Ptolemy's Science of the Stars in the Middle Ages

Ptolemaeus Arabus et Latinus

STUDIES

Volume I

Ptolemaeus
Arabus et Latinus

Ptolemaeus Arabus et Latinus (PAL) is a project of the Bavarian Academy of Sciences and Humanities and the University of Würzburg. As part of the Academies' Programme, PAL is jointly funded by the Federal Republic of Germany and the Free State of Bavaria.

PAL is sponsored by the Union Académique Internationale (UAI).

Ptolemy's Science of the Stars in the Middle Ages

Edited by

David JUSTE
Benno van DALEN
Dag Nikolaus HASSE
Charles BURNETT

BREPOLS

Cover design by Hilde Verhelst, T'Hi, Lier, Belgium.

Cover illustration:
Woodcut (detail) from *La geografia di Ptolemeo Alessandrino*, Venice, 1548.
Copy from Munich, Bayerische Staatsbibliothek.

D/2020/0095/147

ISBN 978-2-503-58639-7

Printed in the E.U. on acid-free paper.

In memory of Paul Kunitzsch (1930-2020)

Contents

Acknowledgements

We would like to thank first the Warburg Institute (London) for hosting the conference (5-7 November 2015) which was the starting point of this book, and especially Jane Fergusson for her care in organising the event. The conference and the preparation of this volume were funded by the project Ptolemaeus Arabus et Latinus (Munich und Würzburg) under the aegis of the Union der deutschen Akademien der Wissenschaften. Our special thanks go to the authors for their contributions. We also would like to thank the scholars who agreed to peer-review one or several articles of this volume, Stephan Heilen for chairing a session at the conference, Claudia Dorl for preparing the indexes and providing assistance throughout the preparation of this book, and all members of the Ptolemaeus Arabus et Latinus team who were always ready to help. This is the second volume of the series 'Ptolemaeus Arabus et Latinus' and the first volume of the subseries 'Studies'. We thank Brepols Publishers for hosting this series, and in particular Bart Janssens for his commitment and efficiency. Finally, our deep gratitude goes to Menso Folkerts and Paul Kunitzsch for their constant support and encouragement since the early days of Ptolemaeus Arabus et Latinus in 2009. Sadly, Paul Kunitzsch did not live to see the publication of this volume. We dedicate this book to him.

The Editors

Introduction

In November 2012, Ptolemaeus Arabus et Latinus (PAL) was established by the Union of the German Academies of Sciences and Humanities as an Academy Project with five researchers for a period of 25 years. This was the result of a long application journey whose idea first originated on a Saturday night of August 2009. In the course of 2013, the project became established at the premises of the Bayerische Akademie der Wissenschaften in Munich.

The initial impetus of the project was the realisation that the reception of Ptolemy was still to a large extent a *terra incognita* of the history of science. This seemed to us a rather odd fact considering that Ptolemy wrote perhaps the two most influential texts both in mathematical astronomy (the *Almagest*) and in astrology (the *Tetrabiblos*). From late Antiquity to the seventeenth century, Ptolemy was to the 'science of the stars' what Aristotle was to logic, natural philosophy and metaphysics, Euclid to geometry and Hippocrates and Galen to medicine in the Western tradition. Both the *Almagest* and the *Tetrabiblos* were translated several times into Arabic and into Latin and were heavily discussed and commented upon in the Islamic world and in Christian Europe. Yet the Arabic and Latin versions of the *Almagest* and the *Tetrabiblos* were unavailable in modern editions, their manuscripts remained largely unexplored and, generally speaking, their history until the seventeeth century had never been systematically studied.

From the outset, it was clear to us that we could not limit ourselves to the *Almagest* and the *Tetrabiblos*. Ptolemy also produced several astronomical works of lesser importance, including the *Planetary Hypotheses*, the *Phaseis*, the *Analemma*, the *Planispherium* and the *Handy Tables*, all of which survive, in one form or another, in Arabic and/or Latin. Moreover, a number of works were falsely attributed to Ptolemy, the most famous of which is the *Karpos* (*Kitāb al-Thamara, Centiloquium*), a collection of one hundred astrological aphorisms which enjoyed extraordinary popularity throughout the Middle Ages and whose ascription to Ptolemy was never questioned until the sixteenth century. Last but not least, the *Almagest*, the *Tetrabiblos* and the *Karpos* gave rise to a large number of commentaries in Arabic and Latin, which were an integral part of the Ptolemaic tradition and hence could not be ignored. As a result, our *Corpus Ptolemaicum* had considerably expanded. In view of the time restriction (25 years being the maximum granted by the Union of the German Academies), we had to limit the scope of the project and to make a number of choices. Already at an early point we had decided to focus on the science of the stars and to exclude from our enquiry the non-astronomical/astrological works, i.e. the *Geography*, the *Optics* and the *Harmonics*. Likewise, we were aware that

Ptolemy's Science of the Stars in the Middle Ages, ed. by David Juste, Benno van Dalen, Dag Nikolaus Hasse and Charles Burnett, PALS 1 (Turnhout, 2020), pp. 1–10
 10.1484/M.PALS-EB.5.120171

Arabic and Latin, however central, represent only two languages of the medieval scientific tradition. The Ptolemaic corpus also has a history in Byzantine Greek, Syriac, Hebrew, Persian, Sanskrit and the European vernaculars, but taking these languages into account would have made the project unrealisable. We do, however, consider works in other languages when they have an immediate bearing on the Arabic or Latin tradition. A good example of this is the Syriac version of the *Tetrabiblos*, which may have served as an intermediary between the Greek and Arabic versions and which is in the process of being edited by Bojidar Dimitrov as part of the project.

Despite these restrictions and choices, we are confident that the project can significantly contribute to several research areas. First, we see it as an essential step towards a fuller and better understanding of medieval and early modern astronomy and astrology, and, more generally, of the worldview that dominated Islamic, Jewish and Christian cultures from the Middle Ages to the seventeenth century. Contrary to received opinion, Arabic and Latin scholars deployed remarkable intelligence and creativity, as well as sharp criticism, in their enquiry into Ptolemy's models and theories, in order to develop a good understanding of them, but also to question, improve on, or refute them. Moreover, research into medieval conceptions of the universe has suffered from the fact that astrology was for a long time neglected by historians of science on the grounds that it is irrelevant to modern science. But we should remember that most medieval astronomers were also astrologers and, in the context of Aristotelian natural philosophy, astrology was conceived as a science, both physical and mathematical, leading to the knowledge of man, nature and God. By considering astrology along with astronomy, this project stands in agreement with Ptolemy's vision, as well as with Greek, Arabic and Latin mainstream science up to the time of Galileo and Kepler inclusively.

The project will also lay the foundations for a fresh approach to the Copernican Revolution. Since the seminal studies of Alexandre Koyré and Thomas Kuhn, the Copernican Revolution has been the subject of intense scholarship around the world. Yet, this scholarship — including, remarkably, the most recent scholarship — has been carried out to a large extent in ignorance of the medieval tradition, partly because of methodological biases, partly because of the lack of studies and editions. The fact that the Copernican Revolution started with an attack launched against the *Almagest* (Copernicus's *De revolutionibus orbium celestium*, 1543) underlines the importance of studying the medieval Ptolemaic tradition, and makes it all the more remarkable that the very Latin text that Copernicus sought to replace is still today unavailable in a modern edition.

Last but not least, the retrieval of the Arabic and Latin Ptolemaic corpus will shed light on the authenticity, form and content of the Greek original texts. As is often the case for ancient scientific and philosophical texts, most extant

Greek manuscripts are late Byzantine copies which sometimes preserve incomplete or corrupt texts. Ptolemy's works are no exception to this rule. Arabic and especially Latin manuscripts pre-date Greek manuscripts for all the texts dealt with in this project, with the exception of the *Almagest*. Moreover, three of them (*Planetary Hypotheses*, *Analemma* and *planispherium*) are for the most part lost in Greek and survive in Arabic and/or Latin only. Even though the Arabic and Latin texts are the result of translation — and hence, inevitably, of alteration —, it may be that some of them preserve a more authentic text than that of the extant Greek manuscripts. A case in point is the *Tetrabiblos*. The Greek manuscript tradition of this text is problematic because the earliest complete witness (out of a total of 47 extant manuscripts) is a Byzantine copy dating from *c.* 1300 (Vatican, BAV, Vat. gr. 1038). The text has received four critical editions, by Franz Boll and Emilie Boer (1940), Frank Robbins (1940), Simonetta Feraboli (1985) and Wolfgang Hübner (1998), all of whom ignored the Syriac, Arabic and Latin traditions. Yet, the *Tetrabiblos* survives in a Syriac translation, in at least three Arabic translations, dating from the ninth to the eleventh century, and thirteen Latin translations, six of which were made before 1300. In 2015, Gudrun Vuillemin-Diem and Carlos Steel published a critical edition of the Latin translation made directly from the Greek by William of Moerbeke between 1266 and 1269. The editors demonstrate that the Greek manuscript used by William of Moerbeke (now lost) is far better than all the surviving copies. As a result, while they confirmed 301 readings adopted or conjectured by Hübner in his edition of the Greek text, they also identified 63 instances which definitely confirm 'a reading not adopted by Hübner', 143 instances where they 'propose to modify Hübner's edition' and 109 instances of a reading that 'seems also possible'.[1] It is also likely that the other Latin translations, and especially the Syriac and Arabic translations, will allow us to improve on the Greek text.

The primary aim of the project is to make the Arabic and Latin Ptolemaic corpus on the science of the stars available to research. The first step towards this aim is to establish the *Corpus Ptolemaicum*, survey the works and catalogue the manuscripts. The works are arranged in three categories as follows:

A. Authentic works: *Almagest*, *Tetrabiblos*, *Planetary Hypotheses*, *Phaseis*, *Analemma*, *Planispherium* and *Handy Tables*.
B. *Pseudepigrapha*: besides the *Karpos*, some thirty astronomical and astrological works falsely attributed to Ptolemy have been identified in Arabic and Latin.

[1] Gudrun Vuillemin-Diem and Carlos Steel, *Ptolemy's Tetrabiblos in the Translation of William of Moerbeke. Claudii Ptolemaei liber iudicialium*, Leuven: Leuven University Press, 2015, pp. 95–129.

C. Commentaries: commentaries are understood in the broad sense, so as to include epitomes, paraphrases, critiques, university lectures, etc., in fact all texts that derive their substance primarily and explicitly from one of the above works (categories A and B). The most influential commentaries include Thābit ibn Qurra's *Tashīl al-Majisṭī* and Jābir ibn Aflaḥ's *Iṣlāḥ al-Majisṭī* (both of which were translated into Latin), the anonymous *Almagesti minor*, Regiomontanus's *Epitome Almagesti*, ʿAlī ibn Riḍwān's commentary on the *Tetrabiblos* and Abū Jaʿfar Aḥmad ibn Yūsuf ibn al-Dāya's commentary on the *Karpos* (both of which were also translated into Latin). Other works that relate to Ptolemy less directly (however important their Ptolemaic component might be) have been excluded, for example al-Farghānī's *Kitāb jawāmiʿ ʿilm al-nujūm wa-uṣūl al-ḥarakāt al-samāwiyya* and the various Latin versions of the *Theorica planetarum*.

Altogether the corpus (A-B-C) amounts to over 80 works preserved in at least 500 manuscripts for the Arabic and to *c.* 170 works preserved in *c.* 670 manuscripts and over 100 early printed editions for the Latin.

While it is not possible to prepare critical editions for all these works, we deemed it important to make the primary material available to scholars at a relatively early stage of the project. The editing of texts has therefore been designed in three steps. In the first place, each work (including each version in cases of multiple translations) receives an online digital reproduction in scanned form from one selected witness (manuscript or early printed edition). Next, the digital images are gradually linked with online standardised transcriptions based on the same selected witness, so as to make each text searchable. Finally, as a third step, the online transcriptions are gradually supplemented by proper critical editions based on examination of all extant witnesses. All authentic works (A) and *pseudepigrapha* (B) are expected to receive a critical edition. In cases of multiple translations, only the most influential versions will be edited. This includes, for example, two versions of the Arabic *Almagest* (al-Ḥajjāj and Isḥāq ibn Ḥunayn revised by Thābit ibn Qurra) and two Latin translations of the *Tetrabiblos* (by Plato of Tivoli and Aegidius de Tebaldis). Likewise, only the most influential commentaries (C) will be edited, including at least those listed above.

Besides the *Corpus Ptolemaicum*, we also take into account a further category of related material, namely astronomical tables, almanacs (or ephemerides) and horoscopes, insofar as these represent the main products of astronomical and astrological activity in the Ptolemaic tradition. The aim here will be to design a database of astronomical tables, almanacs and horoscopes, to produce a critical survey of them and to edit some of the most representative and/or influential ones. To this end, computer programs for editing and analysing astronomical tables, almanacs and horoscopes will be written. First steps in this direction have been made by collaborating in the project of a database of tables and other materials from astronomical and astrological manuscripts led by Matthieu Husson of the Observatoire de Paris.

Another resource developed as part of the project is a glossary of Greek-Arabic-Latin-English mathematical, astronomical and astrological terms. This glossary is designed as a constantly growing tool based on the texts already edited in the project.

Finally, the project will inevitably give rise to new questions, the most important of which are the subject of international conferences and workshops organised by the project. Special attention is paid to three research areas: the reception of Ptolemy in the Arabic world and Western Europe up to 1700 AD; a comparative study of Arabic and Latin astronomy and astrology in their historical contexts; and the place of Ptolemy in the Copernican Revolution.

The progress of the project can be followed on the PAL website (https://ptolemaeus.badw.de), which has been active since December 2016 and updated almost on a daily basis since. The catalogue of texts and manuscripts, the critical editions and the studies will also appear in print in the new series *Ptolemaeus Arabus et Latinus* published by Brepols. The first volume in the subseries *Texts* came out in 2018,[2] whereas the present volume is the first in the subseries *Studies*.

One of the missions of the project is to organise an international conference or a workshop every three years. The first PAL conference was held at the Warburg Institute (University of London) from 5 to 7 November 2015 and this book is the result of it. The title *Ptolemy's Science of the Stars in the Middle Ages* (already the title of the conference) was deliberately kept wide in scope so as to encompass the reception of Ptolemaic astronomy and astrology in the Arabic world and in Western Europe up to 1700 AD. Our aim was to gather together leading scholars and younger researchers in Ptolemaic studies and, while obtaining as broad an overview as possible, we tried to keep some balance between Arabic and Latin, on the one hand, and between astronomy and astrology, on the other. Two speakers at the conference are not represented in this volume, because further research in their respective topics led them to considerable developments which deserve to be published separately: Flora Vafea on the *Dhāt al-kursī* attributed to Ptolemy and Maria Mavroudi on the Greek and Arabic versions of the *Karpos*. At the same time, two contributions included here are by scholars who joined the project after the conference and who offered to contribute to the volume (José Bellver and Paul Hullmeine). With this volume, we therefore hope to present the state of the art of Ptole-

[2] Henry Zepeda, *The First Latin Treatise on Ptolemy's Astronomy: The Almagesti minor (c. 1200)*.

maic studies in the Islamic and Christian cultures in the long Middle Ages. The 15 articles are arranged roughly chronologically.

Alexander Jones establishes the corpus of Ptolemy's authentic works, a task made arduous by our almost complete ignorance of his life and the circumstances of his writings, so much so that 'anything that we can know of him has to come from his writings'. How do we know what Ptolemy wrote? Most texts that are attributed to Ptolemy either refer to the *Almagest* or bear the dedication to Syros, but this is no definite proof of their authenticity (the *Karpos* is also dedicated to Syros). Previous discussions about the authenticity of Ptolemy's works have rested on doctrinal and stylistic comparison with his other works and on testimony of later authors, but these have their limitations too. Jones identifies a number of 'verbal fingerprints', that is, expressions used in two or more works attributed to Ptolemy, but (virtually) nowhere else before the end of the fourth century. This allows him to confirm Ptolemy's authorship for most of the texts commonly attributed to him and to securely confirm the authenticity of *On the Criterion*. Jones also discusses works that are lost in Greek, as well as lost works known to us only through quotations by Ptolemy or others, and concludes with a reconstruction of the chronology of Ptolemy's writings and an assessment of his scientific interests.

Nathan Sidoli reconstructs the mathematical methods found in Ptolemy's *Analemma*, which is extant in full in Latin only, in the context of Greek mathematical practices. The analemma is a plane figure obtained by rotating and orthogonally projecting arcs, lines and points from the heavenly sphere into the plane, basically mimicking the operations that can be carried out with a compass and a set square. In this process the magnitudes of lines and arcs on the sphere are preserved, and the same object may be represented in multiple ways. The analemma thus allows the determination of arc lengths on the sphere by means of plane trigonometry (or by direct measurement). Ptolemy uses it in particular to specify the solar position with respect to the horizon and the local meridian as a function of its declination, the geographical latitude and the time of day. After giving an explanatory table of contents of the 15 sections of Ptolemy's *Analemma*, Sidoli explains in detail Ptolemy's model of the world and his determination of one of the three analogous pairs of angles in which he expresses the local coordinates of the solar position.

Paul Hullmeine discusses the question whether the ninth sphere, which was commonly associated with the Ptolemaic cosmos in medieval Arabic, Latin and Hebrew descriptions of the universe, in fact originated with Ptolemy. The ninth sphere is a starless sphere beyond the eighth sphere of the fixed stars. Since the ninth sphere is not mentioned in the *Almagest*, Hullmeine carries out a detailed analysis of the extant Greek and Arabic versions of the *Planetary Hypotheses*, investigating both the vocabulary related to spheres, orbs and circles and other aspects of Ptolemy's physical descriptions of the planetary models.

He concludes that Ptolemy did not intend to establish a nine-sphere cosmos and finds that John Philoponus (sixth century) was the first Greek scholar to associate the ninth sphere with Ptolemy. Finally, an interesting quotation from *India* (*c.* 1030) by al-Bīrūnī, who was intimately familiar with the *Almagest* and the *Planetary Hypotheses* as well as Philoponus's works, reveals that Philoponus indeed introduced the ascription of the ninth sphere to Ptolemy into the medieval Arabic astronomical tradition.

In his doctoral dissertation, **Bojidar Dimitrov** edits the (incomplete) Syriac translation of the *Tetrabiblos* and compares it in detail with the Greek text as edited by Hübner (1998), the Latin version by William of Moerbeke as edited by Vuillemin-Diem and Steel, and the Arabic versions by ʿUmar ibn al-Farrukhān and Ḥunayn ibn Isḥāq in the preliminary edition by the late Keiji Yamamoto, visiting fellow at PAL in September 2014. The significant variants from all five sources are specified in the apparatus to the Syriac text. As a foretaste, Dimitrov here presents a linguistic comparison of ten cases in which the five versions show significant differences. He finds, among other results, that the Latin version of William of Moerbeke is generally closer to the Greek and Syriac than the two Arabic versions.

Johannes Thomann, who has previously published several articles on the *Almagest* commentary by al-Fārābī (tenth century), here makes use of a short critique by Ibn al-Ṣalāḥ (mid twelfth c.) on al-Fārābī's commentary to identify passages from the otherwise lost early Arabic translation of the *Almagest* made for the caliph al-Maʾmūn before that of al-Ḥajjāj. These passages are quoted by Ibn al-Ṣalāḥ from al-Fārābī's commentary. By comparing the technical vocabulary in these passages with the Greek original of the *Almagest*, with the extant parts of al-Fārābī's commentary, with the Arabic translations by al-Ḥajjāj and Isḥāq ibn Ḥunayn (revised by Thābit ibn Qurra), and with other early ninth-century astronomical works, he concludes that the passages can be assumed to be early and hence to most likely stem from the early Maʾmūnic *Almagest* translation.

In his article, **Dirk Grupe** identifies five Arabic and Persian sources that used the version of the *Almagest* by Thābit ibn Qurra recently identified by him (c. 890, to be distinguished from the translation by Isḥāq ibn Ḥunayn that was revised by Thābit). These include an epitome of the *Almagest* in a manuscript from a private collection (originally in Iran), Ibn Sīnā's *Kitāb al-Shifāʾ*, an abridged reworking of Thābit's *Almagest* extant in the Senate Library in Tehran, the *Talkhīṣ al-Majisṭī* by Quṭb al-Dīn al-Shīrāzī and a commentary on the *Almagest* by Athīr al-Dīn al-Abharī. Furthermore he argues that MS 20 of the Maharaja Sawai Man Singh Museum Library in Jaipur contains an unshortened copy of Books I-V of the Thābit version.

Tzvi Langermann demonstrates the importance of the Islamic commentary tradition by investigating in detail how three important authors from the Islamic world treated Ptolemy's proof of the sphericity of the earth and of the heavenly motion in *Almagest* I.3. In his introduction he discusses in general terms the role of commentaries, which also took the form of epitomes or critiques, as perceived by their Islamic authors. Then he goes on to analyse the way in which Ptolemy's treatment was explained and criticised in three works: Ibn al-Haytham's *Commentary on the Almagest* (*c.* 1000), Jābir ibn Aflaḥ's *Book of Astronomy* or *Correction of the Almagest* (early twelfth century) and al-Bīrūnī's *zīj al-Qānūn al-Masʿūdī*.

José Bellver carries out a detailed comparison of the different extant versions of Jabīr ibn Aflaḥ's *Book of Astronomy*, which was a very important reedition of Ptolemy's *Almagest*. He first establishes that the title by which it is generally known in modern scholarship–*Iṣlāḥ al Majisṭi* (*Correction of the Almagest*)–is probably not the original title, since it only appears on the title page of a manuscript far removed from the original work. Rather, the original title of the work was most likely to be *Kitāb al-Hayʾa* (Book of Astronomy). He then analyses the differences between the four extant Arabic manuscripts in Arabic characters, namely two at the Escorial, one in Berlin and one recently discovered in the Parliament Library in Tehran, also taking into account the Latin translation by Gerard of Cremona from the late twelfth century, to determine the chronological order of the four different recensions and to find that at least three of them most likely stem from Jābir ibn Aflaḥ himself.

In the Arabic tradition, several mathematical methods applied to astrology are attributed to Ptolemy and Hermes. These methods concern the house systems (domification), the projection of rays and the progressions. **Josep Casulleras** reviews the various methods attributed to Ptolemy and Hermes and notes that these attributions are not justified by the extant works of either Ptolemy or Hermes.

It is commonly assumed that the *Almagest* was hardly read in Europe before the time of Peurbach and Regiomontanus. Evidence to the contrary is shown by over ten surviving Latin commentaries on the *Almagest* written between *c.* 1200 and 1450, as well as by glosses present in numerous manuscripts before 1450. In his contribution, **Henry Zepeda** offers the first study of glosses found in the Latin manuscripts of the *Almagest*. After an overview based on an intimate knowledge of the *Almagest* manuscripts, he presents the outstanding case of Paris, BnF, lat. 7256, a thirteenth-century manuscript displaying several layers of glosses, a good deal of which turn out to be by Campanus of Novara.

Carlos Steel examines Henry Bate of Mechelen's views on three astronomical/astrological topics concerning the time for which the horoscope of the revolution of the year should be cast (i.e. for the time of the entry of the Sun into Aries or for the time of the syzygy preceding it), the incertitude of astronomical observations and the commensurability of celestial motions. These views are found in the preface of, and additions to, Bate's translation of Abraham Ibn Ezra's *Book of the World* (completed in 1281), where Bate defends Ptolemy against a number of attacks by Abraham Ibn Ezra. The article is accompanied by an edition and a translation of Bate's preface and additions.

With over 200 extant manuscripts, the *Centiloquium* is by far the most popular Ptolemaic work in the Latin tradition. **Jean-Patrice Boudet**, who is preparing a critical edition of the medieval Latin versions on the basis of the work left unfinished by the late Richard Lemay, offers here a survey of these medieval versions (of which six are identified, all translated from Arabic probably in the twelfth century) and evaluates their quality by comparing selected aphorisms.

Michael Shank explores the vivid controversy that surrounded the study of the *Almagest* between Regiomontanus and George of Trebizond in the second half of the fifteenth century. This controversy concerned the whole *Almagest*, of which George of Trebizond had produced a new translation from the Greek and an extensive commentary in 1451. Sometime after 1461, Regiomontanus embarked on a no less extensive and particularly sharp criticism of George's commentary in his *Defensio Theonis contra Trapezuntium*, a text that has been very little studied. In this article, Shank conducts a detailed analysis of Regiomontanus's discussion of *Almagest* IX.1 on planetary order and distances, a question which had been notoriously problematic — in fact unsolved — since Antiquity, in particular as regards the inferior planets (the Sun, Mercury and Venus). Some of Regiomontanus's developments echo Copernicus's treatment of the planetary order.

Copernicus was not the first to launch a frontal assault against Ptolemy's science of the stars. Some fifty years earlier, Pico della Mirandola produced one of the most devastating refutations of astrology ever written, which was published posthumously in 1496 under the title *Disputationes adversus astrologiam divinatricem*. **Darrel Rutkin** examines Pico's multi-faceted use of Ptolemy, who features 376 times in the *Disputationes*, as the author of the *Almagest*, the *Tetrabiblos* and the *Centiloquium*, whose authenticity was not doubted by Pico. Rutkin shows in particular how Pico rebukes Ptolemy, whom he calls the 'best of the bad ones' ('optimus malorum'), and at the same time uses his silence regarding a particular doctrine (e.g. the decans) to castigate other astrologers who expounded this doctrine, so making Ptolemy his 'anti-astrological ally'.

Ptolemaic astronomy continued to be pursued after Copernicus and even after Kepler well into the seventeenth century. This is perhaps best exhibited by Longomontanus (Christian Sørensen Longberg), a former assistant of Tycho Brahe in Denmark and in Prague, who became professor of mathematics at the University of Copenhagen in 1605 and published in 1622 his *Astronomia*

Danica, a large volume of 550 pages which modern scholarship has called the 'Tychonian Almagest'. In this paper, **Richard Kremer** analyses and reconstructs the theory of Mars developed in several steps by Longomontanus. Proceeding step by step he shows how Longomontanus respectively solved the problems of the first and second anomaly (the latter with a highly original approach), the Mars-Sun and Earth-Sun distances, and finally the mean motions, for which he made use also of observations from Ptolemy's *Almagest*.

The editors

I. The Greek and Near Eastern Traditions

The Ancient Ptolemy

Alexander JONES

1. Introduction

Before the medieval Ptolemy — Ptolemaeus Arabus and Ptolemaeus Latinus, not to forget Ptolemaeus Byzantinus — was the ancient Ptolemy. Or rather, there were ancient Ptolemies, starting with a man who composed a wide range of scientific texts and tables in Antonine Roman Egypt, and trailing after him, the shadowy Ptolemies who were the images of this author as he was known to people of the four remaining centuries of antiquity following his own career. For within a few decades of his time, a process of disintegration of Ptolemy's unity had set in, because even his earliest readers, users, and commentators were unable to mirror the breadth of his scientific interests or grasp the philosophical and didactic agenda that shaped his approach across his individual fields of study; even today, historians tend to specialize according to disciplines whose boundaries cut across Ptolemy's œuvre. My object in the present essay is to explore the extent to which we can know the original, in-the-round Ptolemy, and to identify some aspects of his thought that become more apparent from consideration of the full breadth of his work and that might affect how we receive the specifically astronomical and astrological works that constitute the core of the 'Ptolemaeus Arabus et Latinus' project.

The crucial limitation to our knowledge of the historical Ptolemy is the lack of useful information independent of his writings. This should come as no surprise to any student of antiquity, who knows how rare it is that a Greco-Roman scientific author whose works survived into the medieval manuscript tradition was also a personality traceable in references in literature or in archeologically recovered artifacts and documents from his own time. The case of Archimedes, in which we have on the one hand a corpus of technical mathematical treatises preserved through three early minuscule Byzantine codices and on the other various anecdotes and legends pertaining to his life, is not really an exception, since the biographical reports are only known to us through such later writers as Cicero and Plutarch. A more instructive comparand for Ptolemy is his contemporary Galen. The immense Galenic corpus contains enough autobiographical material for us to reconstruct a detailed if *parti pris* life of a figure whom Bowersock memorably and accurately describes as a lion of

Ptolemy's Science of the Stars in the Middle Ages, ed. by David Juste, Benno van Dalen, Dag Nikolaus Hasse and Charles Burnett, PALS 1 (Turnhout, 2020), pp. 11–34
© BREPOLS PUBLISHERS 10.1484/M.PALS-EB.5.120172

Roman society, yet contemporary and near-contemporary allusions to him are few and scarcely reflect the stature and strong personality conveyed by Galen's self-references.[1] Ptolemy's virtues walked a narrower round, for second-century Alexandria was a provincial intellectual center compared to Galen's Rome — and besides, by the standards of the so-called Second Sophistic movement that served as the gaudy public face of Antonine intellectual life, Ptolemy was an introvert.[2] Perhaps Galen himself came into contact with Ptolemy during his youthful sojourn in Alexandria in the mid-150s (around when Ptolemy published the *Almagest*), and perhaps many years later he included Ptolemy in a list of important but under-read astronomical authors in his commentary on *Airs, Waters, Places* — though the circumstance that in the extant Arabic translation of Galen's commentary he appears as 'Ptolemy king of Egypt' invites suspicion that his presence here is the result of a medieval interpolation.[3] Otherwise any impact Ptolemy the man had on his contemporaries is invisible to us. Effectively, anything that we can know of him has to come from his writings.

2. Establishing the Ptolemaic corpus

How sure are we what were his writings were? The starting point, of course, is the presence of Ptolemy's name at the header or footer of a text as preserved in the extant manuscripts. As we learn from Galen's *On My Own Books*, however, an author — even of technical literature, for which the market was presumably somewhat restricted — could have the disconcerting experience of finding his own name attached fraudulently to a bookseller's wares, to say nothing of false attributions from later times. And unlike Galen, Ptolemy left no catalogue of his literary production.

If we take the *Almagest* as *par excellence* the authentic Ptolemaic text, we can say of several others ascribed to Ptolemy that they must either be his or have been intentionally falsified so as to appear to be his, since they have either an opening address to Syros, the dedicatee of the *Almagest*, or an explicit back-reference to the *Almagest*, or both:

[1] Moraux, *Galien de Pergame*; Bowersock, *Greek Sophists*, p. 66; Nutton, 'Galen in the Eyes'.

[2] It has been vigorously disputed whether Galen qualifies as a figure of the Second Sophistic, e.g. Bowersock, *Greek Sophists*, pp. 59–75; Brunt, 'The Bubble', esp. pp. 43–46; von Staden, 'Galen'. No one, to my knowledge, has associated Ptolemy with the movement.

[3] Toomer, 'Galen on the Astronomers', esp. p. 204; Strohmaier, 'Galen's Not Uncritical Commentary'.

Work	*Primary language of preservation*	*Address to Syros*	*Reference to Almagest*
Planetary Hypotheses	Greek (parts Arabic)	Yes	Yes
Arr. and Comp. Handy Tables	Greek	Yes	Yes
Tetrabiblos	Greek	Yes	Yes
Karpos (*Centiloquium*)	Greek/Arabic/Latin	Yes	No
Geography	Greek	No	Yes
Analemma	Latin (parts Greek)	Yes	No
Planispherium	Arabic/Latin	Yes	Yes

Table 1. Authorship evidence from dedications and cross-references

The remaining texts ascribed to Ptolemy in the manuscripts and that recent scholarship has treated as plausible contenders for authenticity despite the absence of reference to the *Almagest* or dedication to Syros are the *Phaseis, Canobic Inscription, Criterion, Harmonics,* and *Optics.*

On the other hand, there exist many texts that, though ascribed to Ptolemy in the manuscripts, it is unlikely anyone would now make a case for as his work. In Greek, we have the *Karpos* of course, though it has been maintained that the Greek version is a translation of an Arabic original, and *Musica*, a short text partly adapted from the final cosmic-harmonies section of the Canobic Inscription but otherwise devoted to musical terminology unrelated to Ptolemy's *Harmonics.*[4] *Claudii Ptolomei* [*sic*] *de Speculis* is William of Moerbeke's Latin translation of a short treatise on catoptrics that was certainly present in Greek in one of the two lost codices from which William translated several works of Archimedes and Eutocius as well as Ptolemy's (authentic) *Analemma*; the attribution has universally been rejected since the early nineteenth century, and modern editions present it either as a work of Heron of Alexandria or as anonymous.[5] Additionally, numerous manifestly spurious astrological and astronomical texts are extant in Latin or Arabic under Ptolemy's name, for which there is no evidence of Greek originals.[6]

The foregoing discussion has not touched on the *Handy Tables.* In fact what we now understand to be Ptolemy's *Handy Tables* is a modern recon-

[4] On the *Karpos* and the Arabic and Latin traditions of the *Centiloquium* see Juste, 'Pseudo-Ptolemy, *Centiloquium*' and the article by Jean-Patrice Boudet in this volume. *Musica* is edited in von Jan, *Musici Scriptores*, pp. 411–20, and discussed by Swerdlow, 'Ptolemy's Harmonics', esp. pp. 176–78.

[5] Jones, 'Pseudo-Ptolemy *De Speculis*'. The manuscript in which William found the text is listed in the 1311 inventory of the papal Greek manuscripts as 'undecim quaternos... in quibus est liber Tholomei de resumptione [i.e. the *Analemma*], perspectiua ipsius [i.e. the *De Speculis*], perspectiua Euclidis, et quedam figure Arcimenidis [*sic*]'; thus the ascription to Ptolemy was already in this manuscript, not a guess of William's. See Jones, 'William of Moerbeke', esp. p. 19.

[6] For Arabic pseudepigrapha see Sezgin, *Geschichte*, vol. VII, pp. 46–47; for Latin, https://ptolemaeus.badw.de/works/.

struction obtained as a subset of collections of astronomical tables preserved in numerous Byzantine manuscripts, in particular four dating from the ninth and tenth centuries, the selection being guided chiefly by Ptolemy's *Arrangement and Composition of the Handy Tables* and Theon's two commentaries on the *Handy Tables*.[7] Most of the manuscripts in question present the tables anonymously, and the occasional appearance of Ptolemy's name in association with them (which of course would apply on the face of it to all tables in the collection, not just the subset modern scholarship endorses) does not constitute a robust ascription.[8]

Since modern philological methods began to be systematically applied to ancient scientific texts, the works in the Ptolemaic canon whose authenticity has been the subject of serious discussion include the *Tetrabiblos*, the *Karpos* (*Centiloquium*), the *Criterion*, the *Canobic Inscription*, and the *Optics*. Following Boll's 1894 'Studien über Claudius Ptolemäus', in which he argued extensively for the authenticity of the *Tetrabiblos* and more briefly for the spuriousness of the *Karpos*, the status of those two works has effectively been settled.[9] Concerning the *Criterion*, however, Boll writes:[10]

> Dass die Schrift nur dem Mathematiker Claudius Ptolemäus gehören kann, bedarf keines Beweises: Anschauung und Stil zeigen dies selbst dem flüchtigsten Blick.

But the very fact that he felt the need to make this assertion implies that the question of authorship was not entirely straightforward, and in this instance Boll's authority failed to establish a consensus.[11] Any doubts about the *Canobic Inscription* vanished following Hamilton's demonstration that a certain passage in the *Almagest* (4.9) alluding to parts of Ptolemy's lunar and planetary the-

[7] Tihon, 'Les *Tables Faciles*'.

[8] For manuscripts identifying their contents as 'Ptolemy's *Handy Tables*' (Πτολεμαίου πρόχειροι κανόνες) see Heiberg, *Opera astronomica minora*, pp. cxc–cciii. None of the earliest copies has such a heading, and in those that do, it is likely to be a Byzantine scholar's conjecture.

[9] Boll, 'Studien', pp. 111–80 (*Tetrabiblos*) and 180–81 (*Karpos*).

[10] Boll, 'Studien', p. 77.

[11] Rose, *De Aristotelis librorum*, p. 45 had already baldly denied Ptolemy's authorship of the *Criterion* ('ad astronomum certe cui adscribit editor [*scil.* Boulliau] nihil pertinentem'). More recent dissenters include Toomer, 'Ptolemy', esp. p. 201 ('There is nothing in its contents conflicting with Ptolemy's general philosophical position, but the style bears little resemblance to his other works; and the ascription, while generally accepted, seems dubious'.); Taub, *Ptolemy's Universe*, p. 9 ('a work whose attribution to Ptolemy has been questioned'); and Swerdlow, 'Ptolemy's Harmonics', pp. 179–80 ('Concerning the short work on epistemology attributed to Ptolemy, *On the Criterion*, I have nothing to say except to doubt its authenticity, or at least its pertinence to the subjects considered here... It contains not a single reference to the subjects of Ptolemy's other works, *all* in the mathematical sciences, and parallels that have been drawn with the *Harmonics* seem to me vague.').

ories that he had revised refer in fact to the parameters in the inscription.[12] Though the *Optics* has been regarded as authentic by most scholars from the nineteenth century to the present, Rome expressed doubts while more recently Knorr contended that the ascription to Ptolemy was sufficiently insecure that it would be preferable to take as a working hypothesis that it was by a different author.[13]

Arguments for or against the authenticity of writings attributed to Ptolemy have rested chiefly on three types of evidence: comparison of thought with other works accepted to be Ptolemy's, comparison of style, and testimony of later authors. Boll's discussion of the *Tetrabiblos* applies all three, and the abundance and (in large part) the quality of the arguments render his case for Ptolemy's authorship thoroughly persuasive. The *Criterion*'s authorship is supported by no ancient testimony beyond the attribution in the work's manuscript tradition, and Boll backs up his assertion, quoted above, that Ptolemy's authorship is obvious merely by referring to arguments offered by Boulliau in his 1644 *editio princeps*, which are in reality not particularly impressive. The *Optics* is transmitted minus its entire first book, the conclusion of the fifth, and perhaps further books if there were any, and only in an intermittently incoherent Latin translation of an Arabic translation, such that most stylistic traits of the original Greek text can scarcely be discerned. The testimonia do not correspond to any passages in the extant work, leaving the question open whether they refer to material in the lost Book 1 or to another work entirely. Arguments regarding the *Optics*'s authorship have thus operated primarily at the level of thought, which is the most subjective of the criteria, especially considering that the subject matter of the *Optics* has little overlap with the accepted writings of Ptolemy.

Stylistic arguments that are adduced in favor of the common authorship of two or more texts often depend on similarities in vocabulary and idiom, and care must be taken to ensure that the presence of such shared expressions is truly significant. Boll's long list of stylistic features shared by the *Tetrabiblos* and by Ptolemy's acknowledged works includes some that are indeed specially characteristic of Ptolemy as well as others that are not. An example of the latter is the qualifying phrase οὐ τὸ τυχόν (in whatever gender and case is appropriate), meaning 'not just any,' or effectively 'significant'; Ptolemy is fond of it,

[12] Hamilton et al., 'The Canobic Inscription'.

[13] Rome, 'Notes sur les passages', esp. p. 36; Knorr, 'Archimedes', esp. pp. 96–104. (Knorr offers as a potential alternative author Ptolemy's approximate contemporary, the peripatetic Sosigenes.) Ptolemy's authorship of the *Optics* had previously been put in question by Caussin de Perceval, 'Mémoire sur l'*Optique*', esp. pp. 26–29. See also now Siebert, *Die ptolemäische Optik* for an extended argument that the *Optics* is a work from late antiquity.

but it turns up about as frequently in several other authors, including Athenaeus, Galen, and Lucian.

The availability of a near-comprehensive searchable corpus of ancient Greek texts preserved through the medieval manuscript tradition, the *Thesaurus Linguae Graecae* (*TLG*), has made it possible to identify an unexpected trait of Ptolemy's writing that allows a secure test of his authorship applicable to all the texts preserved in Greek that are attributed to him and that include a significant quantity of prose, that is, everything but the *Canobic Inscription* and the *Handy Tables*. Ptolemy's style is, by the standards of his time, not florid, but it is not exactly plain either. In particular, certain words and phrases that he used across multiple works — not specialized technical expressions connected with his subject matter — turn out to be otherwise so rare that, in the *TLG* corpus, they occur in no other author before the fourth century, or in extremely few, and these are often authors who wrote under the strong influence of Ptolemy's writings.

Consider for example the opening sentence of *Almagest* Book 2, a typical example of the transitional passages in which Ptolemy sums up retrospectively the contents of the preceding part of a work before announcing the topic to follow:

> διεξελθόντες ἐν τῷ πρώτῳ τῆς συντάξεως τά τε περὶ τῆς τῶν ὅλων σχέσεως κατὰ τὸ κεφαλαιῶδες ὀφείλοντα προληφθῆναι, καὶ ὅσα ἄν τις τῶν ἐπ' ὀρθῆς τῆς σφαίρας χρήσιμα πρὸς τὴν τῶν ὑποκειμένων θεωρίαν ἡγήσαιτο, πειρασόμεθα κατὰ τὸ ἑξῆς...
>
> Having in the first [book] of the composition gone through *in summary manner* the matters concerning the arrangement of the universe that ought to be assumed beforehand, and all the matters in the *sphaera recta* situation that one would suppose to be useful for the investigation of the subject at hand, we shall next try...

The phrase κατὰ τὸ κεφαλαιῶδες, here translated 'in summary manner', turns up also in the *Tetrabiblos* in the retrospective part of four transitional passages (1.3.20, 2.4.1, 2.14.12, 3.14.9) as well as in one passage (3.4.4) that lists a series of ensuing topics unprefaced by a retrospection. For example, the transition in 2.4.1 is as follows:

> αἱ μὲν οὖν συνοικειώσεις τῶν τε ἀστέρων καὶ τῶν δωδεκατημορίων πρὸς τὰ κατὰ μέρος ἔθνη καὶ τὰ ὡς ἐπίπαν αὐτῶν ἰδιώματα κατὰ τὸ κεφαλαιῶδες τοῦτον ἡμῖν ὑποτετυπώσθωσαν τὸν τρόπον. ἐκθησόμεθα δὲ καὶ...
>
> Let the shared affinities of the stars [i.e. the Sun, Moon, and planets] and the zodiacal signs with respect to the individual peoples and their overall characteristics have been sketched by us *in summary manner* in this way. We shall also set out...

In the *Criterion* it occurs in the prospective part of a transitional passage (15.1):

> τούτων δὲ οὕτως ἐφωδευμένων, ὅτι μὲν ἡγεμονικὸν γίνεται τοῦ σώματος, ἐν ᾧ τὸ ἡγεμονικὸν τῆς ψυχῆς, οὐδὲ εἷς ἂν ἀπορέσειεν, εἰ δ' αὐτὸ τὸ ἡγεμονικὸν οὕτως

ἁπλῶς ληπτέον καὶ οὐχ ὡς τῶν πρός τι ὄν, ὡδί πως κατὰ τὸ κεφαλαιῶδες διοριστέον.

Now that these things have been treated methodically, no one would have difficulty with the fact that there is a *hêgemonikon* of the body, in which is the *hêgemonikon* of the soul, but if this very *hêgemonikon* is to be taken thus absolutely and not as relative to something, one ought to draw distinctions *in a summary manner* in something like the following way.

What makes this instance of the phrase in the *Criterion* significant is the fact that, outside of the *Almagest* and *Tetrabiblos*, it is only attested in authors later than Ptolemy who were heavily influenced by him. In the *TLG* corpus it occurs only in two passages of the astrologer Hephaestion of Thebes (1.20 = Epitome IV 15, and 1.25) which respectively are close paraphrases of the passages from *Tetrabiblos* 2.4 and 2.14 cited above, and in section 6 of the anonymous 'Geographiae expositio compendiaria' (Müller, *Geographi Graeci Minores* 2.494–509), an opuscule of uncertain but definitely late date for which Ptolemy's *Geography* was a major source.[14]

The transitional passage from *Tetrabiblos* 2.4 quoted above also contains the perfect passive imperative verb ὑποτετυπώσθωσαν, following the manuscript reading adopted by Hübner and by Robbins, or ὑποτετυπώσθω following the reading preferred by Boll and Boer (either form is grammatically admissible). This perfect passive imperative of ὑποτυπόω turns out to be another special word for Ptolemy, occurring in the recapitulative parts of transitional passages in the *Tetrabiblos* (2.4 as already mentioned), the *Harmonics* (1.4, 2.3, 2.11, and 3.4), and the *Geography* (1.2.1, 1.18.1, and 2.1.1). And once again, there is an occurrence in the *Criterion* (3.3):

ἐκ πόσων μὲν οὖν καὶ οἵων καὶ τίνα τρόπον συνέστηκεν τὸ κριτήριον ὑποτετυπώσθω διὰ τῶν ἐφωδευμένων. ἐπεὶ δὲ...

Let [the questions] out of how many and what sort of things and in what manner the criterion is composed *have been sketched* by means of the things that have been treated methodically. But since...

The only other occurrence in a text from antiquity in the *TLG* corpus is in Hephaestion's paraphrase (1.20) of *Tetrabiblos* 2.4.[15]

These are just two of many words and phrases that turn up in more than one work attributed to Ptolemy but hardly anywhere else — in some cases

[14] Although the phrase κατὰ τὸ κεφαλαιῶδες does not occur in the *Geography*, Ptolemy's prose description of the known world and its principal features in *Geography* 7.4 is characterized in the chapter title as well as at the end of the preceding chapter as a ὑπογραφὴ κεφαλαιώδης, 'summary caption'; this might have triggered a memory of an expression encountered in other works of Ptolemy's.

[15] There is an instance in the astrological dialogue *Hermippus* (ed. Kroll and Viereck, p. 57 line 23), a work of disputed authorship but definitely of Byzantine date.

nowhere — in texts written up to the end of the fourth century of our era. Table 2 summarizes the patterns of occurrence of fourteen such verbal fingerprints (see the detailed discussion in the appendix to this paper). Every work of continuous prose surviving even partially in Greek in the accepted Ptolemaic corpus (i.e. excluding the *Handy Tables* and the *Canobic Inscription*) is linked by at least one shared expression to at least two other works. As one might expect, the *Almagest* has the largest number of shared expressions — eight of the fourteen — while the *Tetrabiblos* has seven; but the *Criterion* also has seven, making it in this peculiar sense one of the most characteristic works in the Ptolemaic corpus! This is the more remarkable, because the *Almagest* runs to more than 1150 Teubner pages, and the *Criterion* to just 23.

Either the *Criterion* is indeed by Ptolemy, then, or it was composed by someone after Ptolemy using a vocabulary that was strongly influenced, consciously or unconsciously, by Ptolemy's. It is not a work to which Ptolemy's name was accidentally attached, say, merely because it came after genuine works of Ptolemy in a manuscript or because it was written by a different Ptolemy. But the same apparent remoteness of its subject matter from that of Ptolemy's 'scientific' treatises that has led many to doubt its authenticity argues against its being a deliberate forgery or a mistaken ascription of an imitator's composition to the master. The *Criterion* is thus validated as an authentic work of Ptolemy's, and the features of it — such as its very subject matter — that have given rise to doubts about its authorship actually broaden our understanding of Ptolemy's system of thought and perhaps also its development.

The verbal fingerprint test is obviously inapplicable to works that come down to us only in languages other than Greek. In the case of two of the astronomical texts ascribed to Ptolemy, the *Planetary Hypotheses* and the *Analemma*, Greek text whose authenticity is confirmed by verbal fingerprints survives for less than half of each work as represented respectively in Arabic and in Latin. The *Analemma*'s Greek remnants reach us through palimpsest leaves (sixth century?) in the manuscript *Ambrosianus* L99sup that correspond to roughly the middle third of the ostensibly complete text in William of Moerbeke's translation; the parts not covered by the palimpsest were obviously present in the lost Greek manuscript used by William, and there is no reason to suspect that they are inauthentic.[16] The Greek *Planetary Hypotheses*, on the other hand, is roughly the first half of Book 1 as we know it from the Arabic, breaking off in mid-sentence, which suggests descent from a mutilated exem-

[16] William's autograph translation of the *Analemma* in *Ottob. lat.* 1850 cuts off abruptly at the bottom of the last page of a quire, with the first of what the text leads the reader to expect will be a set of tables, and, unlike the other translations in the manuscript, this one lacks a subscription giving the work's title and the date of the translation's completion. There may thus have been a continuation for which we have neither Greek nor Latin.

	Alm	Phas	PH	ACHT	Ana	Tetra	Crit	Harm	Geog	other authors (up to *c.* AD 400)
ἀμετάπιστος	1						2			
ἐπιπολυπραγμονέω						1	1			
εὐκατανόητος	28					1		2		Polybius (1) Hipparchus (2) Pappus*† (1) Theon Alex.*† (3) Porphyry*† (2) Serenus (1)
εὐμεθόδευτον	2		1							Theon Alex.*† (3)
ἐφωδεύμενος (cf. προεφωδευμένος)	7				1	4	2		1	
ἰδιοτροπία	3					35	1	1	1	Aristides Quintilianus (1) Cleomedes (1) Hephaestion*† (18) ps-Galen *De Decubitu* (1)
κατὰ συνεγγισμόν			1						2	Hipparchus (2)
κατὰ τὸ κεφαλαιῶδες	1					5	1			Hephaestion*† (2)
κατὰ τὸ ὁλοσχερές / ὁλοσχερέστερον	5			1		2				Geminus (2) Hephaestion*† (2)
κατὰ τὸν ἁρμόζοντα... λόγον / τρόπον						3		1		
προεντάσσω	4	1								Heron *Metrica* (1) Asclepiodotus (1) Philo Judaeus (1)
προεφωδευμένος	9									Strabo (1)
προσπαραμυθέομαι		1		1			2	1		
προυποτετυπώσθω / προυποτετυπώσθωσαν						2				Hephaestion*† (1)
συνεχέστερα παρατήρησις	1		1							Theon Alex.*† (2)
ὑποτετυπώσθω / ὑποτετυπώσθωσαν (cf. προυποτετυπώσθω)						2	1	4	3	Hephaestion* (1)

* Text reflecting influence of Ptolemy's works. † At least one instance is a paraphrase or quotation from Ptolemy.

Table 2. Some of Ptolemy's verbal fingerprints

plar; and there are changes of subject matter soon after this point as well as between Books 1 and 2 such that no one having just the part existing in Greek would have been able to predict how Ptolemy was going to continue through the rest of the work. But testimonia in Proclus and Simplicius correspond to passages in both Books 1 and 2 that survive only in Arabic, confirming their authenticity (or at a minimum, that these passages existed in Greek in late antiquity).[17] The *Planispherium* seems to be adequately accredited by its references to the *Almagest* and dedication to Syros and by the consistency of its subject matter and mathematical methods with the authenticated astronomical writings, notwithstanding that the only testimonium for it in Greek is its apparent listing in the *Suda* (s.v. Πτολεμαῖος ὁ Κλαύδιος χρηματίσας) by the title ἅπλωσις ἐπιφανείας σφαίρας, 'flattening of a surface of a sphere'.

The *Optics* confronts us with the least satisfactory evidence for its authorship among all the texts whose ascriptions to Ptolemy are not patently spurious. We have testimonia from Simplicius and Damianus in late antiquity and from Symeon Seth in the eleventh century to the existence in Greek of an *Optics* ascribed to Ptolemy, but they do not correspond to passages in the extant, mutilated Latin *Optics*. The Simplicius passage (*In Arist. de Caelo*, ed. Heiberg, *Simplicius*, p. 20) cites both Ptolemy's *Optics* and another work of Ptolemy's 'on the elements' (ἐν τῷ περὶ τῶν στοιχείων βιβλίῳ) for a non-Aristotelian principle that the elements — apparently including both the four 'mundane' elements and the fifth etherial one — have a natural rectilinear motion *only* when they are outside their natural places; the fact that this principle is also found in *Planetary Hypotheses* Book 2 strengthens the case that the *Optics* that Simplicius knew was indeed by Ptolemy.[18] Arguments for the authenticity of the extant *Optics*, however, rest largely on a general sense that it exhibits an intellectual level, engagement with contemporary philosophical concerns, and empirical approach worthy of Ptolemy. Moreover, the extended mathematical discussion of the effect of refraction on observed positions of heavenly bodies in *Optics* 5.23–30 ties one aspect of the treatise's subject to Ptolemy's astronomical interests. A still stronger indication that the author was an astrono-

[17] The end of the authentic Greek text as given, e.g., in *Vat. gr.* 1594 is at Heiberg, *Opera astronomica minora*, p. 104, line 23 after 'ἰσοταχῶς' in the middle of the description of the model for Saturn; the continuation in some manuscripts, which Heiberg retains in his edition, is a mechanical duplication of the preceding description of Jupiter's model with the numerical parameters replaced by blank spaces. Proclus, *In Timaeum* 258a summarizes material from the later part of *Planetary Hypotheses* Book 1, whereas Simplicius, *In Arist. de Caelo* (ed. Heiberg, p. 456) paraphrases a passage in Book 2.

[18] cf. Nix's translation in Heiberg, *Opera astronomica minora*, pp. 112–13. However, since Simplicius also cites the peripatetic Xenarchus and Plotinus in the same context, one cannot maintain that this principle of the rectilinear motion of displaced elements was exclusive to Ptolemy.

mer is that the circular bronze plaque used for the measurements of angles of reflection and refraction in 3.8 and 5.8 is to be inscribed with a division of the circle into degrees, since this appears to be one of only two known ancient instances of use of degrees as a measure of arcs or angles outside of astronomy, astrology, and geography.[19] Though one might wish for something in the *Optics* that marks it specifically as Ptolemy's work, this is barely enough, I believe, to make the transmitted attribution to him convincing.

3. Lost works

It is impossible to be certain how much of Ptolemy's literary production is lost, but from the indications that we have, it is likely to have been significantly less than what survives. Leaving aside the portions known to be missing from extant works — the first book and the conclusion of the fifth of the *Optics*, the last three chapters of *Harmonics* Book 3, and the promised tables at the ends of the *Analemma* and *Planetary Hypotheses* — the only non-extant text explicitly mentioned in any of the surviving ones is the 'dedicated treatise on this subject' (ἡ κατ' ἴδια σύνταξις τῆσδε τῆς πραγματείας) cited at the opening of the *Phaseis*, in which Ptolemy states that he provided a full mathematical treatment of the conditions determining the dates of first and last morning and evening risings of the fixed stars.[20] Since the *Suda* lists among Ptolemy's works 'two books on phases and weather-signs of fixed stars' (περὶ φάσεων καὶ ἐπισημασιῶν ἀστέρων ἀπλανῶν βιβλία β̅), whereas the extant *Phaseis* (transmitted under a slightly different title, φάσεις ἀπλανῶν ἀστέρων καὶ συναγωγὴ ἐπισημασιῶν) is in just one book, it is generally assumed that what we have is Book 2 of a work, the lost Book 1 of which is summarized in its opening sentence, but Ptolemy's wording does not seem to fit a back-reference to a previous part of the same treatise.

A scholion in some manuscripts of the *Almagest* cites a work by Ptolemy 'on paradoxical phases of Venus'.[21] The 'paradoxical' phenomena in question clearly consist of Venus's highly variable intervals of invisibility, in particular around inferior conjunction, which are a topic dealt with in *Almagest* 13.8.[22] The scho-

[19] The other instance is a circular plate graduated in degrees, which forms part of a set of surveyor's instruments of unknown provenance and dating from late antiquity; see Turner, *Mathematical Instruments*, pp. 10–11 and fig. 12d.

[20] Occasionally Ptolemy employs πραγματεία in the sense of 'treatise' (e.g. *Almagest* 13.11), but it can hardly have this meaning here since σύνταξις already designates a composition in its own right.

[21] Jones, 'A Posy', esp. pp. 75–77.

[22] The qualification 'paradoxical' does not appear in the *Almagest*, but is applied by Proclus, *Hypotyposis* 1.17–22 and 7.9–18 to these phenomena and certain visibility phenomena of Mercury also treated in *Almagest* 13.8. Proclus may be making reference to the separate work cited in the scholion as well as to the *Almagest*.

lion is not referring to *Almagest* 13.8, however, since it states that the work in question contained an explanation of positional terminology in Babylonian planetary observation reports, which is in fact not to be found anywhere in the *Almagest*. Like the lost work on stellar visibility, this seems to have been Ptolemy's in-depth handling of a subject that he treated more cursorily in the *Almagest*.

In his commentary on the *Almagest*, Pappus supplements his discussion of the armillary astrolabe whose construction Ptolemy sets out in *Almagest* 5.1 with information about a more complex version of the instrument, with nine rings instead of seven, derived from another work of Ptolemy's that Pappus designates as 'the constructed instrument that is called *meteoroskopeion*' (ἐν δὲ τῷ διακατασκευασμένῳ ὀργάνῳ ὃ καλεῖται μετεωροσκοπεῖον).[23] References to the *meteoroskopeion* also appear in Ptolemy's *Geography* 1.3 and Proclus, *Hypotyposis* 6, though without citation of a specific lost writing.[24]

If the foregoing trio of lost astronomical writings could be classified under the heading, 'more of the same,' others that receive mostly glancing references in later authors hint at facets of Ptolemy's intellectual activity that the extant works represent poorly if at all, especially concerning physics (in the ancient sense). The very first work of Ptolemy's listed in his *Suda* article is *Mechanics* (Μηχανικά) in three books. We know nothing about its contents beyond the implication of its title that it concerned manmade devices and machines, but, presuming it was authentic, it would have counted among Ptolemy's major compositions, and one would imagine that it took at least as theoretical an approach as Heron's *Mechanics* (which interestingly was also in three books). Works *On the Elements* (περὶ τῶν στοιχείων) and *On Weights* (περὶ ῥοπῶν) — or could they be a single work designated by two different descriptive quasi-titles? — are cited by Simplicius, *In Arist. De Caelo* (ed. Heiberg, pp. 20 and 710 for discussions of the behavior of mundane material bodies in and out of their natural places, while Eutocius, *In Archim. De Planorum Equil.* (ed. Heiberg, p. 306) attributes to Ptolemy's *On Weights* a definition of weight.

Simplicius, *In Arist. De Caelo* (ed. Heiberg, p. 9) also refers to a work in a single book (μονόβιβλος) called *On Dimension* (περὶ διαστάσεως), in which Ptolemy presented the same argument as appears in the *Analemma* that there can be only three orthogonal dimensions. Lastly, we have no title for the text ('in some book', ἔν τινι βιβλίῳ) in which, according to Proclus, *In Eucl. Elementa* (ed. Friedlein, pp. 191 and 362–67) and al-Nayrīzī's commentary on the *Elements* (ed. Besthorn & Heiberg, p. 118; ed. Curtze, pp. 65–66), Ptolemy

[23] The verb διακατασκευάω is a hapax legomenon, and perhaps the passage is corrupt, but Pappus clearly has a text attributed to Ptolemy since he follows the phrase quoted above with 'he says' (λέγει).

[24] Rome, 'L'Astrolabe'.

attempted a proof of Euclid's fifth postulate and applied this result to variant proofs of several propositions in *Elements* Book 1.

4. Chronology of the works

Knowing the order in which Ptolemy wrote his works might cast some light on the development of his thought. The cross references to the *Almagest* in the majority of his surviving astronomical writings as well as in the *Tetrabiblos* and *Geography* suffice to show that all these works were completed, if not entirely written, after the *Almagest*. This must also be true of the *Phaseis* and the lost book on the mathematical theory of stellar visibility phenomena summarized in the *Phaseis*'s introduction, since in *Almagest* 8.6 Ptolemy writes of stellar visibility theory as a complex and uncertain undertaking that he has chosen to dispense with 'for the time being' (ἐπὶ τοῦ παρόντος). The *Almagest* in turn cites astronomical observations that Ptolemy asserts that he made over a span of years from 127 through 141, and even if some of these are not genuine and untampered observations, one can safely presume that he would not have claimed to make an observation at a date manifestly before he was capable of doing so. Moreover, the allusion in *Almagest* 4.9 to repudiated earlier astronomical parameters that can be identified in the *Canobic Inscription* establishes that the treatise was not completed in the form we have it before the explicit date of the inscription's erection, the tenth regnal year (according to the Egyptian calendar) of Antoninus Pius, or AD 146–147. Hence almost all Ptolemy's other works on astronomical, astrological, and cartographical subjects are known to have been finished in the period after the *Almagest*, whereas only the *Canobic Inscription*, which is not a writing in the normal sense, can be dated with certainty to the twenty-year interval of Ptolemy's career preceding the *Almagest*'s completion.

Certain developments in Ptolemy's geographical knowledge and astronomical theories make it possible to obtain a plausible sequence for some of the post-*Almagest* works. In *Almagest* 2.6, Ptolemy asserts that the regions around the Earth's equator are 'untrodden' (ἄτριπτοι) by people from his part of the world (ἡ καθ' ἡμᾶς οἰκουμένη) so that one can only guess what the climate there is like. The astrological geography and ethnography of *Tetrabiblos* 2.2 likewise extends southward only as far as the equator. In the *Geography*, however, Ptolemy has learned (from the writings of Marinus of Tyre) of peoples and places located, so he believes, as far south as 16 5/12° south of the equator. Now the core of the *Geography* is a list of several thousand localities with their coordinates in longitude and latitude, grouped by 'provinces and satrapies' and ordered appropriately to provide the basis for systematically drawing a map of the known part of the world; a few hundred of these are singled out as 'noteworthy cities' (πόλεις ἐπίσημοι). The table of Noteworthy Cities in the *Handy Tables* turns out to comprise this same subset from the *Geography*, listed in the same order which had been determined by practical convenience for drawing the map. Turning now to astronomical considerations, the models and parameters built into the *Handy Tables* and presented in the *Planetary Hypotheses* occasionally differ from those of the *Almagest* and each other. In particular, the distinct models for the planets' motions in latitude in the three works make best sense as resulting from a process of simplification in which the *Planetary Hypotheses* represents the final stage.[25]

The concluding section of the *Canobic Inscription* associates the heavenly bodies and the mundane elements one-to-one (or in a few cases two-to-one) with a scale of musical pitches, such that the pitches ascend with increasing distance from the center of the cosmos. This same correlation, which so far as we can tell was devised by Ptolemy himself, was discussed as a harmonic foundation of astrological affinities between the heavenly bodies in *Harmonics* 3.16, one of the lost three closing chapters of the *Harmonics*; the evidence, which is compelling, is a surviving fragment either from the chapter itself or from a scholion or commentary.[26] No trace of the scheme can be found in the *Tetrabiblos*, though a different application of harmonics to astrological relations is introduced in *Tetrabiblos* 1.14. The *Harmonics* thus seems likely to have been a comparatively early work of Ptolemy's, perhaps completed before the *Almagest*. More subjectively, the epistemological discussions of the *Criterion* impress one as both simpler and cruder than those of the *Harmonics*, suggesting that the *Criterion* could belong to the very beginning of Ptolemy's career.[27]

For the remaining major treatise, the *Optics*, three considerations have been adduced as favoring a comparatively late date. First, there is the contrast between the extended discussion of refraction as affecting observed positions of heavenly bodies in *Optics* 5.23–30, which we have already mentioned, and the absence of anything comparable in the *Almagest*.[28] Second, in *Almagest* 1.3 and 9.2 Ptolemy refers to the phenomena that apparent sizes of heavenly bodies, and apparent angular distances separating heavenly bodies, appear larger when they are near the horizon, but in one passage he mistakenly attributes the effect to refraction in the atmosphere while in the other he provides no cause; by way of contrast, in *Planetary Hypotheses* 1B.7 and in *Optics* 3.59 he explains the phenomena psychologically, which is essentially correct.[29] Third, in *Geography* 1.1 Ptolemy seems to invoke the theory (familiar, e.g., from Euclid's

[25] Swerdlow, 'Ptolemy's Theories', pp. 41–71.

[26] Swerdlow, 'Ptolemy's Harmonics'.

[27] Feke, 'Mathematizing the Soul', offers further more or less subjective arguments that the *Criterion* antedated the *Harmonics*.

[28] Smith, *Ptolemy's Theory of Visual Perception*, p. 2.

[29] Smith, *Ptolemy's Theory of Visual Perception*, p. 2.

Optics) that visual perception occurs through rectilinear visual rays fanning out from the eyes with gaps between the rays that enlarge with greater distance, whereas in *Optics* 2.50–51 he rejects the concept of discrete rays.[30] However, none of these considerations constitutes a truly compelling argument for the sequence of the works in question.

Table 3 summarizes what we know or can plausibly guess about the sequence of Ptolemy's writings, where the *Canobic Inscription* and *Almagest* serve as the chronological anchor.

Firmly dated	*Subjectively dated*
	On the Criterion
	Harmonics
Canobic Inscription (AD 147/147)	
Almagest	
Tetrabiblos	
	Treatise on theory of stellar visibility (possibly Book 1 of *Phaseis*)
	Phaseis (possibly Book 2)
Handy Tables	
Arr. and Comp. Handy Tables	
Geography	
Planetary Hypotheses	
	Optics

Table 3. A plausible chronological sequence for some of Ptolemy's works. The *Planispherium* is also firmly dated to after the *Almagest*, and the work describing the *meteoroskopeion* to between the *Almagest* and the *Geography*, but their places in the sequence cannot be further narrowed.

5. Range and connectedness of Ptolemy's interests

Table 4 groups Ptolemy's works, both extant and lost, according to the disciplines by which one would most likely classify them on the basis of their overall subject matter. The primacy of astronomy — defined as the science concerning the nature, movements, and phenomena of the heavenly bodies in their own right — in this list is obvious, both by the number of the works and by their including the *Almagest*, the largest (by a considerable margin) and most highly structured treatise among them.[31] Moreover, significant references to astronomy occur in the *Harmonics, Tetrabiblos, Geography*, and *Optics*. At the same time, taking the lost works into consideration reinforces the realiza-

[30] Berggren and Jones, *Ptolemy's Geography*, p. 57, n. 2.

[31] The *Geography* comes next in bulk, but approximately five of its eight books consist simply of the cartographical data for constructing maps, and most of the eighth book is devoted to captions for the regional maps.

tion that Ptolemy was also deeply interested in phenomena of the sublunary world, whether these phenomena were such as he believed to be amenable to mathematical modelling or not. Conspicuously absent from the list is any work on a strictly biological topic.

Discipline	*Work*	*Preservation*
Astronomy		
	Canobic Inscription	Greek
	Almagest (13 books)	Greek
	Arr. and Comp. Handy Tables	Greek
	Handy Tables	Greek
	Planetary Hypotheses (2 books)	Greek (parts Arabic)
	Phaseis (possibly Book 2)	Greek
	Treatise on theory of stellar visibility (possibly Book 1 of *Phaseis*)	Lost
	On Paradoxical Phases of Venus	Lost
	Description of the *meteoroskopeion*	Lost
	Analemma	Latin (parts Greek)
	Planispherium	Arabic/Latin
Astrology		
	Tetrabiblos (4 books)	Greek
Cartography		
	Geography (8 books)	Greek
Epistemology		
	On the Criterion	Greek
Music theory		
	Harmonics (3 books)	Greek
Optics		
	Optics (5 books)	Latin
Physics and Mechanics		
	Mechanics (3 books)	Lost
	On the Elements	Lost
	On Weights	Lost
Mathematics		
	On Dimension	Lost
	Work related to Euclid's *Elements*	Lost

Table 4. Ptolemy's known works arranged by primary discipline

Cutting across classification by discipline are certain prevailing themes. One that is especially prominent in several of the more ambitious treatises is epistemology. Thus the *Harmonics* and the *Almagest* are both deeply concerned with appropriate strategies for applying sense perception (i.e. empirical observations and measurements with or without specially constructed apparatus) and reason (in particular mathematical analysis) to deduce knowledge of the 'hypotheses' or models underlying the phenomena respectively of musical pitch relations and the apparent behavior of the heavenly bodies; the explicit discussions in the *Harmonics* of the complementary roles of sense perception and reason as criteria (in the Greek philosophical sense) turn out to be highly relevant for grasping the more complex though largely unarticulated deductive structures of the *Almagest*. Book 1 of the *Geography* has an extended discussion of the relative value of different kinds of empirical data for determining absolute and relative locations of terrestrial places, and of methods for evaluating and correcting distorted data. The *Optics*, as a systematic study applying empirical observation, experiment, and deductive analysis to the nature visual perception and the relations (which are often subject to error) between perceived bodies and our perceptions of them, could be described as a study in the scientific epistemology of epistemology itself. In the light of these sophisticated treatments of the processes of acquiring knowledge about the external world, we might be less surprised that Ptolemy wrote a monograph largely devoted to the general topic of criteria than that this part of *On the Criterion* appears comparatively banal and disconnected from scientific applications.

The two central principles of Ptolemy's cosmology are the (originally Aristotelian) four-plus-one elements theory and the division of the cosmos into an inner 'sublunary' sphere in which the four elements earth, water, air, and fire predominate and an outer celestial spherical shell composed of bodies of ether. In the *Almagest* these principles are mostly kept in the background, though Ptolemy does ground his assumption that the heavenly bodies move with eternally uniform circular revolutions in a characterization of etherial bodies as eternal, unchanging, and divine (see for example 13.2). The three-dimensional geometry of these celestial bodies of ether, both visible and invisible, is the chief subject of *Planetary Hypotheses* Book 2, while the *Tetrabiblos* invokes the physical relationship between the celestial outer part of the cosmos and the enclosed sublunary sphere, such that the heavenly bodies are agents of generation and change in the complex, irregularly evolving sublunary world, as the rationale for the viability but inherent inexactness of astrological prediction. Among the non-astral-sciences works, the *Criterion* is particularly interesting for offering a materialistic theory of the composition of human souls, according to which ether is present in the soul and responsible for its intellectual capacity. This would provide a bridge between Ptolemy's notions of the human soul as having mathematical structures (*Harmonics* Book 3) and as having the power to introduce mathematically structured features into the external environment — by making music (*Harmonics* Book 1) and even simply by seeing through the rectilinear emission of a visual ray (*Optics*) — and his belief that the coordinated motions of the celestial etherial bodies are generated by celestial souls (*Planetary Hypotheses* Book 2).

Lastly, didactically appropriate, mathematically defined modes of representation of aspects of the cosmos are a broad concern of Ptolemy's. The *Plani-*

spherium, for example, is about representing celestial circles and revolutions in a single plane through stereographic projection, while in the *Almagest* and *Planetary Hypotheses* Ptolemy writes respectively about the formats of numerical tables and mechanical constructions as means of displaying underlying realities behind astronomical phenomena. Ptolemy's most extensive contribution to this theme, however, is the *Geography*, since this work is more or less entirely concerned with the best ways of displaying geographical information on planar surfaces and globes, in the latter case providing a terrestrial counterpart to the construction of a star globe in *Almagest* Book 8.

Appendix: Words and phrases characteristic of Ptolemy

The fourteen expressions discussed in this appendix are almost certainly not an exhaustive list of those that occur in more than one of Ptolemy's works but rarely or never in other authors; they were found by reading the texts with an eye for candidate expressions, followed by a TLG search. Unless otherwise noted, occurrences of the expressions are according to the editions used in the TLG. Occurrences in authors later than Hephaestion are excluded.

ἀμετάπιστος, 'not subject to change of belief'. While ἀμετάπειστος, thus spelled, is a frequent term in Aristotle and hence also in the Aristotelian commentators (as well as Plutarch and, with one instance, Diodorus), ἀμετάπιστος appears to be distinctive to Ptolemy: *Almagest* 1.1, *Criterion* 2.6 and 12.5. The two words are not identical in meaning, since ἀμετάπειστος is applicable to a belief or a believer that is not subject to alteration, whereas ἀμετάπιστος characterizes an object of thought *about* which belief cannot be altered.

ἐπιπολυπραγμονέω, 'to busy oneself additionally'. Unique to Ptolemy: *Tetrabiblos* 3.6.4, *Criterion* 8.3.

εὐκατανόητος, 'easily comprehended'. Very frequent in the *Almagest* (28 instances); also *Tetrabiblos* 1.11.5, *Harmonics* 1.1 and 1.11. Instances in texts not obviously influenced by Ptolemy: Polybius 18.30.11, Hipparchus *In Arati et Eudoxi phaenomena* 1.1.11 and 2.4.6, Serenus, *De sectione coni* ed. Heiberg p. 250 line 25. In texts influenced by Ptolemy: Porphyry, *Commentary on Harmonics* ed. Düring p. 20 line 9 (quoting *Harmonics*) and p. 133 line 13 (paraphrasing *Harmonics*), Pappus, *Commentary on Almagest* ed. Rome p. 98 line 27 (quoting Ptolemy), Theon of Alexandria, *Commentary on Almagest* ed. Rome p. 502 line 17 (quoting Ptolemy), p. 564 line 7, and p. 569 line 7. Ptolemy may have picked up the word from familiarity with Hipparchus.

εὐμεθόδευτον, 'easily carried out'. *Almagest* 1.10 and 13.4, *Planetary Hypotheses* 1.2. Theon of Alexandria, *Commentary on Almagest* ed. Rome

p. 451 line 2 (quoting Ptolemy), p. 602 line 8, *Great Commentary on Handy Tables* ed. Mogenet and Tihon v. 1 p. 102 line 20.

ἐφωδευμένος, 'worked out' or 'carried out'. Unique to Ptolemy: *Almagest* 2.13, 3.1, 3.4, 6.5, 12.9, 13.8, 13.11, *Analemma* ed. Heiberg p. 195 line 8, *Tetrabiblos* 3.1.1, 3.12.1, 4.9.1, 4.10.1, *Geography* 8.1.2, *Criterion* 3.3, 15.1. The compound προεφωδευμένος, 'previously worked out', occurs in *Almagest* 3.4, 9.9, 12.2, 12.7, 12.9 (3 instances), 13.4 (2 instances), and otherwise only in Strabo 12.8.8.

ἰδιοτροπία, 'characteristic tendency'. Very frequent in the *Tetrabiblos* (35 instances); also *Almagest* 1.1, 8.4, 9.2, *Harmonics* 3.7, *Geography* 2.1.8, *Criterion* 4.3. In texts not obviously influenced by Ptolemy: Aristides Quintilianus 3.26, Cleomedes 2.4, pseudo-Galen, *Prognostica de decubitu*, ed. Kühn (v. 19) p. 538 line 5. Influenced by Ptolemy: Hephaestion (18 instances). In the *Tetrabiblos* (and hence also Hephaestion) the term takes on a quasi-technical status.

κατὰ συνεγγισμόν + genitive, 'by adjustment to fit'. In this usage, unique to Ptolemy: *Planetary Hypotheses* 1.5, *Geography* 1.13.1, 2.1.2. The only other occurrences of κατὰ συνεγγισμόν, without genitive object and with the meaning 'by way of approximation', are in Hipparchus, *In Arati et Eudoxi phaenomena* 1.11.7 and 2.4.6.

κατὰ τὸ κεφαλαιῶδες, 'in summary manner'. *Almagest* 2.1, *Tetrabiblos* 1.3.20, 2.4.1, 2.14.12, 3.14.9, 3.4.4, *Criterion* 15.1. Influenced by Ptolemy: Hephaestion 1.20.1 (paraphrasing Ptolemy), 1.25.25.

κατὰ τὸ ὁλοσχερές, 'in a rough manner'. *Almagest* 9.5, *Tetrabiblos* 3.3.5. In texts not obviously influenced by Ptolemy: Geminus 2.20, 18.14. Influenced by Ptolemy: Hephaestion 1.1.13, 2.2.6 (paraphrasing Ptolemy). κατὰ τὸ ὁλοσχερέστερον, 'in a rougher manner'. *Almagest* 6.11, 8.6, 10.6, 11.5, *Arrangement and Computation of the Handy Tables* 1 (ed. Heiberg p. 161 line 1), *Tetrabiblos* 3.2.6.

κατὰ τὸν ἁρμόζοντα... λόγον/τρόπον, 'in the rationale/manner fitting for...' Unique to Ptolemy: *Tetrabiblos* 1.1.2, 3.7.1, 4.10.27alt,[32] *Harmonics* 2.9.

προσεντάσσω, 'to insert additionally'. *Almagest* 6.11, 8.3 (2 instances), 8.6, *Phaseis* 9 (ed. Heiberg p. 12 line 14). In texts not obviously influenced by Ptolemy: Heron, *Metrica* 2.15, Asclepiodotus 6.1, Philo Judaeus, *In Flaccum* 131.

[32] This refers to the 'alternate' conclusion of the *Tetrabiblos*'s final chapter, which Boll and Hübner did not adopt but is now widely regarded as the authentic version.

προσπαραμυθέομαι, 'to remark additionally'. Unique to Ptolemy: *Phaseis*, ed. Heiberg p. 13 line 21, *Arrangement and Computation of the Handy Tables*, ed. Heiberg p. 185 line 6, *Harmonics* 3.4, *Criterion* 4.2, 6.1.

συνεχεστέρα παρατήρησις, 'more sustained observation'. *Almagest* 1.8, *Planetary Hypotheses* 1.2. Influenced by Ptolemy: Theon, *Commentary on Almagest*, ed. Rome p. 338 line 15, p. 437 line 14 (quoting Ptolemy).

ὑποτετυπώσθω/ὑποτετυπώσθωσαν, 'let there have been sketched'. *Tetrabiblos* 2.4.1 (Boll adopts the variant reading ὑποτυπούσθω), 4.8.6, *Harmonics* 1.4, 2.3, 2.11, 3.4, *Geography* 1.2.1, 1.18.1, 2.1.1, *Criterion* 3.3. Influenced by Ptolemy: Hephaestion 1.25.25. προυποτετυπώσθω/προυποτετυπώσθωσαν, 'let there have been sketched beforehand'. *Tetrabiblos* 1.3.20 (variant reading not adopted by Boll or Hübner), 4.10.13. Influenced by Ptolemy: Hephaestion 2.26.12 (quoting Ptolemy).

Bibliography

Berggren, J. Lennart, and Alexander Jones, *Ptolemy's Geography. An Annotated Translation of the Theoretical Chapters*, Princeton: Princeton University Press, 2000.

Boll, Franz, 'Studien über Claudius Ptolemäus. Ein Beitrag zur Geschichte der griechischen Philosophie und Astrologie', *Jahrbücher für classische Philologie*, suppl. 21 (1894), pp. 49–224.

Bowersock, Glen Warren, *Greek Sophists in the Roman Empire*, Oxford: Oxford University Press, 1969.

Brunt, Peter A., 'The Bubble of the Second Sophistic', *Bulletin of the Institute of Classical Studies* 39 (1994), pp. 25–52.

Caussin de Perceval, Jean Jacques Antoine, 'Mémoire sur l'*Optique* de Ptolémée, et sur le projet de faire imprimer cet ouvrage d'après les deux manuscrits qui existent à la Bibliothèque du Roi', *Histoire et mémoires de l'Institut royal de France* 6 (1922), pp. 1–43.

Feke, Jacqueline, 'Mathematizing the Soul. The Development of Ptolemy's Psychological Theory from *On the Kritêrion and Hêgemonikon* to the *Harmonics*', *Studies in History and Philosophy of Science* 43 (2012), pp. 585–94.

Hamilton, Norman T., Noel M. Swerdlow, and Gerald J. Toomer, 'The Canobic Inscription. Ptolemy's Earliest Work', in J. Lennart Berggren and Bernard R. Goldstein (eds), *From Ancient Omens to Statistical Mechanics. Essays on the Exact Sciences Presented to Asger Aaboe*, Copenhagen: Munksgaard, 1987, pp. 55–73.

Heiberg, Johan L., *Opera astronomica minora*, Leipzig: Teubner, 1907 [*Claudii Ptolemaei opera quae exstant omnia*, vol. II].

Jan, Karl von, *Musici Scriptores Graeci*, Leipzig: Teubner, 1895.

Jones, Alexander, 'William of Moerbeke, the Papal Greek Manuscripts, and the Collection of Pappus of Alexandria in Vat. gr. 218', *Scriptorium* 40 (1986), pp. 16–31.

Jones, Alexander, 'Pseudo-Ptolemy *De Speculis*', *SCIAMVS* 2 (2001), pp. 145–86.

Jones, Alexander, 'A Posy of *Almagest* Scholia', *Centaurus* 45 (2003), pp. 69–78.

Juste, David, 'Pseudo-Ptolemy, *Centiloquium*', *Ptolemaeus Arabus et Latinus. Works*, URL = http://ptolemaeus.badw.de/work/24 (update: 03.10.2017).

Knorr, Wilbur R., 'Archimedes and the Pseudo-Euclidean *Catoptrics*. Early Stages in the Ancient Geometric Theory of Mirrors', *Archives Internationales d'Histoire des Sciences* 35 (1985), pp. 28–105.

Moraux, Paul, *Galien de Pergame. Souvenirs d'un médecin*, Paris: Les Belles Lettres, 1985.

Nutton, Vivian, 'Galen in the Eyes of his Contemporaries', *Bulletin of the History of Medicine* 58 (1984), pp. 315–24.

Rome, Adolphe, 'L'Astrolabe et le Météoroscope d'après le commentaire de Pappus sur le 5e livre de l'Almageste', *Annales de la Société Scientifique de Bruxelles* 47, 2 (1927), pp. 77–102.

Rome, Adolphe, 'Notes sur les passages des Catoptriques d'Archimède conservés par Théon d'Alexandrie', *Annales de la Société Scientifique de Bruxelles* 52, 1 (1932), pp. 30–41.

Rose, Valentin, *De Aristotelis librorum ordine et auctoritate commentatio*, Berlin: Reimer, 1854.

Sezgin, Fuat, *Geschichte des arabischen Schrifttums*, vol. VII: *Astrologie – Meteorologie und Verwandtes bis ca. 430 H*, Leiden: Brill, 1979.

Siebert, Harald, *Die ptolemäische Optik in Spätantike und byzantinischer Zeit: Historiographische Dekonstruction, textliche Neuerschließung, Rekontextualisierung*, Stuttgart: Steiner, 2014.

Smith, A. Mark, *Ptolemy's Theory of Visual Perception. An English Translation of the Optics with Introduction and Commentary*, Philadelphia: American Philosophical Society, 1996.

Staden, Heinrich von, 'Galen and the "Second Sophistic"', *Bulletin of the Institute of Classical Studies* 41 (1997), pp. 33–54.

Strohmaier, Gotthard, 'Galen's Not Uncritical Commentary on Hippocrates' *Airs, Waters, Places*', *Bulletin of the Institute of Classical Studies* suppl. 83, 2 (2004), pp. 1–9.

Swerdlow, Noel M., 'Ptolemy's *Harmonics* and the "Tones of the Universe" in the Canobic Inscription', in Charles Burnett, Jan P. Hogendijk, Kim Plofker and Michio Yano (eds), *Studies in the History of the Exact Sciences in Honour of David Pingree*, Leiden: Brill, 2004, pp. 137–80.

Swerdlow, Noel M., 'Ptolemy's Theories of the Latitude of the Planets in the *Almagest, Handy Tables,* and *Planetary Hypotheses*', in Jed Z. Buchwald and Allan Franklin (eds), *Wrong for the Right Reasons*, New York: Springer, 2005, pp. 41–71.

Taub, Liba, *Ptolemy's Universe. The Natural, Philosophical, and Ethical Foundations of Ptolemy's Astronomy*, Chicago: Open Court, 1993.

Tihon, Anne, 'Les *Tables Faciles* de Ptolémée dans les manuscrits en onciale (IXe–Xe siècle)', *Revue d'Histoire des Textes* 22 (1992) pp. 47–87.

Toomer, Gerald J., 'Ptolemy', in Charles C. Gillispie (ed.), *Dictionary of Scientific Biography*, vol. XI, New York: Scribner, 1975, pp. 186–206.

Toomer, Gerald J., 'Galen on the Astronomers and Astrologers', *Archive for History of Exact Sciences* 32 (1985), pp. 193–206.

Turner, Anthony J., *Mathematical Instruments in Antiquity and the Middle Ages*, London: Vade-Mecum Press, 1994.

Mathematical Methods in Ptolemy's *Analemma*

Nathan Sidoli

1. Introduction

This paper is an attempt to understand the mathematical methods found in Ptolemy's *Analemma* in the context of Greek mathematical practices. I first present an overview of the concepts we will need to read the text, followed by a close reading of select passages, which provide a clear overview of the structure of the argument and give examples of the mathematical methods involved.

In general terms, the *Analemma* provides a method for specifying the location of the sun in three pairs of locally orientated coordinate arcs as a function of three ostensibly empirical variables, namely the declination of the sun as a function of its longitude, $\delta(\lambda)$, the terrestrial latitude, φ, and the hour, η. The key to the approach is to represent the solid configuration in a plane diagram that Ptolemy calls the analemma. The mathematical argument begins with the presentation of a general argument that the analemma figure, or model, can be used to map arcs and lines of the solid configuration, through the example of a proof that this mapping is sound for one of the angles in question. It then proceeds to show that the analemma figure can be used to make computations of arc lengths of the solid sphere through two different methods: chord-table trigonometry, or what we can call analog, or nomographic, computation.[1] The final section of the received text details a series of physical manipulations through which we can compute the values of three pairs of coordinate arcs given the three variables $\delta(\lambda)$, φ, and η.

The *Analemma* has been the subject of a number of important studies, upon which I have drawn. The medieval Latin translation, accompanied by many useful notes that make sense of the mathematical methods, was first printed by F. Commandino in 1562, with no textual apparatus.[2] J. L. Heiberg made use

[1] The terminology of 'analog computation' is common in describing technical devices that use physical manipulation to produce a numerical result—such as a slide rule, or an astrolabe. The usage 'nomographic computation' follows that of Neugebauer, *A History*, pp. 839–856, in denoting a tradition of graphic procedures through which line segments or arcs can be physically measured.

[2] Commandino, *Ptolemaei Liber de analemmate*. This publication also contains Commandino's own work, *Liber de horologiorum descriptione*, which explains how to use the analemma methods set out in Ptolemy's *Analemma* to produce sundials.

Ptolemy's Science of the Stars in the Middle Ages, ed. by David Juste, Benno van Dalen, Dag Nikolaus Hasse and Charles Burnett, PALS 1 (Turnhout, 2020), pp. 35–77

10.1484/M.PALS-EB.5.120173

of this version in his critical edition of the Latin translation and the Greek fragments.[3] An excellent study of the mathematical conceptions underlying the text was made by P. Luckey.[4] All of this material was used by Neugebauer in his overview of ancient analemma methods.[5] D. R. Edwards provided a new critical edition of the Latin translation with an English translation, as well as a careful textual study of the work in his 1984 dissertation.[6] R. Sinisgalli and S. Vastola made an Italian translation, in 1992, accompanied by many useful notes and diagrams.[7] Finally, a full appreciation of the analemma methods must also involve some study of the substantial evidence of the medieval Arabic sources.[8]

In this paper, I make a close reading of key passages of the text, showing how each step of the argument can be understood as justified by other ancient mathematical sources, and how certain arguments are meant to be a justification, or summary of, mathematical practices that are not made explicit in the text. This results in an articulation and development of the approach of Luckey, Neugebauer and Edwards, which fleshes out many of the mathematical details in the context of ancient methods and which, I hope, helps us to understand Ptolemy's claim to be producing a more mathematical natural science.

2. Concepts and terminology

In this section, I introduce the concepts and terminology that we will need to read Ptolemy's text, without showing in detail how they can be derived from the sources.[9] This order of presentation—which may strike some readers as backwards—is motivated by the fact that the analemma approach is unknown to most modern readers, whereas it appears to have been well-known to ancient and medieval readers familiar with the mathematical sciences. Following this, I will make a close reading of key passages of the *Analemma*, arguing along the way that all of these concepts and techniques can be derived directly from the ancient and medieval concepts.

[3] Heiberg, 'Ptolemäus de Analemmate'. Heiberg somewhat revised this version in his *Opera astronomica minora*, pp. 189–223. I have mostly relied on the later version in this study.

[4] Luckey, 'Das Analemma'.

[5] Neugebauer, *A History*, pp. 839–856.

[6] Edwards, *Ptolemy's* Περὶ ἀναλήμματος.

[7] Sinisgalli and Vastola, *L'Analemma*.

[8] An incomplete selection of such material are the following: Schoy, 'Abhandlung'; Id, 'An Analemma Construction'; Kennedy and 'Id, 'A Letter of al-Bīrūnī'; Kennedy, 'Ibn al-Haytham's Determination'; Berggren, 'A Comparison'; Berggren, 'Ḥabash's Analemma'; Carandell, 'An Analemma', and Suzuki, 'A Solution'.

[9] This material takes its point of departure from the work of Luckey, 'Das Analemma'; Neugebauer, *A History*, pp. 839–856, and Edwards, *Ptolemy's* Περὶ ἀναλήμματος. Between when I wrote this paper and when it appeared in press, a useful summary of the text, including excellent diagrams, was made by Guerola Olivares, *El Collegio Romano*, pp. 67–132.

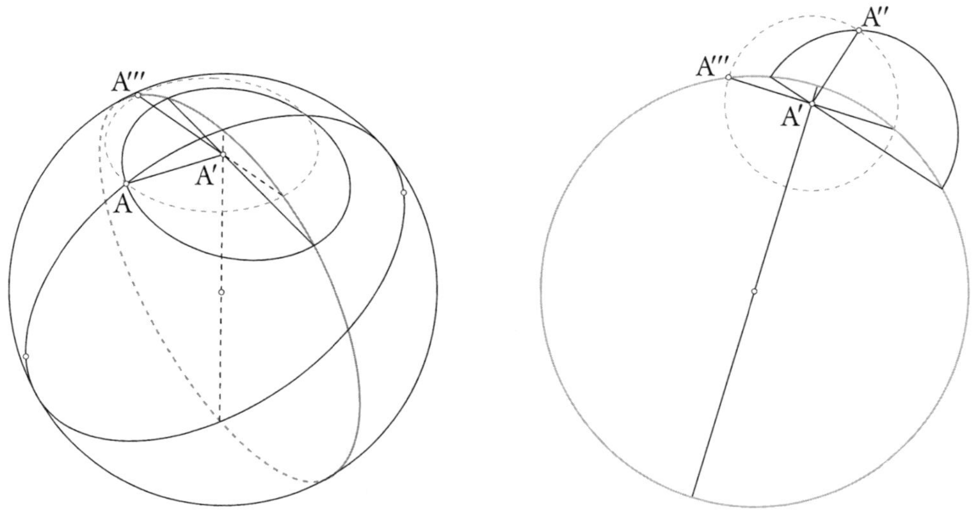

Figure 1: Analemma methods 1: (left) perspective diagram of a point A on a sphere to be mapped to both a lesser circle orthogonal to the analemma and to the great circle joining it with the poles of the analemma circle in the receiving plane, in solid gray; (right) representation in the plane of the analemma of A, which appears in three different representations: (1) as $A \mapsto A'$ in its orthogonal projection into the analemma, (2) as $A \mapsto A''$ in its location on a lesser circle orthogonal to the analemma, and (3) as $A \mapsto A'''$ in its location on the great circle joining point A with the pole of the analemma circle. The original point A does not appear in the analemma representation, because, visually, it coincides with A'.

2.1. Analemma methods

The key to the use of the analemma as a problem-solving device lies in the application of four projective constructions, namely

(M.1) orthogonal projection of individual points into the receiving plane of the analemma,

(M.2) orthogonal projection of great and lesser circles into the lines of their diameters in the receiving plane of the analemma,

(M.3) orthogonal rotation of individual points into the receiving plane of the analemma, and

(M.4) orthogonal rotation of great and lesser semicircles into semicircles in the receiving plane of the analemma.

In all of these geometric transformations, the magnitudes of lines and arcs are preserved, and in the analemma figure we find the same object represented in multiple ways. Some examples will suffice to show the strategy.

A common way of mapping a point in two ways onto the analemma—which we will see Ptolemy perform three times in this account—is seen in Figure 1.

Here we see point A on the sphere in Figure 1 (left), which we will represent both on a lesser circle perpendicular to the analemma, in gray, and on the great circle that joins point A with the poles of the great circle of the analemma. In order to do this, in Figure 1 (right), we represent A by its orthogonal projection in the plane of the analemma, A' (M.1), and draw through A' a line as the diameter of a lesser circle in the sphere that is perpendicular to the analemma, which can also be regarded as the orthogonal projection of the lesser circle into the analemma (M.2). We then rotate this lesser circle into the plane of the analemma by constructing a semicircle on this line and erecting $A'A''$ perpendicular to the diameter of the lesser circle (M.4 and M.3). The length of line $A'A''$ on the analemma will be constant no matter what lesser-circle diameter we draw through A' and it is equal to AA' on the sphere. Next, we effect the mapping of A in its location on the great circle joining A with the poles of the analemma onto the plane of the analemma by a two-stage process. First, (1) we join the orthogonal projection, A' with that of the poles, the center of the sphere (M.1)—that is, by joining A' with the center of the circle of the analemma. This gives us the diameter of this great circle as the orthogonal projection of the great circle into the plane of the analemma (M.2). Next, (2) we find the orthogonal rotation of A on the sphere to A''' on the analemma by rotating point A along the circumference of a circle of radius AA', shown in a gray dotted line, with the axis of rotation being the diameter of the great circle into which we project A (M.3). We do this in the analemma by producing a circle around A' with distance $A'A''$, since this length is equal to AA' on the sphere. This circle, shown in a gray dotted line, however, is not drawn in the plane of the analemma—presumably because arc lengths are not preserved on this circle.[10] In this way, we can exhibit A mapped to A' by orthogonal projection, and to both A'' and A''' by rotation into either a lesser circle or the great circle that joins A with a pole of the analemma.

In Figure 2 (left), if points A and B in the sphere lay on a great circle passing through the poles of the gray analemma circle, they can be projected into the analemma plane orthogonal to the great circle between them by dropping perpendiculars to the plane of the analemma meeting the diameter of the great circle joining them at A' and B' (M.1). When this is represented in the analemma, Figure 2 (right), the line joining A' and B' must pass through the center of the circle, because A and B lie on a great circle that passes through the pole of the analemma (M.2). Arc α of the great circle between A and B is found in the plane of the analemma by constructing perpendiculars at A' and B', mapping them to A'' and B'' (M.1 and 2). In this way, lines $A'A''$

[10] That is, while AA''' is always a quadrant on the sphere, $A''A'''$ is not generally a quadrant on the analemma.

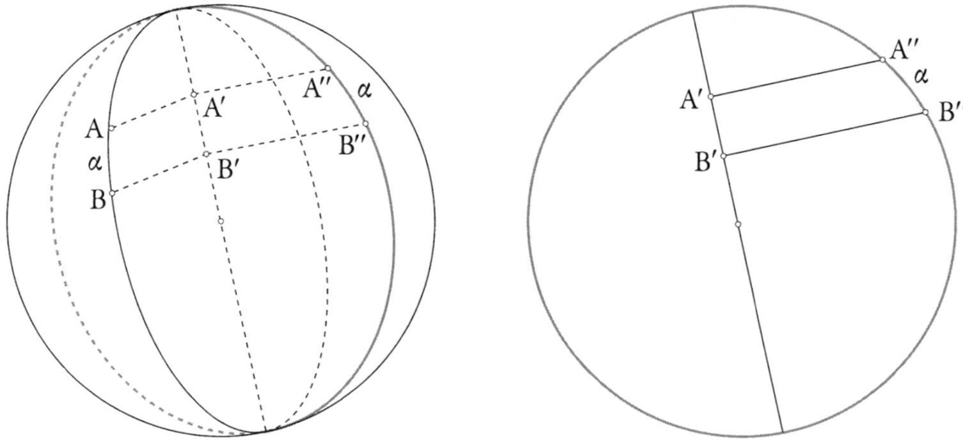

Figure 2: Analemma methods 1: (left) perspective diagram of two points A and B on a sphere, and the great-arc distance between them, α, to be mapped to the receiving plane, in gray; (right) representation in the analemma of A and B as orthogonally projected onto the analemma such that $A \mapsto A'$ and $B \mapsto B'$, along with orthogonal rotation of the great circle into the analemma such that $A \mapsto A''$ and $B \mapsto B''$. The original points A and B do not appear in the analemma representation, because, visually, they coincide with A' and B'.

and $B'B''$, in the analemma, are equal to the perpendiculars dropped from A and B into the receiving plane, in the sphere, and the length of arc α, the great-arc distance between the two points, is preserved in the transformation.

In Figure 3 (left), if points A and B in the sphere lay on a lesser circle perpendicular to the gray analemma circle, they can be mapped into the plane of the analemma circle by rotating the lesser circle into the plane of the analemma—or rather, folding it into two semicircles that are rotated into the same position in the plane of the analemma. This is represented in the analemma, Figure 3 (right), by dropping perpendiculars into the receiving plane, such that A' and B' represent points A and B in the analemma (M.1), and the line joining them is a diameter of the lesser circle and its orthogonal projection into the analemma (M.2). The lesser circle is then folded and rotated into the analemma by erecting a semicircle on the diameter of the lesser circle (M.4). Arc α of the lesser circle is rotated into the plane of the analemma by constructing perpendiculars at A' and B' meeting the semicircle, such that A maps to A'' and B to B'' (M.3).[11] Once again, lines $A'A''$ and $B'B''$, in the analemma, are equal to the perpendiculars, AA' and BB', in the sphere,

[11] The practice, in dealing with a solid configuration, of rotating one plane into another by constructing the objects in the plane to be rotated directly in the receiving plane is common in Greek geometry. See, for example, the solid constructions by Diodorus or Eutocius; Hogendijk, 'The Geometrical Works', pp. 56, 70–71, and Sidoli, 'Review of *The Works of Archimedes*', pp. 160–61.

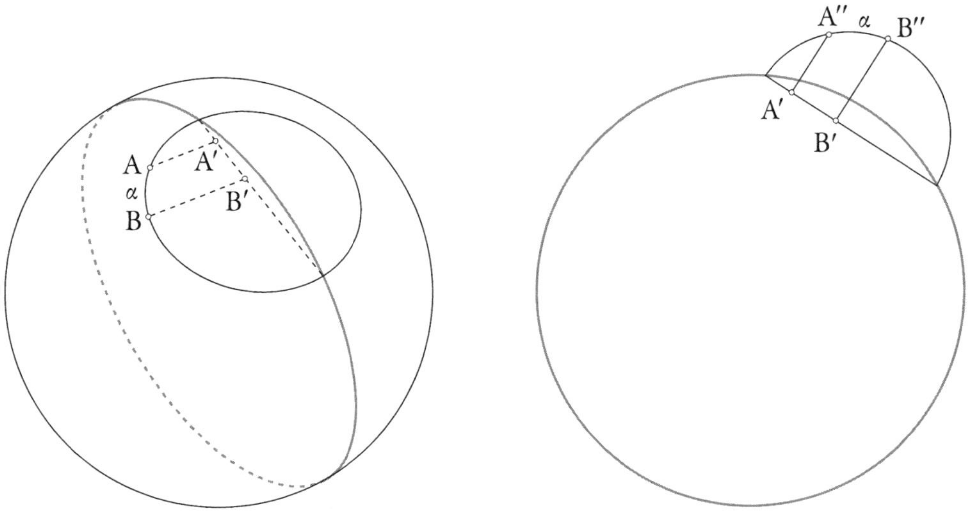

Figure 3: Analemma methods 2: (left) perspective diagram of two points A and B on a lesser circle of a sphere, and the lesser-arc distance between them, α, to be mapped to the receiving plane, in gray, which must be perpendicular to the lesser circle and pass through its poles; (right) representation, in the analemma, of a semicircle of the lesser circle as rotated into the plane of the analemma, such that $A \mapsto A'$ and $B \mapsto B'$, by orthogonal projection, and to $A \mapsto A''$ and $B \mapsto B''$, by rotation.

dropped from A and B into the receiving plane, and arc α is equal to the lesser-arc distance between the two points.

In this section, I have used the term *analemma* as synonymous with the receiving plane of a projection, or mapping. This understanding of the term agrees with Ptolemy's usage in his *Analemma*, and, as Edwards has argued, best conforms with the various functions of the term in ancient sources.[12] Hence, the analemma is the receiving plane of a projection, which is performed by carrying out constructions directly in the plane.[13]

In Ptolemy's *Analemma*, we will see evidence for these various projective operations. Although he explicitly refers to rotating the semicircles of lesser circles, Ptolemy has no special terminology for orthogonal projection of points and circles, and the text speaks only of producing perpendiculars and of taking the diameters of circles in the analemma. Nevertheless, as we read through *Analemma* 6, below, we will see that the underlying solid configuration is essentially that described in this section.

[12] Edwards, *Ptolemy's* Περὶ ἀναλήμματος, pp. 1–10.

[13] This way of producing projective constructions is also found in Ptolemy's *Planisphere*; see Sidoli and Berggren, 'The Arabic Version'.

2.1.1. *Geometrical constructions and instrumental practice*

Although there are almost no synthetic theorems in analemma texts—and only one in Ptolemy's *Analemma*—there are constructions in all uses of the analemma in geometric problem-solving, in both ancient and medieval texts. The types of constructions employed, however, are clearly restricted. In fact, I am not aware of any constructive step in an ancient or medieval analemma problem that uses an operation that cannot be reduced to applications of *Elements* I.posts.1–3, I.11, and 12—that is, the first three postulates of Euclid's *Elements* and the two problems that produce perpendicular lines.[14] Even more, analemma constructions can all be regarded as abstractions of the use of a compass and a set square. Indeed, the three ancient texts that deal with the analemma make explicit mention of various types of instrumental practice, indicating that the mathematical methods of the analemma were closely associated with certain instruments.

As well as referring to operations performed on instruments such as specially prepared plates and hemispheres, analemma texts prescribe the instruments used to carry out geometric constructions. The analemma described by Vitruvius, *Architecture* IX.7, although not addressing a problem, is explicitly produced with a compass,[15] and Ptolemy, as we will see below, explicitly introduces both the compass and the set square. It seems clear that constructions in analemma texts were limited to abstractions of the operations that can be performed with these instruments—that is, a finite set square whose side is just a bit greater than the diameter of the great circle of the analemma, and a finite compass whose radius is just a bit greater than that of the great circle of the analemma, and which can be operated with a given radius.

In order to make this underlying instrumental practice explicit, in the following I will note how each construction on the analemma can be performed with either the compass or the set square.

2.2. Ptolemy's notion of 'model'

Ptolemy's uses of the term *hupothesis* (ὑπόθεσις), and the cognate verb (ὑποτίθεται), are far ranging. These terms often have the sense—implied by the basic meaning of the words—of what is set down in the beginning to be built upon, and, indeed, they are translated literally by Moerbeke, in his Latin translation of the *Analemma*, with *suppositio* and *supponere*. They may also, however, indicate the assumption of a fully elaborated depiction of the structure and function of the objects under investigation.

[14] As we will see below, the proof in *Analemma* 6 also requires *Elements* XI.12, but this is a theorem, not a problem.

[15] See Soubiran, *Vitruve. De l'architecture*, pp. 26–30.

Moreover, a Ptolemaic *hupothesis*, unlike a modern scientific hypothesis, is not subject to testing; in fact, it may sometimes be demonstrated, or even saved—as occasionally in the *Almagest* or in *Harmonics* I.2.[16] G. J. Toomer emphasized that Ptolemy's understanding of a *hupothesis* is often far removed from the modern sense of a hypothesis in scientific discourse, and pointed out that it was closer to our idea of a model.[17]

An *hupothesis* in Ptolemy's writings can be as simple as the assumption of the immobility and sphericity of the earth, or as complicated as the full geometric configurations for the moon or Mercury;[18] either purely arithmetical, as in harmonics, or fundamentally geometric, as in astronomy;[19] an idealized mathematization, as we will see in the *Analemma*, or closely connected with a physical representation, as in *Planetary Models* (commonly *Planetary Hypotheses*) I.1 and 2, or *Almagest* XIII.3.[20]

This broad notion of a conceptual tool for explanation and computation has more overlap with our concept of model than our concept of hypothesis. Hence, in my translation of passages of the *Analemma* I will translate ὑπόθεσις and its cognates with *model* and its cognates.[21] Some readers may find this excessively modern, but I hope the discussion here will help us avoid unwanted anachronism.

Ptolemy's use of modeling can be compared to that of Hellenistic authors working in the exact sciences such as Autolycus, Euclid, Aristarchus, Archimedes and Eratosthenes—all of whom used geometric modeling as the basis of their work in astronomy, mechanics, optics and harmonics.[22] A distinction can be drawn, however, between Ptolemy's use of *hupothesis* and that of Aristarchus in *On the Sizes and Distances of the Sun and the Moon* and that attributed to Eratosthenes by Cleomedes in *On Heavens* I.7.[23] Aristarchus and Eratosthenes used hypotheses both (a) to set out the overall geometrical configuration that

[16] Heiberg, *Syntaxis mathematica*, vol. II, pp. 26, 180, 461, and Düring, *Die Harmonielehre*, p. 5.

[17] Toomer, *Ptolemy's Almagest*, pp. 23–24.

[18] Heiberg, *Syntaxis mathematica*, vol. I, pp. 26, 350; vol. II, p. 255.

[19] Düring, *Die Harmonielehre*, p. 5.

[20] See Heiberg, *Opera astronomica minora*, pp. 70–74; Hamm, *Ptolemy's Planetary Theory*, pp. 72–76; Murschel, 'The Structure'; Heiberg, *Syntaxis mathematica*, vol. II, pp. 532–533. Jones, 'Ptolemy's Mathematical Models', gives an overview of the various ways mathematical modeling functions in Ptolemy's work.

[21] This is also done, for example, by E. Hamm, *Ptolemy's Planetary Theory*, in her translation of Ptolemy's *Planetary Models* I, Part A.

[22] There has been much debate over whether or not Euclid composed the *Division of the Canon*, but there seems to be no objective way to decide the issue; see Barbera, *The Euclidean Division*, pp. 3–29.

[23] Heath, *Aristarchus of Samos*, pp. 352–411, and Todd, *Cleomedis Caelestia*, pp. 35–37.

serves as the basis of the model, and also (b) to set out quantitative assumptions that are, at least in principle, empirically decidable and which serve as a basis for computation.[24] Ptolemy, however, does not use *hupothesis* in this second sense. For Ptolemy the *hupothesis* is only the general geometric configuration of the model, whereas the quantitative parameters to be determined through observation are not referred to by the term *hupothesis*.

Hence, by Ptolemy's time, and probably for a long while before, there was a fairly clear distinction between what we would think of as the model as a basis for computation and the given values that are used in the computation. As we will see in the *Analemma*, although there is no discussion of empirical practice and both the geometric model and the quantitative parameters are simply assumed in the course of the argument, there is a clear linguistic distinction between the two—the model itself is asserted as assumed and the parameters are asserted as fixed, or determined, although they may, of course, vary.

2.3. The two-sphere model

The *two-sphere model* is a name given by modern scholars to the model of the cosmos found in texts such as Autolycus' *Moving Sphere* and *Risings and Settings*, Euclid's *Phenomena*, and Theodosius' *Days and Nights* and *Habitations*. In this model, the sun is taken as located in varying positions on the ecliptic as a great circle in the sphere of the cosmos, which contains the fixed stars. The sphere of the cosmos, carrying the ecliptic, rotates about the celestial poles, P_n and P_s, creating the celestial equator, to which the ecliptic is skew at the angle known as the obliquity of the ecliptic, ε. The sphericity of the earth is only accounted for by the fact that the horizon, which is also a great circle, is generally skew to the ecliptic and the celestial equator, and divides the cosmos into two hemispheres—above and below. In Figure 4, if the eastern point is taken to be in the direction of the viewer, since the horizon is immobile, the sphere of the cosmos is imagined to rotate clockwise. This configuration was used by ancient mathematicians to model the phenomena we associate with spherical astronomy—namely, the solar and stellar phenomena related to local coordinates, determined by the horizon and the local meridian.

Whatever the mathematicians of the Hellenistic period may have thought of this construction, by Ptolemy's time it must certainly have been thought of as a model in the sense discussed above—that is, as a simplified configuration that was known to not be a strict representation of reality, but which could be used mathematically without any significant loss of accuracy. Trivially, the earth is not actually a point, but a sphere—which is what allows us to speak

[24] Berggren and Sidoli, 'Aristarchus's *On the Sizes and Distances*', pp. 231–234; Carman, 'Two Problems', pp. 55–58; Carman and Evans, 'The Two Earths'; Sidoli, 'Mathematical Discourse'.

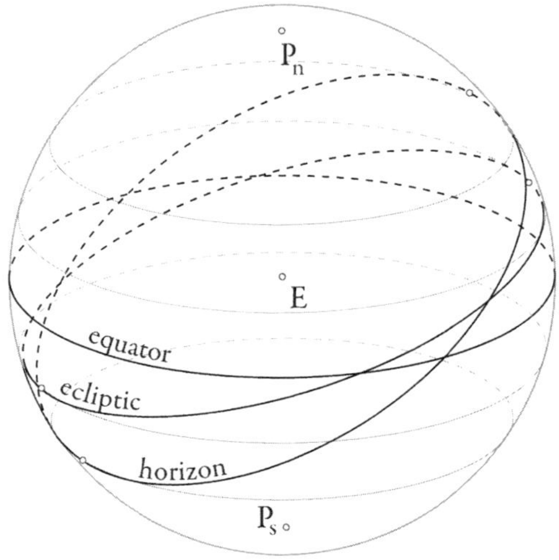

Figure 4: The 'two-sphere' model

of a region above and below the horizon. It is so small in comparison to the cosmos, however, that it can be regarded as a point. That this is a strictly false, but observationally adequate, simplifying assumption must also have been clear to the Hellenistic mathematicians. Secondly, at least by Ptolemy's time, the sun was not actually a point in the celestial sphere, but was rather closer to the earth, below the outer planets, and with a varying distance from the earth. Hence, the two-sphere model could not have been regarded as an accurate depiction of the sun in its relation to the earth, but simply as a mathematical model depicting the perceived location of the sun on the sphere of the cosmos from the perspective of the earth.

There are clear indications in *Analemma* 2 and 3 that Ptolemy thought of the overall model of the cosmos in just these sorts of perceptual terms. In *Analemma* 2, Ptolemy calls this simplified model of the sun in the cosmos, orientated to the local horizon, the 'world sphere', and he says that the great circles in this sphere that can be taken to determine the position of the sun 'move with the sun'.[25] Hence, they can be imagined to be great circles of the world sphere laying in planes that pass through the sun. Furthermore, when he describes the position of the sun in more detail, both in *Analemma* 2, in general terms, and in *Analemma* 3, in setting out the model with letter-names, he refers to the solar position as the 'solar ray'—which we can understand as the line along which we see the sun, drawn from the earth, through the sun, out to the sphere of the cosmos.

[25] These great circles are discussed in detail below.

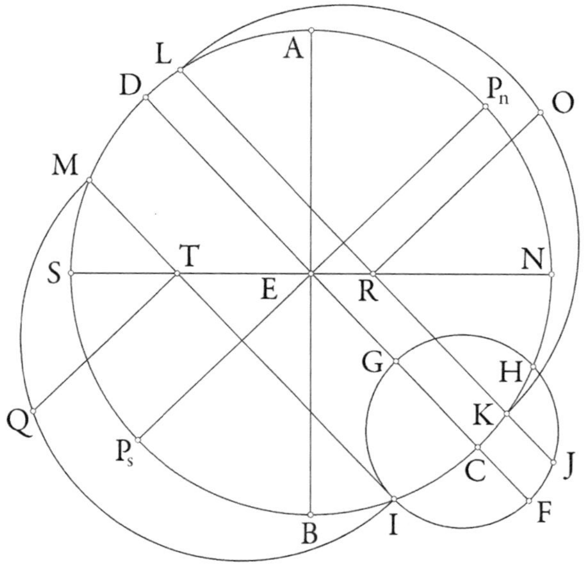

Figure 5: The analemma model

2.4. The analemma model

The analemma model can be found in three ancient sources—Vitruvius' *Architecture* IX.7, Heron's *Dioptra* 35, and Ptolemy's *Analemma*—and a few medieval Arabic sources. Although in medieval sources, analemma methods are used to solve a range of problems in spherical astronomy, in the extant ancient sources, they are almost always used on the analemma model described in this section.[26] The model itself is naturally suited to handle the seasonal hours of Greco-Roman daily life, η_s, and hence was closely associated with *gnomonics* (γνωμονική), the science of sundials. Here, I simply describe the model, with no attempt to derive this description from the sources.

The ancient analemma model is orientated towards a coordinate system of the local horizon and meridian, and can model the motion of the sun, on both its annual and daily paths—that is, the model can be used to specify the location of the sun, relative to local coordinates, given the terrestrial latitude, φ, the declination of the sun as a function of its longitude, $\delta(\lambda)$, and the time of the day, in hours, η.

In Figure 5, the local meridian is the great circle of the analemma *NBSA*, line *NS* is the orthogonal projection of the great circle of the horizon, line *CD* is that of the great circle of the equator, and line P_nP_s is the line joining

[26] An exception is Ptolemy's *Planisphere* 18, which employs an analemma construction; see Sidoli and Berggren, 'The Arabic Version', pp. 132–133.

the celestial poles, perpendicular to *CD*. The orientation of lines *NS* and *CD* determine the terrestrial latitude, φ, because, in great circle *NBSA*, arc NP_n is the elevation of the pole—that is, arc $NP_n = \varphi$.

Since the horizon is motionless, while the cosmos rotates about P_nP_s, the orthogonal projection of both the horizon and the equator into the meridian will remain fixed. The same is not true, however, of the ecliptic. Indeed, the ecliptic will only be orthogonal to the analemma twice daily, when the solstitial colure coincides with the local meridian—and, indeed, the ecliptic is not represented as such in Ptolemy's *Analemma*.

The proper position of the sun on the ecliptic, λ, can, however, be modeled on the analemma with the use of an auxiliary circle, *FHGI*, arranged such that its center lies on the diameter of the equator and it cuts the great circle of the analemma so that the arcs *CH* and *CI* are both equal to the obliquity of the ecliptic, ε. In this case, where *F* represents the vernal equinox, *H* the summer solstice, *G* the autumnal equinox, and *I* the winter solstice, if arc $FJ = \lambda$ is cut off equal to the arc of solar longitude from the vernal equinox at Ari 0°, then arc *CK* of the analemma will be equal to the declination of the sun at this time, $\delta(\lambda)$.[27]

Then, since throughout the course of a day the sun will travel on a course roughly coinciding with the circle of its declination, which we can call its day-circle, the local position of the sun can be modeled on the day-circle folded and rotated into the plane of the analemma.[28] In Figure 5, when the sun is at $FJ = \lambda$, in its annual course, it can be imagined to travel uniformly on semicircle *KOL* throughout the course of the day; or when it is at the winter solstice, $FJI = \lambda$, it will travel on semicircle *IQM*. For example, if a given day in the spring or summer begins at midnight, the sun will start at, say, *K* and move along arc *KO* until sunrise, passing over the horizon at point *O*, and then move up to midday at *L* and return back along *LO* in the afternoon to sunset at *O*, finally returning to *K* at the following midnight. In fact, the second position of *K* will be somewhat altered because of the daily longitudinal movement of the sun of about 1°, but analemma methods do not take this into account.

The final given magnitude, the hour η, is marked off along the day-circle. In the case of the seasonal hours of daily life, η_s, each of the arcs *LO*, *OK*, or *MQ*, *QI* are divided into six equal parts for the six hours between the horizon and the meridian. Although this is not discussed in the ancient sources, it

[27] Neugebauer, *A History*, p. 845, shows this using methods consistent with Greek geometric practice.

[28] Heron, in his *Dioptra* 35, refers to this circle as the 'daily circle' (ἡμερήσιος κύκλος); see Schöne, *Herons von Alexandria Vermessungslehre*, pp. 302–306, or Acerbi and Vitrac, *Metrica*, pp. 103–106.

would also be possible to model the astronomer's equinoctial hours, η_e, by dividing the complete semicircle of the day-circle into twelve equals parts.

In fact, however, Ptolemy's *Analemma* works with further simplifying assumptions. In the first place, the circle *FHGI*, which is called the *menaeus* (from μηναῖος, meaning 'monthly') by Vitruvius,[29] is not found in Ptolemy's presentation. Instead, he simply takes δ as given at one of four solar declinations corresponding to the beginnings of the twelve zodiacal signs—namely,

$\delta = 0°$ for $\lambda \approx 0°, 180°$,
$\delta = 11\,{}^2\!/\!_3{}°(= 11;40°)$ for $\lambda \approx 60°, 120°, 240°, 300°$
$\delta = 20\,{}^1\!/\!_2{}°(= 20;30°)$ for $\lambda \approx 30°, 150°, 210°, 330°$, and
$\delta = 23\,{}^1\!/\!_2\,{}^1\!/\!_3{}°(= 23;50°)$ for $\lambda \approx 90°, 270°$,

and refers to the semicircle constructed at a given declination as the 'monthly circle' (μηνιαῖος κύκλος). Hence, in what follows, for the sake of explicating his text, I will use Ptolemy's terminology and refer to the day-circle of the sun as its month-circle.

These three declinations may have been determined, for example, through a table such as that in *Almagest* I.15, or they may have been values taken by Ptolemy from previous work in gnomonics, having been computed using chord-table trigonometry directly on the analemma model.[30] Whatever the case, although these declinations are those of evenly distributed longitudes of 30°, they are, as declinations themselves, rather unevenly distributed—since their differences are 11;40°, 8;50°, and 3;20° respectively. This is perhaps an indication that, in the *Analemma*, Ptolemy was more interested in the symmetry of his presentation, and the role of symmetry in his instrumental practice, than in the precision of any device that might be made with these methods.

As we have just seen, both Vitruvius and Ptolemy speak of a 'monthly' division of the annual solar cycle, presenting us with a kind of zodiacal month. Of course, there is no discussion of the duration of these months, and given the level of precision evident in Ptolemy's presentation this is probably not important, but these months are clearly a division of the sun's annual progress through the stars into twelve parts. The ancient tool that was used to track the course of the sun through the stars, often noting its passage into each of the twelve zodiacal signs, was the parapegma.[31] Hence, the analemma appears to have been directly related to the two most conspicuous devices used to

[29] Soubiran, *Vitruve. De l'architecture*, p. 29.

[30] It may be significant that Ptolemy states the declination using the proper parts (unit fractions) of standard Greek arithmetical practice, not the sexagesimal fractions of his mathematical astronomy. The values used in the *Analemma* are what we would get if we rounded the values in the *Almagest* to the nearest 0;05°—see *Almagest* I.15, Heiberg, *Syntaxis mathematica*, vol. I, p. 72. We do not know if Ptolemy derived these values in this way.

[31] See Lehoux, *Astronomy*, especially pp. 70–97.

regulate the cycles of daily life in the Greco-Roman world: the sundial and the parapegma.[32]

From the description given in this section, it is clear that, like the two-sphere model, the analemma model functioned as a simplified geometrical configuration that facilitated geometrical and computational problem-solving. We will see below, in reading passages of the *Analemma*, that Ptolemy's mathematical practice clearly distinguished between the assumption of the model itself, as an overall geometrical configuration, and the assumption of given values that may be assumed as parameters of the problem-solving activity.

3. Ptolemy's *Analemma*

The text of the *Analemma* is known to us from two sources—fragments of a 5th–7th century text palimpsested in *Ambrosianus graec.* L 99 sup., **Am**, and a 13th century Latin translation by William of Moerbeke contained in *Vaticanus Ottobonianus lat.* 1850, **O**—which is probably also incomplete. The Ambrosianus codex is an 8th century copy of Isidore's *Etymologiae* that contains eight bifolia that were repurposed from manuscripts that once contained mathematical material, of which twelve pages came from a copy of Ptolemy's *Analemma*.[33] The Ottobonianus codex is an autograph by Moerbeke of his translations of Greek mathematical works, focusing on Archimedes, of which the final three folia contain his translation of the *Analemma* (ff. 62–64)—ending somewhat abruptly with a single mathematical table and no colophon. The Latin text, with many corrections, was first published by Commandino,[34] the Greek fragments and the Latin text were critically edited by Heiberg,[35] and the Latin was reedited by Edwards, who also provided an English translation.[36]

[32] It is worth noting that Ptolemy's own parapegma text, the *Phases of the Fixed Stars*, does not work with zodiacal months, but rather divides a solar year into the twelve 30-day months of the Egyptian calendar, which was also used for astronomical purposes; see Lehoux, *Astronomy*, pp. 261–309. On the other hand, given the low level of precision evidenced in the *Analemma* itself, it is possible that Ptolemy regarded his 'monthly circles' as corresponding to these Egyptian months.

[33] The pages of the codex that contain the *Analemma*, in the order of the Ptolemaic text, are as follows: 119, 120, 139, 140, 137, 138, 143, 144, 129, 130, 117, 118.

[34] Commandino, *Claudii Ptolemaei Liber de analemmate.*

[35] Heiberg, 'Ptolemäus de Analemmate', and Id., *Opera astronomica minora*, pp. 189–223.

[36] Edwards, *Ptolemy's* Περὶ ἀναλήμματος. Since we do not have Greek for the whole text, I will often use the Latin text as the primary source. By relying on Moerbeke's translation of Ptolemy's *Tetrabiblos* and of the Archimedean corpus, I will not always translate the Latin literally, but will make some informed guesses about the original Greek terminology behind the Latin even where we do not have corresponding fragments in the palimpsest. See Clagett, *Archimedes*; and Vuillemin-Diem and Steel, *Ptolemy's* Tetrabiblos.

Although I have made my own translation of the text, I have often been guided by that of Edwards, and in the mathematical passages there is little significant difference.

Although the Greek manuscript numbers the text differently than Heiberg, and the Latin manuscript presents the treatise in continuous prose, I will follow Heiberg's numbering in presenting a short outline of the text:[37]

1: A short dedication to Syrus, explaining that we will take the approach 'of those men in lines' (*uirorum illorum in lineis*),[38] since there is need of a more mathematical conception of natural theory and a more natural-theoretic conception of mathematics.

2, 3: A discussion of the coordinate system from first principles, explaining that three dimensions are used to measure a volume, both in magnitude and in number, and that a point on a sphere can be determined by the motion of three circles of the 'world sphere' (*spera mundi*)—the *horizon*, the *meridian* and the *vertical*—about one of their own diameters as determined by their intersections, producing three pairs of angles—*hectemorius-meridian*, *horarius-vertical*, and *descensivus-horizon*.[39] A description of the analemma model using letter-names.

4: A description of the system of 'the ancients'—which did not use the *hectemorius circle*.

5: A few refinements to the conventions so that no coordinate arc need be taken as greater than a quadrant.

6: (**a**) An introduction of the mathematical goal of the treatise—an instrumental determination, using the analemma, of the six arcs set out in the introduction; (**b**) followed by a synthetic proof that a certain angle in the analemma diagram is equal to the *hectemorius angle* on the sphere.

7: Geometric construction of an analemma diagram containing all six angles, with no proof, for the situation in which the sun is near an equinox.

8: Geometric construction of an analemma diagram containing the same for any other longitudinal position of the sun.

9: (**a**) A short discussion of instrumental practice, in which Ptolemy points out that on the analemma any of the six principal arcs can be determined using 'linear demonstrations'—that is, computational, indeed, trigonometric methods (διὰ τῶν γραμμῶν; *per lineares demonstrationes, per numeros*);[40] (**b**) followed by a *metrical analysis* for the determination of all six arcs in the case in which the sun is at an equinox.

[37] Since not all of the numbers in the Greek fragments have been preserved, it is not possible to be certain where all of the divisions were placed in this version of the text.

[38] Presumably geometers.

[39] These angles are defined below; see page 51.

[40] See note 74 for a discussion of the meaning of the phrase διὰ τῶν γραμμῶν.

10: A *metrical analysis* for the determination of all six arcs for any other longitudinal position of the sun.

11: (**a**) General description of the drawing and the instruments (compass and set square) used to make analog, or nomographic, computations; (**b**) followed by a detailed description of the production of an analemma plate used for carrying out analog calculations.

12: Instructions for producing all six angles on the analemma plate while drawing no new lines, for the situation in which the sun is near an equinox.

13: Instructions for producing all six angles on the analemma plate while drawing no new lines, for any other longitudinal position of the sun.

14: (**a**) Instructions for producing the angles of 'the ancients' on the analemma plate; (**b**) followed by a discussion of which pairs of angles, and taken in which direction, determine the position of the sun.

15: Table of all six angles for the latitude of Meroe, $\varphi = 16;25°$, when the sun is at Can 0°, $\delta(\lambda = 90°) = 23;50°$, for the position of the sun at the horizon, at the end of each of the five pairs of seasonal hours between the horizon and midday, and at the meridian.

It is likely that there were once more tables—perhaps 28 or 49, filling out the seven latitudes mentioned in the text and the four declinations of the beginnings of the twelve signs of the ecliptic.[41] It might be supposed that there was once more text following these tables, but nothing in the extant treatise compels us to this position.

In order to follow Ptolemy's mathematical approach, we will focus on explaining in detail only a few sections of the treatise—in each case, explicating only the *meridian-hectemorius* angle pair, since the three pairs are mathematically analogous and an understanding of one pair will suffice to grasp the overall approach. We will first look at the general exposition of the *world sphere* in *Analemma* 2 and 3. This will be followed by the synthetic proof in *Analemma* 6 that a certain angle of the analemma diagram is equal to the *hectemorius* angle. We will then read passages from the *metrical analysis* in *Analemma* 10, showing that if the terrestrial latitude, φ, solar declination, $\delta(\lambda)$, and seasonal hour, η_s, are given, then the *hectemorius* and *meridian* angles are also given.

[41] Edwards, *Ptolemy's* Περὶ ἀναλήμματος, p. 106, n. 506, states that the total number of tables should have been 49 tables—that is, the seven latitudes by the four declinations, three of which must be taken both to the north and to the south. Neugebauer, *A History*, pp. 854–855, states that there should have been 28 tables—presumably believing that Ptolemy would have made further use of the symmetries between the sets of seasonal hours to reduce the total number of tables.

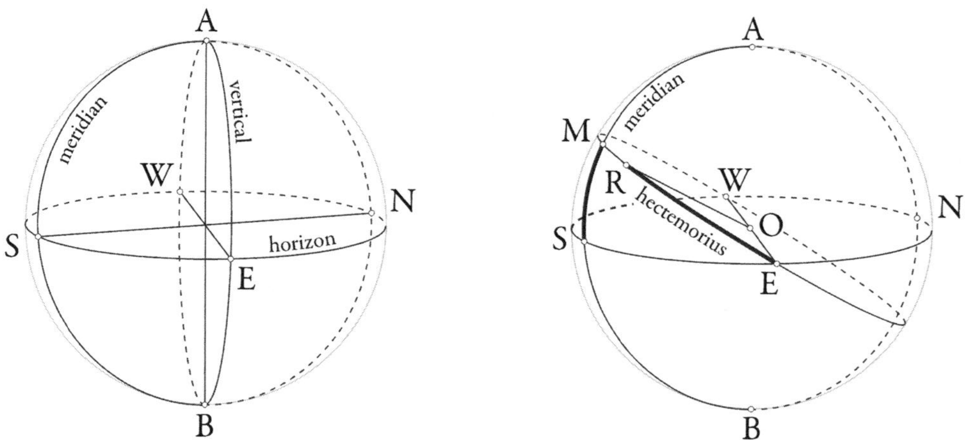

Figure 6: The *world sphere* as described in *Analemma* 2 (modern figures)

Finally, we will follow through the nomographic calculation of the *hectemorius* and *meridian* angles on the analemma plate in *Analemma* 13.

4. Ptolemy's *world sphere*

In *Analemma* 2, Ptolemy explains that three dimensions are sufficient to determine a volume (*moles*) in both magnitude and number, so that, in the *world sphere*, we need only three great circles and their diameters, set at right angles.[42] Because the discussion in *Analemma* 2 concerns the *world sphere* itself, it does not make reference to a lettered diagram—which makes it somewhat difficult to follow. In order to explicate this section, however, we will describe the objects that Ptolemy introduces using a modern diagram, Figure 6, which does not correspond to anything in the manuscript sources.

In Figure 6 (left), the three great circles of the *world sphere* will be taken as (a.1) the *horizon*, *NESW*, dividing the hemisphere above the earth from that below, (a.2) the *meridian*, *NBSA*, dividing the eastern and western hemispheres, and (a.3) the *vertical*, *EBWA*, dividing the northern and southern hemispheres; and the diameters will be (b.1) the *equatorial diameter*, *EW*, (b.2) the *meridional diameter*, *SN*, and (b.3) the *gnomon*, *AB*. Clearly, in this context, we are describing the cosmos using local coordinates, orientated to the position of the observer on the earth.

When these three circles are moved 'with the sun' (*cum sole*) about their diameters, (c.1) the horizon produces the *hectemorius*, rotating about the equatorial diameter, (c.2) the meridian produces the *horarius*, rotating

[42] Moerbeke often uses *moles* to translate ὄγκος, which in this context would mean 'volume'—see Clagett, *Archimedes*, p. 34.

about the meridional diameter, and (c.3) the vertical produces the *descensivus* (καταβατικός), rotating about the gnomon. When any one of these circles is raised 'above the earth with the solar ray' (*cum solari radio super terram*) it will produce two inclinations: (1) an angle contained by lines, between the solar ray and the diameter about which the circle rotates, and (2) an angle contained by planes, between the movable plane and its stationary counterpart—and 'as far as they are given, the position of the [solar] ray is also fixed' (*quibus datis et positio radii determinatur*). The final section of *Analemma* 2 is somewhat obscure because Ptolemy is setting out his own terminology at the same time as that of the ancients, but the main point is that each of the angle pairs *hectemorius-meridian*, *horarius-vertical*, and *descensivius-horizon* can be used to specify the location of the solar ray.

For example, in Figure 6 (right), when the hectemorius, *ERMW*, is inclined with the solar ray, *RO*, the equatorial diameter, *EW*, and *RO* create the rectilinear ∠*EOR*, which can be measured by arc *ER*, while the plane of the hectemorius creates an angle with its stationary counterpart, the horizon, *ESWN*, which can be measured by arc *MS* of the meridian. Hence, the position of the sun can be determined by arc *ER* of the hectemorius and arc *MS* of the meridian. Indeed, it is clear that if arc *ER* has the range 0°–180° and arc *MS* the range 0°–360°, any point on the sphere can be named in these coordinates. In fact, however, Greek geometers did not consider angles greater than 180°, and Ptolemy will introduce conventions in *Analemma* 5 that will insure that these six principal arcs will always have a range of 0°–90°.[43]

Analemma 2 is a discussion of how we can understand the position of the real sun in term of local coordinates—it speaks of the *world sphere* and of movable circles being carried with the sun with no reference to a lettered diagram. The angles that determine the position of the sun are described in terms relative to the position of the observer in the center of the cosmos. The use of the diagrams in Figure 6 helps a modern reader to follow Ptolemy's description, but it is not faithful to his approach, which is to describe the situation as taking place *around us*—with no appeal to the terminology of modeling or supposing.

This changes in *Analemma* 3, which Ptolemy introduces with the following words: 'In order that the sequence (*consequentia*) of the angles and what is modeled (*quod supponitur*) should fall more within our view, in fact, let there be a meridian circle, *ABGD*.'[44] He then proceeds to give a description using a

[43] There may have been practical reasons for this. For example, the graduated quadrants on his analemma plate only measure 0°–90°—which may have to do with the fact that a compass large enough that its radius would be equal to the diameter of his analemma plate would be rather cumbersome.

[44] In translating this and the following Latin passages, I have not attempted to render the

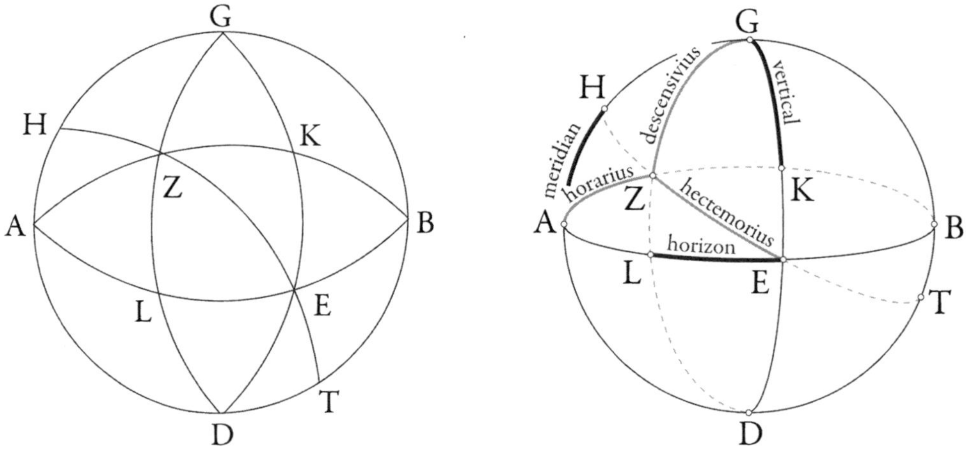

Figure 7: *Analemma* 3: (left) manuscript figure of the *world sphere*; (right) modern figure. All of the elements in the diagram are on the eastern hemisphere, which faces us.

lettered diagram, Figure 7, of the meridian circle and the circles to the east. Hence, Figure 7 depicts the eastern hemisphere of the model, such that all of the lines we see are on the outer surface of the sphere, facing us. Circle *ABGD* is the meridian, semicircle *AEB* is the horizon, and semicircle *GED* is the vertical. The semicircles *HZET*, *AZKB* and *GZLD* are the hectemorius, the horarius and the descensivus, respectively. Thus, the arcs which were mentioned as determining the solar ray are (ZE, AH) as *hectemorius-meridian*, (ZA, GK) as *horarius-vertical*, and (ZG, EL) as *descensivius-horizon*.

Analemma 3 is devoted to the description of a certain geometric object and it does not deal with the *world sphere*, the real sun, or the actual solar ray.[45] Hence, the purpose of *Analemma* 3 is to describe a model—namely the geometric object introduced—that will henceforth stand in for objects in the real world. The rest of the *Analemma* deals only with this geometric model.

5. The mathematical approach

In order to understand the mathematical methods of the *Analemma*, we will follow through the determination of a solar position in local coordinates for one of the three angle pairs, namely the *hectemorius-meridian* pair.

Latin literally but have been guided by the fact that *scilicet* and various forms of *qui*, *ipse*, *idem*, and so on, are used by Moerbeke to translate various functions of the definite article, whose usage in Greek mathematical prose is well known; see Clagett, *Archimedes*, pp. 43–44; Vuillemin-Diem and Steel, *Ptolemy's* Tetrabiblos, p. 14, and Federspiel, 'Sur l'opposition'.

[45] The mention of the 'ray' that appears at the beginning of *Analemma* 3 is actually a reference back to the solar ray introduced in *Analemma* 2, stating that point Z marks its position in the model.

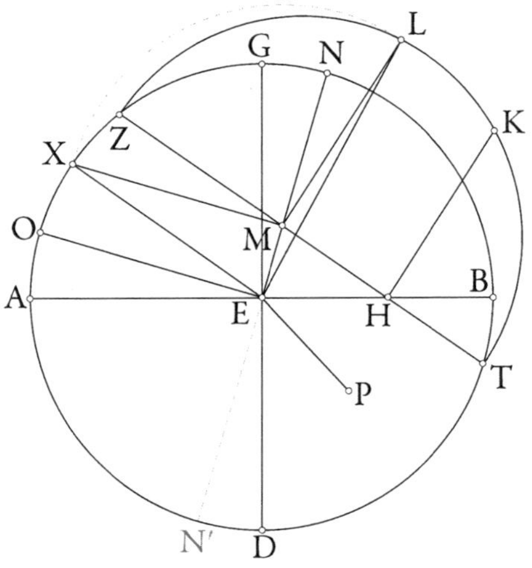

Figure 8: *Analemma* 6. Elements in gray do not appear in the manuscript diagram.

5.1. A synthetic proof

Analemma 6 begins with an introduction to the rest of the work. Ptolemy states that, with the foregoing as preliminaries, we will now set out the 'instrumental determinations' (ὀργανικαὶ λήψεις, *instrumentales acceptiones*)[46] of the coordinate angles. This appears to be a reference to the overall aim of the treatise of producing a physical analemma plate for making analog computations. Indeed, Ptolemy makes it clear that he will only supply a proof (ἀπόδειξις) for a single determination (λῆψις)—that of the new *hectemorius arc*, which he himself has introduced.[47] Hence, if we think of the remaining mathematical sections as establishing the methodological soundness of carrying out analog computations on the analemma plate, we can understand why Ptolemy would refer to this material generally as addressing 'instrumental determinations'.

Brushing off the case in which the sun is at one of the equinoxes as trivial, for the remaining solar positions Ptolemy gives a proof that a certain arc on the analemma is equal to the *hectemorius arc*, as follows (Figure 8 (left)):[48]

Now, as for the remaining monthly [circles],[49] let there be a meridian circle, *ABGD*, in which a diameter of the horizon is *AB*, and at right angles to this along the

[46] The Greek is a conjecture by Edwards, *Ptolemy's* Περὶ ἀναλήμματος, p. 94, n. 454, based on the Greek of the following passage.

[47] Heiberg, *Opera astronomica minora*, pp. 194–195; Edwards, *Ptolemy's* Περὶ ἀναλήμματος, pp. 94–95, 136.

[48] The Greek for this passage is essentially complete, so I have translated Heiberg's text, but kept the letter-names of Moerbeke's Latin. In the palimpsest, this passage concludes Section 1.

[49] That is, besides the month-circles at the equinoxes.

gnomon is *GD*, and the center of the sphere of the sun is *E*,[50] while *ZHT* is a diameter of one of the monthly parallels north of the equator, upon which, in the same plane let an eastern semicircle, *ZKT*, be imagined (νοείσθω).[51]
And let *KH* be produced upright upon *ZT*,[52] such that section (τμῆμα) *ZK* of the parallel [circle] is made to be above the earth, and with arc *KL* being cut off,[53] let a perpendicular, *LM*, be produced from *L* to *ZT*.[54] And, with center *M* and distance *ML*, let a point, *X*, be determined on the meridian,[55] and let *EL*, *MN*, *EX*, and *MX* be joined,[56] and let *EO* be produced upright upon *EN*.[57] I say that angle *OEX* is equal to the sought angle.[58]

For, let semicircle *ZLT* be imagined (νοείσθω) as rotated (ἐπεστραμμένον) to its proper position, that is, the perpendicular to the plane of the meridian. And let a perpendicular, *EP*, be produced, as the equatorial diameter, to the same plane.[59] Then, *LM* being a perpendicular to the meridian, it is obvious that straight lines *EN*, *ML*, and *EP* are in a single plane perpendicular to *ABGD*.[60] Likewise, [it is obvious] that *EN* is the common section of the hectemorius circle and the meridian,[61] while *LE* is in line with the solar ray,[62] and the sought angle, which is contained by the ray and the equatorial diameter, is *LEP*. For, since *EL* is equal to *EX*,[63] and *ML* to *MX*,[64] and *EM* is common, therefore angle *MEL* is equal to angle *MEX*.[65] But angle *MEP* is right, and angle *MEO*,[66] therefore the remaining angle *LEP*[67]

[50] This is a clear indication that Ptolemy thinks of the analemma model as a simplifying assumption, since by his time it was well known that the sun does not orbit the earth in a simple sphere—although the model may have been developed at a time when this was still held to be so.

[51] The MS reads νοείσθαι, Heiberg corrects to νοείσθω.

[52] *Elements* I.11; set square.

[53] This is the arc of the seasonal hour, η_s.

[54] *Elements* I.12; set square.

[55] *Elements* I.post.3; compass. See the discussion of this construction below.

[56] *Elements* I.post.1; side of set square.

[57] *Elements* I.11; set square.

[58] Namely, the *hectemorius arc*.

[59] *Elements* XI.12. That is, the line about which the hectemorius rotates.

[60] *Elements* XI.6, 7 and 18.

[61] *Elements* XI.3.

[62] Note the clear language of modeling here. Ptolemy does not assert that *LE* is the line of the solar ray, but is in line with it—that is, models it.

[63] They are both radii of the sphere.

[64] By construction.

[65] That is, because $\triangle MEX \cong \triangle MEL$, by *Elements* I.8.

[66] Following this, both the Greek and the Latin include the phrase 'and since angle *EML*' (ἐπει καὶ ἡ ὑπὸ τῶν ΕΜΛ, *quoniam et qui sub* EML), which appears to be an interpolation, and has been noted as such by the modern editors.

[67] Here the manuscript includes the phrase 'to angle MEX, that is' (τῇ ὑπὸ ΜΕΞ τουτέστιν, *ei qui sub* MEX *hoc est*), which does not make sense and has been marked as an interpolation by the modern editors.

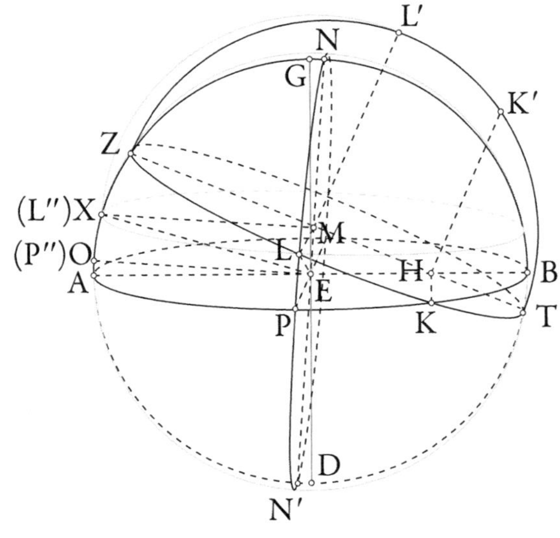

Figure 9: Perspective diagram for *Analemma* 6. Elements in gray do not correspond to any element in the manuscript diagram.

is equal to angle *XEO*. Which was to be shown.[68]

The key to understanding Ptolemy's argument is to consider the analemma figure as a representation of the solid sphere. There are a number of indications that this was Ptolemy's intention. The expression that he uses when he speaks of the monthly parallel, 'let it be imagined' (νοείσθω), is a standard expression in Greek geometric texts used to introduce solid constructions that cannot be fully, or accurately, represented by the plane figure.[69] When the semicircle of the parallel month-circle is introduced in the construction, it is *imagined* to be in the plane of the analemma, because it is, in fact, perpendicular to this plane. Ptolemy makes this clear in the proof when this circle is *imagined* rotated into its 'proper position'—namely, where it is found in the solid configuration. It seems likely that Ptolemy's readers could be expected to know how to view an analemma diagram as a solid configuration, or perhaps to have read the text while working with a solid sphere.

In order to illustrate this point, we will explain the argument with a perspective diagram. Considering Figure 9, let *BPA* be the local horizon and *BGA* the local meridian. Then the terrestrial latitude, φ, and the annual position

[68] Heiberg, *Opera astronomica minora*, pp. 195–198; Edwards, *Ptolemy's* Περὶ ἀναλήμματος, pp. 43–46.

[69] See, for examples, *Elements* XII.13, 16, *Conics* I.52, 54, 56, and Theo. *Spher.* I.19. The expression is also found many times in Archimedes' corpus. For a discussion of the various ways in which this verb is used to introduce objects, see Netz, 'Imagination'.

of the sun, $\delta(\lambda)$, are given by the inclination of the pole and the declination of the month-circle of the sun, *TKLZ*. Ptolemy makes no mention of this, simply assuming that they are fixed by the geometry of the figure. With *TKLZ* as the month-circle, section *KLZ*, being above the horizon, represents the motion of the sun from sunrise in the east to midday. Hence, the seasonal hour, η_s, must be taken as given along this arc, say as arc *KL*. This arc is taken as arbitrary in the construction. Because *Analemma* 6 is a theorem of synthetic geometry, however, there is no mention of any objects being given.[70]

Then, since *L* is the position of the sun and *P* is the east point of the local horizon, the hectemorius is the great circle *PLN* passing through *P* and *L*, and the *hectemorius arc* is *PL*. The construction then amounts to using analemma methods to project these points onto the analemma plane, while the proof amounts to showing that the arc that results from this projection is equal to arc *PL*.

The hectemorius circle is first projected orthogonally onto the analemma as line *NMEN'*, such that *P* maps to *E*, and *L* to *M* (M.1, M.2). Next, the location of *L* on the hectemorius is rotated into the plane of the analemma in two ways, by the method set out in the first example treating analemma methods above, Section 2.1. That is, *L* is mapped to the intersection of the circle about center *M* with distance *ML*, which is perpendicular to line *EN*, with the analemma circle, at point *X*. This construction is effected by taking *X* as the intersection of a circle drawn about center *M* with distance *ML*. In this way, *L* maps to *X*. The circle that produces *X* is not actually drawn in the plane of the analemma—probably because arc lengths are not preserved on it. Then, we project one of the endpoints of the diameter about which the hectemorius rotates into the analemma by erecting a perpendicular to the orthogonal projection of the hectemorius circle, *NEN'*, at *E*, so that *P* maps to *O* (M.3)—effectively, rotating the hectemorius circle into the plane of the analemma (M.4). Ptolemy does not talk about the hectemorius circle as rotated, he simply constructs the points *X* and *O* in the plane, first with the distance of a circle that is not itself drawn and then by constructing a perpendicular. The production of point *X* with distance *ML* is an interesting construction because it differs from any construction performed in the problems of *Elements* I–VI, insofar as the circle about center *M* is not actually drawn—only the point that is cut off by the circle is produced. In the *Elements*, such points are cut off on lines, as justified by *Elements* I.3, but not on circular arcs. We may regard this construction as an analemma construction, and we will see

[70] In general, Greek mathematicians only use the language of givens when treating problems, or in theorems written to facilitate certain problem-solving practices—for example, we do not read of objects being given in the synthetic theorems of the *Elements*; see Acerbi, 'The Language', and Sidoli, 'The Concept'.

that such constructions are often employed in the nomographic procedures on the analemma plate. Finally, with points X and O produced, it is a matter of elementary geometry to show that $\triangle MEL \cong \triangle MEX$, so that $90° - \angle MEL = 90° - \angle MEX$, that is arc LP = arc XO.

The analemma diagram, Figure 8, represents three circles in three different planes superimposed upon one another in the plane of the figure—which is the analemma. The circle of the meridian, $ADBG$, lies in the plane of the figure. The circle of the hectemorius, NPN', is perpendicular to the plane of the figure and intersects it in line NEN'. The month-circle, $TKLZ$, is perpendicular to the plane of the figure and intersects it in line $THMZ$. Although I have spoken of rotations and projections to help explain the solid configuration that motivates the construction, all but one of the constructive steps presented in the proof in *Analemma* 6 are carried out directly in the plane of the analemma—following what appears to have been a common practice among Greek geometers for handling solid configurations.[71] Moreover, because this is ostensibly a purely geometric argument, I have justified each constructive step with a problem, or postulate, from the *Elements* as well as by operations of a set square and compass as described in *Analemma* 11.

The next two sections of the *Analemma* set out the constructions for the remaining five angles with no proofs.[72] For our purposes, here, we will simply note that the arc of the meridian, which completes the angle pair with the hectemorius, is equal to arc AO in Figures 8 and 9.

This synthetic proof—which makes explicit reference to the solid sphere—provides the background to understanding analemma methods. Since we will not return to the solid configuration in this discussion, it may be helpful to summarize the analemma construction of the *hectemorius-meridian* angle pair. The *hectemorius arc* is found as follows:

Hec.1: The diameter of the hectemorius circle is found by taking the orthogonal projection of the solar position, L, onto the meridian plane, M (M.1); and then joining this with the center, E, extended to produce $NMEN'$ (M.2).

Hec.2: The *hectemorius arc* is found by rotating the hectemorius circle into the plane of the meridian circle about its diameter, $NMEN'$ (M.4), such that P, or E, the east point, maps to O, and L, the solar position, maps to X. Using analemma methods, we find X by taking the intersection of a

[71] See note 11, above.

[72] Luckey, 'Das Analemma', cols 25–26, gives a clear account of how the remaining analemma constructions are related to the solid sphere. Following these descriptions it would be a relatively simple matter to reconstruct proofs along the lines of *Analemma* 6. For a summary of the constructions of all the arcs in the analemma, see Guerola Olivares, *El Collegio Romano*, pp. 81–101.

circle of radius ML about center M, or a perpendicular erected to NEN' at M, with the analemma, and we find O by taking the intersection of a perpendicular to NEN' erected at E with the analemma (M.3).

Because the meridian circle is in the same plane as the figure, which is the analemma plane, the analemma construction of the *meridian arc* is somewhat simpler:

Mer.1: The diameter of the hectemorius circle is found, as before, by taking the orthogonal projection of the solar position, L, onto the meridian plane, M (M.1); and then joining this with the center, E, extended to produce $NMEN'$ (M.2).

Mer.2: The *meridian arc* is found by rotating the hectemorius circle about $NMEN'$, such that, as before, P, or E, maps to O, and the *meridian arc* is cut off on the meridian circle between O and A—the south point (M.4). Again, using analemma methods, we simply erect the perpendicular from E to NEN' (M.3).

5.2. A metrical analysis

In *Analemma* 9, after introducing the methods of nomographic computation discussed above, Ptolemy explains that each of the six principal arcs can also be calculated using geometric, indeed trigonometric, means:[73]

> Such a determination, for those who prefer, would also exist precisely by means of lines (διὰ τῶν γραμμῶν),[74] although it would be easily brought about through

[73] Again, I have translated the Greek for this passage.

[74] Διὰ τῶν γραμμῶν is a technical expression in Ptolemy's writings; see Heiberg, *Syntaxis mathematica*, vol. I, pp. 32, 42, 251, 335, 380, 383, 416, 439; vol. II, pp. 193, 198, 201, 210, 321, 426, 427, 429; Heiberg, *Opera astronomica minora*, pp. 202, 203. It designates the geometric means through which a computation can be, or has been, carried out, either by elementary geometry, or by chord-table trigonometry. When it is used in the *Almagest*, it is in reference to either an actual calculation or to a metrical analysis, where the later is understood as showing that the former is, in principle, possible. As Edwards, *Ptolemy's* Περὶ ἀναλήμματος, p. 107, n. 512, suggests, the phrase is closely related to chord-table trigonometry, and in a number of places in the *Almagest* it clearly must indicate computation through chord-table trigonometry; see Heiberg, *Syntaxis mathematica*, vol. I, pp. 251, 335, 380, 439; vol. II, pp. 321, 426, 427. In a number of other places it refers to a computation, which, considering the context, was almost certainly carried out through chord-table trigonometry; see Heiberg, *Syntaxis mathematica*, vol. I, pp. 383, 416, 439; vol. II, p. 429. And, finally, in some cases it refers to a metrical analysis that is intended to justify a computation; see Heiberg, *Syntaxis mathematica*, vol. II, pp. 193, 198, 201, 426.

We find two uses of the phrase διὰ τῶν γραμμῶν in reference to the same metrical analysis, which, taken together, make it clear that we must understand this metrical analysis as justifying, or standing in for, trigonometric calculation; see Heiberg, *Syntaxis mathematica*, vol. II, pp. 426, 427, and Nathan Sidoli, 'Mathematical Tables', pp. 25–26.

> the analemma itself, even if it is not exactly the same as that through geometrical demonstrations (διὰ γραμμικῶν ἀποδείξεων), but rather near enough for theory in agreement with the senses, to which the practical goal of the proposed task leads.[75]

He then provides what I call a metrical analysis to show that given the declination of the sun, $\delta(\lambda)$, terrestrial latitude, φ, and seasonal hour, η_s, each of the six angles is also given—for $\lambda = 0°$ and $\lambda = 180°$ in *Analemma* 9, and for all other λ in *Analemma* 10. A metrical analysis is a type of argument about what is *given* that is found in the writings of both Heron and Ptolemy, in which each step can generally be justified by reference to theorems of Euclid's *Data*, but which itself justifies, or establishes the possibility of, a computational procedure involving arithmetical operations—adding, subtracting, multiplying, dividing, and taking square roots—and, in the case of Ptolemy, entries into a chord table.[76] After reading one of Ptolemy's metrical analyses, we will discuss the significance of this type of argument.

In the foregoing passage, the distinction between producing the final determinations 'by means of lines', or 'through geometrical demonstrations', on the one hand, and those brought about 'through the analemma itself', on the other, addresses the fact that there will probably be slight differences in the values obtained through chord-table trigonometry, as justified by the metrical analyses of *Analemma* 9 and 10, and those obtained through the analog, or nomographic, methods, that will be provided in *Analemma* 12–14. Here, Ptolemy claims that the minor differences in these values will not undermine the overall validity of the analog calculations, which will produce values good enough for the practical goals of the treatise.

After giving the metrical analysis for the six angles in the case where the sun is at one of the equinoxes, in *Analemma* 9, the passages of metrical analysis in *Analemma* 10 that concern the *hectemorius-meridian* pair read as follows (Figure 10):[77]

[75] Heiberg, *Opera astronomica minora*, pp. 202–203; Edwards, *Ptolemy's* Περὶ ἀναλήμματος, p. 50.

[76] *Metrical analysis* is my terminology, but this type of argument is called an 'analysis' by Heron throughout his *Measurements*, and by Pappus in his commentary on Ptolemy's *Almagest* V; see Rome, *Commentaires*, p. 35. I have not found a passage where Ptolemy himself refers to this type of argument as an 'analysis'. I have discussed elsewhere the role of this type of argument with respect to mathematical tables in the *Almagest*; see Sidoli, 'Mathematical Tables', pp. 25–26. See also the discussion by Acerbi, 'I codici stilistici', pp. 201–208; note, however, that his attempt to rewrite Ptolemy's prose should be treated with caution—there is good reason why Ptolemy does not include the passages that Acerbi adds to the text (see n. 107, below).

[77] Only the beginning of this passage is preserved in Greek; I have translated first from the Greek and then from the Latin. In the palimpsest this passage begins Section 5. Here I translate only those passages necessary for the determination of the *hectemorius-meridian* angle pair.

As for the remaining monthly [circles],[78] let meridian *ABGD* be set out along with the diameters upright upon one another and axis *EZ*. And let a diameter, *HTK*, of any of the southern monthly parallels to the equator be produced through, upon which let semicircle *HLK* imagined (νοούμενον)[79] to the east be drawn.[80] And let axis *EZL* be extended,[81] obviously bisecting diameter *HTK* at *T* and semicircle *HK* at *L*.[82] And let line *MN* be produced through [as a perpendicular] to *HT*,[83] dividing the section, *HN*, of the semicircle above the earth from that below the earth. And with arc *NX* being determined as the given hours,[84] let a perpendicular, *XO* be produced from *X* to *HM*.[85] And through *O* let perpendiculars upon *AE*, *POR*, and upon *GE*, *SOC*, be produced through.[86]

Then, since arc *HTK* of the meridian is given,[87] and the double of line *ET* subtends its remainder in the semicircle, the ratio of *HTK* and *ET* to the diameter of the meridian will be given.[88] Likewise, since arc *AZ*, of the elevation [of the pole], is given,[89] angle *MET* of right-triangle *MET* will also be given. So, the ratio of *ET* to each of *EM* and *MT* will also be given,[90] and, moreover, that of diameter *HK* to each of them.[91] But, the double of line *MT* subtends the double of arc *LN*, so arc *LN* will

[78] That is, besides the month-circle at the equinoxes.

[79] The use of 'imagined' is a reminder that, in order to understand the analemma diagram, we must consider it as a representation of a solid configuration.

[80] The assumption of this configuration produces the first two given magnitudes: φ and $\delta(\lambda)$.

[81] *Elements* I.post.2; side of the set square.

[82] *Elements* III.3, I.def.17.

[83] *Elements* I.11; set square. The location of *M* on the horizon is specified in the phrase that follows. The fact that this line must be perpendicular is made explicit in Moerbeke's Latin.

[84] This is the final given magnitude, η_s.

[85] *Elements* I.12; set square.

[86] *Elements* I.12; set square.

[87] In terms of the *Data*, this is so because both meridian *ADBG* and line *HK* are assumed to be given in position, so that, by *Data* 25 and 26, line *HK* is given in position and in magnitude, so that, by *Data* Def.8, arc *HTK* is given in position and magnitude. Edwards, *Ptolemy's* Περὶ ἀναλήμματος, p. 111, n. 537, points out that computationally this follows since arc $HZK = 180° - 2\delta(\lambda)$.

[88] That is $(HTK : \text{diameter}_m)$ and $(ET : \text{diameter}_m)$ are given. The first ratio is given by *Data* 1. The second ratio is given because, by *Data* 88, $\angle HEK$ is given, so that, by *Data* 2, half of it, $\angle HET$ is given. Then, by *Elements* I.32 and *Data* 4, the angles of $\triangle HET$ are given, so that, by *Data* 40, $\triangle HET$ is given in form. Then, by *Data* Def.3, $(ET : EH) = (ET : \text{radius}_m)$ is given, so that, by *Data* 8, $(ET : \text{diameter}_m)$ is given. In terms of computation, the use of *Data* 88 would involve entering into a chord-table, as would the computation of ratios based on angles implied by the use of *Data* 40.

[89] This is φ—assumed to be given by the geometric configuration, or by taking the meridional altitude of the pole, in degrees.

[90] That is, since by *Elements* I.32 and *Data* 4, the angles of $\triangle MET$ are given, by *Data* 40, $(ET : EM)$ and $(ET : MT)$ are given. The computation of these ratios would involve entering into a chord-table.

[91] That is, by *Data* 8, $(EM : \text{diameter}_m)$ and $(MT : \text{diameter}_m)$ are given.

also be given,[92] as well as the remainder *NXH* from a quadrant.[93] But *NX* is given,[94] therefore both *LX* and *XH* will be given.[95] But, the double of line *XO* subtends the double of arc *HX*, and the double of line *OT* [subtends] the double of arc *XL*,[96] so the ratio of *XO* and *OT* to diameter *HK* is given,[97] and because of this also to that of the meridian.[98] But, since the ratio of *TM* [to diameter$_m$] is also given,[99] the ratio of *MO* [to diameter$_m$] will be given.[100] And it is that as *EM* to *MO*, so is *TM* to *MP* and *ET* to *OP*,[101] for the triangles *EMT* and *OPM* are equiangular.[102] Therefore, the ratio of *MP* and *OP* to the diameter of the meridian is given.[103] On account of this, also the ratio of *ES* [to diameter$_m$], and of the whole of *EMP*, which is *OS* [to diameter$_m$, will be given].[104]

With these things demonstrated, with center *O* and distance *OX* let a point of the meridian, *Y*, be determined.[105] [...] And let *EY* [...] and *EO* [...] be joined.[106]

[92] That is, since by *Data* 2, 2*MT* is given, arc 2*LN* is given by *Data* 88. Computationally, we enter into a chord-table.

[93] *Data* 4.

[94] This is determined by η_s. Geometrically, it is simply assumed as given by taking the seasonal hour from sunrise to noon going from *N* through *X* to *H*, or from noon to sunset going back from *H* to *N*. Arc *NH* can be divided into six parts using one of the various trisections of an angle preserved in Greek sources; see Heath, *A History*, vol. I, pp. 235–244. Computationally, arc $NX = \eta_s \cdot$ arc $NH/6$; see Edwards, *Ptolemy's* Περὶ ἀναλήμματος, p. 112, n. 542.

[95] *Data* 3 and 4.

[96] That is, in semicircle *HLK*, by *Data* 87, 2*XO* and 2*OT* are given; so by *Data* 2, *XO* and *OT* are given.

[97] *Data* 1.

[98] That is, by *Data* 8, $(XO : \text{diameter}_m)$ and $(OT : \text{diameter}_m)$ are given. The Greek fragment ends here—we continue with Moerbeke's Latin. With regard to my translation choices, see note 44, above.

[99] That is, $(TM : \text{diameter}_m)$ is given, as shown above.

[100] That is, since $(TO : \text{diameter}_m)$ and $(TM : \text{diameter}_m)$ are given, by *Data* 8, $(TO : TM)$ is given. Hence, by *Data* 5, $(TO : (TO - TM))$ is given. Hence, again by *Data* 8, $((TO - MT) : \text{diameter}_m) = (MO : \text{diameter}_m)$ is given.

[101] *Elements* VI.4.

[102] *Elements* I.15 and 32.

[103] That is, $(EM : TM) = (MO : MP)$, and $(EM : ET) = (MO : OP)$, and each of $(EM : TM)$ and $(EM : ET)$ are given, so by *Data* 8, $(MP : \text{diameter}_m)$ and $(OP : \text{diameter}_m)$ are given.

[104] That is, since $(ME : \text{diameter}_m)$ and $(MP : \text{diameter}_m)$ are given, by *Data* 8, $(ME : MP)$ is given. Hence, by *Data* 6, $((ME + MP) : (ME))$ is given. Hence, again by *Data* 8, $((ME + MP) : \text{diameter}_m) = (OS : \text{diameter}_m)$ is given.

[105] *Elements* I.11; compass.
O reads *G* in place of *Y*—which error was noted by the modern editors.

[106] *Elements* I.post.1; side of the set square. Heiberg, *Opera astronomica minora*, pp. 206–209, Edwards, *Ptolemy's* Περὶ ἀναλήμματος, pp. 54–56.

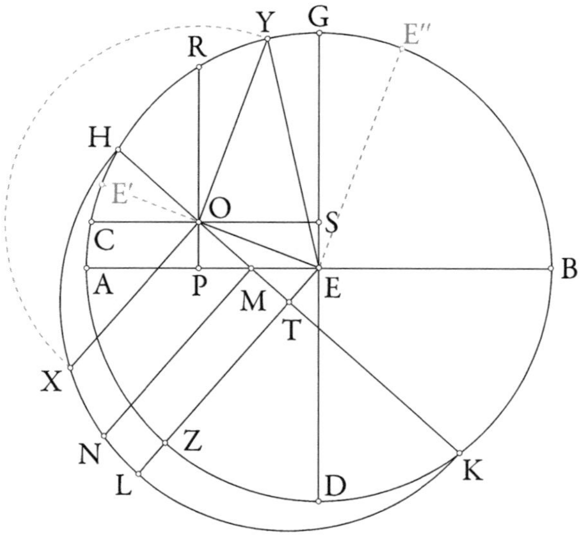

Figure 10: *Analemma* 10: Partial diagram of the analemma. Objects which do not concern the *hectemorius-meridian* angle pair have been omitted. Elements in grey do not appear in the manuscript diagram.

As before, in Figure 10, the solar position is projected into the analemma in two ways. Since O is the orthogonal projection of the solar position, X, onto the plane of the meridian—produced by dropping a perpendicular from X to the diameter of the solar month-circle, HK (M.1)—the line OE, joining O with the orthogonal projection of the east point, E, will be the diameter of the hectemorius circle (M.2). Hence, the *hectemorius arc*, YE'', will be produced by rotating the hectemorius circle into the meridian (M.4)—that is, by taking Y on the analemma circle such that $OY = OX$, and erecting EE'' perpendicular to OE (M.3). Since, by *Elements* I.29, $\angle YEE'' = \angle EYO$, Ptolemy will simply work with $\angle EYO$—probably to avoid having to produce EE'' in an already cluttered diagram. The *meridian arc* is found by extending the diameter of the hectemorius circle, OE, to meet the meridian at E', and taking the arc between this intersection and the south point, arc AE'. Since arc $AE' = \angle PEO$, Ptolemy simply works with this angle—again to avoid producing any unnecessary lines.

Up to this point in the metrical analysis, Ptolemy has dealt with all lines in terms of ratios to other lines. This practice agrees with that found in the *Almagest* for plane trigonometry, and derives from the fact that when we enter into a chord table with an angle, the resulting chord is always given in terms of the radius of the circumscribing circle—that is, as a ratio.[107] In what

[107] This is the reason why Acerbi, 'I codici stilistici', pp. 204–208, has gone too far in attributing to Ptolemy claims about given lines in his rewriting of the ancient text.

follows, however, he will assert that the radius of the meridian is given, and he will then state the other lengths as also given—that is, given in terms of the diameter of the meridian.

The foregoing passage continues to argue that the six principal arcs are all given. We read only those passages pertaining to the *hectemorius-meridian* pair:[108]

> Since, then, in the preceding, the angle *EOY* was shown to be right,[109] while hypothenuse *EY*, being a radius of the meridian, is given,[110] as well as *OY*, being equal to *OX*,[111] angle *EYO*, containing that of the hectemorius circle, will be given.[112] [...] Then, since both *OP* and *EP*, of right-angled [triangle] *EOP* are given,[113] hypothenuse *EO* and angle *OEP*, which makes the *meridian arc*, will be given.[114] [...][115]

The foregoing metrical analysis constitutes a general argument that, where $\delta(\lambda)$, φ, and η_s are assumed as given, the two arcs of the *hectemorius-meridian* angle pair are also given—that is fixed, or determined. The argument, as all extant metrical analyses, works on two levels: (1) as a purely geometrical proof, in which each step can be justified by theorems of the *Data*, and (2) as the articulation of an effective computational procedure, involving the basic arithmetic operations, taking square roots and entries into a chord-table.[116] That is, in the *Analemma*, for Ptolemy, *given* means both *geometrically given*—that is, producible using the constructive methods of *Elements* I–VI—and *numerically given*—that is, computable as some numerical value.[117]

As he stated at the beginning of *Analemma* 9, this argument constitutes a demonstration that these arcs can be determined 'by means of lines' (διὰ τῶν γραμμῶν)—that is, they are computable through geometric, or rather

[108] There is no Greek for this passage.

[109] This is a reference to the argument in *Analemma* 6, in which it was shown, in Figure 8, that $\triangle LME \cong \triangle XME$ and *LM* was imagined as constructed perpendicular to the plane of the meridian.

[110] That is, assumed as given—given by the geometry of the figure, or taken, as say 60^p, following the practice of the *Almagest*.

[111] *Data* 2, since $(OX : \text{diameter}_m)$ was shown to be given.

[112] *Data* 43. Computationally, this involves entering a chord-table.

[113] *Data* 2, since $(OP : \text{diameter}_m)$ and $(OS : \text{diameter}_m)$ were shown to be given.

[114] *EO* is given by *Data* 52, 3 and 55; so that, by *Data* 39, $\angle OEP$ is given. Calculating this angle would involve Ptolemy's usual convention of taking *OP* when *OE* is assumed as given, and entering a chord-table.

[115] Heiberg, *Opera astronomica minora*, p. 209; Edwards, *Ptolemy's* Περὶ ἀναλήμματος, p. 56.

[116] Edwards, *Ptolemy's* Περὶ ἀναλήμματος, pp. 115–117, gives an example calculation following Ptolemy's methods that proceeds along the same lines as that established by Ptolemy's metrical analysis.

[117] Ptolemy does not express any concern with the fact that the numerical value used to express certain geometric objects will not be perfectly precise.

trigonometric, methods.[118] In the context of this treatise, this is contrasted with the analog computations that will be outlined below. The determinations, and likewise computations, 'by means of lines' are said to be more precise than those of the nomographic procedures to which we now turn.

5.3. An analog computation

In *Analemma* 11, Ptolemy describes the production of an analemma plate, on which 'with only the compass and the set square' (διὰ μόνου τοῦ τε καρκίνου καὶ τοῦ ὀρθογωνίου πλατύσματος)[119] constructions of the six principal arcs can be carried out.[120] The plate is a 'drum-shaped plane' (τυμπανοειδὲς ἐπίπεδον) on which certain, permanent lines are drawn. Ptolemy gives instructions for how this plate should be made using a lettered diagram that serves as a sort of mathematical recipe for a physical construction. The finished plate, however, would probably not have had labels on it, as in Figure 11.

The plate may be made of inscribed lines on bronze or stone, or colored lines drawn on wood, which is then covered with wax so that the horizon and gnomon can be drawn in the wax.[121] The wooden tablet—which we will treat here—is inscribed with red lines for the meridian and the diameter of the equator and black for three month-circles. Quadrants, graduated at 1° intervals, are produced on one or both sides of the equator, as well as in one of the quadrants of the outer circle. In each quadrant of the outer circle a set of seven marks is drawn for the elevation of the pole at seven well-known latitudes, φ, of Greco-Roman geography: 16 1/3 1/12°, 23 1/2 1/3 °, 30 1/3 °, 36°, 40 1/2 1/3 1/12°, 45°, and 48 1/2°.[122] Three month-circles are inscribed corresponding to the following solar declinations, $\delta(\lambda)$: 23 1/2 1/3°, 20 1/2°, and 11 2/3°.[123] The first month-circle is used when the sun is at the tropics, in the signs of Cancer or Capricorn ($\lambda \approx 90°$, 270°); the second is used when the sun is in Gemini, Leo, Sagittarius or Aquarius ($\lambda \approx 60°$, 120°, 240°, 300°); the third is used when the sun is in Taurus, Virgo, Scorpio or Pisces ($\lambda \approx 30°$, 150°, 210°, 330°); while the equator is used when the sun is in Aries or Libra ($\lambda \approx 0°$, 180°). The wooden plate is orientated by rotating it such that the elevation corresponding to the given terrestrial latitude, φ, is in the zenith and drawing the horizon and the gnomon in the wax. In the operations to be described below, arcs will be set

[118] See note 74 for a discussion of the meaning of the phrase 'by means of lines' (διὰ τῶν γραμμῶν).

[119] Literally, 'crab' and 'rectangular plate'.

[120] Heiberg, *Opera astronomica minora*, p. 212.

[121] Presumably working with bronze or stone involved having a number of different sets of plates for the different latitudes.

[122] These latitudes should be compared with those in *Almagest* II.6 and *Geography* I.23.

[123] These should be compared with the declinations in *Almagest* I.5.

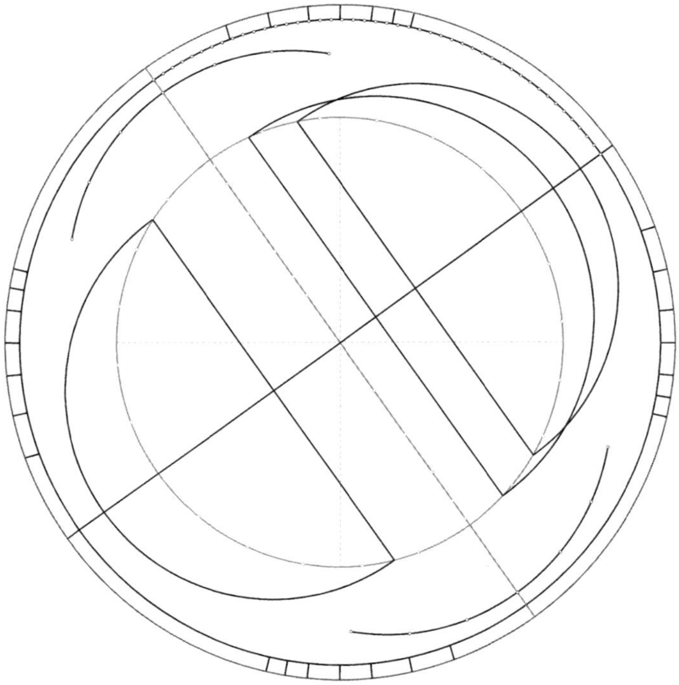

Figure 11: *Analemma* 11: The wooden analemma plate, rotated to carry out computations at the latitude of Rhodes, $\varphi = 36°$. Red lines are shown in gray, dotted lines are to be drawn in the wax.

out on the meridian. These arcs can then be measured by carrying them, with the compass, to one of the graduated quadrants at the side, which have the same diameter as the meridian.

In order to follow the method of the analog computation, we will read *Analemma* 13 closely, with a new diagram for each step of the procedure.[124] *Analemma* 13, in which Ptolemy describes the analemma-plate computation for the six principal angles, begins as follows:[125]

> Again, [1] let a diameter of any one of the monthly circles be modeled, and let it be *ZHTK*, upon which is the eastern semicircle *ZLK*.[126] [2a] And with center *T* and

[124] Luckey, 'Das Analemma', cols 32–39, gives a complete account of the nomographic computation for all of the angles. See also the account by Guerola Olivares, *El Collegio Romano*, pp. 122–131.

[125] There is no Greek text for this part of the treatise; I translate Moerbeke's Latin—omitting those passages unnecessary to the computation of the *hectemorius-meridian* pair.

[126] *L* has not actually been produced yet, so at this point it serves as an unspecified name for the semicircle. The positioning of *L* and *M* in the diagram provided by Heiberg, *Opera astronomica minora*, p. 219, which accurately reflects that on **O** f. 64v, is incorrect.

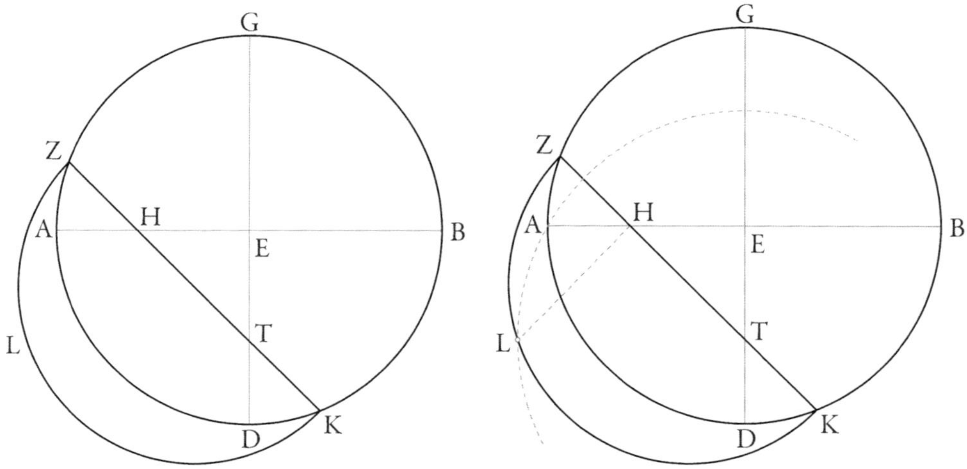

Figure 12: *Analemma* 13: Computing on the analemma plate, steps [1] and [2]: (left) initial set up, determination of $\delta(\lambda)$ and φ; (right) determination of L, the position of the horizon on the month-circle.

distance TA, let a point of the meridian, L, be determined,[127] by which ZL—the semicircle above the earth—and LK—below the earth—are separated. [2b] But point L is determined with the set square (*per platinam rectangulam*) if the angle will have been brought to H such that the other side is fitted to ZK—for what the remaining side cuts on the semicircle will be the point [L], because the perpendicular produced from H of HK will be the [common] section of the planes of the horizon and the monthly circle.[128]

In the first step, [1], we orientate the plate to the given latitude—say $\varphi = 36°$, the latitude assumed for Rhodes—and draw the horizon, AB, and the gnomon, GD, into the wax on the plate. In the actual procedure, we would not need to label these lines, but we label them in Figure 12 for the sake of clarity. We then chose one of the month-circles, which will determine H and T—the plate can be rotated 180° so that any month circle can be taken in the northern or southern direction.

The second step, [2], can be carried out in two ways. We determine the point on the month-circle that divides between day- and night-time, L, by [2a] either using the compass to take the intersection of a length $TL = TA$ on the month-circle, [2b] or using the set square to take the perpendicular from H, the intersection of the diameter of the month-circle and the horizon. That we must set $TL = TA$ is clear from considering the solid configuration—as hinted

[127] The fact that $TA = TL$ can be shown by considering the solid configuration—see below.

[128] Heiberg, *Opera astronomica minora*, p. 219; Edwards, *Ptolemy's* Περὶ ἀναλήμματος, pp. 68–69.

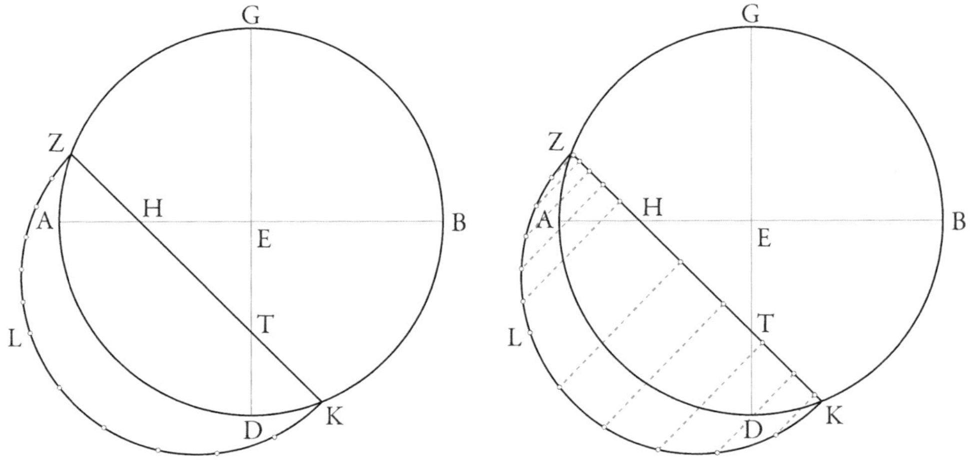

Figure 13: *Analemma* 13: Computing on the analemma plate, steps [3] and [4]: (left) determination of the day- and night-time hours; (right) projection of the hours onto the diameter of the month-circle.

at in the text. If *ZLK* is rotated into its proper position, *A* and *L* will both lay on the great circle of the horizon and *T* is some point on the gnomon. Since the gnomon is perpendicular to the horizon and passes through its center, by an argument similar to that in Theodosius' *Spherics* I.1, the distances from *T* to every point on the great circle of the horizon are equal.

Analemma 13 continues, as follows:

> Then, [3] let each section [*ZL* and *LK*] be divided equally in 6, and with these points, [4] by an application of the set square let points on *ZK* made by perpendiculars to it from the divisions determined on the semicircle be determined.[129]

It is not stated, in step [3], how to perform the division of the daytime arc into six parts. Various possibilities come to mind. We could use one of the purely geometrical solutions to this problem that are extant in the ancient sources—for example, one from among those treated by Pappus in *Collection* IV.[130] Indeed, Pappus tells us that he showed how to trisect an angle in his lost commentary to the lost *Analemma* of Diodorus.[131] Alternatively, the plate itself could be used to perform this division as follows:

- We draw an auxiliary circle with the same radius as the month-circle, concentric with the meridian,

[129] Heiberg, *Opera astronomica minora*, p. 219, Edwards, *Ptolemy's* Περὶ ἀναλήμματος, p. 69.

[130] See Hultsch, *Pappi Collectionis*, pp. 272–288, and Sefrin-Weis, *Pappus. Book 4*, pp. 146–155. Heath, *A History*, vol. I, pp. 235–244, gives an overview of the ancient solutions to trisecting an angle.

[131] See Hultsch, *Pappi Collectionis*, pp. 244–246.

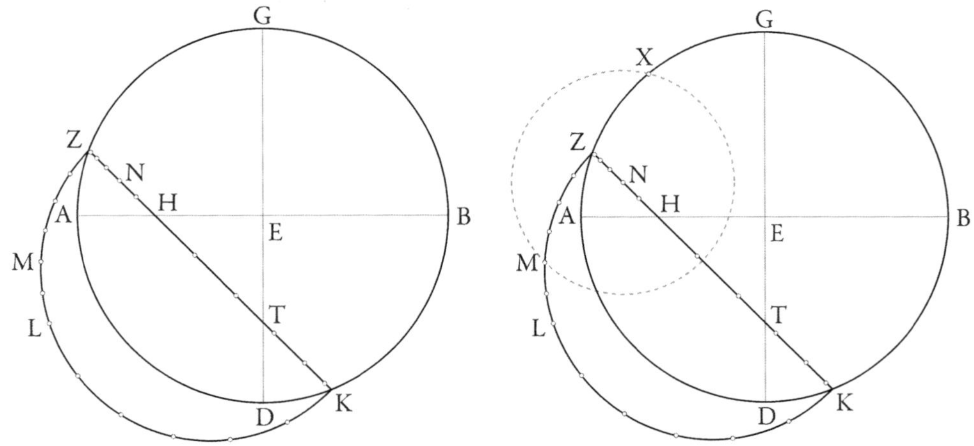

Figure 14: *Analemma* 13: Computing on the analemma plate, steps [5] and [6]: (left) determination of the solar position, *M*, and its projection onto the diameter of the month-circle; (right) determination of the solar position on the hectemorius circle in the plane of the meridian, *X*.

- we transfer the daytime arc to this auxiliary circle with the compass, such that one endpoint falls on the axis that bounds the outer graduated quadrant,
- we project the other endpoint onto the outer graduated quadrant with the set square,
- we read off the angle measure on the graduated quadrant and divide this value by six,
- we mark this value off on the graduated quadrant and project this arc back onto the auxiliary circle, and
- we transfer this arc back to the month-circle with the compass and mark it off six times.

Since both the geometrical and analemma plate methods of producing the hours are non-trivial, it seems likely that Ptolemy took his readers to have some familiarity with these sorts of constructive procedures.

Step four, [4], is carried out by lining up one side of the set square on the diameter of the month-circle such that the other side passes through the hour points—as is made explicit in the text. The points on the diameter of the month-circle are then marked at the angle of the set square.

The text continues, as follows:

But, [5] let one of them that is above the earth be that at *M*,[132] and the ordinate

[132] This is the given hour, η_s.

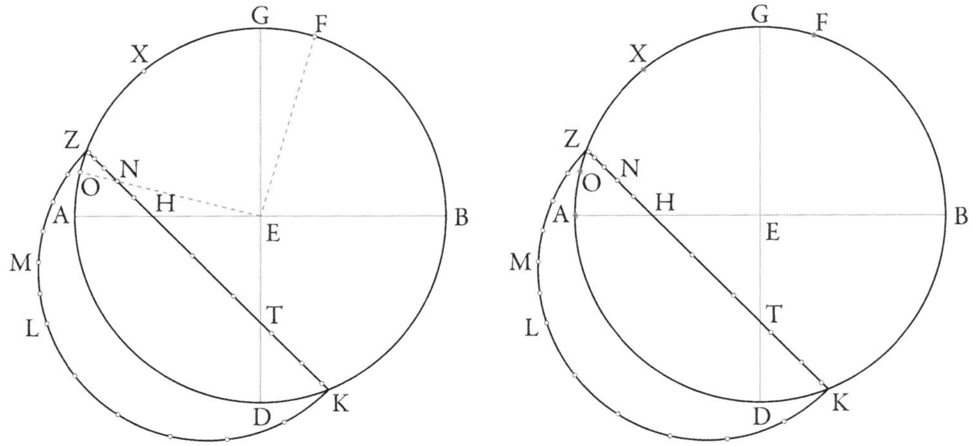

Figure 15: *Analemma* 13: Computing on the analemma plate, steps [7] and [8]: (left) determination of the diameter of the hectemorius circle and of the east point of the hectemorius circle in the plane of the meridian; (right) determination of the *hectemorius-meridian* pair, *XF* and *OA*.

with it, *N*, of those on *ZH*. Then, [6] with center *N* and distance *NM*, let point *X* be determined on the meridian.[133]

Step five, [5], simply consists in choosing a pair of corresponding points along arc *LZ* and line *HZ* for the solar position of the sun at the given hour, *M*, and its projection onto the diameter of the month-circle, *N*.

In step six, [6], we find the projection of the solar position onto the rotation of the hectemorius into the plane of the meridian—that is, the plane of the analemma. Following the first example of the analemma methods in Section 2.1 and the construction provided in *Analemma* 6, this is found by setting the stationary end of the compass on *N*, the mobile end on *M*, and then marking the intersection of the mobile end with the meridian at *X*.

The material from *Analemma* 13 that concerns the *hectemorius-meridian* pair concludes as follows:

And, [7] with the side of the set square brought to points *E* and *N* such that it cuts the meridian at *O*, [8] arc *XO* will make the complement of the hectemorius,[134] and that from *X* to the other intersection of the set square and the meridian, [*F*,] is the hectemorius [...] Again, arc *AO* will make that of the meridian [...][135]

[133] Heiberg, *Opera astronomica minora*, p. 219; Edwards, *Ptolemy's* Περὶ ἀναλήμματος, p. 69.

[134] **O** reads *ZO* in place of *XO*—which error was noted by the editors.

[135] Heiberg, *Opera astronomica minora*, pp. 219–220; Edwards, *Ptolemy's* Περὶ ἀναλήμματος, p. 69.

In step seven, [7], we determine the diameter of the hectemorius circle by placing the angle of the set square at the center of the figure and passing one side over point *O* so that the other side falls above the earth, at point *F*.

In the final step, [8], we note that *XF* is the *hectemorius arc* and *AO* is the *meridian arc*. These can be measured by placing the compass points at their endpoints and then transferring them to the graduated quadrant at the equator.

By following a series of physical manipulations of this sort, each of the six principal arcs can be computed nomographically.

6. Conclusion

Following the details of Ptolemy's presentation of the *Analemma*, as we have done in this paper, has made it clear that the analemma methods, as a loose collection of problem-solving methods in the science of *gnomonics*, were closely associated with various instrumental practices. We have seen both a restriction to operations that can be performed by abstractions of realizable instruments and explicit instructions for the production and use of an analemma plate as a tool for analog computations. This basis in instrumental practice, and its justification through metrical analysis, appears to have been a significant part of Ptolemy's mathematical bequest to scholars of the mathematical sciences in the late ancient and medieval periods. This explicitly instrumental approach, which is not found in the extant writings of authors like Euclid or Diophantus, was, nevertheless, an important aspect of the Greek mathematical sciences.[136]

Although there is one theorem in the *Analemma*, the analemma methods, as they are preserved in ancient and medieval sources, were clearly focused around problem-solving—based on operations that can be performed with a real compass and set square. This provides us with an interesting example of a mathematical practice that is clearly the articulation and abstraction of an actual instrumental practice. In fact, the contrast between the constructive methods of gnomonics and those of Euclid's *Elements*, allows us to cast the Euclidean problems in a new light. It is often claimed that Euclid's postulates derive from the operations of a compass and a straightedge,[137] but in fact they are more abstract than this. For example, *Elements* I.post.1 can be used to join points that are any distance apart, such as in *Elements* I.2, which a straightedge cannot do. Of course, one could argue that the postulates in the *Elements* suppose an indefinitely long straightedge—but there is no such thing. Again, *Elements* I.16 requires that *Elements* I.post.2 be used to extend a line to any assumed length, which a straightedge cannot do, since every actual straightedge is finite.

[136] I have used this interpretation of Ptolemy's *Analemma* as both computational and instrumental to give an interpretation of the mathematical methods underlying Heron's *Dioptra* 35 as an application of analemma methods; see Sidoli, 'Heron's *Dioptra* 35'.

[137] See, for one of many examples, Mueller, *Philosophy*, pp. 15–16.

Elements I.post.3 is used to produce a circle about a given point as center and passing through another given point, which can be at any distance from the center. Again, this is not possible with a real compass—since every compass has a fixed finite radius. Moreover, as its application in *Elements* I.2 shows, *Elements* I.post.3 cannot be used to produce a circle about a given point with a given radius—but any actual compass can perform this operation. Hence, it has sometimes been argued that *Elements* I.post.3 concerns the operation of a compass that closes when it is no longer in contact with the plane—but again, there is no such compass.[138] In this way, we can contrast the level of abstraction allowed in the *Elements* with that allowed in the *Analemma*, in which every problem can be carried out with constructive operations that are direct abstractions of the physical manipulations of a compass that can operate with given radii and a set square—both of some preassigned, definite size.

Another interesting feature of the *Analemma* is its concern with providing multiple methods for computing the same value. It is clear from the way in which Ptolemy presents his procedures that a primary goal of the text is to provide nomographic techniques, but this is proceeded by a full argument that the values in question are both geometrically determined and computable through chord-table trigonometry. This presentation constitutes a multilevel argument that the procedure—geometrical, computational, and nomographical—is complete. At the most basic level, the theorems of the *Data* implied by the steps of the metrical analysis insure that the geometric magnitudes are fixed; at the next level, the metrical analysis itself provides confirmation that there is an effective procedure for computing the value; and at the final level, the articulation of a geometric and computational procedure assures us that the physical manipulations of the analog computation will produce results that, although perhaps not terribly precise, are, in principle, sound.

Although there is no evidence that Ptolemy's *Analemma* was translated into Arabic, the gnomonics of his predecessor Diodorus certainly was, and there is clear evidence that mathematicians working in Arabic were familiar with analemma methods already at the time of Ḥabash al-Ḥāsib in the early 9th century.[139] These methods must have come directly or indirectly from Greek sources—since there is no evidence of analemma methods in other ancient cultures.[140] Hence, since Ptolemy's *Analemma* presents the most complete

[138] Heath, *A History*, vol. I, p. 246, recounts this interpretation of Euclid's postulate by Augustus De Morgan.

[139] Edwards, *Ptolemy's* Περὶ ἀναλήμματος, pp. 152–182, gives a full study of what is known of Diodorus' life and work. For evidence of the translation of, at least parts of, Diodorus' work into Arabic, see Kennedy, *The Exhaustive Treatise*, pp. 157–166, and Hogendijk, 'Geometrical Works'.

[140] It used to be argued that analemma methods provide the best explanation for certain

explanation we have of the mathematical conceptions underlying analemma methods, and since these methods were so fruitfully applied in the classical Islamic period, if we want to fully understand the medieval development of analemma methods, we should begin with a firm basis in Ptolemy's text. The key features of the *Analemma* that should inform our reading of the medieval sources are (1) its essentially projective approach and (2) its interest in the mathematical justification of the methods of analog computation.

Finally, the *Analemma* provides a well-contained example of the approach to mathematical astronomy developed in the Hellenistic period and still practiced by Ptolemy and others in the Roman Imperial period.

- A geometric model is posited, with relatively little attempt to argue that it is a sound representation of the physical world.
- The model itself becomes the object of geometrical investigation and geometrical claims that can be made about the model are assumed, without comment, to apply also to the world.
- Numerical values, which are ostensibly empirical, enter into the model as given parameters for computation.
- The mathematical methods of computation (λογιστική) are mixed with the constructive methods of geometry, with no evidence for the division of these two areas of mathematics that we find, for example, in the *Elements* and certain philosophical authors.
- The geometrical methods of the *Data* are used as a theoretical basis for a computational practice that is understood as producing measurements of various aspects of the underlying geometric model.

Mathematical scholars of the late ancient and medieval periods, who read Ptolemy as a mathematician, found in these aspects of his approach various methods to articulate, critique and revise in their effort to further develop the mathematical sciences.

Acknowledgements

I presented the details of Ptolemy's mathematical procedures in the *Analemma* at workshops in Paris at the invitation of the SAW Project, under the direction of Karine Chemla. I would like to thank the participants in these workshops for helping me to clarify my presentation. During the time that I was a guest of the SAW Project in Paris, some of the research leading to these results received funding from the European Research Council under the European Union's Seventh Framework Programme (FP7/2007–2013)

Indian sources, but there is no Indian source that clearly contains an analemma construction and the spherical methods in Indian sources can also be explained through other approaches to spherical geometry; see Mimura, *The Tradition*, pp. 36–55.

/ ERC Grant agreement n. 269804. The structure of the argument in this paper was motivated, in part, by a desire to clarify issues that arose in the discussion following my talk at the conference on 'Ptolemy's Science of the Stars in the Middle Ages' at the Warburg Institute, November 2015, organized by the *Ptolemaeus Arabus et Latinus* project. I thank the organizers of this project for the invitation to participate. I presented some topics from this talk at the Excellence Cluster Topoi, Humboldt-Universität zu Berlin, and benefitted from the discussion after this talk. I would like to thank Topoi for hosting me in Berlin and giving me the chance to make this presentation. Fabio Guidetti read the paper carefully and caught a number of errors. This paper has also benefited from the close attention of an anonymous referee and the editors of this volume.

Bibliography

Medieval Manuscripts

Am: Milan, *Ambrosianus graec.* L 99 sup. 8th CE (palimpsest, 5th–7th CE).

O: Vatican, *Vaticanus Ottobonianus lat.* 1850. Autograph, 13th CE.

Modern Scholarship

Acerbi, Fabio, 'The Language of the "Givens": Its Forms and its Use as a Deductive Tool in Greek Mathematics', *Archive for History of Exact Sciences* 65 (2011), pp. 119–153.

Acerbi, Fabio, 'I codici stilistici della matematica greca: dimostrazioni, procedure, algoritmi', *Quaderni Urbinati di Cultura Classica* NS 101/2 (2012), pp. 167–214.

Acerbi, Fabio, and Bernard Vitrac, *Metrica. Héron d'Alexandrie*, Pisa-Roma: Fabrizio Serra, 2014.

Barbera, André, *The Euclidean Division of the Canon*, Lincoln: University of Nebraska Press, 1991.

Berggren, J. Lennart, 'A Comparison of Four Analemmas for Determining the Azimuth of the Qibla', *Journal for the History of Arabic Science* 4 (1980), pp. 69–80.

Berggren, J. Lennart, 'Ḥabash's Analemma for Representing Azimuth Circles on the Astrolabe', *Zeitschrift für Geschichte der arabisch-islamischen Wissenschaften* 7 (1991–1992), pp. 23–30.

Berggren, J. Lennart, and Nathan Sidoli, 'Aristarchus's *On the Sizes and Distances of the Sun and the Moon*: Greek and Arabic Texts', *Archive for History of Exact Sciences* 61 (2007), pp. 213–254.

Carandell, Juan, 'An Analemma for the Determination of the Azimuth of the *qibla* in the *Risāla fī 'ilm al-ẓilāl* of Ibn al-Raqqām', *Zeitschrift für Geschichte der arabisch-islamischen Wissenschaften* 1 (1984), pp. 61–72.

Carman, Christián, 'Two Problems in Aristarchus's Treatise *On the Sizes and Distances of the Sun and Moon*', *Archive for History of Exact Sciences* 68 (2014), pp. 35–65.

Carman, Christián, and James Evans, 'The Two Earths of Eratosthenes', *Isis* 106 (2015), pp. 1–16.

Clagett, Marshall, *Archimedes in the Middle Ages*, vol. II: *The Translations from the Greek by William of Moerbeke*, Philadelphia: The American Philosophical Society, 1976.

Commandino, Federico, *Claudii Ptolemaei Liber de analemmate*, Rome: Paulum Manutium Aldi F., 1562.

Düring, Ingemar, *Die Harmonielehre des Klaudios Ptolemaios*, Göteborg: Wettergren & Kerbers, 1930 [Göteborgs Högskolas Årsskrift 36/1; reprinted New York: Garland, 1980].

Edwards, Don R., *Ptolemy's* Περὶ ἀναλήμματος – *An Annotated Transcription of Moerbeke's Latin Translation and of the Surviving Greek Fragments, with an English Version and Commentary*, PhD dissertation, Brown University, Department of Classics, 1984.

Federspiel, Michel, 'Sur l'opposition défini/indéfini dans la langue des mathématiques grecques', *Les études classiques* 63 (1995), 249–293.

Guerola Olivares, Joaquim, *El* Collegio Romano *i els orígens de la trigonometria: de l'*Analemma *de Ptolemeu a la* Gnomonica *de Clavius*, PhD dissertation, Universitat Autònoma de Barcelona, Centre d'Història de la Ciència, 2018.

Hamm, Elizabeth Anne, *Ptolemy's Planetary Theory: An English Translation of Book One, Part A of the* Planetary Hypotheses *with Introduction and Commentary*, PhD dissertation, University of Toronto, Institute for the History and Philosophy of Science and Technology, 2011.

Heath, Thomas L., *Aristarchus of Samos. The Ancient Copernicus*, Oxford: Clarendon Press, 1913.

Heath, Thomas L., *A History of Greek Mathematics*, 2 vols, Oxford: Oxford University Press, 1921 [reprinted New York: Dover, 1981].

Heiberg, Johan L., 'Ptolemäus de Analemmate', *Abhandlungen zur Geschichte der mathematischen Wissenschaften* 7 (1895), pp. 1–30.

Heiberg, Johan L., *Syntaxis mathematica*, 2 vols, Leipzig: Teubner, 1898–1903 [*Claudii Ptolemaei opera quae exstant omnia*, vol. I].

Heiberg, Johan L., *Opera astronomica minora*, Leipzig: Teubner, 1907 [*Claudii Ptolemaei opera quae exstant omnia*, vol. II].

Hogendijk, Jan, 'The Geometrical Works of Abū Saʿīd al-Ḍarīr al-Jurjānī', *SCIAMVS* 2 (2001), pp. 47–74.

Hultsch, Friedrich, *Pappi Alexandrini Collectionis quae supersunt*, Berlin: Weidmann, 1876.

Id, Yusif, 'An Analemma Construction for Right and Oblique Ascensions', *Mathematics Teacher* 62 (1969), pp. 669–672.

Jones, Alexander, 'Ptolemy's Mathematical Models and their Meaning', in Glen Van Brummelen and Michael Kinyon (eds), *Mathematics and the Historian's Craft*, New York: Springer, 2005, pp. 23–42.

Kennedy, Edward S., *The Exhaustive Treatise on Shadows by Abu al-Rayḥān Muḥammad b. Aḥmad al-Bīrūnī*, 2 vols, Aleppo: University of Aleppo, 1976.

Kennedy, Edward S., 'Ibn al-Haytham's Determination of the Meridian from One Solar Altitude', *Zeitschrift für Geschichte der arabisch-islamischen Wissenschaften* 5 (1989), pp. 141–144.

Kennedy, Edward S., and Yusuf 'Id, 'A Letter of al-Bīrūnī. Ḥabash al-Ḥāsib's Analemma for the Qibla', *Historia Mathematica* 1 (1974), pp. 3–11.

Lehoux, Daryn, *Astronomy, Weather, and Calendars in the Ancient World.* Parapegmata *and Related Texts in Classical and Near Eastern Societies*, Cambridge: Cambridge University Press, 2007.

Luckey, Paul, 'Das Analemma von Ptolemäus', *Astronomische Nachrichten* 230 (1927), cols 17–46.

Mimura, Taro, *The Tradition of Problem-Solving through the Idea of Spherics in Hellenism, India and the Medieval Islamic World* (in Japanese), MA thesis, University of Tokyo, Graduate Department of History and Philosophy of Science, 2002.

Mueller, Ian, *Philosophy of Mathematics and Deductive Structure in Euclid's* Elements, Cambridge (Mass.): MIT Press, 1981.

Murschel, Andrea, 'The Structure and Function of Ptolemy's Physical Hypotheses of Planetary Motion', *Journal for the History of Astronomy* 26 (1995), pp. 33–61.

Netz, Reviel, 'Imagination and Layered Ontology in Greek Mathematics', *Configurations* 17 (2009), pp. 19–50.

Neugebauer, Otto, *A History of Ancient Mathematical Astronomy*, 3 vols, New York: Springer, 1975.

Rome, Adolphe, *Commentaires de Pappus et de Théon d'Alexandrie sur l'Almageste*, 3 vols, Vatican: Biblioteca Apostolica Vaticana, 1931–1943.

Schöne, Hermann, *Herons von Alexandria Vermessungslehre und Dioptra*, Leipzig: Teubner, 1903 [Opera quae supersunt omnia III].

Schoy, Carl, 'Abhandlung über die Ziehung der Mittagslinie, dem Buche über das Analemma entnommen, samt dem Beweis dazu von Abū Sa'īd aḍ-Ḍarîr', *Annalen der Hydrographie und maritimen Meteorologie* 10 (1922), pp. 265–271.

Sefrin-Weis, Heike, *Pappus of Alexandria. Book 4 of the* Collection, London: Springer, 2010.

Sidoli, Nathan, 'Review of *The Works of Archimedes: Volume I. The Two Books On the Sphere and the Cylinder* by Reviel Netz', *Aestimatio* 1 (2004), pp. 148–162.

Sidoli, Nathan, 'Heron's *Dioptra* 35 and Analemma Methods: An Astronomical Determination of the Distance Between Two Cities', *Centaurus* 47 (2005), pp. 236–258.

Sidoli, Nathan, 'Mathematical Tables in Ptolemy's *Almagest*', *Historia Mathematica* 41 (2014), pp. 13–37.

Sidoli, Nathan, 'Mathematical Discourse in Philosophical Authors: Examples from Theon of Smyrna and Cleomedes on Mathematical Astronomy', in Christián Carman and Alexander Jones (eds), *Instruments – Observations – Theories: Studies in the History of Early Astronomy in Honor of James Evans*, forthcoming.

Sidoli, Nathan, 'The Concept of *Given* in Greek Mathematics', *Archive for History of Exact Sciences* 72 (2018), pp. 353–402.

Sidoli, Nathan, and J. Lennart Berggren, 'The Arabic Version of Ptolemy's *Planisphere* or *Flattening the Surface of the Sphere*: Text, Translation, Commentary', *SCIAMVS* 8 (2007), pp. 37–139.

Sinisgalli, Rocco, and Salvatore Vastola, *L'Analemma di Tolomeo*, Firenze: Edizioni Cadmo, 1992.

Soubiran, Jean, *Vitruve. De l'architecture, Livre IX*, Paris: Les Belles Lettres, 1969.

Suzuki, Takanori, 'A Solution of the *qibla*-Problem by Abu 'l-Qāsim Aḥmad ibn Muḥammad al-Ghandajānī', *Zeitschrift für Geschichte der arabisch-islamischen Wissenschaften* 4 (1987–1988), pp. 139–149.

Todd, Robert B., *Cleomedis Caelestia*, Leipzig: Teubner, 1990.

Toomer, Gerald J., *Ptolemy's Almagest*, London: Duckworth, 1984 [reprinted Princeton: Princeton University Press, 1998].

Vuillemin-Diem, Gudrun, and Carlos Steel, with Pieter De Leemans, *Ptolemy's* Tetrabiblos *in the Translation of William of Moerbeke*, Leuven: Leuven University Press, 2015.

Was there a Ninth Sphere in Ptolemy?*

Paul HULLMEINE

An important part of ancient and medieval cosmology was the determination of the total number of spheres needed for any given astronomical model to account for the celestial phenomena.[1] Notorious in this respect is Aristotle's conclusion in *Metaphysics* XII.8 that there are up to 55 spheres, exceeding the amount reckoned by his predecessors Callippus and Eudoxus. Aristotle arrives at this large number for two reasons. First, he needs to account for the complex motions of the stars and planets. Second, he strives in his cosmological setup to adhere to the presuppositions and foundations of his physics.[2] Without delving too deeply into the various pre-Ptolemaic systems, the core idea of the Aristotelian model is the following. All celestial bodies, namely, the Sun, Moon, and the five planets visible to the naked eye — Mercury, Venus, Mars, Jupiter, and Saturn — are fixed in a sphere homocentric with the centre of the universe, the immobile earth. Since one sphere through itself can only move with one

* I am grateful to the editors for allowing me to publish this article in the present volume although I did not attend the conference in 2015. I want to thank Charles Burnett and Peter Adamson for their careful reading of a previous version of this text and for their ongoing support. My gratitude goes also to the anonymous referee for helpful comments and references.

[1] See Lloyd, 'Saving the Appearances', and especially the discussion of Ptolemy's *Planetary Hypotheses* on pp. 215–17. There, Lloyd discusses the question whether Ptolemy aims at establishing a functioning physical picture of the universe in addition to his mathematical calculations. He concludes (p. 217): 'But the fundamental point remains that to represent Ptolemy in general as interested purely in the mathematics of his problems cannot be right given first the appeal to physical arguments in chs. 3 and 7 of the *Syntaxis* and second in the straightforwardly realist account offered in the second book of the *Planetary Hypotheses*.'

[2] The main passage on the number of spheres and celestial movers is *Metaph.* XII.8, 1073b3–1074a18. See for example Lloyd, 'Metaphysics Λ 8', and Beere, 'Counting the Unmoved Movers'. The physical foundations underlying his astronomical system are spread throughout the *De caelo*, for example the nature of the fifth element, the aether (I.3), the nature of the circular motion (I.4), the spherical motion of the heavens and the stars (II.3–8), and the sphericity of the earth (II.14). The fragments by Eudoxus are collected in Eudoxus of Cnidus, *Die Fragmente*. For the relationship between the astronomical models by Eudoxus, Callippus and Aristotle and their different calculations of the total numbers of spheres and celestial movers, see Neugebauer, *A History*, pp. 675–89; Heglmeier, 'Die griechische Astronomie'; Yavetz, 'On the Homocentric Spheres'; Mendell, 'The Trouble with Eudoxus'. Additionally, a brief but concise overview is offered by Alberto Jori in his commentary added to his German translation of the *De caelo*, see Aristotle, *Über den Himmel*, pp. 296–301.

Ptolemy's Science of the Stars in the Middle Ages, ed. by David Juste, Benno van Dalen, Dag Nikolaus Hasse and Charles Burnett, PALS 1 (Turnhout, 2020), pp. 79–96
© BREPOLS PUBLISHERS 10.1484/M.PALS-EB.5.120174

motion, further orbs need to be added to account for the complex motion of the planets. Above all these spheres lies the sphere of the fixed stars, which is moved by the Prime Mover. This sphere, in turn, transmits the diurnal westwards rotation to every planet below it. In order to make certain that only this diurnal rotation is transmitted to, for example, Jupiter, the combined motion for Saturn must be cancelled, which happens by means of Aristotle's notorious 'counter-acting' spheres. This results in a model that assigns multiple spheres to every planet, thereby generating the 55 or 49 spheres posited by Aristotle.

However, one could also reduce the number of spheres by assigning one 'main sphere' to every planet, which then consists of a variable number of orbs responsible for the seemingly irregular motion of the planets. This approach results in assuming eight 'main spheres', one for each of the five planets, one for the Sun, one for the Moon, and one for the fixed stars. Such a reduced version can be found in Alexander of Aphrodisias.[3] In this context, István M. Bodnár has posed the question whether Alexander actually argued only for the existence of eight spheres in total or of eight 'bundles' of spheres (i.e. what I called 'main spheres'), deciding for the latter option.[4] Subsequently, Damien Janos argued that a model similar to that of Alexander can be found in two of the most important representatives of the medieval Arabic philosophical tradition, al-Fārābī and Avicenna. Interestingly, al-Fārābī as well as Avicenna adopt a model that consists not of eight, but rather of nine such 'bundles' of spheres, each of these carrying within itself a certain number of eccentrics and epicycles.[5] In fact, this model of nine concentric main spheres was widespread in medieval Arabic philosophy.[6] To put forward one telling example, the metaphysical part of Avicenna's *Kitāb al-Šifāʾ* (*Book on the Healing*) addresses the number of celestial movers. Already for Aristotle and Alexander the question of the number of spheres was not only a purely astronomical one to account for the phenomena, but rather had some severe philosophical consequences regarding the further question on the number of Unmoved Movers, since they assigned one celestial mover to every celestial motion. Leaving aside the philosophical background of this question, Avicenna tells us that there

[3] Alexander of Aphrodisias, *On the Cosmos*, pp. 10 and 93–95. Genequand refers to a similar passage in Alexander's *Quaestiones*, I.25, see Alexander of Aphrodisias, *Quaestiones*, p. 40: 23–26, tr. Sharples, *Quaestiones 1.1–2.15*, p. 85.

[4] See Bodnár, 'Alexander on Celestial Motions', pp. 196–98.

[5] For al-Fārābī, see Janos, *Method, Structure, and Development*, pp. 119–28 (and for a possible influence by Alexander, pp. 152–53); for Avicenna, see Janos, 'Moving the Orbs', p. 174.

[6] Another example from the philosophical side are the Iḫwān al-Ṣafāʾ, see Iḫwān al-Ṣafāʾ, *On the Natural Sciences*, epistle 16, ch. 3, where they pose a system of nine spheres and connect this system to the Qurʾānic report that eight angels, which they compare to the eight spheres of the planets and the fixed stars, carry God's throne, which they compare to the outermost ninth, starless sphere (*Qurʾān* 69:17). Thereby, they provide the ninth sphere with a theological fundament.

are two opinions about the nature of the first sphere (*al-kura al-ūlā*). Before Ptolemy, according to Avicenna, the first sphere was thought to be the sphere of the fixed stars (*kurat al-ṯawābit*). However, the scientists following Ptolemy claimed the existence of a starless sphere outside that of the fixed stars.[7] Thus, Avicenna reports that it was Ptolemy who first introduced an additional sphere above the fixed stars, a move which the majority of the later tradition followed. The fact that the departing point of this report is the debate on the number of celestial movers highlights the philosophical importance of this question. In addition, the acceptance of a starless outermost sphere contradicts Aristotelian cosmology. Aristotle emphasizes that there is no body outside the realm of the cosmos, which means outside the sphere of the fixed stars.[8] In addition, he poses the question why it is that the single primary motion carries a huge mass of stars, whereas the inner spheres carry only one planet each. Of the two arguments Aristotle puts forward, the stronger is that the fact that the outermost sphere carries all the fixed stars is an indication for its superiority above the others.[9] This argument leads Averroes to reject a ninth sphere, thereby consciously pitting himself against Ptolemy, since Ptolemy described the precession that is additional to the diurnal motion.[10] This critique stands in the context of a development in al-Andalus that tried to replace Ptolemaic cosmology with a system that was compatible with Aristotelian physics.[11] Finally, the discussion about the number of spheres is also found in Jewish thought, where we likewise find traces of the ascription of a ninth sphere to Ptolemy.[12]

While these implications belong to the philosophical reception of a ninth sphere, we also have evidence for its mathematical-astronomical reception

[7] Avicenna, *The Metaphysics*, p. 317:9–13. Marmura's translation reads as follows: '[The first sphere,] for those who preceded Ptolemy, is the sphere of the fixed stars; and for those who learned the sciences that became manifest to Ptolemy, [it] is a sphere outside the [former] which surrounds it and is without stars'. On pp. 328:28 and 331:8, this sphere is called *al-ǧirm al-aqṣā* and *al-falak al-aqṣā*, i.e. 'the utmost body/sphere'. It may be noted here that Avicenna does not enumerate the different spheres or movers. He does not call this 'the ninth sphere'. Nevertheless, since the topic of the ninth sphere was so common in the Islamic world from the early ninth century onwards, Avicenna surely must have an independent main sphere in mind.

[8] *De caelo* I.9, 279a11-b3, II.10, 291a34-b9 and *Metaph.* XII.6–7.

[9] *De caelo* II.12, 292a10–15 and 292b25–293a12. See also *Metaph.* XII.8, 1074a26–28, where Aristotle writes that all motions exist for the sake of the stars.

[10] See his account in his *Epitome* of Aristotle's *Metaphysics*, in Averroes, *On Aristotle's Metaphysics*, p. 146, and further Endress, 'Averroes' *De caelo*', pp. 43–44.

[11] See Sabra, 'The Andalusian Revolt'.

[12] See Tanenbaum, 'Nine Spheres or Ten?', esp. p. 310. The author addresses mostly the philosophical and theological implications. The ascription of nine spheres to Ptolemy can be found in a gloss on Ibn Ezra, which also gives evidence for a reception of Ptolemy's *Planetary Hypotheses*. See also Sela, 'Maimonides and Māshaʾallāh' on the ninth sphere in a more astrological context in medieval Jewish thought.

already in the early ʿAbbāsid period. A very early, ninth-century critique of the ninth sphere that deals with the problem of how two concentric orbs interact with each other, survives in a citation by Quṭb al-Dīn al-Šīrāzī.[13] It is interesting that the author of this cited treatise, whom Saliba identified with one of the Bānū Mūsā, namely Muḥammad ibn Mūsā, seems to consider himself as belonging to the Ptolemaic tradition. The citation, as it survives in al-Šīrāzī, begins with two references to Ptolemy's *Almagest* that Muḥammad ibn Mūsā uses to strengthen his point that the first universal motion is produced by a deity and not by a bodily ninth sphere.[14] The fact that al-Šīrāzī draws on this treatise provides evidence that the astronomical tradition in Marāġa also dealt with the question of the ninth sphere's existence.[15]

So far, I have shown that the doctrine of a ninth sphere was, firstly, adopted as well as rejected in medieval Arabic and Jewish philosophy and astronomy and, secondly, that its introduction is sometimes ascribed to Ptolemy. However, the puzzle is that there is no information on a ninth sphere in Ptolemy's most important astronomical work, the *Almagest*. In *Almagest* I.8, Ptolemy discusses the first two primary motions, the precession of the equinoctial points and the diurnal rotation of the heavens.[16] Although from antiquity onwards the discovery of precession was considered as the reason for introducing a ninth sphere, as we will see, Ptolemy does not mention anything along these lines in the *Almagest*. Instead, modern scholars have referred to the *Planetary Hypotheses*, which is extant in its entirety only in an Arabic and a Hebrew version,[17] in

[13] This citation is edited and translated in Saliba, 'Early Arabic Critique', pp. 130–37.

[14] Saliba, 'Early Arabic Critique', p. 131.

[15] At this point one can refer to Naṣīr al-Dīn al-Ṭūsī, al-Šīrāzī's teacher, who included a ninth sphere in his description of the physical arrangements of the celestial spheres in his *al-Taḏkira fī ʿilm al-hayʾa* (*Memoir on Astronomy*), see Ragep, *al-Ṭūsī's Memoir*, vol. I, pp. 108–11 and 124–25, and vol. II, pp. 389–90 (in the same book, Ragep gives some valuable insights into questions of the relationship between the eighth and ninth sphere and the transmission of motion in his commentaries, see Ragep, *al-Ṭūsī's Memoir*, vol. II, pp. 400–10). On the reception of the ninth sphere within the Marāġa-tradition and its dependence on *falsafa*, see Morrison, 'Falsafa and Astronomy', pp. 313–16. But also astronomers outside of Marāġa did employ a ninth sphere in their astronomical works, for example the Andalusian al-Biṭrūǧī (see Mancha, 'Al-Biṭrūjī's Theory', pp. 147–61) who became well-known in the Latin tradition. Thus, Pierre Duhem's account that this conception was adopted by nearly all Islamic astronomers of the Middle Ages might generally speaking still be true, see Duhem, *Le système du monde*, vol. II, p. 204.

[16] For Ptolemy's report of the discovery of precession, see Ptolemy, *Almagest*, VII.2 and 3. Further, see Neugebauer, *A History*, pp. 292–98.

[17] The extant Greek text of the *Planetary Hypotheses* only covers the first part of the first book, see Ptolemy, *Hypotheseōn* (an English translation is available in Hamm, *Ptolemy's Planetary Theory*, pp. 44–64). An English translation (from the Arabic) of the second part of Book I can be found in Goldstein, *The Arabic Version*, pp. 5–9, and an edition of the Arabic version of the entire Book I with a French translation in Morelon, 'La version arabe'.

order to show that Ptolemy might have argued for such a ninth sphere.[18] Interestingly, there are indeed some passages that seem to justify this course of the story.

In order to look for the textual basis of this claim, in what follows I will investigate the terminology Ptolemy uses throughout the *Planetary Hypotheses* for referring to the different kinds of spheres, orbs, and circles.[19] The first part of the first book, which is also extant in Greek, deals with the mathematical models and the basic data of the motions of the planets. Before he gives detailed numerical values for the revolutions of the different planets, Ptolemy provides short introductory statements. In chapter I.3,[20] he introduces different spheres and circles with their specific names. For example, he starts as follows:

> νοείσθω μέγιστος κύκλος περὶ τὸ κέντρον τῆς τοῦ κόσμου σφαίρας μένων καὶ καλείσθω ἰσημερινός
>
> Let there be imagined a stationary great circle that is centered on the center of the sphere of the cosmos and let it be called the 'equator' [...][21]
>
> فلنتوهّم دائرة من الدوائر العظام مخطوطة على مركز العالم ثابتة ولتسمّ فلك معدّل النهار
>
> Let us imagine one of the great circles drawn about the center of the world, fixed, and it shall be called the 'circle of the equator' [...][22]

Besides the minor point that the Arabic does not translate the Greek *sphaira* here, it is interesting to point at the way in which the Arabic translator renders the Greek *kyklos* ('circle'). As one would expect, the corresponding Arabic term is *dāʾira*, which is the usual term for a geometrical circle. However, where the Greek omits a reiteration of the term *kyklos* as reference for the adjective *isēmerinos* ('equator'), the translator choses to render this as *falak muʿaddal al-nahār* (which can mean both, 'orb' or 'circle of the equator').[23] Thus, *falak*

[18] See, for instance, Saliba, 'Early Arabic Critique', pp. 118 and 121, and Janos, *Method, Structure, and Development*, p. 120, n. 16. On the other hand, scholars did point out that there might be some problems with this ascription, for example Richard Sorabji, see Sorabji, 'Adrastus', p. 588, and also Damien Janos in Janos, 'Moving the Orbs', p. 168, n. 7, where he writes that 'the remarks on the ninth orb in this work [i.e. the *Planetary Hypotheses*] are somewhat ambiguous'.

[19] The most accurate analysis of the *Planetary Hypotheses* can be found in Murschel, 'Structure and Function'.

[20] I follow the chapters suggested in the edition and German translation in Ptolemy, *Hypotheseōn*.

[21] Ptolemy, *Hypotheseōn*, p. 74:3–4; English translation in Hamm, *Ptolemy's Planetary Theory*, p. 46.

[22] Morelon, 'La version arabe', p. 19:4–5, the English translation is my own. Morelon, 'La version arabe', p. 18 translates as follows: 'Représentons-nous l'un des grands cercles traces autour du centre du monde, fixe, appelé orbe de l'équateur'.

[23] One finds a parallel passage in Ptolemy, *Almagest*, I.8. There, Ptolemy writes that 'the greatest of these circles is called the equator' (tr. Toomer, *Ptolemy's Almagest*, p. 45). The Greek

is used here as in the rest of the first seven chapters to describe circles. The same pattern, namely translating *falak* when the Greek text does not specify an adjective any further, is also used in the case of a moving circle, which is simply *pherōn* in the Greek and *al-falak al-muḥarrik* in the Arabic.[24] Also, in these first seven chapters, the Arabic term *kura* ('sphere') is only used for the sphere of the fixed stars, translating literally *sphaira*.[25]

From I.8 onwards, Ptolemy discusses the geometrical models and the mathematical data of the revolutions in more detail with respect to every planet. In these chapters, he describes the sphere (*sphaira* in Greek and *kura* in Arabic) of the planet, which in its entirety contains all the various circles needed for describing the course of the planet. Take I.10 on Mercury as an example:

> ἐπὶ δὲ τῆς τοῦ Ἑρμοῦ σφαίρας νοείσθω κύκλος ὁμόκεντρος τῷ ζῳδιακῷ [...] φερέτω δὲ οὗτος ὁ κύκλος ἕτερον κύκλον ἐγκεκλιμένον πρὸς αὐτὸν
>
> Concerning the sphere of Mercury, let there be imagined a concentric circle with the zodiac circle [...] Let this circle carry another circle inclined to it [...][26]
>
> حال أفلاك عطارد وأمّا عطارد فإنّا نتوهّم في كرته فلكًا مركزه مركز فلك البروج [...] وليحرّك هذا الفلك بحركته فلكًا آخر مائلًا عنه
>
> The situation of the circles of Mercury. Regarding Mercury, we imagine in its sphere a circle, the center of which is the center of the zodiac [...] By its motion, this circle shall move another circle that is inclined to it.[27]

The Arabic manuscripts add the phrase 'the situation of the spheres of' at the beginning of the discussion of every planet. In the case of the Sun, this is *ḥāl falak al-šams*, where the singular is used, since in what follows only one eccentric circle is described to account for the Sun's motion.[28] For all other planets, the Arabic has the plural, *aflāk*, just as in the above cited example. Afterwards, in the Greek phrase stating 'concerning the sphere of' *sphaira* is always translated as *kura*, and the various circles, Greek *kyklos*, are rendered as *falak*. We

text has *hōn ho megistos kyklos isēmerinos kaleitai* (Ptolemy, *Syntaxis*, p. 26:19–20), which is analogous to the text from the *Planetary Hypotheses*. However, the Arabic reads *wa-yusammā aʿẓam hāḏihi l-dawāʾir muʿaddal al-nahār* (cited from the translation by Isḥāq ibn Ḥunayn and Ṯābit ibn Qurra, extant in the manuscript Tunis, National Library, 7116, fol. 5v:12). There, the translator did not use the additional *falak* as in the *Planetary Hypotheses*.

24 See Ptolemy, *Hypotheseōn*, p. 74:11 and Morelon, 'La version arabe', p. 19:8.

25 Compare, for example, the edition in Ptolemy, *Hypotheseōn*, pp. 76:28, 78:3–4 and 78:17–18 with Morelon, 'La version arabe', p. 21:17, p. 23:2, and p. 23:16.

26 Ptolemy, *Hypotheseōn*, p. 84:24–29; English translation in Hamm, *Ptolemy's Planetary Theory*, p. 53.

27 Morelon, 'La version arabe', p. 33:1–4, the French translation (ibid., p. 32) reads as follows: 'Situation des orbes de Mercure. Quant à Mercure, nous nous représentons dans sa sphère un orbe dont le centre est celui de l'écliptique [...]; que, par son mouvement, cet orbe meuve un autre orbe incliné'.

28 See Morelon, 'La version arabe', p. 27:1.

can infer two important details from these descriptions. First, Ptolemy (in this case the Greek version, which is available to us for this chapter) indicates that there is a *sphere* for every planet, in which the reader is supposed to imagine the various geometrical *circles*. Second, the Arabic translator now choses to translate *kyklos* as *falak* rather than *dāʾira*.[29] For the five planets Mercury, Venus, Mars, Jupiter and Saturn, there is another expression that might deserve our attention. When it comes to the construction of the epicycle, he calls this the 'sphere of the epicycle', both in Greek as well as in the Arabic, *hē epikyklou sphaira* and *kurat falak al-tadwīr*. This indicates that Ptolemy uses the term *sphaira*, rendered in Arabic as *kura*, to describe the entirety of the epicycle, which then consists of different circles.[30]

From this point onwards, the remainder of the Greek text has not survived.[31] Consequently, the possibility to compare the Arabic translation to the Greek text also comes to an end here. The upshot of the comparison drawn so far is the following: In these first chapters of the Arabic translation of Ptolemy's *Planetary Hypotheses*, in which Ptolemy deals with the geometrical representation of the model of the planets and with the numerical values of their motions, the term *falak* is exclusively used for geometrical circles. These circles, on the other hand, are integrated into the spheres of the planets. For the spheres, the Arabic translator uses *kura*, always corresponding to the Greek *sphaira*. *Kura* is used in a similar way in the following, last chapters of the first book.[32] In I.16 to 18, Ptolemy turns to the order and distances of the planets.[33] He signifies their order by referring to 'the sphere of the Moon', 'the sphere of the Sun', etc. Thereby, he means the main sphere, carrying with it the planet, whereas it may contain all these aforementioned circles. It is interesting to highlight that Ptolemy does not mention the ninth sphere at this point, although this might easily be explained by the fact that it was supposed to be starless.[34]

[29] On the different meanings of *falak* in Ibn al-Hayṯam, see Langermann, 'A Note on the Use'.

[30] A discussion of the mathematical values given in the description of these models can be found in Neugebauer, *A History*, pp. 901–13.

[31] In fact, the text stops in the middle of a description of Saturn. Some extant Greek manuscripts complete the text for Saturn so that it is analogous to the previous models. However, they leave empty spaces for the numerical data, which Heiberg fills in with the data that are found in the Arabic manuscripts. See Ptolemy, *Hypotheseōn*, pp. 104–07.

[32] These chapters, which are neither part of Heiberg's edition nor of Nix's German translation, were first discovered and translated by Bernard R. Goldstein, see Goldstein, *The Arabic Version*. They are included in Morelon's edition, see Morelon, 'La version arabe', pp. 56–85.

[33] See Morelon, 'La version arabe', pp. 62–81. In Goldstein, *The Arabic Version*, these are chapters 2 to 4.

[34] The only hints towards the first motion are at Morelon, 'La version arabe', pp. 57:4 and 69:8.

Ptolemy devotes the second book of the *Planetary Hypotheses* to the question of how to physically represent the mathematical model of the heavens. The first nine chapters discuss the physical and metaphysical foundations of Aristotelian cosmology, such as the existence of aether and Aristotle's counter-acting spheres.[35] In II.10, Ptolemy lays the ground for the following presentation of his own physical models by clarifying his terminology.[36] Among other things, he explains how the term 'mover' appears:[37]

> ويسمّى ما كان من الأجسام متحرّكًا من المشرق إلى المغرب على أقطاب فلك معدّل النهار وكان يذهب بجميع ما يحيط به إلى ناحية حركة الكلّ بالضرورة باسم عامّ له وهو المحرّك وأوّل هذه الأجسام هو الذي يحرّك كرة الكواكب الثابتة والثاني الذي يحرّك كرة زحل الخارجة والثالث الذي يحرّك كرة المشتري الخارجة وكذلك ما يتلوا هذا على الوِلاء[38]
>
> The bodies that move from east to west on the poles of the equator and that necessarily carry along with them everything that they encompass in the direction of the motion of the cosmos, are called by a general name, namely 'mover'. The first of these bodies is that which moves the sphere of the fixed stars, the second that, which moves the outer sphere of Saturn, the third that which moves the outer sphere of Jupiter, and so likewise [for the rest of the spheres] according to [their] sequence.

At first glance, one might take the Arabic Ptolemy here as arguing for a sphere above that of the fixed stars, which one could consider as the (in)famous ninth sphere. However, this passage unambiguously contradicts such a reading. As Ptolemy points out, every sphere belonging to a planet (or to the fixed stars, respectively) has a body that moves it from east to west, i.e. which transmits the diurnal rotation to it. Each of these movers is called *jism*, 'body', whereas *kura* is the term used to signify the outer spheres of the planets. As he describes later and makes clear with every particular model for each planet, all the orbs[39] below this outer sphere, are referred to by their specific terms, such as *falak al-tadwīr* for epicycle (a fixed technical term).[40] Following this, in the next chapter he explains his model for the first motion of the cosmos. This model includes three spheres (*ukar*, the plural of *kura*) that are concentric with the

[35] Valuable summaries can be found in Murschel, 'The Structure and Function', pp. 37–41; Taub, *Ptolemy's Universe*, pp. 112–23, and Feke, *Ptolemy's Philosophy*, pp. 187–200.

[36] The fact that he again turns to an explanation of terminology indicates the start of a new topic within the *Planetary Hypotheses*, namely the transmission from the geometrical models in Book I to their physical representation in Book II.

[37] For the unedited passages of Book II, I refer to the facsimile of the London manuscript (London, British Library, Add. 7473, fols 81v–102v) in Goldstein, *The Arabic Version*. The second book can be found on pp. 36–55, and descriptions of the Arabic manuscripts on p. 5. Translations of these passages are my own.

[38] Goldstein, *The Arabic Version*, p. 42:9–13.

[39] Instead of 'circle', which I used with regard to the geometrical representation, I now shift my translation of *falak* to 'orb'. Thereby, I intend to capture the three-dimensional figure of these *aflāk* when it comes to the physical functioning of Ptolemy's models.

[40] See for example Goldstein, *The Arabic Version*, p. 42:20.

centre of the cosmos: 1) the body moving the sphere of the fixed stars, 2) the sphere of the fixed stars itself, and 3) the body moving the outer sphere of Saturn.[41] Thus, Ptolemy uses here the term *kura* for the movers as well as for the sphere of the fixed stars and the outer sphere of Saturn itself. All of them share the same centre, which is the centre of the cosmos, and are thus concentric.[42] This is important to highlight, because it is the first indication that Ptolemy did not introduce a system of only nine spheres that are designated by the term *kura*.

The following chapters, II.12 to II.16, describe these physical models for each of the planets. By doing so, he occasionally reminds the reader of the hierarchy and number of the moving spheres. In the chapter about the model of Saturn, for example, he writes that the moving sphere for the model of Saturn is the second of the moving spheres (after the one that moves the fixed stars) and that the third of these moving spheres belongs to Jupiter.[43] These moving spheres are similar to the first one, which means that they are responsible for the diurnal rotation of every set of spheres belonging to a planet. Ptolemy keeps counting until he arrives at the physical model of Mercury, below which is the eighth moving sphere belonging to the Moon.[44]

This means not only that there is a sphere above that of the fixed stars, which is responsible for its diurnal rotation, but also that there is an outer moving sphere at the level of every planet. This is again emphasized towards the end of the *Planetary Hypotheses*, where Ptolemy provides us with a summary of the motions and spheres:

> فجميع الأكر على الوجه الأوّل إحدى وأربعون كرة من ذلك ثماني أكر محرّكة وكرة للكواكب الثابتة وكرة الشمس وأربع للقمر ولكلّ واحد من زحل والمشتري والمرّيخ والزهرة خمس أكر وفي هذه الأكر في كلّ واحد من الكواكب كرة مقارنة وكرة[45] تتحرّك على خلافها ولعطارد سبع أكر فيها واحدة مقارنة وواحدة[46] تتحرّك على خلافها فجميع ذلك إحدى وأربعون كرة[47]

> Thus, on account of the first way[48] all the spheres are 41: Of these, eight are moving spheres, one is for the fixed stars, one for the Sun, four for the Moon, and five spheres each for Saturn, for Jupiter, for Mars, and for Venus. Among these spheres, for every one of the planets there is one accompanying sphere and one that moves

41 See Goldstein, *The Arabic Version*, pp. 42:26–43:2.

42 In his analysis of this chapter, Neugebauer, *A History*, p. 923, does not mention that this moving sphere was interpreted as ninth sphere. This strengthens my point that a literal analysis of the extant material does not indicate that Ptolemy argued for nine main spheres.

43 cf. Goldstein, *The Arabic Version*, p. 45:16–18.

44 cf. Goldstein, *The Arabic Version*, p. 50:1–4.

45 The London manuscript printed in Goldstein's article has *wa-kayfa*. The manuscript from Leiden (Leiden, UB, Or. 180, fol. 42r:5) has *wa-kura*, the reading here adopted.

46 The London manuscript omits *muqārina wa-wāḥida*, which is included in Leiden, UB, Or. 180, fol. 42r:6.

47 Goldstein, *The Arabic Version*, p. 53:14–18.

48 By this, Ptolemy refers to the configuration of the cosmos by means of complete spheres.

> contrary to it. Mercury has seven spheres, among them one accompanying and one that moves contrary to it. Therefore, all of these are 41 spheres.

From all the passages collected here, it becomes clear that the moving spheres cannot be counted as 'individual spheres', which must be added to the number of the eight main spheres. Otherwise, we would not end up with nine, but sixteen spheres or sets of spheres, respectively. For counting the outermost moving sphere as independent from the sphere of the fixed stars would lead to the need of counting every moving sphere for every planet independently as well. However, there is one additional problem that we should address here. In the second book of the *Planetary Hypotheses*, Ptolemy expounds his theory of the so-called sawn-off pieces (*manšūrāt*). He contrasts the Aristotelian cosmological model with his assumption of a combination of two different kinds of spheres. The first kind are the complete spheres, which we know from the classical pre-Ptolemaic system, which are assigned to the outer spheres of the cosmos, since the fixed stars are dispersed throughout the heaven. The second kind are only rings or sawn-out pieces of the inner spheres for the five wandering planets and the Sun and the Moon, since they only move against a part of the heavens.[49] Since the space between the inner surface of the outer complete spheres and the inner rings are filled with aether, which transmits the diurnal rotation to every one of the rings, Ptolemy needs fewer spheres in this second model than before:

> وأمّا على الوضع الثاني فإنّ جميع الأجسام تكون تسعة وعشرين جسمًا من ذلك ثلاث أكر مجوّفة وهي الكرة المحرّكة للكواكب الثابتة وكرة الكواكب الثابتة[50] وكرة ما يبقي من الأثير وستة وعشرون منشورًا من منشورات الأكر وكذلك أيضًا يكون للشمس منشور واحد وللقمر أربع منشورات ولكلّ واحد من زحل والمشتري والمرّيخ والزهرة أربعة ولعطارد خمسة فجميع ذلك تسعة وعشرون جسمًا[51]

> On account of the second model, all bodies are [only] 29. Of these, three spheres are hollow, namely the sphere moving the fixed stars, the sphere of the fixed stars [itself], and the sphere of what remains of the aether. 26 are sawn-off pieces of the spheres. Likewise, the Sun also has one sawn-off piece, the Moon four, and every one of Saturn, Jupiter, Mars, and Venus have four, and Mercury has five. Therefore, the total of these are 29 bodies.

But also in this picture, if we decided to think of the first moving sphere as the ninth sphere, we would need to consider the 'rest of aether' as well, and would arrive at 10 sets of spheres. In short, the discussion revolves around the problem of the status of the outermost sphere, for which Ptolemy undoubtedly argued. However, it is not the case (contrary to what the Arabic tradition supposes)

[49] For the best analysis up to this day, see Murschel, 'The Structure and Function', esp. pp. 41–52.

[50] The London manuscript omits *wa-kurat al-kawākib al-ṯābita*, which is included in MS Leiden, UB, Or. 180, fol. 42r:9.

[51] Goldstein, *The Arabic Version*, p. 53:18–22.

that he conceived the outermost starless sphere as being independent from the others below it. Rather, it simply induces the diurnal westwards rotation for the sphere of the fixed stars, which by itself moves eastwards to account for precession. As the model for the complete spheres reveals, there is an analogous moving sphere for every planet, which no one (neither in Neoplatonic, Arabic, or modern scholarship) added to the number of main spheres or sets of spheres.

The investigation of the Greek and Arabic terminology underlying the *Planetary Hypotheses* reveals, most importantly, that *kura* or *sphaira* do not indicate these eight superordinate spheres specifically. On the one hand, in the mathematical part of Book I of the *Planetary Hypotheses* a set of spheres belonging to a planet is called *kura*, which translates the Greek *sphaira*.[52] On the other hand, the translator chose to use *falak* to translate circles, which are the tools for computing the motion of the planets. In the second book, Ptolemy turns to the physical representation of these spheres. The Arabic translator continued to use *daʾira* (probably *kyklos*) for describing the geometrical figure. When it comes to describing the physical explanation of the motion of the planets, the eccentrics and epicycles are referred to as *falak al-tadwīr* and *al-falak al-ḫāriǧ al-markaz*. On the other hand, the term *kura* receives a wider meaning than in the mathematical part before. It indicates concentric spheres, whether they are just 'movers' for the spheres of the planets or not. Thus, *kura* also has, at least in the *Planetary Hypotheses*, a twofold meaning. First, it can describe a 'set of spheres', of which there are only eight, and second, concentric spheres within this set of spheres. This means that it is not wrong to think of sets of spheres or main spheres, carrying a certain number of concentric, eccentric, and epicyclic spheres within them, as already Alexander of Aphrodisias had done. Since Ptolemy only enumerates the moving spheres together with the overall number of spheres needed for the motions of the planets, he never calls the outermost sphere the ninth. This *argumentum ex silentio* taken together with the textual evidence gathered from the *Planetary Hypotheses* shows that Ptolemy did not intend to establish a nine-sphere-cosmos.

The outcome of this investigation is that we can definitely state that Ptolemy was only concerned with enumerating the various orbs in order to show that his model of sawn-off pieces is more economic than the model of com-

[52] This meaning can be found in Ṯābit ibn Qurra as well. This is remarkable, since Ṯābit is said to have revised the translation of the *Planetary Hypotheses* (though Murschel, 'The Structure and Function', p. 34 doubts that on the basis of the poor quality of the translation). The work in question is the short treatise *Presentation of the Orbs of the Heavenly Bodies*, in Morelon, *Thābit ibn Qurra*, pp. 18–25. There, on p. 19, Ṯābit writes: 'These spheres (*ukar*) that belong to the planets contain (*fīhā*) various circles (*aflāk muḫtalifa*), which move in various ways, …' Since this treatise is usually referred to as being a reprise to the first part of the first book of the *Planetary Hypotheses*, this should be read in the same way as being a description of the geometrical figures.

plete spheres. However, he does not enumerate what we have called 'bundles of spheres' above, because it is not important for the physical functioning of his model whether the outermost orb belongs to the sphere of the fixed stars as a mover or whether it is regarded as an individual ninth one. Since it is now safe to say that Ptolemy himself did not argue for a ninth sphere, the question can be raised where this report, which seems to have been widespread in the Middle Ages, ultimately comes from. From what we have seen before, at the basis there seems to have been a misunderstanding. To my knowledge, it is not before John Philoponus and Simplicius in the sixth century AD that Ptolemy is reported to have argued for a ninth sphere. For example, Proclus argued against Ptolemy regarding precession, but did not mention the theory of a ninth sphere.[53] Thus, John Philoponus is the first author who wrote about the existence of a ninth sphere with reference to Ptolemy.[54] In his *De opificio mundi*, he defends the Biblical report about a heaven above the fixed stars by alluding to Hipparchus and Ptolemy.

> If someone does not believe the prophet [i.e. Moses], who supposes another heaven outside the so-called sphere of the fixed stars, because there would be no proof of its existence, he shall have in mind that no one of the mathematicians before Ptolemy and Hipparchus knew the ninth and starless sphere, the utmost of all. Plato, together with the others, believed that only eight exist, but on the basis of some observations, about which we do not need to speak now, Hipparchus and Ptolemy introduced the ninth and starless [sphere]. It is not necessary that things that are completely unknown to some do not exist. So far, I only pointed out here that Ptolemy and before him Hipparchus were in agreement with Moses in supposing the utmost of all spheres as starless. Even more so, they took the beginning of [their] findings from him [i.e. Moses].[55]

Philoponus connects the hypothesis of a ninth sphere with some newly undertaken observations by Hipparchus and Ptolemy, by which he alludes to the discovery of precession. However, he does not directly refer to a specific Ptolemaic work. Although Philoponus argues that there cannot be any proof for the

[53] Proclus, *Hypotypōsis*, p. 234:7–23. See Siorvanes, *Proclus*, pp. 290–93. Proclus knew the *Planetary Hypotheses*, as is evident from his commentaries, see Proclus, *In rep.*, II, p. 230:14–15 and Proclus, *In tim.*, III, p. 60:31–63:30, and see further Hartner, 'Medieval Views', pp. 260–61.

[54] The evidence for a similar thought in Asclepius, as pointed out by Maróth, 'The Ten Intellects Cosmology', pp. 110–11, is rather vague. Commenting on Aristotle *Metaph.* V.11, 1018b9–29, he adds to the example of prior in the sense of prior in space (*topos*) the example that the sphere of the fixed stars is prior since it is close to the starless (*anastrō*) sphere. See Asclepius, *In metaph.*, p. 323:26–27. In addition, Asclepius does not say from where he takes the teaching of a starless sphere.

[55] Philoponus, *De opificio mundi*, I.7, pp. 15:17–16:8, the translation is my own from the Greek. Compare to the German translation by Scholten, *Über die Entstehung der Welt*, p. 103.

astronomers' models,[56] he nevertheless seems to favour the nine-sphere-model, since it is in agreement with the Biblical report of the first heaven. Thus, Philoponus' motivation to accept the existence of a sphere above the fixed stars and to give it the status of an independent main sphere derives most importantly from theological, not astronomical reasons. He does not add an astronomical proof for this theory, but only singles out that one could find support for the Biblical text in astronomical sources. He even writes that Moses might have been a motive for Ptolemy's discovery, as he calls it. Simplicius, as well, favoured the theory of a sphere above that of the fixed stars, which he thinks to be Ptolemaic. He writes:

> So perhaps it would be truer to say that the starless sphere which contains all [the spheres], of which it seems there was no knowledge at the time of Aristotle, carries around all the other [spheres] with its single simple motion from the east; that the sphere which we call fixed has two motions, one which is from the east and is that of the universe, the other which is from the west and is its own.[57]

Like Philoponus, Simplicius also regards the discovery of precession to be the reason for the postulation of a sphere above the fixed stars. He recognizes this to be a departure from the Aristotelian model and tries to excuse Aristotle for not considering its existence, by saying that the starless sphere was a mathematical discovery after the time of Aristotle. However, Simplicus' remark refers only to a starless sphere above the fixed stars. This is a correct rendition of Ptolemy's account, since Ptolemy's first moving sphere was indeed no part of the Aristotelian cosmos. It is important to note that Simplicius does not call it 'the ninth sphere', as Philoponus did, because it might indicate that also Simplicius did not consider this outermost sphere to be a ninth 'main sphere'. Furthermore, one should bear in mind that also the Neoplatonic sources did not allude to the *Planetary Hypotheses* while ascribing the ninth sphere or, generally speaking, the starless sphere to Ptolemy. Therefore, the vague way in which the Arabic tradition commented upon the question on the number of spheres in Ptolemy is also reminiscent of the Neoplatonic commentators, with the difference that it was more common to speak of a ninth sphere. This opens up the possibility that the Arabic tradition was acquainted with post-Ptolemaic

[56] See the more detailed passage in Philoponus, *De opificio mundi*, III.3, pp. 113:15–116:17. Already Duhem pointed at this passage, see Duhem, *Le système du monde*, vol. II, pp. 496–97. See also Philoponus, *De aeternitate mundi*, XIII.18, p. 537:7–10. Since the *De opificio mundi* was not translated into Arabic, the latter work is more important in terms of Philoponus' impact on the Arabic tradition. See further Philoponus, *In meteor.*, p. 110:14–15, where he writes about the whole substance moving in a circle as including 'eight or perhaps nine spheres, as Ptolemy thinks'.

[57] Simplicius, *In de caelo*, p. 462:12–31, tr. Mueller, *On the Heavens 2.1–9*, p. 119. Another reference can be found in Simplicius, *In phys.*, p. 643:31–36, where he ascribes this theory to the astronomers in general.

sources ascribing this theory to Ptolemy, but were themselves unable to detect the exact Ptolemaic passage.

I want to conclude by presenting one more passage from the Arabic tradition which also questions Ptolemy's connection to the argument for the existence of a ninth sphere. In his astronomical *opus magnum*, *al-Qānūn al-masʿūdī* (*Masudic Canon*), Abū al-Rayḥān al-Bīrūnī engages with Ptolemaic astronomy and its reception in his time. He also makes use of the *Planetary Hypotheses*, criticizing Ptolemy's conception of the above mentioned sawn-off pieces (*manšūrāt*).[58] However, in his compendium on Indian beliefs and their relationship to ancient Greek sources, one can find an argument against the belief in a ninth sphere which he connects with the Hinduistic belief of the *braḥmānda*, the world-egg that encompasses the entire world. After a complicated refutation of the existence of the ninth sphere, for the sake of which al-Bīrūnī borrows arguments from Aristotelian physics and metaphysics, al-Bīrūnī gives the following report:

> Thus, the theory of the ninth sphere leads to an impossibility. To the same effect are the words of Ptolemy in the preface of his Almagest: 'The first cause of the first motion of the universe, if we consider the motion to be simply, is according to our opinion an invisible and motionless god, and we call the study of this subject a divine one. We perceive his action in the highest heights of the world, but as an altogether different one from the action of those substances, which can be perceived by the senses'.[59] These are the words of Ptolemy on the first mover, without any reference to the [ninth] sphere, on which John the Grammarian[60] reports in his refutation of Proclus. He says: 'Plato did not know the ninth sphere, which has no stars and the conception of which Ptolemy claimed'.[61]

In this passage, al-Bīrūnī gives us the following picture of how he got acquainted with this topic. He read Philoponus' *De aeternitate mundi contra Proclum*, where he found the remark that Ptolemy argued for the existence of the ninth sphere. Obviously surprised by this, he went through the *Almagest* but was not

[58] See al-Bīrūnī, *Kitāb al-Qānūn*, vol. II, pp. 633–35.

[59] cf. Ptolemy, *Syntaxis*, I.1, p. 5:13–19, tr. Toomer, *Ptolemy's Almagest*, pp. 35–36. Al-Bīrūnī cites the Arabic translation by Isḥāq ibn Ḥunayn and Ṯābit ibn Qurra literally.

[60] Of course, this refers to John Philoponus.

[61] Al-Bīrūnī, *Kitāb fī Taḥqīq*, pp. 183:16–184:15, tr. Sachau, *Alberuni's India*, vol. I, pp. 225–26, modified. Sachau translates the last sentence as follows: '"Plato did not know a ninth, starless sphere". And, according to John, it was this, i.e. the negation of the ninth sphere, which Ptolemy meant to say'. This is obviously wrong, both from the Arabic text as well as from the source from which al-Bīrūnī cites, namely Philoponus, *De aeternitate mundi*, XIII.18, p. 537:7–10.

It may be noted that in what precedes the cited passage, al-Bīrūnī argues against the corporeality of the mover of the sphere of the fixed stars. This is an approach similar to that described above by Muḥammad ibn Mūsā, as cited by al-Šīrāzī, see again Saliba, 'Early Arabic Critique', p. 131.

able to find any clue in support of Philoponus' claim. It is important to highlight that al-Bīrūnī knew the *Planetary Hypotheses* quite well and referred to it in his *Qānūn*.[62] To emphasize once more, the *Planetary Hypotheses* are often cited as being the ultimate source for the ascription of the theory of the ninth sphere to Ptolemy. The above cited passage taken together with the evidence that al-Bīrūnī was familiar with the *Planetary Hypotheses* clearly shows that he was unable to find any argument for the existence of this ninth sphere either in Ptolemy's *Almagest*, as he explicitly informs us, or in the *Planetary Hypotheses*. Thus, in the special case of al-Bīrūnī, Philoponus is the origin of the ascription of a ninth sphere to the Ptolemaic universe. Therefore, it may be either the case that Philoponus was the source for other Arabic authors as well, or that there is some other unknown line of transmission. It might be that Ptolemy's ninth sphere was already such a common *topos* in Late Antiquity that it easily entered the Arabic debates through many different sources. Be that as it may, the concept of a ninth sphere, to which a tenth or even an eleventh sphere was added in the Latin and Jewish tradition, is not of genuine Ptolemaic origin. Rather, I have clearly identified one concrete source for this ninth sphere, namely, John Philoponus, who compares this sphere to the first heaven from the Old Testament and who, therefore, had no astronomical reason to make the outermost moving sphere a ninth independent one. This is in line with the impact Philoponus had on the medieval Arabic tradition regarding other issues such as the eternity of the world as well.[63] Thus, al-Bīrūnī's above cited account adds another indication not only of how Ptolemy's works were read in the Middle Ages, but also of how Philoponus altered the way in which medieval astronomers and philosophers thought about the Ptolemaic cosmos.

Bibliography

Alexander of Aphrodisias, *Quaestiones*, ed. Ivo Bruns, in *Praeter commentaria scripta minora. Quaestiones, De fato, De mixtione*, Berlin: Georg Reimer, 1892, pp. 1–163; tr. Robert W. Sharples, *Quaestiones 1.1–2.15*, London: Duckworth, 1992.

Alexander of Aphrodisias, *On the Cosmos*, ed. and tr. Charles Genequand, Leiden/Boston/Köln: Brill, 2001.

Aristotle, *Über den Himmel*, tr. and comm. Alberto Jori, Berlin: Akademie Verlag, 2009.

Asclepius, *In Aristotelis metaphysicorum libros A – Z commentaria*, ed. Michael Hayduck, Berlin: Georg Reimer, 1888.

[62] See n. 58 above and Hartner, 'Medieval Views', pp. 260–61.

[63] See most famously the contributions by Herbert A. Davidson, in Davidson, 'John Philoponus as a Source', and Davidson, *Proofs for Eternity*. Most recently, Andreas Lammer wrote a comprehensive study on Philoponus' physics and its reception by Avicenna, see Lammer, *The Elements of Avicenna's Physics*.

Averroes, *On Aristotle's Metaphysics. An Annotated Translation of the so-called Epitome*, tr. Rüdiger Arnzen, Berlin/New York: De Gruyter, 2010.

Avicenna, *The Metaphysics of the Healing*, tr. Michael E. Marmura, Chicago: Chicago University Press, 2005.

Beere, Jonathan, 'Counting the Unmoved Movers: Astronomy and Explanation in Aristotle's *Metaphysics* XII.8', *Archiv für Geschichte der Philosophie* 85 (2003), pp. 1–20.

Al-Bīrūnī, *Kitāb fī Taḥqīq mā li-l-hind min maqūla maqbūla fī l-ʿaql aw marẕūla*, Hyderabad: Osmania Oriental Publications Bureau, 1958; tr. Eduard Sachau, *Alberuni's India*, 2 vols, London: Kegan Paul, Trench, Trübner & Co, 1910.

Al-Bīrūnī, *Kitāb al-Qānūn al-masʿūdī*, 3 vols, Hyderabad: Osmania Oriental Publications Bureau, 1954–1956.

Bodnár, István M., 'Alexander of Aphrodisias on Celestial Motions', *Phronesis* 42 (1997), pp. 190–205.

Davidson, Herbert A., 'John Philoponus as a Source of Medieval Islamic and Jewish Proofs of Creation', *Journal of the American Oriental Society* 89 (1969), pp. 357–91.

Davidson, Herbert A., *Proofs for Eternity, Creation and the Existence of God in Medieval Islamic and Jewish Philosophy*, New York: Oxford University Press, 1987.

Duhem, Pierre, *Le système du monde. Histoire des doctrines cosmologiques de Platon à Copernic*, 10 vols, Paris: Hermann et fils, 1913–1959.

Endress, Gerhard, 'Averroes' *De caelo*. Ibn Rushd's Cosmology in his Commentaries on Aristotle's *On the Heavens*', *Arabic Sciences and Philosophy* 5 (1995), pp. 9–49.

Eudoxus of Cnidus, *Die Fragmente*, ed. Francois Lasserre, Berlin: De Gruyter, 1966.

Feke, Jacqueline, *Ptolemy's Philosophy. Mathematics as a Way of Life*, Princeton: Princeton University Press, 2018.

Goldstein, Bernard R., *The Arabic Version of Ptolemy's Planetary Hypotheses*, Philadelphia: American Philosophical Society, 1967.

Hamm, Elizabeth A., *Ptolemy's Planetary Theory: An English Translation of Book One, Part A of the Planetary Hypotheses with Introduction and Commentary*, PhD dissertation, University of Toronto, 2011.

Hartner, Willy, 'Medieval Views on Cosmic Dimensions and Ptolemy's *Kitāb al-Manshūrāt*', in I. Bernard Cohen and René Taton (eds), *Mélanges Alexandre Koyré*, Paris: Hermann et fils, 1964, vol. I, pp. 319–48.

Heglmeier, Friedrich, 'Die griechische Astronomie zur Zeit des Aristoteles. Ein neuer Ansatz zu den Sphärenmodellen des Eudoxos und Kallippos', *Antike Naturwissenschaft und ihre Rezeption* 6 (1996), pp. 51–71.

Iḫwān al-Ṣafāʾ, *On the Natural Sciences. An Arabic Critical Edition and English Translation of Epistles 15–21*, ed. and tr. Carmela Baffioni, Oxford: Oxford University Press, 2013.

Janos, Damien, 'Moving the Orbs: Astronomy, Physics, and Metaphysics, and the Problem of Celestial Motion According to Ibn Sīnā', *Arabic Sciences and Philosophy* 21 (2011), pp. 165–214.

Janos, Damien, *Method, Structure, and Development in al-Fārābī's Cosmology*, Leiden/Boston: Brill, 2012.

Lammer, Andreas, *The Elements of Avicenna's Physics. Greek Sources and Arabic Innovations*, Berlin: De Gruyter, 2018.

Langermann, Y. Tzvi, 'A Note on the Use of the Term *orbis* (*falak*) in Ibn al-Haytham's *Maqūlah fī hay'at al-'ālam*', *Archives Internationales d'Histoire des Sciences* 32 (1982), pp. 112–13.

Lloyd, Geoffrey E. R., 'Saving the Appearances', *The Classical Quarterly* 28 (1978), pp. 202–22.

Lloyd, Geoffrey E. R., 'Metaphysics Λ 8', in Michael Frede and David Charles (eds), *Aristotle's Metaphysics Lambda. Symposium Aristotelicum*, Oxford: Clarendon Press, 2000, pp. 245–73.

Mancha, José L., 'Al-Biṭrūjī's Theory of the Motions of the Fixed Stars', *Archive for History of Exact Sciences* 58 (2004), pp. 143–82.

Maróth, Miklós, 'The Ten Intellects Cosmology and Its Origin', in Alexander Fodor (ed.), *Proceedings of the 14th Congress of the Union Européenne des Arabisants et Islamisants*, Budapest: Eötvös Loránd University, Chair for Arabic Studies, 1995, pp. 103–11.

Mendell, Henry, 'The Trouble with Eudoxus', in Patrick Suppes, Julius M. Moravscik, and Henry Mendell (eds), *Ancient & Medieval Traditions in the Exact Sciences. Essays in Memory of Wilbur Knorr*, Stanford: CSLI Publications, 2000, pp. 59–138.

Morelon, Régis, *Thābit ibn Qurra. Œuvres d'astronomie*, Paris: Les Belles Lettres, 1987.

Morelon, Régis, 'La version arabe du *Livre des hypothèses* de Ptolémée', *MIDEO* 21 (1993), pp. 7–85.

Morrison, Robert, '*Falsafa* and Astronomy after Avicenna: An Evolving Relationship' in Y. Tzvi Langermann (ed.), *Avicenna and his Legacy. A Golden Age of Science and Philosophy*, Turnhout: Brepols, 2009, pp. 307–26.

Murschel, Andrea, 'Structure and Function of Ptolemy's Physical Hypotheses of Planetary Motion', *Journal for the History of Astronomy* 26 (1995), pp. 33–61.

Neugebauer, Otto, *A History of Ancient Mathematical Astronomy*, 3 vols, Berlin/Heidelberg: Springer, 1975.

Philoponus, John, *De opificio mundi*, ed. Walther Reichardt, Leipzig: Teubner, 1897; German tr. Clemens Scholten, *Über die Entstehung der Welt*, 3 vols, Freiburg: Herder, 1997.

Philoponus, John, *De aeternitate mundi contra Proclum*, ed. Hugo Rabe, Leipzig: Teubner, 1899.

Philoponus, John, *In Aristotelis meteorologicorum librum primum commentarium*, ed. Michael Hayduck, Berlin: Georg Reimer, 1901.

Proclus, *Hypotypōsis tōn astronomikōn hypotheseōn*, ed. Karl Manitius, Stuttgart: Teubner, 1974.

Proclus, *In Platonis timaeum commentaria*, ed. Ernst Diehl, 3 vols, Leipzig: Teubner, 1903–1906.

Ptolemy, *Syntaxis Mathematika*, ed. Johan L. Heiberg, *Syntaxis mathematica*, 2 vols, Leipzig: Teubner, 1898-1903 [*Claudii Ptolemaei opera quae exstant omnia*, vol. I]; tr. Gerald J. Toomer, *Ptolemy's Almagest*, London: Duckworth, 1984, 2nd ed. Princeton: Princeton University Press, 1998.

Ptolemy, *Hypotheseōn tōn planōmenōn*, ed. Johan L. Heiberg, *Opera astronomica minora*, Leipzig: Teubner, 1907 [*Claudii Ptolemaei opera quae exstant omnia*, vol. II], pp. 69–107; German tr. Ludwig Nix, ibid., pp. 71–145.

Ragep, F. Jamil, *Naṣīr al-Dīn al-Ṭūsī's Memoir on Astronomy*, 2 vols, New York: Springer, 1993.

Sabra, Abdelhamid I., 'The Andalusian Revolt against Ptolemaic Astronomy', in Everett Mendelsohn (ed.), *Transformation and Tradition in the Sciences. Essays in Honor of I. Bernard Cohen*, Cambridge: Cambridge University Press, 1984, pp. 133–53.

Saliba, George, 'Early Arabic Critique of Ptolemaic Cosmology: A Ninth-Century Text on the Motion of the Celestial Spheres', *Journal for the History of Astronomy* 25 (1994), pp. 115–41.

Sela, Shlomo, 'Maimonides and Māshā'allāh on the Ninth Orb of the Signs and Astrology', *Aleph* 12 (2012), pp. 101–34.

Simplicius, *In Aristotelis de caelo commentaria*, ed. Johan L. Heiberg, Berlin: Georg Reimer 1894; tr. Ian Mueller, *On Aristotle. On the Heavens 2.1–9*, London: Duckworth, 2004.

Simplicius, *In Aristotelis physicorum libros quattuor priores commentaria*, ed. Hermann Diels, Berlin: Georg Reimer, 1882.

Siorvanes, Lucas, *Proclus. Neo-Platonic Philosophy and Science*, Edinburgh: Edinburgh University Press, 1996.

Sorabji, Richard, 'Adrastus: Modifications to Aristotle's Physics of the Heavens by Peripatetics and Others, 100 BC to 200 AD', *Bulletin of the Institute of Classical Studies* 94 (2007), pp. 575–94.

Tanenbaum, Adena, 'Nine Spheres or Ten? A Medieval Gloss on Moses Ibn Ezra's "Be-Shem El Asher Amar"', *Journal of Jewish Studies* 46 (1996), pp. 294–310.

Taub, Liba Chaia, *Ptolemy's Universe. The Natural Philosophical and Ethical Foundations of Ptolemy's Astronomy*, Chicago: Open Court, 1993.

Yavetz, Ido, 'On the Homocentric Spheres of Eudoxus', *Archive for the History of Exact Sciences* 52 (1998), pp. 221–78.

'Fort. recte': Witnesses to the Text of Ptolemy's *Tetrabiblos* in Its Near Eastern Transmission

Bojidar DIMITROV

Claudius Ptolemy's *Tetrabiblos* is among the fortunate cases of Graeco-Arabic textual transmission where extant versions of the text in Greek, Syriac, Arabic, Latin and Hebrew present scholars with the opportunity to examine in detail the history of an extremely influential astrological treatise's dissemination in the Near East and Europe.

The manuscript Paris, BnF, syr. 346 contains a fragmentarily preserved, possibly pre-Islamic,[1] Syriac translation of the *Tetrabiblos*. In the context of the project Ptolemaeus Arabus et Latinus, my research aims at editing the Syriac translation and establishing its significance for the *Tetrabiblos*' chain of transmission. This objective is to be achieved by comparing the content of the Syriac text with that of the Arabic translations and the Greek source text.

There are two major Arabic versions of the *Tetrabiblos*.[2] According to the introductory paragraph in the Uppsala manuscript, the older version was prepared by the Persian astrologer ʿUmar b. al-Farruḫān in 812 AD.[3] ʿUmar was involved in the translation of astrological material available in Pahlavi into Arabic during the early ʿAbbāsid period.[4] However, according to the tenth-century bibliographer Ibn an-Nadīm, the *Tetrabiblos* was not translated by ʿUmar, but on his behalf, by al-Biṭrīq.[5] Al-Biṭrīq was a contemporary primarily known for translations of medical texts from Greek into Arabic.[6] David Pingree disagrees with Ibn an-Nadīm on the strength of the indication provided by the Uppsala manuscript, and argues that the older Arabic *Tetrabiblos* is a paraphrase which was probably derived from a Pahlavi source.[7] As we shall see, the content of ʿUmar's version appears to support Pingree's hypothesis to a certain

[1] Villey, *Les textes astronomiques*, p. 350.

[2] The late Prof Keiji Yamamoto generously shared his draft editions of the two texts with me.

[3] cf. Uppsala, Universitätsbiblioteket, MS 203, fol. 2r.

[4] Dodge, *The Fihrist*, vol. II, p. 589; Gutas, *Greek Thought*, pp. 108–10.

[5] Dodge, *The Fihrist*, vol. II, pp. 649–50.

[6] Müller, *ʿUyūn ul-anbāʾ*, vol. I, p. 205; Dunlop, 'The Translations', pp. 140–50; Pormann, 'The Development', pp. 143–62.

[7] Pingree, ''Umar Ibn al-Farrukhān'.

Ptolemy's Science of the Stars in the Middle Ages, ed. by David Juste, Benno van Dalen, Dag Nikolaus Hasse and Charles Burnett, PALS 1 (Turnhout, 2020), pp. 97–113

10.1484/M.PALS-EB.5.120175

extent, because it often differs significantly from the Greek text and the other translations.

The second Arabic translation of the *Tetrabiblos* is said to have been made by Ibrāhīm b. aṣ-Ṣalt and subsequently improved by Ḥunayn b. Isḥāq[8] (808–873 AD). It is important to note that Ibrāhīm translated from Greek into both Syriac and Arabic, belonged to Ḥunayn's circle and worked under his direct supervision.[9] This fact implies that Ibrāhīm and Ḥunayn are likely to have used the same source text(s), but leaves open the question as to what the language of the source text(s) may have been, and respectively, whether the *Tetrabiblos* was translated in two stages (Greek-Syriac, Syriac-Arabic), as was often the case,[10] or in one (Greek-Arabic, possibly Syriac-Arabic).

The evidence contained in the Syriac and the Arabic translations of the *Tetrabiblos* can also provide vital clues for our understanding of the Greek source text.[11] Moreover, the recent publication of William of Moerbeke's *Tetrabiblos* translation from Greek into Latin by Gudrun Vuillemin-Diem and Carlos Steel,[12] and their comparison of Moerbeke's *Tetrabiblos* with Wolfgang Hübner's authoritative edition of the Greek source text, have posed many questions.[13] William of Moerbeke had access to a Greek manuscript that was older than the ones which Hübner used for his edition.[14] Just like William of Moerbeke's translation, the Syriac and the Arabic texts are much older than Hübner's witnesses, and might therefore contain valuable variant readings that support some of Vuillemin-Diem and Steel's conclusions.

This essay aims to compare a number of significant readings, which have been identified by Vuillemin-Diem and Steel, with corresponding material from the above-mentioned Near Eastern translations of the *Tetrabiblos*. Furthermore, this essay demonstrates the relevance of the Semitic translations for the overall transmission history of the treatise, and for the re-assessment of Hübner's Greek text in particular. For the reader's convenience, passages or phrases con-

[8] Dodge, *The Fihrist*, vol. II, p. 640; Müller and Lippert, *Taʾrīkh al-ḥukamāʾ*, p. 98.

[9] Bergsträsser, *Ḥunain ibn Isḥāq*, pp. 25 and 28–29.

[10] ibid., p. 27 – Ḥunayn relates, for instance, how he revised with the help of a Greek manuscript and afterwards translated into Arabic a Galenic treatise, which had initially been translated from Greek into Syriac by one of his associates.

[11] Hübner, *Ἀποτελεσματικά*.

[12] Vuillemin-Diem and Steel, *Ptolemy's Tetrabiblos*.

[13] Vuillemin-Diem and Steel's thorough analysis groups the variant readings offered by Hübner's and Moerbeke's sources in several categories. The majority of the readings discussed in this essay belong to a category the scholars have designated with the symbol (-*) ' = the reading confirmed by G (i.e. Moerbeke's main witness – BD) seems better to us: we propose to modify Hübner's edition'. – cf. Vuillemin-Diem and Steel, *Ptolemy's Tetrabiblos*, p. 95.

[14] ibid., pp. 49–58.

taining the variants in question will be presented in synoptic tables (where G designates Hübner's edition, with book and line number; L, William of Moerbeke's Latin translation with the book and line number of Vuillemin-Diem and Steel's edition; S, the Syriac version in my edition-in-preparation, with the folio and line number of the Paris manuscript; F, the earlier Arabic version attributed to ʿUmar b. al-Farruḫān; and H, the version attributed to Ḥunayn b. Isḥāq, both with the page and line number. The comparison and the evaluation of the readings will follow below the tables. The tables also contain transcriptions (for the Semitic languages) and literal renditions provided for analytical purposes, with Robbins' standard English translation[15] of the relevant passages added in the footnotes where this is deemed necessary.

Reading 1

G	L	S	F	H
III.9	III.10	7v, ll. 6–7	p. 266, l. 17	p. 266, l. 13
περὶ ἀρρενικῶν καὶ θηλυκῶν γενέσεων	De masculinis et femininis	ܥܠ ܕܟܪܐ ܘܢܩܒܬܐ *ʿal dekrē w-neqbāṯā*	في الذكور *fi-d-ḏukūri*	في الذكور والإناث *fi-d-ḏukūri wa-l-ʾināṯi*
('of the nativities of males and females')	('of males and females')	('of males and females')	('of males')	('of males and females')

The seventh chapter in the Greek text's contents of Book III reads περὶ ἀρρενικῶν καὶ θηλυκῶν γενέσεων ('of the nativities of males and females'). According to Hübner's *apparatus criticus*, '*γενέσεων* om. **YM** fort. recte'.[16] In other words, Hübner supposes that the omission of γενέσεων ('nativities') in the contents may be correct, because some witnesses provide the same variant for the title of the actual chapter, and he accordingly adopted it.[17] Vuillemin-Diem and Steel's text also supports this omission.[18] Moreover, the two authors point out that 'the text of chapter eight speaks of ἀρρενογονία and θηλυγονία (*procreatio*)', that is to say 'male and female procreation', and 'not of γένεσις' (i.e. 'nativity').[19]

As we can see on the above table, the Syriac translation confirms the evidence obtained from William of Moerbeke's Latin rendering, and so does Ḥunayn's ninth century Arabic text. Oddly enough, the earlier Arabic translation associated with ʿUmar b. al-Farruḫān omits *ʾināṯ* ('females') — this

15 Robbins, *Ptolemy. Tetrabiblos*.

16 Hübner, *Ἀποτελεσματικά*, III.9.

17 ibid., III.392: 'περὶ ... θηλυκῶν **αC** Heph.'

18 Vuillemin-Diem and Steel, *Ptolemy's Tetrabiblos*, p. 115.

19 ibid.

omission requires further examination because it does not occur in any of the witnesses, on which the transmission of the other versions of the *Tetrabiblos* is based. It might well be due to a scribal error, illegible handwriting or, perhaps, a damaged manuscript, and not an actual variant reading.

Reading 2

G	L	S	F	H
III.23	III.24	7v, ll. 18–19	–	p. 270, ll. 5–7
ὧν τὸ προγ- νωστικὸν μέρος γενεθλιαλογίαν καλοῦμεν	quorum pronosticam partem uocamus geneaticum sermonem	ܢܐܡܪ ܕܡܢܬܐ ܓܝܪ ܕܡܩܕܡܘܬ ܝܕܥܬܐ ܕܕܐܟ ܗܕܐ ܡܡܠܠܘܬ ܡܘܠܕܐ *nemar da-mnāṯā* *ger da-mqaḏmūt* *ida'ṯā ḏ-d-aḵ* *hāḏē* *mmalālūṯ* *mawlāḏā*		وهذا الجزء من هذه الصناعة تسمّى تقدمة المعرفة بالأمور المواليد *wa hāḏā l-ǧuzʾu min* *hāḏihi ṣ- ṣināʿati* *tusammā taqdimatu* *l-maʿrifati bi-* *ʾumūri l-mawālīdi*
('the prognostic part of which we call genethlialogy')	('the prognostic part of which we call genethlialogy')	('we say that a prognostication part such as this is the study of nativity')		('this part from this art is called prognostication of the matters of nativities' [i.e. studies dealing with nativities' genethlialogy])[20]

With the next entry we have an instance where William of Moerbeke's editors agree with Hübner's choice. The Greek relative clause ὧν τὸ προγνωστικὸν μέρος γενεθλιαλογίαν καλοῦμεν ('the prognostic part of which we call genethlialogy') includes the neuter noun τὸ μέρος ('part') that appears to have led some scribes to adopt a congruent neuter adjective instead of γενεθλιαλογίαν ('genethlialogy') – γενεθλιαλογιὸν (i.e. 'the prognostic part of which we call *genethlialogical*').[21] Hübner opted for γενεθλιαλογίαν, yet he did, however,

[20] cf. صناعة المواليد (*ṣināʿatu l-mawālīdi*, 'the art of nativities') and γενεθλιαλόγοι in Ullmann, *Wörterbuch*, p. 970.

[21] Hübner, *Ἀποτελεσματικά*, III.23.

honour the alternative variant with another 'fort. recte',[22] which Vuillemin-Diem and Steel have taken into consideration as well.[23]

The evidence presented by the transmission of the *Tetrabiblos* in the Near East speaks against the adjective γενεθλιαλογιόν and Hübner's 'fort. recte'. The Syriac translation follows the word order of the Greek sentence quite closely. It renders γενεθλιαλογία with the genitive construction *mmalālūṯ mawlāḏā* ('study of nativity'), where the term *mmalālūṯā* is the exact semantic pendant of λογία, designating a scientific discipline (cf. *mmalālūṯ kawkbē* – 'study of the stars', i.e. astrology).[24] The translator resorted to the same approach in order to tackle the other *terminus technicus* in the sentence — the adjective προγνωστικόν ('prognostic'). Instead of an adjective, the Syriac employs *mqaḏmūt idaʿṯā* — a genitive construction with the meaning 'preceding knowledge', functioning as an adjective, which the genitive particle *d-* links to *mnāṯā* ('part').

This passage is translated quite freely in the earlier Arabic version, and provides no conclusive evidence concerning the variant reading in question. Ḥunayn's translation, on the contrary, is very close to the Syriac and also has a genitive construction for γενεθλιαλογία, as opposed to an adjective. What it adds to the passage is the phrase *min hāḏihi ṣ- ṣināʿati* ('from this art'), apparently for the purpose of greater clarity — this detail can be interpreted as a case of *explicitation*, i.e. the 'spelling out' of implicit source text content in the target text, a feature that has been ascribed to the translation style of Ḥunayn and his school.[25] The other genitive construction, *taqdimatu l-maʿrifati*, is the usual Arabic astrological term for 'prognostication', and as far as its semantic value is concerned, the construction is etymologically identical with the Syriac *mqaḏmūt idaʿṯā* ('preceding knowledge'). Unlike the Syriac, however, Ḥunayn's pendant for προγνωστικόν is not intended to define *ǧuzʾ* ('part') as an adjective. Instead, it is linked with the other genitive construction (*ʾumūru l-mawālīdi*, i.e. γενεθλιαλογία) by way of the preposition *bi*. Thus we end up with the phrase *tusammā taqdimatu l-maʿrifati bi-ʾumūri l-mawālīdi* ('... it is called prognostication of the matters of nativities'), which represents a slight deviation from the Greek and the Syriac.

[22] ibid.

[23] Vuillemin-Diem and Steel, *Ptolemy's Tetrabiblos*, III.24.

[24] Sokoloff, *A Syriac Lexicon*, p. 777.

[25] cf. Pormann, 'The Development', p. 154.

Reading 3

G	L	S	F	H
III.53–54	III.46	8r, l. 15	p. 274, ll. 2–3	p. 274, ll. 3–6
Ἀρχῆς δὲ χρονικῆς ὑπαρχούσης τῶν ἀνθρωπίνων τέξεων φύσει μὲν τῆς κατ' αὐτὴν τὴν σποράν	Temporali autem principio generationum hominum existente per naturam quidem eo quod secundum seminationem	ܟܕ ܕܝܢ ܪܝܫܝܬܐ ܟܠܢܝܬܐ ܕܙܒܢܐ ܥܠ ܙܪܥܐ ܟܝܬ ܘܙܒܐ* ܕܡܘܠܕܐ ܐܝܬܘܗ̇ ܕܡܘܠܕܐ ܕܒܢܝܢܫܐ ܒܟܝܢܐ ܡܢ ܡܢ ܙܒܢܐ ܕܛܪܡܝܬܗ ܕܙܪܥܐ *kaḏ den rīšīṯā* *kullānāyṯā ḏ-zaḇ-* *nā ʿal zrāʿā kēṯ* *w-zabā** [*sic* = zabnā] *ḏ-maw-* *lāḏā iṯ-eh* *ḏ-mawlāḏā* *ḏa-bnay-nēšē* *ḇ-ḵyānā man men* *zabnā ḏ-ṯarmīṯā* *ḏ-zarā*	طبائع المولود وقوّة كيانه يعرف من ساعة مسقط النطفة *ṭabāʾiʿu* *l-mawlūdi wa-* *quwwatu kiyāni-* *hi yuʿrafu min* *sāʿati masqaṭi* *n-nuṭfati*	لمّا كان لكون للناس ابتداء زماني وكان هذا الابتداء إمّا بالطبع فالذي يكون عند سقوط النطفة *lammā kāna li-* *kawni li-n-nāsi* *ʾibtidāʾun* *zamāniyyun* *wa-kāna hāḏā* *l-ʾibtidāʾu ʾimmā* *bi-ṭ-ṭabiʿi* *fa-l-laḏī yakūn* *ʿinda suqūṭi* *n-nuṭfatin*
('since the chronological beginning of human procreations is by nature according to conception itself')[26]	('the temporal beginning of the procreations of men being by nature that which is according to the conception)	('since the total chronological beginning is upon conception [lit. *sowing*, cf. σπορά], that is to say the time of nativity, the nativity of men is, indeed, by nature from the time of the seed's sowing')	('the natures of the newborn and the strength of his [pl.!] being are known from the hour of conception [lit. *fall of sperm*]')	('since being born into mankind has a chronological beginning, this beginning is, by nature, that which is according to the conception [lit. *fall of sperm*]')

[26] I would like to express my gratitude to Prof Charles Burnett and Prof Martin Heide for their valuable suggestions concerning this and other passages; cf. Robbins, *Ptolemy. Tetrabiblos*, p. 223: 'Since the chronological starting-point of human nativities is naturally the very time of conception'.

Vuillemin-Diem and Steel propose to replace Hübner's semantically more concrete τέξεων (lit. 'child-bearings / procreations') with the proper astrological *terminus technicus* γενέσεων ('nativities'), which is somewhat closer to the Latin text's *generationum*.[27] The Syriac version employs the standard term *mawlāḏā* ('nativity') that we already encountered above.

A problem in the Syriac rendition arises from the substitution of the Greek adjective χρονικῆς ('chronological') for the noun *zaḇnā* ('time'). The resulting genitive construction *rīšīṯā* ... *ḏ-zaḇnā* can thus mean both 'the beginning of time' and 'chronological beginning', the former being the more natural interpretation in Syriac. It is very similar to the adjectival genitive construction S has in the previous entry: cf. *mnāṯā* ... *da-mqaḏmūt ida'ṯā* (lit. 'part of the prognostication', intended to mean 'prognostication part', i.e. 'prognostic part' = μέρος γενεθλιαλογιὸν). The addition of the adjective *kullānāyṯā* ('universal, total, absolute') appears to be an attempt to render the Greek ὑπαρχούσης. In this particular case Ptolemy uses the verb ὑπάρχω in its secondary meaning, as a more sophisticated way to designate existence, i.e. 'the chronological beginning *is*', cf. 'existente' in the Latin text. Since the primary meaning of ὑπάρχω is 'to begin, to make a beginning', the Syriac translator may have understood ὑπαρχούσης to designate a total, absolute beginning.

If there are variant readings in this passage that the Latin, the Syriac and the Ḥunayn translation unanimously agree against, we ought to point at the Greek's use of an adjective deriving from ἄνθρωπος ('human') and not ἄνθρωπος itself.[28] All the other texts (except the early Arabic version) have genitive constructions with a noun: cf. generationum hominum ('of humans' nativities'), *ḏ-mawlāḏā ḏa-bnay-nēšē* ('of humans' nativity'), *li-kawni li-n-nāsi* ('to the nativity into mankind').

All translations except the early Arabic one exhibit an awareness of, and an effort to preserve the syntactical structure of the inverted Greek μέν ... δέ clause ('on the one hand...on the other hand' or 'indeed...but'): cf. autem ... quidem (L), *kaḏ den* ... *men* (S), *lammā* ... *ʾimmā* (H).

[27] Vuillemin-Diem and Steel, *Ptolemy's Tetrabiblos*, p. 115.

[28] cf. Hübner, *Ἀποτελεσματικά*, III.53: '**53** *ἀνθρωπίνων* **VΣDγ** *ἀνθρωπείων* **YMS**'.

Reading 4

G	L	S	F	H
III.308–310	III.242	9v, ll. 13–14	p. 298, ll. [9]1–3	p. 298, ll. [9]1–3
θανάτους τοὺς μητρικοὺς αἰφνιδίους καὶ τὰ σίνη περὶ τὰς ὄψεις ποιεῖ	mortes matrum efficit repentine et lesuram circa oculos	ܡܘܡܐ ܥܒܕ ܒܥܝܢܐ ܕܐܡܗܬܐ: ܘܒܥܓܠ ܥܐܛ ܠܗܝܢ *mūmā ʿāḇeḏ b-ʿaynē d'emhāṯā wə-ḇ-ʿaġal ʿāṭe l-hen*	دلّ على داء في عيون الأمّهات ويهلك بصرها سريعاً *dalla ʿalā dāʾin fī ʿuyūni l-ʾummahāti wa-yuhlik baṣara-hā sarīʿan*	حدث عن ذلك موت الأمّهات فجأة أو آفة تعرض لهنّ في الأعين *ḥadaṯa ʿan ḏalika mawtu l-ʾummahāti faǧʾatan ʾaw ʾāfatun tuʿriḍa lahunna fī l-ʾaʿyuni*
('makes sudden *motherly* deaths and injuries around the eyes')	('suddenly effectu-ates mothers' deaths and injury around the eyes')	('makes infirmity to the eyes of the mothers and swiftly destroys them')	('leads to illness in the eyes of the mothers and swiftly destroys their sight')	('from this suddenly occurs the death of the mothers, or illness befalls them in the eyes')

An adverbial reading of the adjective αἰφνιδίους ('sudden') would present a *lectio difficilior* in this passage,[29] i.e. 'suddenly (αἰφνιδίως) makes / causes the deaths of mothers', as opposed to 'sudden deaths'. This adverbial reading is confirmed by all Semitic versions: S has *ḇ-ʿaġal* ('swiftly'), F – *sarīʿan* ('promptly'), H – *faǧʾatan* ('suddenly'), the latter being particularly close to αἰφνιδίως.

The adjective μητρικοὺς (lit. 'motherly / maternal' deaths) in G technically presents another *lectio difficilior*, which no other text supports — L and H offer the closest renditions and both employ genitive constructions, cf. 'mortes matrum' ('mothers' deaths') and *mawtu l-ʾummahāti* ('mothers' death').

The relationship of the 'mothers' and the 'eyes' is of critical importance for the passage. S and F do not mention death in connection with the mothers. And while the Syriac phrase *ʿāṭe l-hen* ('he destroys them', the accusative object being feminine) is somewhat ambiguous and can refer to both the mothers and the eyes, ʿUmar's translation clearly mentions the mothers' sight (*baṣara-hā*). Ḥunayn, on the other hand, retains the two elements separately, just as the Greek and the Latin texts do, but uses the conjunction *ʾaw* ('or') between them

29 cf. Vuillemin-Diem and Steel, *Ptolemy's Tetrabiblos*, p. 117: '...αἰφνιδίους ... α βγ] ... αἰφνιδίως ... V + G (mortes ... efficit repentine ...) Procl.'

instead of 'and' (cf. καὶ, et). This creates the impression that Mars can bring about either sudden death or eye illness, but not *both* — a possibility we do not find in G and L. S, F, and H specify that it is the mothers' eyes that are concerned, something left unspecified in GL.

Reading 5

G	L	S	F	H
III.419	III.323–4	10v, l. 22	p. 308, l. [1]3	p. 308, l. [8.1]4
παρακολουθεῖν δὲ εἴωθε τοῦτο τὸ σύμπτωμα παρὰ τὰς συγκράσεις ('it is usual for this event to follow along with the minglings[30] together')	consequi enim consueuit tale simptoma circa commix-tiones ('such an event is accustomed to follow around minglings')	ܢܩܦ ܕܝܢ ܓܕܫܐ ܕܐܝܟ ܗܢܐ ܠܘܬ ܡܘܙܓܐ *nāqeph den gedšā d-ak hānā lwāt mūzāğā* ('an event like this follows along with the mixture')	لأنّ أكثر ذلك يكون في المزاج *li-'anna 'aktara dālika yakūna fī l-mizāği* ('because that occurs frequently in the mixture')	هذا الأمر إن يعرض عند التركيبات *hādā l-'amr'inna ya'ridu 'inda t-tarkībāti* ('indeed, this matter occurs along with the compoundings')

The Latin and the Syriac confirm that in the earlier stages of the transmission the Greek σύμπτομα ('event') was probably defined by the adjective τοιοῦτον ('such as this', cf. tale, *d-ak hānā*), as opposed to the demonstrative pronoun τοῦτο ('this').[31] It is important to note that the Latin also uses 'simptoma' as a loan word and an astrological *terminus technicus*. William of Moerbeke's rendering also retains the plural form of συγκράσεις – commixtiones ('minglings together'). The Near Eastern transmission seems to have introduced the singular form early on, because it is attested by the Syriac as well as the earlier Arabic version: cf. S – *mūzāğā* ('mixture') and F – *mizāği* ('mixture') as opposed to H – *tarkībāti* ('compoundings'). While Ḥunayn's peculiar use of a noun based on the verbal root **rkb V** deviates from the semantic pendants presented by S and F, his text has also preserved the plural form *tarkībāti*, which corresponds to the Latin and the Greek and was thus chosen by Yamamoto as the better reading.

Of course, the singular form in Syriac could have resulted from a common phenomenon: the plural form *mūzāğē* is virtually homographous with the singular *mūzāğā*, the only difference being the addition of the diacritical *seyāmē*

[30] Robbins, *Ptolemy. Tetrabiblos* prefers to translate the Greek plural form τὰς συγκράσεις as 'intermixture', cf. p. 257; he renders the whole phrase as follows: 'For such an event is apt to attend the intermixture'.

[31] cf. Vuillemin-Diem and Steel, *Ptolemy's Tetrabiblos*, p. 117: 'τοῦτο τὸ σύμπτωμα V] τὸ τοιοῦτον σύμπτωμα α βγ + G (tale simptoma), cf. Heph.'

points, designating the plural form (cf. ܡܕܒܪܢܐ and ܡܕܒܪ̈ܢܐ). The occasional omission of the *seyāmē* points is one of the most common features occurring in Syriac manuscripts. It would be interesting to speculate whether the reading variants with the singular form in the Arabic versions of the text have resulted from this Syriac translation.

The Syriac text is the only Semitic source to retain the verb 'to follow' (cf. *nāqeph*). Moreover, the Syriac also provides a very precise pendant for σύμπτομα / simptoma, i.e *gedšā* ('event') – something F (*ḏālika*, i.e. 'that') and H (*hāḏā l-ʾamr*, i.e. 'this matter') do not.

Reading 6

G	L	S	F	H
III.464–465	III.464–465	11v, ll. 3–4	p. 312, ll. [2]13–14	p. 312, ll. [9.2]8–9
μηδενὸς μὲν μαρτυροῦντος τοῖς φωσὶν ἀγαθοποιοῦ, ἀλλὰ τῶν κακοποιῶν	sed nullo quidem benefico luminaribus testimonium reddente	ܐܠܐ ܟܕ ܛܒܐ ܟܕ ܢܣܗܕ ܠܢܗܝܪ̈ܐ *ʾelā kaḏ ṭāḇā kaḏ nsaheḏ l-nahīrē*	فإن لم يكن شيء من النحوس يلي النيّرين *fa-ʾin lam yakun šayʾun min n-nuḥūsi yalī n-nayyirayni*	فإن لم يشهد شيء من المسعدة النيّرين وشهدت لهما المنحسة *fa-ʾin lam yašhad šayʾun min l-musʿadatin li n.-nayyirayni wa-šahidat lahumā l-munḥisatu*
('if no beneficent one [i.e. planet] bears witness to the luminaries, but the maleficent ones do')	('but if no beneficent one bears witness to the luminaries')	('for unless a good one is to bear witness to the luminaries')	('but if none of the misfortunate ones follows the two luminaries')	('but if none of the fortunate ones bears witness to the two luminaries, and the misfortunate one bears witness to them')

According to Vuillemin-Diem and Steel, 'The reading of the archetype (ω) in 464°-5° could have been simply: ἀλλὰ μηδενὸς μαρτυροῦντος τοῖς φωσὶν ἀγαθοποιοῦ',[32] i.e. 'but if no beneficent one bears witness to the luminaries'. The Syriac translation confirms this conclusion with the conditional clause *ʾelā* ('unless, if not') which, in combination with the positive adjective *ṭāḇā* ('good'), more or less corresponds to the Latin 'sed nullo ... benefico'.

The earlier Arabic text introduces a negative conditional clause, *fa-ʾin lam yakun* (lit. 'if it is not'), whose meaning is similar to the one in the Syriac. The verbal root **wly**, from which the finite form *yalī* derives, can mean anything

[32] Vuillemin-Diem and Steel, *Ptolemy's Tetrabiblos*, p. 118.

from 'to be close' or 'to follow' to 'to govern / manage' or 'to protect'.[33] At face value, none of **wly**'s meanings is similar to the Greek μαρτυροῦντος (lit. 'bearing witness'), the key verb in the sentence, but this may have been F's intended usage, particularly if we consider the sense 'to follow'. Ḥunayn's text, on the other hand, not only employs **šhd** (cf. *lam yašhad*, 'it does not bear witness') — the verbal pendant of the Syriac **shd** (cf. *nsahed*, 'it bears witness') — but also preserves, as does the Syriac, the positive definition characterising the planet (cf. *šay'un min mus'adatin*, lit. 'something fortunate'). Ḥunayn's version also contains the clarification *wa-šahidat lahumā l-munḥisatu* ('the misfortunate one bears witness to them'), which corresponds to the Greek phrase ἀλλὰ τῶν κακοποιῶν ('but the maleficent ones do') that Vuillemin-Diem and Steel consider to be a redundant later addition.[34]

Reading 7

G	L	S	F	H
III.476–477	III.366	11v, 11–12	p. 314, [3]7–8	p. 314, [9.3]8–9
μηδὲ εἷς τῶν ἀγαθοποιῶν ἀστέρων προσμαρτυρῇ ('not one of the beneficent planets bears witness')	nullus beneficorum contestificetur ('no one of the beneficent ones calls to witness')	ܡܢ ܥܒ̈ܕܝ ܛܒ̈ܬܐ ܠܐ ܬܗܘܐ ܡܢ ܣܗܕܘܬܐ *men 'āḇday ṭāḇāṯā lā tehwē men sahdūṯā* ('from the beneficent ones will be no testimony')	ولم ينظر ... شيء من السعود *wa-lam yunẓar … šay'un min as-su'ūdi* ('if none of the fortune [planets] is observed')	فإن لم يشهد... شيء من المسعدة *fa-'in lam yašhad … šay'un min al-mus'adatin* ('if none of the fortune [planets] bear witness')

The first problem in the above passage is the combination of the indeclinable particle μηδὲ ('and not' <μή + δέ) and the masc. numeral εἷς ('one'). The Latin variant 'nullus' ('nobody, not one, not even one') prompts Vuillemin-Diem and Steel to favour its Greek pendant μηδεὶς ('no one' or 'nobody') as the more sensible option.[35] The Arabic versions address this reading by combining partitive constructions (cf. *šay'un min*, lit. 'something of …', i.e. 'something disastrous', 'something fortunate' etc.) with negated finite verbs.[36] The two texts confirm Vuillemin-Diem and Steel's conclusion that the 'addition of ἀστέρων

[33] cf. Kazimirski, *Dictionnaire arabe-français*, vol. 2, p. 1606.

[34] Vuillemin-Diem and Steel, *Ptolemy's Tetrabiblos*, p. 118.

[35] Vuillemin-Diem and Steel, *Ptolemy's Tetrabiblos*, III.366: 'nullus: μηδεὶς **Σ β γ** Heph.] μηδὲ εἷς **V** (*Hüb.*)'.

[36] cf. Ullmann, *Wörterbuch*, p. 854.

is not needed',[37] and so does the Syriac. Of course, ἀστέρων could have been a gloss — a possibility that should not be excluded.

The Syriac rendition *ʿāḇday ṭāḇāṯā* (lit. 'doers of good things') provides a semantic equivalent of both ἀγαθοποιῶν and 'beneficorum'. Ḥunayn's use of the verbal root **šhd** ('to witness') brings his text closer to the Syriac, the Latin and the Greek, as opposed to the passive form of **nẓr**[38] ('to observe, perceive, look at') employed by F.

Reading 8

G	L	S	F	H
III.482–486	III.370–372	11v, 16–17	p. 314, [9.4]6–8	p. 314, [8.4]6–7
εἰ δὲ καὶ ὁ τοῦ Ἑρμοῦ μαρτυρήσειε ... μόνος δὲ ὁ τοῦ Ἑρμοῦ ... ἀπεργάζεται ('if Mercury should bear witness ...but when alone, Mercury makes [them] ...')	si autem Mercurius testificetur ... solus autem ... efficit ('if Mercury should bear witness ... but when alone, he makes [them] ...')	ܐܢܕܝܢ ܐܪܡܝܣ ܢܣܗܕ ... ܢܗܘܘܢ *ʾen-den ʾermīs* *neshad ... nehwon* ('but if Mercury bears witness ... they will become')	فإن نظر عطارد ...جعله *fa-ʾin nuẓira* *ʿuṭārid ǧaʿala-* *hu ...* ('but if Mercury is observed, he makes him ...')	فإن شهد عطارد ... كان *fa-ʾin šahida* *ʿuṭārid ... kāna...* ('but if Mercury bears witness ... he is')

This entry examines another variant reading where Hübner thought it necessary to add a 'fort. recte' comment.[39] William of Moerbeke's version confirms the omission of Mercury's name in the last phrase of the sentence.[40] Adding the name, perhaps for the sake of clarity, could have been a measure dictated by the Greek language's propensity for verbal indulgence.

The Syriac and the two Arabic texts support the omission of the name. In all of them Mercury is mentioned only once, at the beginning of the sentence. A key difference is exhibited in the way the texts present Mercury's influence on human fates — the earlier Arabic version follows the Greek and the Latin by using an active verb, i.e. the planet 'makes' people dumb, deaf, etc. (cf. ἀπεργάζεται, efficit and *ǧaʿala-hu*, the latter's literal meaning being 'he makes him'). The Syriac translation and Ḥunayn choose a different approach

[37] Vuillemin-Diem and Steel, *Ptolemy's Tetrabiblos*, p. 118: 'ἀγαθοποιῶν ἀστέρων V D γ Procl.] ἀγαθοποιῶν α M S + G (beneficorum) Heph.'

[38] However, this verb can mean 'being in aspect' in the astrological sense — cf. Dozy, *Supplément*, vol. 2, p. 685.

[39] Vuillemin-Diem and Steel, *Ptolemy's Tetrabiblos*, p. 118: 'ὁ τοῦ Ἑρμοῦ γ] ὁ Ἑρμοῦ V D: *om.* α M S + G ('*fort. recte*')'; cf. Hübner, Ἀποτελεσματικά, III.484.

[40] ibid.

— people 'become' (cf. *nehwon*, *kāna*) various things as a result of Mercury bearing witness. Moreover, the two Arabic versions refer to the subject of the planet's influence in the singular (cf. *ǧaʿala-hu*, 'he [i.e. Mercury] makes him / it' and *kāna*, 'he / it becomes').

The earlier Arabic text's choice of the verbal root **nẓr** (cf. n. 38 supra) as a pendant of the Greek μαρτυρέω ('to bear witness') and its derivatives appears to be persistent. In fact, this version stands out as the only one which does not, at least in the cases that have been examined so far, employ a closer semantic equivalent of μαρτυρέω.

Reading 9

G	L	S	F	H
III.512–513	III.393–394	12v, ll. 12–13	p. 318, ll. [10.3]5–8	p. 318, ll. [9.3]7–8
ἢ ὁ μὲν ἕτερος διαμετρῶν, ὁ δὲ ἕτερος ἐπανα-φερόμενος ('or the one is opposing, the other one is ascending')	uel alter quidem diametralis sit, alter autem superallatus ('or one would be opposite, the other one is exalted [lit. brought up]')	ܐܘ ܟܕ ܗܘ ܚܕ ܡܢ ܕܝܐܡܛܪܘܢ. ܐܚܪܢܐ ܕܝܢ ܒܐܢܐܦܘܪܐ *ʾaw kad̲ haw ḥad men diāmeṭron* [<διάμετρος] *ḥrinā d̲en b-ʾanaphorē* [<ἀναφορά, cf. ἐπαναφερόμενος] ('or when this one is [diametrically] opposed, the other one is on the ascent')	a very vague paraphrase	أوكان أحدهما مقابلا على القطر والآخر صاعدا إلى موضع النيّر *ʾaw kāna ʾaḥadhumā muqābilan ʿalā l-quṭri wa-l-ʾāḫar ṣāʿidan ʾilā mawḍiʿi n-nayyiri* ('or one of them is opposite to the diameter and the other one ascending to the location of the luminary')

Hübner's edition prefers the participle διαμετρῶν ('opposing') to the adjective 'διάμετρος' which Vuillemin-Diem and Steel propose on the strength of the evidence provided by the Latin translation.[41] The Syriac text presents a rendition almost identical with that of Moerbeke and confirms the use of an adjective, as opposed to a participle. What can be somewhat misleading is that the loan word used in Syriac is derived from the accusative form of the Greek term, i.e. *diāmeṭron* corresponds to the adjective διάμετρος, not to the homophonous διαμετρῶν.[42]

[41] cf. Vuillemin-Diem and Steel, *Ptolemy's Tetrabiblos*, p. 118: 'διαμετρῶν α βγ] διάμετρος ἢ V + G (diametralis sit), cf. Heph.'

[42] cf. Villey, *Les textes astronomiques*, p. 405.

Ḥunayn makes an effort to improve and clarify the passage which results in a more scientific language. The phrase *muqābilan ʿalā l-quṭri* ('opposite according to the diameter') is another case of *explicitation*,[43] building upon the Greek διάμετρος. Unlike the Syriac translator, who simply transcribed the term, Ḥunayn not only employs a literal Arabic pendant but also augments it with the explanation *ʿalā l-quṭri* (lit. 'according to the diameter') in order to arrive at the meaning 'diametrically opposed', which διάμετρος already contains, in a mathematical sense. The same approach is applied in the case of ἐπαναφερόμενος ('ascending'), too. We encounter another augmented phrase where the participle *ṣāʿidan* ('ascending') is accompanied by *ʾilā mawḍiʿi n-nayyiri* ('to the location of the luminary'), which specifies the implicit direction of the ascent.

Reading 10

G	L	S	F	H
III.824	III.617	17r, 8	–	not applicable
τῶν προκει-μένων ('above mentioned')	presuppositarum	ܕܩܕ̈ܡܝ ܐܬܬܣܝܡܘ *d-qaḏmay ʾetsim* ('which were established beforehand')		

The last variant reading the present survey examines concerns Ptolemy's use of the peculiar double prefix προϋπο-.[44] Hübner's decision to choose προκειμένων instead of προϋποκειμένων was informed by the scholar's philological acumen and experience. The participle προκειμένων is derived from the medio-passive verb πρόκειμαι and can mean 'set before' or 'set forth' in the sense of 'proposed' or 'established' (hence Robbins' translation 'above mentioned'). The fact that William of Moerbeke made an effort to translate Ptolemy's double prefix into Latin speaks for meticulous attention to detail. The Latin reading 'presuppositarum' appears to emphasise the temporal aspect of the participle, and maybe this was the reason why Ptolemy resorted to the unusual double prefix in the first place. The Syriac translator appears to have made an effort to capture the sense of προϋποκειμένων as well, thus confirming the reading found in Moerbeke's text. He combined a finite medio-passive verb (*ʾetsim*, i.e. 'they were established') and a participle (*qaḏmay*, i.e. 'they precede') in order to do so.

43 cf. Pormann, 'The Development'[, p. 154; n. 17 supra.

44 cf. Vuillemin-Diem and Steel, *Ptolemy's Tetrabiblos*, p. 119: 'προκειμένων α βγ, cf. Procl. (προστιθέντος)] προϋποκειμένων V + G (presuppositarum). The double prefix προϋπο- is used several times III.106 135 IV.712 790: see also προϋφέστηκα IV.691 and προϋποτυπόομαι IV.788'.

Conclusion

Comparing the variant readings suggested by Vuillemin-Diem and Steel with the Syriac and the Arabic texts has shown that in almost all cases at least one of the Near Eastern translations confirms the evidence provided by Moerbeke's Latin version. The Syriac translation backs the Latin readings in eight out of ten instances, Ḥunayn does so in five. Of particular interest are the occasions where Ḥunayn's readings confirm Hübner against the Syriac version and Moerbeke (cf. the omission of τῶν κακοποιῶν, i.e. 'the maleficent ones' in Reading 6). The earlier Arabic text, despite its tendency to paraphrase, nevertheless follows the Syriac and the Latin on three occasions, two of them being important omissions.[45]

Keeping in mind the fact that Moerbeke's version practically serves as the earliest extant recension of the Greek text, its proximity to the Syriac, both in terms of context as well as particular variant readings, speaks for a general affiliation with the older stages of the *Tetrabiblos*' transmission. The Syriac remains closer to the Latin and the Greek than the Arabic translations in the majority of the cases, generally confirms the improvements proposed by Vuillemin-Diem and Steel, tends to use Greek astrological terminology or lexical pendants, displays awareness of Greek syntax and in some instances attempts to emulate it. The Semitic nature of the language, along with minor omissions and deviations, may account for the impression that the Syriac translation somehow compresses the Greek text, but cases such as Readings 4, 9 and especially 10 clearly appear to refute the possibility of the Syriac text being a mere paraphrase, 'and a poor one at that'.[46] The overall quality of this version of the *Tetrabiblos*, its author's adequate solutions and attention to detail evade the somewhat polarised concepts of 'free' (i.e. paraphrastic) and 'literal' translation methods.[47] The Syriac passage in Reading 3 resorts to additional explanations in order to clarify the Greek content, and does that in a fashion which is not all that different, albeit less refined, from Ḥunayn's *explicitation* approach (cf. Readings 9 and 2).

The readings from the older Arabic text appear to confirm Pingree's hypothesis of ʿUmar b. al-Farruḫān as translator.[48] The omissions and the tendency to simplify the technical nature of Ptolemy's treatise and its astrological terminology are, in fact, more deserving of a definition such as 'paraphrase'. Moreover, the content of the older translation renders Ibn an-Nadīm's attribution

[45] cf. Reading 6, mentioned above, and Reading 8, where the addition of Mercury's name is concerned; Reading 8 is also significant because F is the only Near Eastern translation which translates ἀπεργάζεται / efficit ('to make'), cf. *ǧaʿala-hu* ('he makes him').

[46] cf. Saliba, *Islamic Science*, p. 12.

[47] cf. Gutas, *Greek Thought*, pp. 142–44; Pormann, 'The Development', p. 145.

[48] cf. n. 7 supra.

to al-Biṭrīq[49] rather problematic, because Pormann's analysis of another translation attributed to al-Biṭrīq[50] points out the latter's frequent use of transliterated Greek terms — a feature we have encountered frequently in the Syriac text, but not in F. Of course, a scholar's translation skills can evolve, and the quality of his work may vary substantially, depending on the circumstances surrounding each particular translation.

Since ʿUmar b. al-Farruḫān's text, by virtue of being a probable paraphrase, differs from Ḥunayn's translation to a significant degree, it would be tempting to deduce that the two are not directly related. H's readings contain many reference points which can be associated with the Greek, Latin and Syriac versions, and may indicate the availability of good source texts. Reading 9, in particular, attests that the translator had at his disposal a more precise technical vocabulary in Arabic, a fact which speaks for a more sophisticated translation technique.

The comparisons and analyses presented in this essay demonstrate that the three Near Eastern translations of the *Tetrabiblos* have the potential to make significant contributions to the research on the treatise's complex transmission, which the evidence from William of Moerbeke's Latin translation has clearly advanced to a new level.

Bibliography

Bergsträsser, Gotthelf, *Ḥunain ibn Isḥāq. Über die syrischen und arabischen Galen-Übersetzungen*, Leipzig: F. A. Brockhaus, 1925.

Dodge, Bayard, *The Fihrist of al-Nadīm. A Tenth-Century Survey of Muslim Culture*, 2 vols, New York/London: Columbia University Press, 1970.

Dozy, Reinhart, *Supplément aux dictionnaires arabes*, 2nd ed., 2 vols, Leiden: Brill, 1927.

Dunlop, Douglas, 'The Translations of Al-Biṭrīq and Yaḥyā (Yuḥannā) b. Al-Biṭrīq', *The Journal of the Royal Asiatic Society of Great Britain and Ireland* 3/4 (1959), pp. 140–50.

Gutas, Dimitri, *Greek Thought, Arabic Culture. The Graeco-Arabic Translation Movement in Baghdad and Early ʿAbbāsid Society (2nd-4th/8th-10th centuries)*, London/New York: Routledge, 1998.

Hübner, Wolfgang, *Ἀποτελεσματικά. Post Franz Boll et Ae. Boer secundis curis*, Stuttgart/Leipzig: Teubner, 1998 [*Claudii Ptolemaei opera que extant omnia*, vol. III.1].

Kazimirski, Albert de Biberstein, *Dictionnaire arabe-français contenant toutes les racines de la langue arabe*, 2 vols, Paris: Maisonneuve, 1860.

Müller, August, *ʿUyūn ul-anbāʾ fī ṭabaqāt ul-aṭibbāʾ li-ibn Abī Uṣaybiʿa*, 2 vols, Königsberg: Selbstverlag, 1882–1884.

[49] cf. n. 5 supra.

[50] Pormann, 'The Development', pp. 148–49.

Müller, August, and Julius Lippert, *Ibn al-Qifṭī's Taʾrīkh al-Ḥukamāʾ*, Leipzig: Theodor Weicher, 1903.

Pingree, David, "Umar Ibn al-Farrukhān aṭ-Ṭabarī', *Complete Dictionary of Scientific Biography* (digital format, accessed via www.encyclopedia.com, 23 December 2018).

Pormann, Peter, 'The Development of Translation Techniques from Greek into Syriac and Arabic: The Case of Galen's *On the Faculties and Powers of Simple Drugs*, Book Six', in Rotraud Hansberger, Afifi Al-Akiti and Charles Burnett (eds), *Medieval Arabic Thought. Essays in Honour of Fritz Zimmermann*, London: Warburg Institute, 2012, pp. 143–62.

Robbins, Frank E., *Ptolemy. Tetrabiblos*, Cambridge (Mass.): Harvard University Press, 1940.

Saliba, George, *Islamic Science and the Making of the European Renaissance*, Cambridge (Mass.): The MIT Press, 2007.

Sokoloff, Michael, *A Syriac Lexicon. A Translation from the Latin, Correction, Expansion, and Update of C. Brockelmann's Lexicon Syriacum*, Piscataway: Gorgias Press, 2009.

Ullmann, Manfred, *Wörterbuch zu den griechisch-arabischen Übersetzungen des 9. Jahrhunderts. Supplement Band II*, Wiesbaden: Harrassowitz, 2007.

Villey, Emily, *Les textes astronomiques syriaques (VIe et VIIe siècles). Établissement d'un corpus et de critères de datation. Édition, traduction et lexique*, PhD dissertation, Université de Caen Basse-Normandie, 2012.

Vuillemin-Diem, Gudrun, and Carlos Steel, *Ptolemy's Tetrabiblos in the Translation of William of Moerbeke. Claudii Ptolemaei Liber iudicialium*, Leuven: Leuven University Press, 2015.

II. The Arabic Tradition

The Oldest Translation of the *Almagest* Made for al-Maʾmūn by al-Ḥasan ibn Quraysh: A Text Fragment in Ibn al-Ṣalāḥ's Critique on al-Fārābī's Commentary

Johannes THOMANN

1. Life and times of Ibn al-Ṣalāḥ (d. 1154 CE)

The first half of the twelfth century was a pivotal time in Western Europe. In that period translation activities from Arabic into Latin became a common enterprise on a large scale in recently conquered territories, of which the centres were Toledo, Palermo and Antioch. This is a well known part of what was called the Renaissance of the Twelfth Century.[1] Less known is the situation in the Islamic World during the same period. Traditionally it was denounced as post-classical, implying some kind of decadence. Politically it was the time when the Seljuqs had surpassed their apogee of power, but still dominated the Islamic East from Syria to Central Asia. The Christian kingdom of Jerusalem in the recently conquered territories was a zone of permanent conflicts, but formed only part of the periphery. The territory of the Fatimids was reduced to Egypt. In the West the Almoravids were about to extend their empire in the Maghreb towards al-Andalus.[2]

Concerning the mathematical disciplines, the first half of the twelfth century has been called the age of Omar Khayyam.[3] His works on geometrical solutions of algebraic problems are famous, and a number of other treatises document a broad field of scientific activities.[4] He was active in Central Asia in the Eastern part of the Seljuk Empire.[5] In this area a great number of lesser-known mathematicians were active, and it must be seen as one of the two main centers of mathematical science at the time.[6] The other center was al-Andalus, where an even greater number of mathematicians were active.[7] Among

[1] Haskins, *The Renaissance*, pp. 278–302.

[2] Kennedy, *An Historical Atlas*, p. 10.

[3] Sarton, *Introduction*, vol. I, pp. 738–83.

[4] *MAOSIC*, pp. 168–70 (No. 420).

[5] Aminrazavi, *The Wine of Wisdom*, pp. 18–31.

[6] *MAOSIC*, pp. 168–86 (Nos 420, 423–26, 435, 437–39, 443, 450, 453, 458–59, 461, 467, 469, 471, 473–76, 484, 489).

[7] *MAOSIC*, pp. 168–86 (Nos 422, 428, 431, 433–34, 436, 440–42, 448–49, 452, 455, 462, 464, 468, 477, 479–80, 483, 486).

Ptolemy's Science of the Stars in the Middle Ages, ed. by David Juste, Benno van Dalen, Dag Nikolaus Hasse and Charles Burnett, PALS 1 (Turnhout, 2020), pp. 117–138

10.1484/M.PALS-EB.5.120176

these only Jābir ibn Aflaḥ became famous, since his commentary on the *Almagest* was translated into Latin.[8]

Baghdad had lost its position as the primary place of learning in the Islamic world. However, it attracted still some students of the sciences. Even though it was not the home of eminent scholars, there must have been exceptionally rich and valuable treasures of books available. One of those who took profit of these treasures was Abū l-Futūḥ Aḥmad ibn Muḥammad ibn al-Sarī, called Ibn al-Ṣalāḥ.[9] According to his biographers he was a Persian, born in Hamadān in Western Iran, who came to Baghdad and had gained a reputation as a physician.[10] In this quality he went to the court of Temür Tāsh ibn Īl Ghāzī, the Artuqid ruler at Mārdīn (r. 1122–1154 CE). Towards the end of his life he moved to Damascus, which was ruled by the Börid Atabeg Abaq (r. 1140–1154 CE).[11] There are different statements concerning the date of Ibn al-Ṣalāḥ's death in the sources. According to al-Qifṭī he died at the end of the year 548 (March 1154 CE), and according to Ibn Abī Uṣaybiʿa in the year '540 odd'.[12] A manuscript of the *Conics* of Menelaos at the British Library contains a colophon with the date 'Monday 4 Rabīʿ II 548' (29 June 1153 CE), in which Ibn al-Ṣalāḥ is mentioned.[13] The formula *aṭāla llāhu baqāhu* ('may God make his life long') after his name indicates that he was still alive at that date. This corroborates al-Qifṭī's date March 1154 CE for his death. In the same colophon Ibn al-Ṣalāḥ is called *al-zāhid* ('the ascetic'), which might explain his surname, since *ibn al-ṣalāḥ* ('son of salvation') points to a pious lifestyle.

Ibn al-Ṣalāḥ was a somewhat unusual scholar. Among his preserved works there are only very few which are of his own creation. Almost all of them are critiques directed against the works of others. The targets of his critical attacks were the most famous scholars of the past: Aristotle, Euclid, Ptolemy, Galen, Ibn al-Haytham, Abū Sahl al-Kūhī, Jābir ibn Ibrāhīm al-Ṣābiʾ and al-Fārābī.[14]

The work by Ibn al-Ṣalāḥ which is best known among scholars working on the history of astronomy is his critique of the transmission of coordinates in the star catalogue of the *Almagest*. This is a meticulous analysis of the values of coordinates in a Syriac and four Arabic translations of the *Almagest* and other

[8] *MAOSIC*, p. 176 (No. 448).

[9] *MAOSIC*, pp. 177–78 (No. 458).

[10] Lippert, *Taʾrīḫ al-ḥukamā*, p. 428; Müller, *ʿUyūn al-anbāʾ*, vol. II, pp. 164–67.

[11] For the life of Ibn al-Ṣalāḥ see Lorch, 'Ibn al-Ṣalāḥ's Treatise', p. 401.

[12] Lippert, *Taʾrīḫ al-ḥukamā*, p. 428; Müller, *ʿUyūn al-anbāʾ*, vol. II, p. 164.

[13] MS London, British Library, Or. 13127, fol. 51r, lines 6–14; see the online catalogue at http://searcharchives.bl.uk (search for 'Or 13127'; retrieved 21 April 2016); digital images are available at http://www.qdl.qa/en/archive/81055/vdc_100000038406.0x000001 (retrieved 21 April 2016).

[14] For a list of Ibn al-Ṣalāḥ's works see Thomann, 'Al-Fārābī's Kommentar', pp. 101–02; the marginal glosses by Ibn al-Ṣalāḥ to the text of Menelaos in the MS London, British Library, Or. 13127 are to be added to this list.

works containing a star catalogue. It was edited, translated and commented upon by Paul Kunitzsch in 1975.[15]

2. Ibn al-Ṣalāḥ's critique on al-Fārābī's commentary on the *Almagest*

In the focus of the present paper is another work by Ibn al-Ṣalāḥ on the *Almagest*, namely a critique of al-Fārābī's commentary on the *Almagest*. This work is preserved in a single manuscript in the library of the Holy Shrine in Mashhad (MS 5593).[16] The manuscript was written in 1462 and the work by Ibn al-Ṣalaḥ, contained on pages 81 to 92, is entitled 'Reasoning on Proof of the Error Made by Abū Naṣr al-Fārābī in his Commentary on the Seventeenth Section of the Fifth Book of the Almagest and the Explanation of this Section'.[17]

The passage of the *Almagest* on which Ibn al-Ṣalaḥ writes is not in Chapter V.17 but in Chapter V.19 as we know it from the Greek text and the extant Arabic translations. The topic of the work is a small passage in the section on parallax. At the beginning of Chapter V.19 Ptolemy explains how to find the lunar parallax in altitude.[18] This is the change in the lunar position in vertical direction for an observer at a distance from the centre of the earth. After that Ptolemy explains how to split up this parallax in altitude into two components, the parallax in ecliptical longitude and the parallax in ecliptical latitude.[19] This second part of Chapter V.19 is the topic of Ibn al-Ṣalāḥ's critique.

Ptolemy's approach is rather crude. First he makes an approximation by transforming the spherical problem into a plain one and assuming that the two circles of measuring the ecliptical latitude are straight parallel lines. In doing so, the problem is reduced to a trivial geometrical case. Later in the chapter he criticizes this method, invented by Hipparchos, and proposes another solution, allegedly operating 'in a [mathematically] sound way' (κατὰ τὸν ὑγιῆ τρόπον), but this is an approximation too.[20] It seems that he could not find an exact solution by his mathematical means. Otto Neugebauer's verdict was that 'The chapter on parallax is undoubtedly one of the most unsatisfactory sections in the whole Almagest'.[21]

[15] Kunitzsch, *Ibn aṣ-Ṣalāḥ*.

[16] Maʿānī, *Fihrist-i kutub-i ḫaṭṭī*, pp. 344–48; Sezgin, *Geschichte des Arabischen Schrifttums*, p. 195; Thomann, 'Al-Fārābī's Kommentar', pp. 102–04.

[17] *MAOSIC*, p. 178 (No. 458).

[18] Heiberg, *Syntaxis mathematica*, vol. I, pp. 444–45; Toomer, *Ptolemy's Almagest*, pp. 265–66.

[19] Heiberg, *Syntaxis mathematica*, vol. I, pp. 446–50; Toomer, *Ptolemy's Almagest*, pp. 266–67.

[20] Heiberg, *Syntaxis mathematica*, vol. I, pp. 450–55; Toomer, *Ptolemy's Almagest*, pp. 269–71; Pedersen, *A Survey*, pp. 218–19, 471; Neugebauer, *A History*, vol. I, pp. 116–17.

[21] Neugebauer, *A History*, vol. I, p. 116; but see Toomer, *Ptolemy's Almagest*, p. 273, note 87 for a different view.

The first astronomer who was able to provide an exact and valid solution of the same problem was Ḥabash al-Ḥāsib in the mid ninth century.[22] He based his calculations not on Greek trigonometry with chords but on Indian trigonometry with sine and cosine and used for his solution both the cosine rule and the sine rule for spherical triangles. In this case at least Indian style trigonometry was superior to Greek style trigonometry.

Ibn al-Ṣalāḥ writes at the beginning of his treatise:[23]

> I had a look at a book by the outstanding Abū Naṣr al-Fārābī called *Commentary on the Book by Ptolemy Known as the Almagest*. I studied it thoroughly in full clarity and understanding of its concepts up to the information in Chapter 17 of Book V. I found that he wanted to establish the proof based on the relation which was there in connection with a complete commentary on the chapter. But the premises which he used in the composition of his proof were impossible and fallacious.

Thus the critique of Ibn al-Ṣalāḥ is not directed towards Ptolemy himself but towards al-Fārābī's *Commentary on the Almagest*. This work has only recently been discovered, and some information is appropriate here.

3. Al-Fārābī's commentary on the *Almagest*

The great philosopher al-Fārābī (d. 950 CE), who had the honorary title of 'the Second Teacher' (sc. after Aristotle), is most famous for his works on logic, metaphysics and political philosophy. But he wrote also on mathematical disciplines. Since the times of Moritz Steinschneider it has been known that al-Fārābī wrote a commentary on the *Almagest*.[24] It is mentioned in the biographies in al-Qifṭī, Ibn Abī Uṣaybiʿa and al-Ṣafadī, and it appears in a list of commentaries on the *Almagest* by al-Nasawī (eleventh century CE).[25] A supposed copy in the British Library turned out to be the *Talkhīṣ* by Ibn Sīnā,[26] and the work was considered to be lost.[27] In 2011 the discovery of a part of a comprehensive commentary on the *Almagest*, probably al-Fārābī's commentary, was announced.[28] The MS Tehran, Majlis Library, 6531 has a modern title-page with the name of al-Fārābī. The beginning of the original manuscript is missing, and at the end it has no colophon. Thus the text is transmitted

[22] Kennedy, 'Parallax Theory', pp. 42–43.

[23] MS Mashhad, Holy Shrine Library, 5593, p. 81; for the Arabic text see Appendix II.

[24] Steinschneider, *Al-Farabi*, p. 78.

[25] Lippert, *Taʾrīḫ al-ḥukamā*, p. 279; Müller, *ʿUyūn ul-anbāʾ*, vol. II, p. 138; Ritter, *Kitāb al-Wāfī*, vol. I, p. 108; for al-Nasawī see Lorch, *Thābit ibn Qurra*, p. 348.

[26] Goldstein, book review of Sezgin, p. 342.

[27] Janos, 'Al-Fārābī', p. 239; Janos, *Method*, pp. 22–26.

[28] Paper presented at the conference 'Contexts of Learning in Baghdad from 8th–10th centuries', University of Göttingen, September 12–14, 2011, published later as: Thomann, 'From Lyrics', pp. 500–02; first publication: Thomann, 'Ein al-Fārābī zugeschriebener Kommentar', pp. 48–53.

anonymously. It contains a commentary on the *Almagest* based on the Isḥāq translation, and covers parts of Book IX and all of Books X to XIII. In 2012 another manuscript with the same text was found (MS Tehran, Majlis, 6430), but again with no indications of the author.[29] At the beginning several pages are missing, but it covers slightly more text than the first manuscript. Further investigations made an attribution of this commentary to al-Fārābī more and more likely. It is evident that it was written by a philosopher rather than by a professional astronomer.[30] This limits the number of candidates for being the author of the Tehran commentary considerably. Further, there are some characteristics in the vocabulary which coincide with Fārābīan usage.[31]

The identification of the Tehran manuscripts as al-Fārābī's commentary on the *Almagest* finally became beyond doubt when the treatise of Ibn al-Ṣalāḥ on the critique of al-Fārābī's commentary was studied for the first time.[32] The text of Ibn al-Ṣalāḥ consists for a large part of literal quotations from al-Fārābī's commentary. For the first time documented original parts of al-Fārābī's work were at hand. Since the quoted texts belong to Book V of the *Almagest* a direct comparison with the two Tehran manuscripts, which cover Books IX to XIII, was not possible. But the relative quantity of text of al-Fārābī's commentary in comparison to related text of Ptolemy could be estimated and conspicuous terminology could be compared. There is one noteworthy abnormality in the parts quoted by Ibn al-Ṣalāḥ. In the text of al-Fārābī the term for parallax is always *inḥirāf al-manẓar*, while the standard term, also found in the translations of the *Almagest*, is *ikhtilāf al-manẓar*.[33] The reason why al-Fārābī chose this non-standard term may be his propensity to be philologically precise, and indeed, *inḥirāf* 'deviation' is semantically closer to Greek *parallaxis* than *ikhtilāf*, which means simply 'difference'.[34] In any case, the occurrence of this abnormality in the text of the two Tehran manuscripts would provide a perfect terminological test. There is only one passage in Books IX to XIII of the *Almagest* where parallax is mentioned.[35] The corresponding commentary is only preserved in the second Tehran manuscript, where parallax is indeed called *inḥirāf al-manẓar*.[36] Therefore there can hardly be any doubt that the passages quoted by Ibn al-Ṣalāḥ and the text in the two Tehran manuscripts are parts of

[29] Paper presented at the 26th Congress of the Union Européenne des Arabisants et Islamisants (UEAI 26), Basel, September 12–16, 2012; see now Thomann, 'Terminological Fingerprints', pp. 304–05.

[30] Thomann, 'Ein al-Fārābī zugeschriebener Kommentar', pp. 58–59.

[31] Thomann, 'Terminological Fingerprints', pp. 305–10.

[32] Thomann, 'Al-Fārābī's Kommentar'.

[33] Thomann, 'Al-Fārābī's Kommentar', pp. 110–11; see the text in Appendix II.

[34] Eckhard Neubauer, personal communication (July 26, 2015).

[35] Heiberg, *Syntaxis mathematica*, vol. II, p. 207; Toomer, *Ptolemy's Almagest*, p. 419.

[36] MS Tehran, Majlis Library, 6430, fol. 22r; see the text in Appendix II.

the same work, and that in the twelfth century this work was regarded by the attentive and well-informed Ibn al-Ṣalāḥ as the work of al-Fārābī.

4. An anonymous translation of the *Almagest* and its terminology

At the very beginning of his critique on al-Fārābī's commentary, after the introductory phrase, Ibn al-Ṣalāḥ quotes literally the passage of the *Almagest* upon which al-Fārābī comments.[37] Ibn al-Ṣalāḥ does not say anything about the authorship of the quoted translation, therefore in the following it will be called provisorily 'Anonymous'. In a first step, the text will be compared with the two well-known Arabic translations of the *Almagest* by al-Ḥajjāj and Isḥāq/Thābit.[38] The three Arabic translations, the Greek text and the Latin translation of Gerard of Cremona are given in Appendix I. Words and expressions which differ in the three Arabic translations are listed in the four following tables. The first table contains words and expressions which differ in all three translations:

Greek	*Anonymous*	*Al-Ḥajjāj*	*Isḥāq/Thābit*
ἵνα	فإذا أردنا أن	ولكي	ولكيما / وكيما
διακρίνωμεν	ونفصل ...	نعدل	نقوّم
ἐπισκεψόμεθα	ونأخذ	ونطلب	ننظر
σελιδίῳ	السطر	الجدول	الصفّ
τοσούτων	فإنّا إذا فعلنا ذلك	وذلك	فإن هذا
ἐπειδήπερ	فلمّا	لأنّ	من قبل أن
γραφομένου	الّتي تمرّ	المخطوط على	ترسم مادّة

The second table contains words and expressions which are identical or similar in al-Ḥajjāj and Isḥāq/Thābit but different in the Anonymous:

Greek	*Anonymous*	*Al-Ḥajjāj*	*Isḥāq*
ἀπέχει	بين ... وبين	بعد ... من	بعد ... من
μεσημβρινοῦ	وسط السماء	فلك نصف النهار	دائرة نصف النهار
μεσημβρινοῦ	توسّط القمر السماء	بعد نصف النهار	بعد دائرة نصف النهار
ἀπογραψόμεθα	وكتبناه	أثبتناها	أثبتناها
ἐκκειμένην	اللتين تليان	الّتي على هذه	اللتين في هذا
ἐν κύκλῳ εὐθειῶν κανόνιον	جداول القسي والأ[و]تار	في جدول أوتار القسي	في جدول الأوتار الّتي في الدائرة
εὑρισκομένην	حصلناه	الموجود	يوجد
μερίζοντες	قسمناه	نقسم	ونقسم
συναγόμενα	بما يخرج	ما اجتمع	ما اجتمع
ἕξομεν	علمنا	فهو	فهو

[37] MS Mashhad, Holy Shrine Library, 5593, pp. 81–82.

[38] Other translations of the *Almagest* will be discussed in Section 5.

The third table contains words and expressions which are identical or similar in al-Ḥajjāj and the Anonymous but different in Isḥāq/Thābit:

Greek	*Anonymous*	*Al-Ḥajjāj*	*Isḥāq*
ἰσημερινὰς	المعتدلة	المعتدل	الاستوائيّة
κανόνος	جداول	جداول / جدول	جدول
εἰς τὸ αὐτὸ μέρος	إلى الموضع الّذي كنّا أدخلناه به فيما تقدّم	ذلك الموضع	في ذلك القسم بعينه
λειπούσας	ما نقص	الّتي تنقص	ما يبقى بعدها
τομὴν	قطعة	القطعة	التقاطع
καὶ ὃν ἂν ἔχῃ	فيكون لنا	فيكون	فأيّ
πολυπλασιάζοντες	فضربناه	فيضرب	فيضاعف
κατὰ κορυφὴν	بسمت الر[ؤ]وس	سمت الرؤوس	سمت الرأس

The fourth table contains words and expressions which are identical or similar in Isḥāq/Thābit and the Anonymous but different in al-Ḥajjāj:

Greek	*Anonymous*	*Al-Ḥajjāj*	*Isḥāq*
παρακειμένας	بحياله	الّتي تقابل	حياله / بحياله
οὖν	ثمّ	ف	ثمّ
ἀδιαφοροῦσιν	كان لا فرق	⟨لا⟩ تكون ... مختلفة	فليس بينه وبين ... فرقان

The fact that the second table is the largest indicates that the Anonymous differs more from the two other translations than al-Ḥajjāj and Isḥāq/Thābit differ from each other. This leads to the question if the Anonymous version is a genuine translation from the Greek, or a paraphrase of one of the two other Arabic translations.[39] There are three cases in the Anonymous where knowledge of the Greek original is evident. In the Anonymous, Greek διακρίνωμεν ('we distinguish, we set apart') is at first translated by *nufaṣṣilu* ('we divide'). Later in the sentence it is specified by the expression *wa-nafṣila kulla wāḥidin minhumā ʿani l-ākhari* ('and we separate each of them from the other'). This is a precise paraphrase of the litteral meaning of διακρίνω and could not have been derived from one of the two other translations. Al-Ḥajjāj writes *naʿdilu* or *nuʿaddilu* 'we normalize the parallax ...', and Isḥāq *nuqawwimu* 'we arrange the parallax'. This suggests that the anonymous translation is based on the Greek text, and that it is not just a paraphrase of one of the two other translations. A second case is the translation *wa-katabnāhu* ('we have written it') of Greek ἀπογραψόμεθα ('we have written off'). Al-Ḥajjāj and Isḥāq/Thābit translate it with *athbatnāhā* ('we have made it fixed'), which does not preserve

[39] There is no need to consider a translation from the Syriac since according to Ibn al-Ṣalāḥ all Arabic translations were made from the Greek; cf. Kunitzsch, *Ibn aṣ-Ṣalāḥ*, p. 155, lines 12–19 (Arabic text) and p. 40 (German translation).

the meaning of 'writing'.[40] A third case is the Greek word ἐκκειμένην ('lying outside'), which is translated by the Anonymous as *allatayni taliyāni* ('which are adjacent'). The other translations are less precise: Al-Ḥajjāj translates as *allatī ʿalā* ('which are on') and Isḥāq/Thābit *allatayni fī* ('which are in'). These three examples show clearly that the Anonymous is based on the Greek text independently from the translations of al-Ḥajjāj and Isḥāq/Thābit.

In a next step, some conspicuous expressions used by the Anonymous will be compared to other astronomical texts in order to derive arguments for a chronological classification.

The Greek adjective μεσημβρινός means literally 'belonging to noon', composed of the adjective μέσος ('middle'), the substantive ἡμέρα ('day') and the suffix -ινος (for building adjectives). In an astronomical context ὁ μεσημβρινός κύκλος ('the circle belonging to noon') is the technical term for 'meridian', and μεσημβρινός can be used alone as a noun to denote 'meridian', as is the case in the text here. Al-Ḥajjāj uses *falak niṣf al-nahār* ('sphere of half day') and Isḥāq/Thābit *dāʾirat niṣf al-nahār* ('circle of midday'). In the translation of the *Anaphorikos* by Hypsikles, made either by Qusṭā ibn Lūqā or Isḥāq ibn Ḥunayn, μεσημβρινός is translated also as *niṣf al-nahār*. The expressions *falak niṣf al-nahār*, *dāʾirat niṣf al-nahār* and *khaṭṭ niṣf al-nahār* ('line of midday') became standard and were used interchangeably in astronomical texts of different epochs. Ḥabash al-Ḥāsib (d. *c*. 864 CE) uses *falak niṣf al-nahār* and *khaṭṭ niṣf al-nahār* as technical terms for 'meridian'.[41] Al-Bīrūnī (973–1048) uses *falak niṣf al-nahār* for 'meridian' in his introductory work on astronomy and astrology.[42] The term *dāʾirat niṣf al-nahār* is found in the terminological dictionary by al-Tahānawī (eighteenth century).[43] Different from these common translations, the Anonymous translates μεσημβρινός as *wasaṭ al-samāʾ* ('middle of the heaven'). It is conspicuous that in one of the oldest extant Arabic astronomical texts, *On the Use of the Astrolabe* by al-Khwārizmī, *khaṭṭ wasaṭ al-samāʾ* ('line of the middle of heaven') is used as the technical term for 'meridian'.[44] Besides that the expression *wasaṭ al-samāʾ* is used for a different notion. In contrast to *khaṭṭ wasaṭ al-samāʾ*, which denotes a line, *wasaṭ*

[40] No example for *athbata* in Lane's Lexicon refers to 'writing', see Lane, *An Arabic-English Lexicon*, p. 329.

[41] MS Istanbul, Süleymaniye Library, Yeni Cami 784, fols 130v, 149r, 150r, 156r–v, 161v, 162v, 164v, 190r–v (*falak niṣf al-nahār*), fols 130v, 151v, 167v, 168v–170v, 172r–v, 176r, 190r, 191r, 194v, 195v, 196v, 197v, 198v, 208r, 219v, 220r (*khaṭṭ niṣf al-nahār*).

[42] Wright, *The Book of Instruction*, p. 49 (§ 129).

[43] Daḥrūj, *Kashshāf*, p. 241.

[44] Charette and Schmidl, 'Al-Khwārizmī', p. 115 (§ 2c), p. 116 (§ 2d), p. 116 (§ 3) et passim.

al-samāʾ denotes a point defined by the intersection of the meridian with the ecliptic. This becomes evident when al-Khwārizmī writes:[45]

> [Then look at which degree] is cut by the line of midheaven (*khaṭṭ wasaṭ al-samāʾ*), and this will be the degree of midheaven (*darajat wasaṭ al-samāʾ*).

The expression *darajat wasaṭ al-samāʾ* in the sense of '(ecliptical) degree of the meridian' is used by Ḥabash too.[46] Later the meaning of *wasaṭ al-samāʾ* became restricted to 'the point of intersection of the ecliptic with the meridian'. But obviously the Anonymous imitates the Greek expression ὁ μεσημβρινός as an abbreviated form of ὁ μεσημβρινός κύκλος by writing *wasaṭ al-samāʾ* as an abbreviated form of *khaṭṭ wasaṭ al-samāʾ*.

Another abnormality concerns the translation of the Greek conjunction ἵνα ('that, in order that'). Al-Ḥajjāj translates it as *wa-lākin* ('however, yet, but'), and Isḥāq/Thābit more literally as *wa-likaymā* ('that, in order that'). The Anonymous departs considerably from the Greek text and starts the sentence by *wa-idhā aradnā an naʿrifa* ('when we want to know'). A similar expression is found only once in the Greek *Almagest*: Chapter III.8 begins with the expression Ὁσάκις οὖν ἂν ἐθέλωμεν ... ἐπιγιγνώσκειν ('So whenever we want to know').[47] There must thus have been another source of inspiration for the Anonymous to use this expression. Indeed, in al-Khwārizmī's treatise 'On the Use of the Astrolabe' 42 paragraphs out of 53 (79%) start either with *idhā aradta an taʿrifa* ('when you want to know'), *idhā aradta an taʿlama* (ditto), *in aradta an taʿrifa* (ditto), or *idhā aradta* ('when you wish') followed by a noun in the accusative. The second person singular was based on the style of Sanskrit astronomical works, while the first person plural was the style of Greek works.[48] The Anonymous keeps the first person plural from the Greek text, but uses the conditional phrase that was the standard start of a paragraph in astronomical treatises of his time. The phrase *idhā aradta an taʿrifa* ('when you want to know') and its synonyms are found in later astronomical texts too, but never again as rigorously as in the astronomical writings of al-Khwārizmī. In the *Zīj* of Ḥabash al-Ḥāsib it occurs only twelve times.[49] In the *Zīj* of al-Battānī still 40 chapters and subchapters out of 65 (62%) start with such a phrase,[50] and in

45 Charette and Schmidl, 'Al-Khwārizmī', p. 116 (§ 3).

46 MS Istanbul, Süleymaniye Library, Yeni Cami 784, fols 160r, 161r, 169r–185v, 205r–222v.

47 Heiberg, *Syntaxis mathematica*, vol. I, p. 259, lines 12–14; Toomer, *Ptolemy's Almagest*, p. 169.

48 Thomann, 'From Lyrics', pp. 510–14.

49 MS Istanbul, Süleymaniye Library, Yeni Cami 784, fols 74v, 78r, 101v (2x), 102v, 124r, 224v, 225r (3x), 228v.

50 Nallino, *Al-Battānī*, vol. III, p. 20 line 6, p. 29 line 7, p. 30 line 11, p. 31 line 23, p. 31 line 23, p. 33 line 33 et passim.

contrast, Thābit ibn Qurra uses the phrase rarely.[51] Al-Bīrūnī uses the phrase only occasionally. For example, in Book V of his *Qānūn* the phrase occurs at the beginning of three chapters out of 21 (14%).[52]

A third noteworthy case is the terminology for 'table', 'row' and 'column'. In the *Almagest* the Greek expressions are κανών (literally 'straight rod, bar'), στίχος ('row of soldiers', also 'line of poetry'), and σελίδιον, diminutive of σελίς ('cross-beam', also 'column in a papyrus or a mathematical table'). Al-Ḥajjāj translates these terms as *jadāwil*,[53] plural of *jadwal* (litteraly 'creek, brook'), *saṭr* ('line')[54] and *jadwal*. In Isḥāq/Thābit they are translated as *jadwal*, *saṭr*[55] and *ṣaff* ('row, line'). In the terms for 'table' a shift from the plural *jadāwil* to the singular *jadwal* is seen. If the plural is used for 'table', it is logical to use the singular for 'column'. However, the Anonymous calls the table *jadāwil*, but uses *saṭr* for 'column' instead, the same term which Isḥaq/Thābit use in the sense of 'row'. The same use of *saṭr* in the sense of 'column' is found in al-Khwārizmī's *On the Construction of the Astrolabe*.[56] It is also found in Yaḥyā ibn Abī Manṣūr's *al-Zīj al-Mumtaḥan*.[57] Al-Battānī uses *saṭr* still in the same sense.[58] But otherwise *saṭr* was used predominantly for 'row'. This holds for Thābit ibn Qurra,[59] for the *Mafātīḥ al-ʿulūm* (tenth c. CE),[60] and also for Ibn al-Ṣalāḥ.[61]

These examples suggest that the translation of the Anonymous was made at an early time, probably at the beginning of the ninth century CE. At least, nothing in the terminology speaks against such an early date.

5. Translations of the *Almagest* known to Ibn al-Ṣalāḥ

Ibn al-Ṣalāḥ mentions in his work on the star catalogue of the *Almagest* explicitly which translations he had at hand:[62]

51 Lorch, *Thābit ibn Qurra*, pp. 42–111: no occurrences; Morelon, *Thābit ibn Qurra*, pp. 65, 135, 137; shorter expressions as *wa-in aradta* and the like: pp. 96, 101, 105, 135, 138, 139, 141, 145, 146, 148, 149, 150, 160.

52 al-Bīrūnī, *al-Qānunu'l-Masʿūdī*, vol. II, pp. 516 line 3, 522 line 7, 526 line 3.

53 This is the reading in MS Leiden, Universiteitsbibliotheek, Or. 680, fol. 85v. In MS London, British Library, Add. 7474, fol. 150r the singular *jadwal* is found.

54 *Almagest* I.10, final paragraph; see MS London, British Library, Add. 7474, fol. 14r, line 3.

55 *Almagest* I.10, final paragraph; see MS Tunis, National Library, 7116, fol. 9v, line 4.

56 Charette and Schmidl, 'Al-Khwārizmī', p. 110, line 6.

57 Sezgin, *Al-Zīj al-Maʾmūnī*, p. 125, line 4.

58 See the glossary in Nallino, *Al-Battānī*, vol. III, p. 337.

59 Morelon, *Thābit ibn Qurra*, p. 55, line 7 and p. 106, line 18.

60 van Vloten, *Liber Mafâtîh al-olûm*, p. 55, line 8.

61 Kunitzsch, *Ibn aṣ-Ṣalāḥ*, p. 131, line 21.

62 Kunitzsch, *Ibn aṣ-Ṣalāḥ*, p. 155, lines 12–20 (Arabic text) and p. 40 (German translation).

> Five copies (*nusakh*) of the Book al-Majisṭī, different in language and translation had come about (*kāna qad ḥasala*), a Syriac copy, translated from the Greek, a second copy in the translation of al-Ḥasan ibn Quraysh for al-Maʾmūn, from Greek into Arabic, a third copy in the translation of al-Ḥajjāj ibn Yūsuf ibn Maṭar and Hilīyā ibn Sarjūn, also for al-Maʾmūn from Greek into Arabic, a fourth copy in the translation of Isḥāq ibn Ḥunayn for Abū al-Ṣaqr ibn Bulbul, from Greek into Arabic, and this [copy] is the original archetype (*dustūr*) of Isḥāq and in his handwriting, and a fifth copy with the correction of Thābit ibn Qurra of this translation of Isḥāq ibn Ḥunayn for Abū al-Ṣaqr ibn Bulbul. It agrees (*muwāfiq*) with Isḥāq's translation except for the pieces of information which were in the margin of the version of Isḥāq, such as doubts (*tashakkuk*) [concerning variant readings]. These pieces of information were not in the copy of Thābit. All these copies were differing and faulty.

According to this statement, Ibn al-Ṣalāḥ had four Arabic translations at his disposal, which he lists in chronological order: A translation by al-Ḥasan ibn Quraysh, the translation by al-Ḥajjāj, the original translation of Isḥāq in an autograph with marginal notes, and the Isḥāq/Thābit translation. The last three translations are well known, and the translations of al-Ḥajjāj and of Isḥāq/Thābit exist in a number of manuscripts.[63] Later on in the text, Ibn al-Ṣalāḥ calls the translation by al-Ḥasan ibn al-Quraysh 'the Maʾmūnic translation by al-Ḥasan' (*al-maʾmūnī bi-naql al-Ḥasan*),[64] or simply 'al-Ḥasan's translation' (*naql al-Ḥasan*),[65] or occasionally also 'the Maʾmūnic [translation]' (*al-maʾmūnī*).[66] There is a passage in Ibn al-Nadīm's *Fihrist* on a translation of the *Almagest* made before al-Ḥajjāj, but al-Ḥasan ibn Quraysh is not mentioned there,[67] nor is he mentioned in Ibn al-Nadīm's list of translators from Greek into Arabic.[68] The only biographical source which makes a reference to him is Ibn Abī ʿUṣaybiʿa in his biography of the physician Sahl al-Kawsaj, where al-Ḥasan ibn Quraysh is listed among the colleagues of Sahl.[69]

Sahl al-Kawsaj died shortly before the Caliph al-Maʾmūn (d. 833 CE). Despite the lack of further evidence of a translation of al-Ḥasan ibn Quraysh, the account of Ibn al-Ṣalāḥ has to be taken seriously. He must have had a manuscript of this translation at hand, from which he quoted as often as from the other translations. Most of the quotations concerned numerical values of star coordinates. Ibn al-Ṣalāḥ did not explicitly evaluate the different translations in general. There are approximately equally many cases in which he judges the numerical values in the Maʾmūnic translation to be correct against some of

[63] Kunitzsch, *Claudius Ptolemäus*, vol. I, pp. 3–4; Kunitzsch, 'A Hitherto Unknown', pp. 31–32.

[64] Kunitzsch, *Ibn aṣ-Ṣalāḥ*, p. 149, line 12 (Arabic text) and p. 49 (German translation).

[65] Kunitzsch, *Ibn aṣ-Ṣalāḥ*, p. 139, lines 9–10 (Arabic text) and p. 63 (German translation).

[66] Kunitzsch, *Ibn aṣ-Ṣalāḥ*, p. 149, line 15 (Arabic text) and p. 49 (German translation).

[67] Dodge, *The Fihrist*, vol. II, p. 639.

[68] Dodge, *The Fihrist*, vol. II, pp. 586–88.

[69] Müller, *ʿUyūn ul-anbāʾ*, vol. I, p. 160, line 23; cf. Kunitzsch, *Der Almagest*, p. 23, note 33.

the other translations, as cases in which he judges them to be wrong. Often the Ma'mūnic translation agrees with the Syriac translation against those of al-Ḥajjāj and Isḥāq (or Isḥāq/Thābit).

Besides the critique of numerical values, there are also a few remarks on different translations of star names. In one case Ibn al-Ṣalāḥ criticized al-Ḥasan, since he translated Greek βέλος ('arrow') with *nawl* ('loom').[70] In another case concerning the translation of Greek ὁ θύρσος ('the wand of Thyrsos') he wrote:[71]

> This star (= b Cen), and the eighth, ninth and tenth [star] (= ψac[1] Cen) stand according to the translation of Isḥāq on the 'branches of vine' (*ʿalā quḍbān al-karm*), but according to the Syriac on the 'shield' (*ʿalā l-turs*), which is called in Syriac *sakrā*, and according to the version of al-Ḥasan ibn Quraysh on the 'lance' (*ʿalā l-ḥarba*). Similarily I saw them in the form of a lance (*ṣūrat ḥarba*) on a celestial globe made by the Ḥarranians. The lance appears to me as the most likely [translation], since Centaur is holding a wild beast of prey at its forefoot, and it is mentioned in the commentary to Aratos that Centaur wanted to sacrifice the animal to the God, and to fumigate it with the nearby incense burner.

In the manuscript of Isḥāq's translation the star is called 'branch of vine' in the singular (*qaḍīb al-karm*), and never in the plural.[72] The Syriac translater read ὁ θυρεός ('oblong shield') instead of ὁ θύρσος.[73] The translation of al-Ḥajjāj is not quoted, but it agrees with the Syriac translation by rendering the star name as *al-turs* ('the shield'). Thus we see that in this case Ibn al-Ṣalāḥ prefers the Ma'mūnic translation against all others. Considering this judgment, it would seem perfectly reasonable if he would quote the *Almagest* in the Ma'mūnic translation at other occasions as well.

6. Authorship of the translation quoted by Ibn al-Ṣalāḥ in his critique on al-Fārābī

It seems reasonable to assume that the anonymous translation of the *Almagest* quoted by Ibn al-Ṣalāḥ in his critique on al-Fārābī's commentary was one of the four Arabic translations which he used in his work on the star catalogue. It has been shown that the Anonymous differs considerably from al-Ḥajjāj and Isḥāq/Thābit. In view of the fact that the Anonymous has even less in common with Isḥāq/Thābit than with al-Ḥajjāj, the Anonymous could hardly be identical with the original translation of Isḥāq. There are only few cases where

[70] Kunitzsch, *Ibn aṣ-Ṣalāḥ*, p. 145, lines 1–2 (Arabic text) and p. 54 (German translation); cf. Kunitzsch, *Der Almagest*, pp. 184–85.

[71] Kunitzsch, *Ibn aṣ-Ṣalāḥ*, p. 134, line 20 – p. 133, line 1 (Arabic text) and pp. 70–71 (German translation).

[72] Kunitzsch, *Der Almagest*, p. 339.

[73] Kunitzsch, *Der Almagest*, p. 339, note 191.

Ibn al-Ṣalāḥ reported differences between Isḥāq/Thābit and the original Isḥāq translation in numerical values, and none in verbal expressions. Therefore, the Maʾmūnic translation remains as the only candidate among the translations used by Ibn al-Ṣalāḥ in his work on the star coordinates.

Two more possibilities have to be taken into consideration. Ibn al-Ṣalāḥ had some knowledge of Greek, and he might have translated the passage of V.19 himself. But it has already been demonstrated that the terminology used by Ibn al-Ṣalāḥ in his own works does not correspond to the Anonymous.[74]

Besides the translations mentioned by Ibn al-Ṣalāḥ, there was another translation made by Thābit ibn Qurra after having finished his corrections for the Isḥāq translation.[75] Even though unlikely, it cannot be excluded that Thābit's own translation became available to Ibn al-Ṣalāḥ only after he had finished his work on the star coordinates, and then he used it in his critique on al-Fārābī. However, there are examples which show that the terminology in the Anonymous does not correspond to Thābit's terminology in his own works.[76]

At this point, the only option remains to identify the Anonymous with the old Maʾmūnic translation. This is compatible with the observations concerning its terminology, which point rather to an early epoch, when technical terms in astronomy were not yet as standardized as they became later. Moreover, there is nothing in the text which precludes from assuming an early date in the first third of the ninth century CE.

In the former section on the terminology of the Anonymous it was observed that some of its peculiarities are found also in the *Zīj* of al-Battānī. This can be explained now, since Paul Kunitzsch found that al-Battānī's star catalogue was mainly based on the Maʾmūnic translation.[77] Therefore it is likely that al-Battānī adopted some of the terminology of the Maʾmūnic translation too.

A final problem remains to be discussed. The statement of the authorship of the Maʾmūnic translation does not correspond to the passage on the early translation of the *Almagest* in Ibn al-Nadīm's *Fihrist*:[78]

> The first person to become interested in translating it and issuing it in Arabic was Yaḥyā ibn Khālid ibn Barmak. A group of people explained it for him but, as they did not understand it perfectly, he was not satisfied with it, so he called upon Abū Ḥassān and Salm, the director of the Bayt al-Ḥikmah, for its explanation. They made sure [of its meaning] and persevered in making it accurate, after having sum-

[74] See Section 4.

[75] See Lorch, *Thābit ibn Qurra*, pp. 355–57; Grupe, 'The Thābit-Version', and Grupe's article in this volume.

[76] See Section 4.

[77] Kunitzsch, *Ibn aṣ-Ṣalāḥ*, pp. 97–108.

[78] Dodge, *The Fihrist*, vol. II, p. 639.

> moned the best translators, testing their translation, and making sure of its good literary style and accuracy.

The name of al-Ḥasan ibn Quraysh, to whom Ibn al-Ṣalāḥ attributed the Maʾmūnic translation, is not mentioned here. However, this is no contradiction, since the text, taken at face value, does not mention the names of the translators, but only those of the supervisors, who did not translate themselves. The date of the translation indicated by Ibn al-Nadīm differs from the one indicated by Ibn al-Ṣalāḥ, who wrote that the translation was made for al-Maʾmūn (d. 833 CE). According to Ibn al-Nadīm the initiator was Hārūn al-Rashīd's famous Vizier Yaḥyā ibn Khālid ibn Barmak (733 or 737–805 CE). He was responsible for translations of literary and scientific texts into Arabic, but his main focus was on works in Sanskrit.[79] This orientation towards Indian works was a consequence of his Buddhist family background from Balkh. Greek works were translated too, but not from Greek, but from Middle Persian or Syriac, and this would also hold for a translation of the *Almagest*.[80]

There is a sharp contrast between the reports of Ibn al-Nadīm and of Ibn al-Ṣalāḥ on the earliest Arabic translation of the *Almagest*. Paul Kunitzsch characterized this in the following way:[81]

> This witness [of Ibn al-Ṣalāḥ] is of the utmost importance because of its authenticity, and it merits to be placed on the same level as the direct transmission. With its brief objectivity and unambiguity it distinguishes itself impressively from the vague or verbose bibliographical notes of the other authors, which in general do nothing else then to quote second-hand information without verifying it, and to carry it on from book to book.

Even if this may be somewhat exaggerated, Ibn al-Ṣalāḥ has been proven to be a meticulous and scrutinizing scholar who based his judgment on first-hand investigation. Therefore it is the preferable option to accept his attribution of the Maʾmūnic translation to al-Ḥasan ibn Quraysh, and to consider Ibn al-Nadīm's narrative with great caution. The attribution of authorship to al-Ḥasan ibn Quraysh can claim to be based on the most trustworthy source, and to be at present without alternatives.

[79] van Bladel, 'Barmakids', p. 35.

[80] van Bladel, 'The Bactrian Background', p. 85.

[81] Kunitzsch, *Der Almagest*, p. 23: 'Dieses Zeugnis ist wegen seiner Authentizität von allergrößter Bedeutung und verdient es, mit der direkten Überlieferung auf eine Stufe gestellt zu werden. Es hebt sich in seiner knappen Sachlichkeit und Eindeutigkeit eindrucksvoll von den vagen oder weitschweifigen bibliographischen Notizen der übrigen Autoren ab, die im allgemeinen nichts anderes tun, als Angaben zweiter Hand ohne eigene Nachprüfung zu zitieren und von Buch zu Buch weiterzuschleppen'.

7. Conclusions

The translation of the *Almagest* quoted by Ibn al-Ṣalāḥ in his critique of al-Fārābī's commentary contains knowledge of the Greek text which could not have been derived from the translations of al-Ḥajjāj and Isḥāq/Thābit. It shows more differences from both the translation of al-Ḥajjāj and the translation of Isḥāq/Thābit than the latter two among themselves. Besides that, it has more in common with al-Ḥajjāj than with Isḥāq/Thābit. Its terminology agrees best with some of the earliest preserved Arabic astronomical texts by al-Khwārizmī, and therefore an early chronological classification, possibly at the beginning of the ninth century CE, is probable. From the four Arabic translations of the *Almagest* which Ibn al-Ṣalāḥ used in his work on the star coordinates only the Maʾmūnic translation by al-Ḥasan ibn Quraysh could be the one which he quoted in his critique of al-Fārābī's commentary, an attribution which is supported by our terminological analysis. Only scattered splinters of this translation have hitherto been available. Now a small, but intact window into its text has been opened.

Appendix I: Text and Translations of *Almagest* V.19.2

Greek text:[82]

ἵνα οὖν καὶ τὴν πρὸς τὸν διὰ μέσων τῶν ζῳδίων τότε γινομένην παράλλαξιν διακρίνωμεν κατά τε μῆκος καὶ κατὰ πλάτος, τὰς αὐτὰς πάλιν ἰσημερινὰς ὥρας, ἃς ἀπέχει τοῦ μεσημβρινοῦ ἡ σελήνη, εἰσενεγκόντες εἰς τὸ αὐτὸ μέρος τοῦ τῶν γωνιῶν κανόνος ἐπισκεψόμεθα τὰς παρακειμένας τῷ ἀριθμῷ τῶν ὡρῶν μοίρας, ἐὰν μὲν πρὸ τοῦ μεσημβρινοῦ ᾖ ἡ σελήνη, τὰς ἐν τῷ γʹ σελιδίῳ, ἐὰν δὲ μετὰ τὸν μεσημβρινόν, τὰς ἐν τῷ δʹ, κἂν μὲν ἐντὸς τῶν ϟ μοιρῶν ὦσιν, αὐτὰς ἀπογραψόμεθα, ἐὰν δ' ὑπὲρ τὰς ϟ, τὰς λειπούσας εἰς τὰς ρπ· τοσούτων γὰρ ἔσται ἡ ἐλάσσων τῶν περὶ τὴν ἐκκειμένην τομὴν γωνιῶν, οἵων ἡ μία ὀρθὴ ϟ. τὰς ἀπογεγραμμένας οὖν μοίρας διπλώσαντες εἰσοίσομεν εἰς τὸ τῶν ἐν κύκλῳ εὐθειῶν κανόνιον αὐτάς τε καὶ τὰς λειπούσας εἰς τὰς ρπ, καὶ ὃν ἂν ἔχῃ λόγον ἡ τὴν τῶν δεδιπλωμένων μοιρῶν περιφέρειαν ὑποτείνουσα εὐθεῖα πρὸς τὴν ὑποτείνουσαν τὴν λείπουσαν εἰς τὸ ἡμικύκλιον, τοῦτον ἕξει τὸν λόγον ἡ κατὰ πλάτος παράλλαξις πρὸς τὴν κατὰ μῆκος, ἐπειδήπερ αἱ τηλικαῦται τῶν κύκλων περιφέρειαι ἀδιαφοροῦσιν εὐθειῶν. πολυπλασιάζοντες οὖν τὸν ἀριθμὸν τῶν παρακειμένων εὐθειῶν ἐπὶ τὴν εὑρισκομένην ὡς ἐπὶ τοῦ διὰ τοῦ κατὰ κορυφὴν σημείου γραφομένου κύκλου παράλλαξιν καὶ τὰ γινόμενα μερίζοντες εἰς τὸν ρκ χωρὶς τὰ ἐκ τοῦ μερισμοῦ συναγόμενα μόρια ἕξομεν τῆς οἰκείας παραλλάξεως.

[82] Heiberg, *Syntaxis mathematica*, vol. I, pp. 446–47.

English translation:[83]

Now, in order to determine the parallax with respect to the ecliptic, in both longitude and latitude, at the given time, we again enter, with the same distance of the moon from the meridian in equinoctial hours [as before], into the same part of the Table of Angles [II.13], and take the number of degrees corresponding to that hour, in the third column if the moon is to the east of the meridian, or in the fourth column if it is to the west of the meridian. We examine the result, and if it is less than 90° we write down the number itself; but if it is greater than 90°, we write down its supplement, since that will be the size in degrees of the lesser of the two angles at the intersection [of ecliptic and altitude circle] in question. We double the number written down, and enter with this [doubled] number, and also with its supplement, into the Table of Chords [I.11]. The ratio of the chord of the doubled number to the chord of the supplement will give the ratio of the latitudinal parallax to the longitudinal parallax (for circular arcs of such small size are not noticeably different from straight lines). So we multiply the amounts of the chords in question by the parallax determined with respect to the altitude circle, and divide the products, each separately, by 120. The results of the division give us the separate components of the parallax.

Old Maʾmūnic translation:[84]

فإذا أردنا أن نعرف اختلاف المنظر ‹الّذي ينحرف به يعني القمر عن النقطة الّتي هو فيها من فلك البروج› في الطول والعرض ونفصل كلّ واحد منهما عن الآخر فإنّا نأخذ أيضا الساعات المعتدلة الّتي بين القمر وبين وسط السماء فندخلها إلى جداول الزوايا إلى الموضع الّذي كنّا أدخلناه به فيما تقدّم ونأخذ ما بحياله في السطر الثالث إن كانت الساعات في قبل توسّط القمر السماء وإن كانت الساعات بعد توسط القمر السماء مما في السطر الرابع فما وجدنا أي السطرين أخذناه وكتبناه إن كان ما فيه أقلّ من تسعين جزءا وإن لم يكن أقلّ من تسعين جزءا كتبنا ما نقص عن مائة وثمنين جزءا فإنّا إذا فعلنا ذلك كنّا قد أخذنا الزاوية الصغرى من الزاويتين اللتين تليان قطعة القوس الّتي بين سمت الر[ؤ]وس وموضع القمر بالمقدار الّذي تكون به الزاوية القائمة تسعين جزءا ثمّ نضـ[ـعـ]ـف الأجزاء الّتي كتبنا و نأخذ وتر ما يجتمع من جداول القسّي والأ[و]تار ووتر ما نقصت هذه الأجزاء عن مائة وثمانين جزءا فيكون لنا نسبة وتر الأجزاء الّتي أضعفت إلى وتر ما نقصته تلك الأجزاء المضعفة على مائة وثمانين جزءا كنسبة اختلاف المنظر الّذي في العرض إلى اختلاف المنظر الّذي في الطول فلمّا كان لا فرق بين استعمال القسي وبين استعمال أوتارها عند هذه الحال لأنّ القسي ههنا صغار جدّا كنّا متّي أخذنا كلّ واحد من هذين الوترين فضربناه في اختلاف منظر القمر الّذي قد حصلناه من الدائرة الّتي تمرّ بسمت الر[ؤ]وس فما اجتمع من كلّ واحد منهما قسمناه على مائة وعشرين جزءا علمنا بما يخرج من القسمة كم اختلاف المنظر في الطول والعرض

[83] Toomer, *Ptolemy's Almagest*, p. 266.

[84] MS Mashhad, Holy Shrine Library, 4493, pp. 81–82.

English translation:

If we want to know the parallax⟨, with which it deviates, that is to say the moon from the point on which it is on the ecliptic,⟩ in longitude and latitude, and to split apart each of the two from the other, we take again [as before] the equinoctial hours which are between the moon and midheaven. We enter with them the tables of angles at the [same] place at which we had entered with them in what was already mentioned before. If the hours are before [the time when] the moon is in midheaven, we take in the third column [the value] that is opposite. If the hours are after [the time when] the moon is in midheaven, we take in the fourth column [the value] that is opposite. We take what we find in either of the two columns and write it down, if it is less then ninety degrees. If it is not less than ninety degrees, we write down [the amount by] which it is less than hundred eighty degrees. When we have done this, we have taken the smaller of the two angles which are adjacent to the division by the arc between the zenith and the position of the moon, using a measure in which a right angle has ninety degrees. Next we double the degrees which we have written down, and we take the chord which is collected in the tables of arcs and chords, and [we take] the chord of [the amount by] which it is less than hundred eighty degrees. Thus we will have the ratio of the chord of the degrees which were doubled to the chord of the complement of these doubled numbers from hundred and eighty degrees. [This ratio] is like the ratio of the parallax in latitude to the parallax in longitude, since there is no difference in the use of angles and the use of their chords in this situation, because the arcs here are very small. When we take each of these two chords, multiply them with the parallax of the moon, which we have already obtained on the circle through the zenith, and divide each of the two results by hundred and twenty parts, then we know from the results of the division how much the parallax is in longitude and in latitude.

Translation of al-Ḥajjāj:[85]

ولكي نعدل اختلاف المنظر الّذي يكون في ذلك الوقت في الطول والعرض نأخذ تلك الساعات المعتدل أيضا[86] الّتي هي بِعد القمر من فلك نصف النهار فندخلها في ذلك الموضع من جداول[87] الزوايا ونطلب الِأجزاء الّتي تقابل عدد الساعات[88] فإن[89] كان موضع القمر قبل نصف النهار أخذنا الأجزاء الّتي في الجدول الثالث وإن كان موضعه بعد نصف النهار أخذنا الأجزاء الّتي

[85] Text of MS Leiden, Universiteitsbibliotheek, Or. 680, fol. 85v; variant readings of MS London, British Library, Add. 7474, fol. 150r–v in the footnotes.

[86] فنأخذ أيضا الساعات.

[87] جدول.

[88] الساعات الّتي هي بعد القمر من فلك نصف النهار.

[89] خوضع القمر قبل.

في الجدول الرابع فإن كانت الأجزاء أقلّ من التسعين[90] أثبتناها وإن كانت أكثر من التسعين[91] أثبتنا الأجزاء الّتي تنقص عن تمام مائة وثمانين[92] جزءا وذلك هو قدر الزاوية الصغرى من الزوايا[93] الّتي على هذه القطعة بالمقدار الّذي به تكون الزاوية القائمة تسعين[94] جزءا فنأخذ الأجزاء الّتي أثبتنا فنضعّفها وندخل ما اجتمع في جدول أوتار القسي ندخل تلك الأجزاء بعينها وما[95] نقص من[96] تمام مائة وثمانين جزءا[97] فيكون نسبة وتر القوس الّتي هي ضعف هذه الأجزاء إلى الوتر الّذي توتّر[98] القوس الناقصة عن تمام نصف الدائرة كنسبة اختلاف منظر القمر في العرض إلى اختلافه في الطول لأنّ أقدار مثل هذه القسي من الأفلاك <لا> تكون أوتارها مختلفة فيضرب[99] عدد هذه الأوتار في عدد أوتار قسي اختلاف المنظر الموجود كمثل اختلاف المنظر الّذي يكون في الفلك المخطوط على نقطة سمت الرؤوس ثمّ نقسم ما اجتمع على مائة وعشرين فما خرج من القسمة من الأجزاء فهو اختلاف ذلك المنظر

Translation of Isḥāq/Thābit:[100]

وكيما[101] نقوّم[102] أيضا اختلاف[103] النظر الّذي يكون[104] عند ذلك بالقياس إلى فلك البروج في الطول وفي العرض فإنّا ندخل أيضا تلك الساعات الاستوائيّة بأعيانها الّتي هي بعد القمر من دائرة نصف النهار في ذلك القسم بعينه من جدول الزوايا ثمّ ننظر ما حيال[105] ذلك العدد من الساعات من أجزاء[106] أمّا إن كان القمر قبل دائرة نصف النهار فما كان من الاجزاء حياله[107] في الصفّ الثالث وأمّا إن كان بعد دائرة نصف النهار فما كان من الأجزاء حياله[108] في الصفّ الرابع فإن كانت الأجزاء تسعين وما[109] دون ذلك أثبتناها[110] وإن كانت مجاوزة للتسعين أثبتنا ما يبقى بعدها إلى مائة وثمانين فإن هذا مبلغ أصغر الزاويتين اللتين في هذا التقاطع بالأجزاء الّتي بها زاوية واحدة قائمة تسعون جزءا ثمّ نضعّف ما أثبتناه من الأجزاء بعينه وما أثبتناه مما[111] هو ما يبقى بعد التسعين إلى مائة وثمانين وندخله في جدول الأوتار الّتي في الدائرة فأيّ نسبة كانت للخط المستقيم الّذي يوتّر قوس الأجزاء المضعّفة إلى الخط الّذي يوتّر القوس الباقية إلى نصف الدائرة فبقي نسبة اختلاف المنظر في العرض إلى اختلاف المنظر في الطول من قبل أن ما كان هذا مقداره من قسي الدولية[112] فليس بينه وبين الخطوط المستقيمة فرقان[113] فيضاعف عدد الخطوط المستقيمة الّتي بإزائها باختلاف المنظر الّذي يوجد في الدائرة العظمى الّتي ترسم مادّة بالنقطة الّتي على سمت الرأس[114] ويقسم ما اجتمع على مائة وعشرين كلّ واحد من العددين على حياله فما حصل من[115] الأجزاء عند القسمة فهو اختلاف المنظر على حسب العدد المقسوم

Appendix II: Arabic texts translated in the main text

Ibn al-Ṣalāḥ's critique:[116]

كنت نظرت كتابا للفاضل أبي نصر الفارابي موسوما بشرح كتاب بطلميوس المعروف بالمجسطي فتصفّحته مستوفيا حقّ الإصفاء والتفهّم بمعانيه بحيث انتهيت إلى انباء الفصل السابع عشر في المقالة الخامسة وجدته يروم إقامة البرهان على النسب التي هناك مع شرح للفصل مستوفى الّا أنّ تلك المقدّمات الّتي يستعملها في تركيب برهانه ممتنعة مغالطيّة

[90] تسعين [91] تسعين [92] المائة والثمانين [93] الزاوية [94] ص [95] وندخل ما [96] عن
[97] ø [98] فتكون [99] فنضرب

[100] Text of MS Tunis, National Library, 7116, fol. 88r; variant readings of MS Philadelphia, Penn Libraries, LJS 268, fol. 65r in the footnotes.

[101] ولكيما [102] يُقَوِمُ [103] اختلافات [104] الّتي تكون [105] بحيال [106] الأجزاء
[107] بحياله [108] بحيَاله [109] بما [110] أثبتناها بعين [111] فما [112] Read الدورية?
[113] فرق. [114] وبمركز القمر [115] فما حصل من

[116] MS Mashhad, Holy Shrine Library, 5593, p. 81.

Al-Fārābī's commentary on *Almagest* V.19:[117]

وقد بقي الآن أن نستخرج في هذا الانحراف المعلوم انحراف المنظر في العرض وانحراف المنظر في الطول والسبيل إلى ذلك أن أخذ تلك الساعات المعتدلة بعينها او الساعات وما اتّفق فيها وهي بعد القمر من دائرة نصف النهار في ذلك الجدول بعينهٔ الّذي كنا أدخلنا تلك الساعات فيه

English translation:

Now it remains to extract for that known parallax (*inḥirāf al-manẓar*) the latitudinal parallax (*inḥirāf al-manẓar fī l-ʿarḍ*) and the longitudinal parallax (*inḥirāf al-manẓar fī l-ṭūl*). The means for that are that we take those equinoctial hours themselves, or the [seasonal] hours and what [difference] occurs in them. These are the distance of the moon from the circle of the meridian in that same table in which we had entered those hours [already before].

Al-Fārābī's commentary on *Almagest* V.19:[118]

وقد صحّحنا موضع القمر كلّ جهاته في تلك الساعة وبانحراف منظره

English translation:
We confirmed the position of the moon in all directions in that hour and with its parallax (*wa-bi-nḥirāfi manẓarihi*).

Acknowledgement

I thank Sonja Brentjes for her valuable suggestions and Natalia Bachour for her help in reading a difficult passage in the Mashhad MS.

Bibliography

Aminrazavi, Mehdi, *The Wine of Wisdom. The Life, Poetry and Philosophy of Omar Khayyam*, Oxford: Oneworld, 2007.

al-Bīrūnī, Abū Rayḥān Muḥammad ibn Muḥammad, *al-Qānunu'l-Masʿūdī (Canon Masudicus). An Encyclopaedia of Astronomical Sciences*, 3 vols, Hyderabad: Osmania Oriental Publications Bureau, 1954–1956.

Charette, François, and Petra G. Schmidl, 'Al-Khwārizmī and Practical Astronomy in Ninth-Century Baghdad. The Earliest Extant Corpus of Texts in Arabic on the Astrolabe and Other Portable Instruments', *SCIAMVS* 5 (2004), pp. 101–98.

Daḥrūj, ʿAlī, *Muḥammad Aʿlā ibn ʿAlī al-Tahānawī. Kashshāf iṣṭilaḥāt al-funūn*, Bayrūt: Maktabat Lubnān Nāshirūn, 1996.

Dodge, Bayard, *The Fihrist of al-Nadīm. A Tenth-Century Survey of Muslim Culture*, 2 vols, New York/London: Columbia University Press, 1970.

[117] MS Mashhad, Holy Shrine Library, 5593, p. 82.

[118] MS Tehran, Majlis Library, 6430, fol. 46v.

Goldstein, Bernard R., book review of Fuat Sezgin, *Geschichte des arabischen Schrifttums*, vol. VI: *Astronomie bis ca. 430 H.*, *Isis* 71 (1980), pp. 341–42.

Grupe, Dirk, 'The "Thābit-Version" of Ptolemy's *Almagest* in MS Dresden Db.87', *Suhayl* 11 (2012), pp. 147–53.

Haskins, Charles Homer, *The Renaissance of the Twelfth Century*, Cambridge (Mass.): Harvard University Press, 1927.

Heiberg, Johan Ludvig, *Syntaxis mathematica*, 2 vols, Leipzig: Teubner, 1898–1903 [*Claudii Ptolemaei opera quae exstant omnia*, vol. I].

Janos, Damien, 'Al-Fārābī on the Method of Astronomy', *Early Science and Medicine* 15 (2010), pp. 237–65.

Janos, Damien, *Method, Structure, and Development in al-Fārābī's Cosmology*, Leiden; Brill, 2012.

Kennedy, Edward S., 'Parallax Theory in Islamic Astronomy', *Isis* 47 (1956), pp. 33–53.

Kennedy, Hugh, *An Historical Atlas of Islam*, Leiden: Brill, 2002 [second, revised edition of the atlas by William C. Brice of 1981].

Kunitzsch, Paul, *Der Almagest. Die Syntaxis mathematica des Claudius Ptolemäus in arabisch-lateinischer Überlieferung*, Wiesbaden: Harrassowitz, 1974.

Kunitzsch, Paul, *Ibn aṣ-Ṣalāḥ. Zur Kritik der Koordinatenüberlieferung im Sternkatalog des Almagest. Arabischer Text nebst deutscher Übersetzung, Einleitung und Anhang*, Göttingen: Vandenhoeck & Ruprecht, 1975.

Kunitzsch, Paul, *Claudius Ptolemäus. Der Sternkatalog des Almagest. Die arabisch-mittelalterliche Tradition*, 3 vols, Wiesbaden: Harrassowitz, 1986–1991.

Kunitzsch, Paul, 'A Hitherto Unknown Arabic Manuscript of the Almagest', *Zeitschrift für Geschichte der arabisch-islamischen Wissenschaften* 14 (2001), pp. 31–37.

Lane, Edward William, *An Arabic-English Lexicon*, London: Williams and Norgate, 1863–1893.

Lippert, Julius, *Ibn al-Qifṭī's Ta'rīḫ al-ḥukamā*, Leipzig: Dieterich, 1903.

Lorch, Richard, 'Ibn al-Ṣalāḥ's Treatise on Projection: a Preliminary Survey', in Menso Folkerts and Richard Lorch (eds), *Sic itur ad astra. Studien zur Geschichte der Mathematik und Naturwissenschaften. Festschrift für den Arabisten Paul Kunitzsch zum 70. Geburtstag*, Wiesbaden: Harrassowitz, 2000, pp. 401–08.

Lorch, Richard, *Thābit ibn Qurra. On the Sector-Figure and Related Texts*, Frankfurt am Main: Institute for the History of Arabic-Islamic Science, 2001 [2nd ed. Augsburg: Rauner, 2008].

Ma'ānī, Aḥmad Gulčīn, *Fihrist-i kutub-i ḫaṭṭī-i kitābḫāna-i markazī-i Āstān-i Quds-i Raḍawī*, vol. VIII, Mashhad: Āstan-i Quds-i Raḍawī, 1971.

MAOSIC = Boris A. Rosenfeld and Ekmeleddin İhsanoğlu, *Mathematicians, Astronomers, and Other Scholars of Islamic Civilization and Their Works (7th-19th c.)*, Istanbul: Research Centre for Islamic History, Art and Culture, 2003.

Morelon, Régis, *Thābit ibn Qurra. Œuvres d'astronomie*, Paris: Les Belles Lettres, 1987.

Müller, August, *Ibn Abī Usaybiʿa. ʿUyūn al-anbāʾ fī ṭabaqāt al-aṭibbāʾ*, 2 vols, Cairo: al-Maṭbaʿa al-Wahbiyya, 1882.

Nallino, Carlo Alfonso, *Al-Battānī sive Albatenii opus astronomicum*, 3 vols, Milan: Hoepli, 1899–1907 [reprinted Hildesheim: Olms, 1977].

Neugebauer, Otto, *A History of Ancient Mathematical Astronomy*, 3 vols, Berlin: Springer, 1975.

Pedersen, Olaf, *A Survey of the Almagest*, Odense: Odense University Press, 1974 [revised edition, with annotation and new commentary by Alexander Jones, New York: Springer, 2010].

Ritter, Hellmut et al., *Khalīl ibn Aybak al-Ṣafadī. Kitāb al-Wāfī bi-l-wafāyāt*, Leipzig: Brockhaus, 1931–2013 [Bibliotheca Islamica 6].

Sarton, George, *Introduction to the History of Science*, 3 vols, Baltimore: Williams & Wilkins, 1927–1948.

Sezgin, Fuat, *Geschichte des arabischen Schrifttums*, vol. VI: *Astronomie bis ca. 430 H.*, Leiden: Brill, 1978.

Sezgin, Fuat, *Yaḥyā ibn Abī Manṣūr. Al-Zīj al-Maʾmūnī al-mumtaḥan. The Verified Astronomical Tables for the Caliph al-Maʾmūn. Reproduced from MS árabe 927, Escorial Library*, Frankfurt am Main: Institute for the History of Arabic-Islamic Science, 1986.

Steinschneider, Moritz, *Al-Farabi (Alpharabius). Des arabischen Philosophen Leben und Schriften, mit besonderer Rücksicht auf die Geschichte der griechischen Wissenschaft unter den Arabern*, St.-Pétersburg: Académie Impériale des Sciences, 1869 [Mémoires de l'Académie Impériale des Sciences de St.-Pétersbourg. Série 7; 13, 4].

Thomann, Johannes, 'Ein al-Fārābī zugeschriebener Kommentar zum Almagest (Hs. Tehran Mağlis 6531)', *Zeitschrift für Geschichte der arabisch-islamischen Wissenschaften* 19 (2010–11), pp. 37–76.

Thomann, Johannes, 'From Lyrics by al-Fazārī to Lectures by al-Fārābī: Teaching Astronomy in Baghdād (750–1000 C.E.)', in Jens Scheiner and Damien Janos (eds), *The Place to Go. Contexts of Learning in Baghdād, 750–1000 C.E.*, Princeton: The Darwin Press, 2014, pp. 503–25.

Thomann, Johannnes, 'Al-Fārābī's Kommentar zum Almagest in sekundärer Überlieferung bei Ibn aṣ-Ṣalāḥ. Ein vorläufiger Bericht', *Asiatische Studien* 69 (2015), pp. 99–113.

Thomann, Johannes, 'Terminological Fingerprints in a Commentary on the Almagest Attributed to al-Fārābī', in Maurus Reinkowski and Monika Winet (eds), *Arabic and Islamic Studies in Europe and Beyond*, Leuven: Peeters, 2016, pp. 301–12.

Toomer, Gerald. J., *Ptolemy's Almagest*, London: Duckworth, 1984 [2nd edition Princeton: Princeton University Press, 1998].

van Bladel, Kevin, 'Barmakids', in Kate Fleet et al., *The Encyclopaedia of Islam. Three*, Leiden: Brill, 2007–, vol. III, pp. 32–38.

van Bladel, Kevin, 'The Bactrian Background of the Barmakids', in Anna Akasoy, Charles Burnett and Ronit Yoeli-Tlalim (eds), *Islam and Tibet. Interactions along the Musk Routes*, Farnham: Ashgate, 2011, pp. 43–88.

van Vloten, Gerlof, *Liber Mafâtîh al-olûm explicans vocabula technica scientiarum tam arabum quam peregrinorum auctore Abû Abdallah Mohammad ibn Ahmed ibn Yûsof al-kâtib al-Khowarezmi*, Leiden: Brill, 1895.

Wright, R. Ramsay, *The Book of Instruction in the Elements of the Art of Astrology by Abu'l-Rayḥān Muḥammad ibn Aḥmad al-Bīrūnī*, London: Luzac, 1934.

Thābit ibn Qurra's Version of the *Almagest* and Its Reception in Arabic Astronomical Commentaries

(based on the presentation held at the Warburg Institute, London, 5th November 2015)

Dirk Grupe*

1. Introduction

Since Paul Kunitzsch's ground-breaking study of the Arabic and Latin transmission of Claudius Ptolemy's *Syntaxis mathematica*, better known by its Arabic title, *Almagest*, it has been the common belief of historians of Islamic science that a total of four different Arabic versions of the *Almagest* with notable influence were produced during the ninth century as part of the Arabic acquisition of Hellenistic astronomy.[1] The key witness for this assumption has been a report from the twelfth century by the physician Ibn al-Ṣalāḥ (d. 1154 CE, Damascus). Although other, partly deviating accounts on Arabic versions of the *Almagest* were known, Kunitzsch gave the most credibility to Ibn al-Ṣalāḥ. One reason was Ibn al-Ṣalāḥ's diligent expertise and the relative proximity of

* The essentials of this article were presented in 2015, in my conference paper with the same title, at *Ptolemy's Science of the Stars in the Middle Ages*, The Warburg Institute (London), 5–7 November 2015, and on various occasions thereafter. I wish to express my thanks to Dr Mª José Parra. Her well-organised collection of Arabic manuscript reproductions that she assembled during a postdoc period at the project *Ptolemaeus Arabus et Latinus* and her expertise with Arabic and Persian astronomical works were important prerequisites for the findings presented in that paper and in the present article. Since 2013, Dr Parra has been keeping her eyes open for a Thābitian influence in the manuscripts that passed through her hands, and several of the texts included here would be missing without her help. The database of Ptolemaic manuscripts and their classification, which is hosted by the Bavarian Academy of Sciences and Humanities (Munich), provides a useful tool for studies on the dissemination and the reception of texts related to Ptolemy's work.

A first version of this article was submitted in spring 2016, with the provisional title 'Further witnesses of Thābit ibn Qurra's version of the *Almagest*'. A late editing stage of this volume (December 2018) allows me to briefly add that some of my conjectures from 2015 require adjustment, since MS Jaipur, Maharaja Sawai Man Singh II Museum Library, 20, discussed under 2.4, has become accessible to me. I am grateful to the owners and the custodians of the Jaipur manuscript for enabling Dr Parra and me to undertake a thorough study of the text and its transcription during a stay in India between September and November 2018. An article on some of our observations about the Jaipur manuscript is presently forthcoming in a different volume.

[1] Kunitzsch, *Der Almagest*.

Ptolemy's Science of the Stars in the Middle Ages, ed. by David Juste, Benno van Dalen, Dag Nikolaus Hasse and Charles Burnett, PALS 1 (Turnhout, 2020), pp. 139–157

10.1484/M.PALS-EB.5.120177

his lifetime to the events in question. Another reason was that the extant manuscripts related to the *Almagest* in the Arabic language seemed to agree with Ibn al-Ṣalāḥ's account.[2]

According to Ibn al-Ṣalāḥ, the first translation of Ptolemy's *Almagest* into Arabic was made for the caliph al-Maʾmūn in the early ninth century. Except for fragments, apparently from this version, in al-Battānī (ninth c.), the work is lost.[3] It was followed by a second translation made in 827/8 by al-Ḥajjāj ibn Yūsuf ibn Maṭar. Al-Ḥajjāj's version is extant in one complete copy, in MS Leiden, UB, cod. or. 680, and in a large fragment, in MS London, BL, add. 7474.[4] A third Arabic translation of the *Almagest* was produced, and completed between 879 and 890, by Abū Yaʿqūb Isḥāq ibn Ḥunayn. Corrections to that third translation were made by Thābit ibn Qurra (d. 901), who thereby produced the fourth and latest of the Arabic versions listed in Ibn al-Ṣalāḥ's report. No surviving copy of Isḥāq ibn Ḥunayn's translation of the *Almagest* without Thābit ibn Qurra's corrections has been identified, whereas ten manuscripts of Isḥāq's text after correction by Thābit are known, the oldest complete one being MS Tunis, National Library, 07116, from 1085.[5] The rich evidence of the Isḥāq/Thābit version of the *Almagest* corresponds to a general estimate according to which this version became the preferred translation of Ptolemy's work in large parts of the Arabic-speaking world.[6] It was also from a combined use of the two surviving Arabic versions, Ḥajjāj and Isḥāq/Thābit, that the most influential Latin translation of Ptolemy's *Almagest* was produced, by Gerard of Cremona in Toledo (Spain) some time before 1175 and with continued modifications possibly until Gerard's death in 1187.[7]

[2] Kunitzsch, *Der Almagest*, ch. I.A, pp. 15–82, see esp. pp. 22–24, including a transliteration and a German translation of the passage from Ibn al-Ṣalāḥ, and pp. 59–71, with an analysis of preserved witnesses. An English translation of Ibn al-Ṣalāḥ's report can be found in the article by Johannes Thomann in this volume.

[3] Kunitzsch, *Der Almagest*, pp. 60–64. For further citations of the Maʾmūnic translation of the *Almagest* in the work of Ibn al-Ṣalāḥ, see the article by Johannes Thomann in this volume.

[4] On preserved manuscripts of the *Almagest* in Arabic see Kunitzsch, *Der Almagest*, pp. 34–46, successively updated and partly revised in Kunitzsch, *Der Sternkatalog*, vol. I, pp. 3 ff., and Kunitzsch, 'A Hitherto Unknown', pp. 31–37. New information on most of these manuscripts was contained in the article by Dr Mª José Parra, 'Making a Catalogue of Arabic Ptolemaic Manuscripts', which was prepared for publication together with the present paper, as an essential complement. Although Dr Parra's work is currently uncitable, it is referred to herein on various occasions (as Parra, 'Making a Catalogue'), due to its outstanding importance to the present paper.

[5] See Kunitzsch, *Der Almagest*, pp. 38–41, where the discovery of the Tunis manuscript is attributed to Fuat Sezgin; cf. Sezgin, *GAS* VI, pp. 88–89.

[6] See e.g. Kunitzsch, *Der Almagest*, pp. 35 and 71.

[7] Kunitzsch, *Der Almagest*, pp. 83–112. An introductory survey of the Greek-Arabic-Latin transmission of mathematical works and their translators, along with a helpful bibliography, is given in Lorch, 'Greek-Arabic-Latin'.

Different from the above commonly held belief, I was recently able to prove the existence of another, previously unknown Arabic version of Ptolemy's *Almagest* and its translation into Latin.[8] My argument was founded mainly on three sources; (a) a Latin translation from the twelfth century (extant in MS Dresden, SLUB, Db. 87, fols 1r–71r) of the first four books of that Arabic version, (b) the appearance of distinctive passages from the same Arabic text, in the form of quotations from the *Almagest*, in al-Nasawī's *Commentary on the Sector Figure* (eleventh c.), and (c) detailed remarks about this version of the *Almagest* in Naṣīr al-Dīn al-Ṭūsī's famous *Taḥrīr of the Almagest* (thirteenth c.).[9] Al-Ṭūsī refers to the text as 'Thābit's version', *scil.* of the *Almagest.* This is in agreement with remarks by other historic witnesses, from the tenth to the fifteenth centuries, who accredit the ninth-century translator and mathematician Thābit ibn Qurra with the production of a revised version of the *Almagest,* aside from his corrections to Isḥāq ibn Ḥunayn's translation.[10] Although the references reveal a reception of Thābit's *Almagest* across several centuries, the work seems to have remained less known than other Arabic versions of Ptolemy's work or not to have reached the same authoritative status.[11] In particular, evidence of a circulation of Thābit's text was found only in the eastern parts of the Islamic world.[12]

Since my first publications on the subject I have been able to identify further texts related to Thābit's *Almagest.* They reveal a greater popularity of Thābit's

[8] Grupe, 'The Thābit-Version'. The newly found version of the *Almagest,* made by Thābit ibn Qurra, is also the subject of Grupe, *The Latin Reception,* pp. 90–134 and Appendix B. Concerning the previous scepticism about an Arabic version of the *Almagest* by Thābit ibn Qurra alone see the analysis in Kunitzsch, *Der Almagest,* pp. 25–34. See also Carmody, *Astronomical Works,* p. 19; Morelon, *Thābit ibn Qurra,* and, as a recent example, Pedersen, *A Survey,* pp. 14–16, who does not consider a text of this kind among Thābit's astronomical works.

[9] Correspondences between the Latin Dresden *Almagest* and al-Nasawī's *Commentary on the Sector Figure* were known before from Lorch, *Thābit ibn Qurra,* pp. 355–75. Transcriptions of al-Ṭūsī's remarks are given in Grupe, 'The Thābit-Version', p. 151.

[10] For a collection of these remarks see Kunitzsch: *Der Almagest,* pp. 17–34, supplemented in Kunitzsch, *Ibn aṣ-Ṣalāḥ,* pp. 115–23 passim. References to a version of the *Almagest* by Thābit ibn Qurra were known, especially, from the writings of Abū ʿAlī al-Muḥassin al-Ṣābiʾ (tenth c.), Ibn al-Nadīm (2nd half tenth c.; relating a version by Thābit based on the 'old' translation) and Qāḍīzāde al-Rūmī (late fourteenth to early fifteenth c.).

[11] Thābit's text seems to have been unknown to Ibn al-Ṣalāḥ or not considered by him to be a 'translation' of Ptolemy's work. I am grateful to Paul Kunitzsch for fruitful discussions about this aspect. Similarly, an explanatory note in several manuscripts of al-Ṭūsī's *Taḥrīr* indicates that later recipients of the *Taḥrīr* felt the need to explain al-Ṭūsī's repeated references to a 'Thābit's version' of the *Almagest*; a transcription of the note is given in Kunitzsch, *Der Almagest,* p. 26. The tradition of Thābit's *Almagest* also appears less homogeneous than others, as some of Thābit's insertions to the original content of the *Almagest* became intentionally removed again in some transmission branches; cf. Grupe, 'The Thābit-Version', pp. 150 f.

[12] cf. Grupe, 'The Thābit-Version', p. 152.

text than I had previously thought. At the same time, they confirm a reception mostly in the Persian area. In the present paper a brief account of these texts and their significance as witnesses of Thābit ibn Qurra's *Almagest* is given.[13]

2. Further texts and manuscripts related to Thābit ibn Qurra's version of the *Almagest*

2.1. MS (formerly) Tehran, private collection Naṣīrī, 789 (epitome of the *Almagest*)

This manuscript contains on 127 folios a text that once was described as an autograph by the thirteenth and early-fourteenth-century Persian astronomer Quṭb al-Dīn Maḥmūd b. Masʿūd al-Shīrāzī (1236–1311).[14] For some time the text was wrongly believed to be a copy of the Isḥāq/Thābit version of the *Almagest*.[15] Paul Kunitzsch later observed that the manuscript does not as a whole provide the 'authentic' text of the *Almagest*; only *Alm.* I, 1 and some of the chapter titles appear in the wording of Isḥāq/Thābit. As for the remaining text, Kunitzsch believed this to be an epitome of the *Almagest* which al-Shīrāzī had produced based on Isḥāq/Thābit's translation.[16] Kunitzsch also noticed that *Alm.* I, 1 has been turned into a preface which precedes a general table of contents and which has been excluded from the chapter count. The rearrangement caused the numbering of Ptolemy's subsequent chapters to be reduced by one. Parts of the private collection where the manuscript was kept have been sold in the meantime, leaving the manuscript's present whereabouts unknown.[17] However, from existing paper copies of some of its pages it is evident that al-Shīrāzī did not compose the text as freely as has been commonly believed.[18] Correspondences with the Latin translation of Thābit's *Almagest* in MS Dresden, SLUB, Db. 87 show that for the first chapters of his treatise al-Shīrāzī took substantial parts from Thābit's *Almagest*. At least from *Alm.* I, 10 onwards, however, al-Shīrāzī seems to have continued his epitome mainly based on Isḥāq/Thābit's alternative version of the *Almagest*.[19]

[13] Recently a reference by the Persian scholar al-Harawī in his edition of Menelaus' *Spherics* has also been plausibly related to Thābit's *Almagest* by Sidoli and Kusuba, 'Al-Harawī's Version', p. 164. Traces of Thābit's translation in the Islamic West, in marginal notes in MS Tunis, National Library, 7116, containing the Isḥāq/Thābit version of the *Almagest*, have been found by Mª José Parra, 'Making a Catalogue' (see above, note 4).

[14] Sezgin, *GAS* VI, p. 89. For a brief account on al-Shīrāzī's life and work see Ragep, 'Shīrāzī'.

[15] cf. Sezgin, *GAS* VI, p. 89.

[16] Kunitzsch, *Der Sternkatalog*, vol. III, p. 200, referring back to vol. II, p. 171. See also Kunitzsch, 'A Hitherto Unknown', p. 31.

[17] I owe this information to Mohammad Bagheri.

[18] I am grateful to Paul Kunitzsch for providing me with these copies in 2014, which also enabled me to prove that the Jaipur *Almagest* stems from Thābit's translation; see below 2.4.

[19] In the talk on which this article is based, I argued that al-Shīrāzī gave up on Thābit's *Almagest* before *Alm.* I, 10 and for the remainder of his text used only the Isḥāq/Thābit ver-

The following transcriptions of the chapter titles and the opening sentences of *Alm.* I, 3 and I, 5 according to MS (formerly) Tehran, Naṣīrī, 789 and MS Dresden, SLUB, Db. 87 show the dependency of both texts on the same Arabic tradition. The correspondences become clearer from a simultaneous comparison with the different wording in Isḥāq/Thābit's version of the *Almagest* (transcribed from MS Tunis, National Library, 07116).[20]

Alm. I, 3.

MS (formerly) Tehran, Naṣīrī, 789, fol. 4v:5 f.:

الباب الثاني في ان السماء كرية الشكل وحركتها كرية الدور

ان اول ما دعا القدماء الى ان قالوا بان السماء وحركتها كريين هو ما ظهر لهم من القياسات والارصاد التي راوها

English translation:

Chapter 2: On that the heaven is spherical in shape and its motion is spherical in rotation.

The first [thing] that called the ancients to say that the heaven and its motion are spherical is what was perceptible for them from the analogies and the observations that they saw.

MS Dresden, SLUB, Db. 87, fols 1r:3 and 2r:24 f. (trl. of Thābit's *Almagest*):[21]

Quia celum est sperale et suus motus speralis motus.

Primum igitur qui vocavit antiquos ut dicerent quia motus celi est speralis est hoc quid fuit ipsum visum ab illis in consideracione sua.

sion. This estimate needs to be revised, since a microfilm of al-Shīrāzī's epitome reappeared in Munich in October 2016 in Menso Folkerts' collection (noticed by Mª José Parra). I am thankful to Menso Folkerts, who in 2011 offered me the use of his microfilm collection and in 2016 took it upon himself to bring his microfilms back to Munich. A brief inspection of the microfilm of the Naṣīrī manuscript has revealed an influence of Thābit's *Almagest* also in later passages of al-Shīrāzī's epitome, especially in the discussion of the sector figure. Al-Shīrāzī's creative treatment of the sector figure in his epitome, which includes two sets of numbered diagrams for twelve modes and their respective 'reverses' for each of the planar and the spherical versions of the theorem, has in the meantime been the subject of my paper 'Geometric Reasoning in Arabic Works on the Almagest', presented at *From Pseudo-Bede to Duarte De Sande*: *Arts and Sciences in East and West*, Würzburg, 22 November 2016.

[20] Transliterations of the opening sentences of *Alm.* I, 3 and I, 5, also according to the earlier Arabic version of the *Almagest* by al-Ḥajjāj as well as Gerard of Cremona's Latin translation, are given in Kunitzsch, *Der Almagest*, pp. 134 f.

[21] In the Dresden *Almagest*, a copy from *c.* 1300, the use of Latin pronouns is often incorrect and tense-agreement is lacking. This may reveal a corresponding insecurity on the part of the (presumably oriental) translator of the text, ʿAbd al-Masīḥ Wittoniensis. I have abstained from emending these characteristics of the text.

MS Tunis, NL, 7116, fol. 2v:17–19 (Isḥāq/Thābit's translation):

في ان السماء كرية وحركتها ايضا كرية

ان اول توهم القدماء لما ذكرنا انما كان لانهم راوا الشمس والقمر وسائر النجوم متحركات ابدا من المشارق الى المغارب

(Trl.: On that the heaven is spherical and its motion is also spherical.
The first imagination of the ancients of what we have mentioned was because they saw the sun and the moon and the passing of the stars always moving from the east to the west.)

Alm. I, 5.

MS (formerly) Tehran, Nasīrī, 789, fol. 6v:1–3:

الباب الرابع في ان الارض في وسط السماء

فاذ كنا قد بينا ما قلنا فينبغي ان نعلم انه لا يمكن ان تعرض الاشياء التي نشاهدها ونراها من امر السماء [و]الارض الا والارض في وسط السماء بمنزلة المركز

English translation:

Chapter 4: On that the earth is in the middle of the heaven.

Since we have demonstrated what we said, we need to know that there cannot occur the things that we witness and see concerning the heaven and the earth, if the earth is not in the middle of the heaven, in the status of the centre.

MS Dresden, SLUB, Db. 87, fols 1r:5 and 3v:32–34, (trl. of Thābit's *Almagest*):

Quia terra est in medio celi.

Ostensis igitur quem prediximus sciendum quia non possunt esse accidentalitates quas videmus in celo et in terra si non sit terra in medio celi quasi centrum.

MS Tunis, NL, 07116, fol. 3v:27–4r:1 (Isḥāq/Thābit's translation):

في ان الارض في وسط السماء

ومن بعد علمنا بهذا ان طلبنا ان نعلم موضع الارض وجدنا انه انما يكون ما ظهر لنا فيها كما نرى اذا نحن اثبتنا موضعها وسط السماء كالمركز في الكرة فقط

(Trl.: On that the earth is in the middle of the heaven.
After we have become aware of this, if we seek to know the position of the earth, we find that what is perceivable for us about it, as what we see, occurs only if we assert its position to be the middle of the heaven, like the centre in the sphere.)

The Dresden translation has not preserved a chapter count and therefore does not allow for a comparison in this regard. However, the Dresden text contains *Alm.* I, 1 (fols 1r:14–2r:8) in a form different from the Arabic traditions Isḥāq/Thābit and al-Ḥajjāj, which suggests that this variant of Ptolemy's introduction was an original part of Thābit's *Almagest.* A reworking of the *Almagest,* discussed below under 2.3, confirms the shifted numbering of the chapters in Thābit's tradition, while Ptolemy's introductory chapter *Alm.* I, 1 is again missing. An explanation could be that Thābit rearranged Ptolemy's chapters to present *Alm.* I, 1 as an uncounted preface that was followed by a general table of contents. An inspiration for this can be found in the *Almagest* itself, where Ptolemy's philosophical introduction does not connect closely to the technical content of the work, whereas the following chapter, *Alm.* I, 2, already provides a catalogue of the different subjects of the *Almagest*. Thus separated from the rest, the introduction was much more at risk of becoming lost or omitted. It may thus have survived only in the Dresden translation, whereas al-Shīrāzī compensated for its loss by putting in its place *Alm.* I, 1 from Isḥāq/Thābit's alternative version.

2.2. Ibn Sīnā, *Kitāb al-shifā'*

Similarities between Ibn Sīnā's (Latinised: Avicenna; *c.* 980–1037) discussion of the sector figure, following *Alm.* I, 13, and the treatment of the same theorem in al-Nasawī's (fl. mid-eleventh c.) *Commentary on the Sector Figure* as well as in the Dresden *Almagest* were found by Richard Lorch.[22] The passage in al-Nasawī's *Commentary* has meanwhile been identified as being copied from Thābit's *Almagest*.[23] The correspondences between Ibn Sīnā and al-Nasawī can therefore be explained if one accepts that Ibn Sīnā, too, took parts of his information from Thābit ibn Qurra's version of the *Almagest*.

Agreement between al-Nasawī's quotations from the *Almagest* and Ibn Sīnā's *Kitāb al-shifā'* can also be found in other passages. Of particular interest is an extension to another theorem, again from *Alm.* I, 13,[24] which both authors present in a similar way, as follows (the transcription and the diagrams largely follow the Cairo edition of Ibn Sīnā;[25] elements that are not found in al-Nasawī, based on a collation of the MSS Istanbul, Topkapı, Ahmet III 3464, fol. 215v:4–16, id., Hazine 455, fols 50v:17–49r:6, and Leiden, UB, Or. 556/4, fol. 56v:3–12, are indicated by angle brackets < >, whereas additional elements in al-Nasawī have been inserted in square brackets [ ]; among further differences, in the manuscripts of al-Nasawī the points on the circle arc in the right one of the following two diagrams are labelled in reverse order, and the second arc in the first argument reads بـ جـ instead of بـ ح; minor variants are not considered):

> [*Preceded by a discussion of Ptolemy's first case, corresponding to an intersection of the extended diameter and the secant on the side of the points* ا *and* بـ, *where it is shown that, if the arc* بـ جـ *and the relation sin(*ا جـ*)/sin(*ا بـ*) are known, the arc* ا بـ *can be determined.*]

[22] Lorch, *Thābit ibn Qurra*, pp. 355 f.

[23] Grupe, 'The Thābit-Version', p. 151.

[24] The theorem is a rider to what Toomer counts as Ptolemy's Theorem 13.4.

[25] Riḍā Madwar et al., *Ibn Sīnā*.

واما ان كان الالتقاء من الجهة الاخرى فانا نعلم قوسي جـ ح، بـ ح بمثل <ما> علمنا في الشكل الاول قوس ا بـ فتصير جميع قوس <بـ ح معلومة لكن جميع قوس> بـ جـ معلومة لكن جميع نصف دائرة ح جـا معلومة يبقى بـ ا معلوما

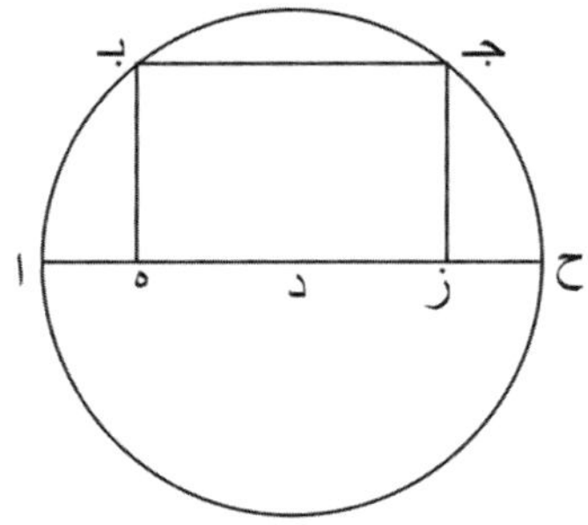

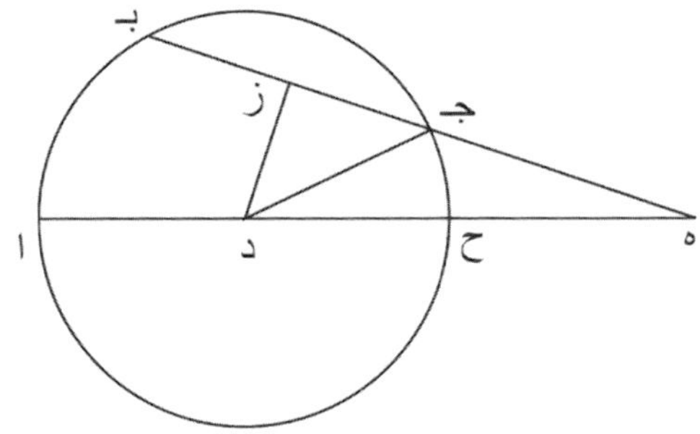

واما ان كان موازيا لا يلتقي فليكن بـ ه جيب [قوس] ا بـ وهو لا محالة عمود على قطر ا [بـ] ح و: جـ ز جيب [قوس] ا جـ وهو ايضا عمود على [قطر] ا [بـ] ح تبقى زاويتا بـ ، جـ بين المتوازيين قائمتين ويكون سطح جـ ه متوازي الاضلاع فيكون بـ ه ، جـ ز متساويين [كما تبين في شكل لد من مقالة ا من س] لكن جـ ز ايضا جيب [قوس] جـ ح فـ[خرج] جـ ح ، بـ ا متساويان و: جـ بـ معلوم فنصف ما يبقى الى تمام نصف الدائرة معلوم وهو بـ ا [وذلك ما اردنا]

This partly corrupted passage may be interpreted as follows:

If the intersection occurs on the other side, we know [the relation of the sines of] the two arcs جـ ح, بـ ح [from what is given, using the identity sin(ا بـ) = sin (بـ ح), etc.] and, analogously to what we found in the first case, the arc [جـ ح, by corresponding application of the 'first' case]. Then the whole arc بـ ح becomes known, as the arc بـ جـ is known [from what is given]. But the whole semicircle ح جـ ا is known, hence there remains بـ ا known.

And if they are parallel and do not intersect, بـ ه is the sine of the arc ا بـ and that, of course, is a column on the diameter ا ح, and جـ ز is the sine of the arc ا جـ and it is also a column on the diameter ا ح, and بـ [ه], جـ [ز] equally join the two parallels perpendicularly, and the area جـ ه is a parallelogram, then بـ ه, جـ ز are equal, as is shown in Proposition 3[3] of Book 1 of Euclid's *Elements*. But جـ ز is also the sine of the arc جـ ح, hence the result is that جـ ح, بـ ا are equal, and جـ بـ is known [from what is given], hence half of what remains as the complement to the semicircle is known, and it is بـ ا, and that is what we wanted.

The passage supplements one of Ptolemy's theorems in a manner very similar to Thābit's inserted comments on the sector figure at the end of *Alm.* I, 13.[26] Analogous to Thābit's treatment of the sector figure, explicit proofs of the validity of Ptolemy's theorem are given for constellations which Ptolemy had ignored. Moreover, the constellations correspond again to an opposite and a parallel case relative to the first case discussed by Ptolemy, and proving these cases involves again a completion of the semicircle and the identity

[26] Thābit's insertion, according to the quotation by al-Nasawī, is edited in Lorch, *Thābit ibn Qurra*, pp. 362–70.

sin a = sin(180°–a).[27] The occurrence of the sine in this passage differs from Ptolemy's general use of the chord of the doubled arc; however, it is in line with other Arabic witnesses of Thābit's *Almagest* which indicate the possibility that Thābit systematically replaced Ptolemy's chords of doubled arcs with the sine.[28] Though not preserved by Ibn Sīnā, a reference to Euclid's *Elements* also appears in al-Nasawī's quotation of the passage. Such references were found to be another characteristic of Thābit's *Almagest*.[29]

The present theorem is of interest with regard to the transmission of Thābit's text. The known witnesses of Thābit's *Almagest* differ systematically in how much of the above extension they include or confirm. The most comprehensive account is the one cited above, as found in Ibn Sīnā and al-Nasawī. By contrast, al-Ṭūsī speaks in the same context of only one additional diagram in Thābit's *Almagest*, viz. the one related to the opposite case. This agrees with Qāḍīzāde al-Rūmī, who says that in total *Alm.* I, according to Thābit's version, contained four, not five, diagrams more than the other Arabic versions, while typically three of the additional diagrams pertain to the sector figure.[30]

An omission of the diagram for the parallel case may have been decided on in some parts of the tradition, because the validity of the theorem is trivial in this case and can be easily demonstrated without a dedicated illustration. However, also the diagram for the opposite case might be considered dispensable, especially if al-Nasawī's labels are used, which render the diagram hardly more than a mirror image of Ptolemy's own diagram of the first case. This may have led to a third group of texts, represented by the Latin Dresden *Almagest* and a reworking of the *Almagest* kept in Tehran, which will be discussed next. They contain neither of the two additional diagrams, while also the discussion of the corresponding cases has been identically shortened in both texts (cf. below 2.3).

It is the two oldest witnesses, Ibn Sīnā and al-Nasawī, both from the first half of the eleventh century, that give the most comprehensive version of the supplement. It is therefore plausible that the above reading resembles Thābit's insertion most closely, whereas later recipients of Thābit's text occasionally shortened the argument. Such abbreviations could have been motivated by the fact that this theorem is not used later on in the *Almagest*.

[27] For the correspondences in Thābit's discussion of the sector figure cf. the summaries in Lorch, *Thābit ibn Qurra*, pp. 155 f. and 374 f.

[28] The use of the sine in al-Nasawī's quotations in contrast to the chords in the Latin Dresden translation was noticed by Lorch, *Thābit ibn Qurra*, p. 357. Until now, the Dresden translation has remained the only witness of Thābit's *Almagest* that does not have the sine but uses the chord of the doubled arc. Ptolemy's chords therefore may have been restored in some parts of the tradition of Thābit's *Almagest*.

[29] See Grupe, 'The Thābit-Version', pp. 149 ff.

[30] Transliterations of al-Ṭūsī's and Qāḍīzāde al-Rūmī's statements are given in Kunitzsch, *Der Almagest*, pp. 31 f. (Qāḍīzāde al-Rūmī) and Grupe, 'The Thābit-Version', p. 151 (al-Ṭūsī).

2.3. MS Tehran, Majlis-Senate, 1231 (abridged reworking of Thābit's *Almagest*)

The manuscript no. 1231 of the Parliament (Majlis-Senate) Library in Tehran preserves on 157 folios the most extensive surviving derivative of Thābit's text.[31] The text is a systematically reshaped and shortened version of Thābit's *Almagest*. All tables, including all chapters that consist of tables, such as Ptolemy's star catalogue in *Alm*. VII and VIII, as well as any references to tables in Thābit's text, as they are known from the Dresden translation, have been removed. Furthermore, the division of the *Almagest* into books, which is also documented for Thābit's version, has been abolished in favour of a continuous sequence of ninety-two numbered chapters. The numbering of the chapters has been adapted to the omission of the table chapters, whereas it still reflects the shifting of the chapter numbers due to Thābit's separation of Ptolemy's introduction as well as further changes in the chapter division.[32] As in the texts discussed before, under 2.1 and 2.2, Ptolemy's introduction is again not contained in the Majlis manuscript, which instead begins immediately with an adapted variant of *Alm*. I, 2.

Further shortenings in the Majlis text consist in the removal of references to Euclid's *Elements* and of cross-references to numbered books and theorems of the *Almagest*, which Thābit had originally inserted.[33] Whereas such cross-references would in any case have required an adaptation to the abandoned division into books, Thābit's relative or less distinct cross-references, to a respective 'preceding theorem' etc., partly survived the revision.[34] Apart from this systematic removal of specific elements, also Thābit's prose, especially in longer narrative or descriptive parts of the *Almagest*, has occasionally been brought into a more concise form.[35]

[31] I am grateful to Mª José Parra, who located this manuscript in 2015 and drew my attention to it as a possible representative of the Thābit tradition; Parra, 'Making a Catalogue' (see above, note 4).

[32] Thābit himself had already applied a different chapter division in his version; cf. Grupe, 'The Thābit-Version', p. 150.

[33] cf. Grupe, 'The Thābit-Version', p. 149.

[34] Cf., e.g., MS Tehran, Majlis-Senate Library, 1231, fol. 20r:13: وقد تبين فيما يقدم, corresp. to MS Dresden, SLUB, Db. 87, fol. 22r:1: 'demonstracione figure huic preposite', or MS Tehran, Majlis-Senate Library, 1231, fol. 22r:24: فانه تبين من المقدمة الاولى, corresp. to MS Dresden, SLUB, Db. 87, fol. 26r:10 f.: 'manifestum est ex prima proposicione'.

[35] On fol. 26r:25 of MS Tehran, Majlis-Senate Library, 1231, relating to *Alm*. III, 1, the Greek names 'Meton and Euctemon' are transliterated as فاطونا وافطيمونا. Apart from a probable corruption of the letters *mīm* and *qāf* to *fā* at the beginning of the two names, the transliterations differ from what is found in the manuscripts of other Arabic versions of the *Almagest*, by the added terminal *alif*s and by rendering the Greek letter omega by *wāw*. As a terminal *alif* is a common ending of nouns in Syriac, Dr Mohammad Mozaffari conjectured in a private conversation that the transliterations possibly indicate a Syriac influence. This could be of relevance to the production process of Thābit's *Almagest* version. Regarding that conjecture,

Except for the aforementioned modifications in the Majlis reworking, a comparison with Books I to IV of the Dresden translation reveals close correspondence in the wording. This makes the Majlis text a valuable source of information, especially on Books VI to XIII of Thābit's *Almagest*, which are not preserved in any other known manuscript.[36]

The following examples from the Majlis reworking of Thābit's *Almagest* demonstrate its relation to the previous witnesses. Corresponding to the epitome in the (former) Naṣīrī manuscript no. 789 and the Latin Dresden translation, the same chapters in the Majlis text start as follows (cf. above 2.1).

Alm. I, 3.

MS Tehran, Majlis-Senate, 1231, fol. 1r:15–17 (the folio numbers have been determined based on the order of the pages in the digital scan as published online by the Parliament Library in Tehran):

الباب الثاني في ان السماء كرية الشكل والحركة

ان اول ما دعا القدماء الى ان قالوا ان حركة السماء كرية هو ما ظهر لهـ[م من ا]لقياسات والارصاد التي راوها

Alm. I, 5.

MS Tehran, Majlis-Senate, 1231, fol. 2r:15–17:

الباب الرابع في ان الارض في وسط السماء

ليس يمكن ان تعرض الاشياء التي نشاهدها ونراها من امر السماء والارض الا والارض في وسط السماء بمنزلة المركز

Similarly, a passage from Thābit's *Almagest* that I have discussed elsewhere on the basis of the Dresden translation[37] has a direct correspondence in the Majlis reworking. Except for Thābit's second cross-reference, which has been removed in the Majlis text as part of the revision, both texts also show the same insertions by Thābit (underlined in the Latin):

Alm. I, 10.

MS Tehran, Majlis-Senate, 1231, fol. 7r:8–21:

اذا علمنا وتري قوسين من دائرة فان وتر فضل ما بين القوسين يكون لنا معلوما

فلتكن نصف دائرة عليه ابجد وليكن القطر اد وليكن وترا ابـ اجـ معلومين

فنصل خط بجـ

Prof. Kunitzsch noted that it was rather untypical of Greek-Syriac translations that an *alif* became added to a Greek personal name.

[36] Apparently only several coordinates from the star catalogue in Books VII and VIII of Thābit's version are preserved in the work of Qāḍīzāde al-Rūmī; cf. Kunitzsch, *Ibn aṣ-Ṣalāḥ*, pp. 115–23 passim.

[37] cf. Grupe, 'The Thābit-Version', pp. 152 f., Appendix.

فاقول ان بجـ معلوم

برهان ذلك انا نصل خطي بد جد وهو بين ان هذين الخطين يكونان معلومين لانا قلنا انه اذا كان وتر قوس ما معلوما فان وتر ما تنقص تلك القوس عن نصف الدائرة يكون معلوما وقد احاطت دائرة بشكل ذي اربعة اضلاع وهو شكل ابجد فالذي يكون من ضرب ابـ في جد مع الذي يكون من ضرب اد في بجـ مساو للذي يكون من ضرب اجـ في بد وليكن الذي يكون من ضرب ابـ في جد معلوم والذي يكون من ضرب اجـ في بد ايضا معلوم فيبقي المجتمع من ضرب اد في بجـ معلوما وقطر اد معلوم فيصير بجـ معلوما

فاذا كانا وترا قوسين من دائرة معلومين فان وتر فضل ما بين القوسين معلوم

وذلك ما اردنا ان نبين

MS Dresden, SLUB, Db. 87, fol. 7v:8–24 (trl. of Thābit's *Almagest*):

Ut scierimus duorum arcuum duas cordas circuli alicuius, corda superhabundantis quod est inter duos arcus nobis scietur.

Sit medietas circuli ABCD, *sitque diametrum ipsum* AD, *sintque prescite due corde* AB AC.

Excopulemus lineam BC.

Dico quia BC *scietur.*

Racio. Copulabimus enim duas lineas BD CD, *sed est manifestum quia ipse due linee sunt prescite. Diximus namque quia quando fuerit alicuius arcus corda prescita, corda minoritatis illius arcus a medietate circuli erit scita. Et ipse circulus est circuicio figure quatuor laterum et est figura* ABCD *unde quod est ex multiplicacione* AB *in* CD *et quod est ex multiplicacione* AD *in* BC *est equale ei quod ex multiplicacione* AC *in* BD *hoc quod est ostensum in figura huic precedenti. Et quod est ex multiplicacione* AB *in* CD *est prescitum et quod est ex multiplicacione* AC *in* BD *est prescitum. Igitur remanet quod est ex multiplicacione* AD *in* BC *scitum et diametrum* AD *est prescitum, igitur* BC *est scita.*

Igitur cum fuerint due corde duorum arcuum alicuius circuli prescite, corda superhabundantis quod est inter ipsos duos arcus erit scita.

Et hoc est <quod> demonstrare voluimus.

Furthermore, as mentioned above in connection with Ibn Sīnā's *Kitāb al-shifāʾ*, Thābit's supplement to Ptolemy's Theorem 13.4 (Toomer's count) appears shorter and without diagrams in the Majlis text. This again corresponds to the Dresden translation (cf. the shortened 'proofs' of Thābit's additional cases underlined in the Latin):

Alm. I, 13; extension to theorem 13.4.

MS Tehran, Majlis-Senate, 1231, fol. 11r:23–11v:14:

برهان ذلك انا نجعل مركز الدائرة نقطة د ونصل خط [جـ]بـ وخط اد ونخرجهما على استقامة

فاما ان يكونا متوازيين واما ان يلتقيا

فان كانا متوازيين فان قوس ابـ سيكون نصف ما ينقص قوس بجـ التي هي معلومة عن تنصف دائرة ولذلك تكون معلومة

وان لم يكن خطا اد بجـ متوازيين فيلتقيا على نقطة ه...

[continues with a discussion of the 'first' case in accordance with Ptolemy]

وبمثل هذا المسلك نعلم قوس ابـ اذا التقي خطا بجـ دا في الجهة الاخرى

وذلك ما اردنا ان نبين

MS Dresden, SLUB, Db. 87, fols 12v:23–13r:13 (trl. of Thābit's *Almagest*):

Racio. Sit centrum circuli D *et copulemus* CB DA *et faciamus exire rectas utrasque.*

Aut enim erunt paralellice nusquam convenientes aut erunt sibi obviantes.

Si vero sunt paralellice erit arcus AB *medietas diminucionis arcus* BC *presciti de medietate circuli, ideoque est scitus.*

Sed si non sunt due linee DA BC *paralellice, sint sibi obviantes super punctum* E ...

[continues with a discussion of the 'first' case in accordance with Ptolemy]

Eademque via sciemus arcum AB *cum fuerint iuncte due linee* BC AD *alia in parte.*

Et hoc est quod demonstrare voluimus.

2.4. MS Jaipur, Maharaja Sawai Man Singh II Museum Library, 20 (incomplete copy of Thābit's *Almagest?*)

The Jaipur *Almagest* was made known to a wider public in 1980, by David King, in his 'Handlist' of astronomical manuscripts that he had seen on the occasion of a visit to the Maharaja Sawai Man Singh II Museum Library in Jaipur in 1978.[38] Copied around 1600, the manuscript no. 20 of this library contains on *c.* 150 folios Books I till the beginning of Book VI of what King described as 'the Arabic version by Thābit ibn Qurra of Ptolemy's *Almagest*'. King indicated that the text was a copy of the well-known Isḥāq/Thābit version of the *Almagest,* which Thābit ibn Qurra created by making corrections to Isḥāq ibn Ḥunayn's Arabic translation. Accordingly, King concluded in his report that the astronomical works in Jaipur were well known from copies in other collections and added little to the corpus of material available for the further study of the history of Islamic astronomy in general.[39]

A different view was taken after 1991, when George Saliba and Richard Lorch had the opportunity to examine the manuscript for a second time and also to transcribe some of its chapter beginnings at the request of Paul Kunitzsch. The transcriptions corresponded in the comparable parts to what was found in the epitome of the *Almagest* in the former Naṣīrī manuscript

[38] King, 'A Handlist'.

[39] King, 'A Handlist', pp. 81 f., describes all the texts in the list to be already known and also preserved in other libraries. The discussion of manuscript no. 20, on p. 82, is followed by a reference to other manuscripts of the Isḥāq/Thābit version of the *Almagest.* This identification of manuscript Jaipur 20 became accepted. However, the mention of Thābit ibn Qurra alone in King's report left some uncertainty about the Jaipur *Almagest* on the part of Paul Kunitzsch; see Kunitzsch, *Der Sternkatalog*, vol. I, p. 4, where King's estimate is followed, though with some reservation.

no. 789 (cf. above 2.1).[40] It thus became clear that the previous assumption about the Jaipur text had been incorrect. However, the Arabic origin of the Latin Dresden *Almagest*, and also the reworking in MS Tehran, Majlis-Senate, 1231 (cf. above 2.3), both having similar chapter beginnings as the Naṣīrī text, were still unknown at the time.[41] Kunitzsch thus inferred that the Jaipur text was not an *Almagest* at all but a copy of the same epitome as preserved in the Naṣīrī manuscript.[42]

This second assumption contradicted the reference to Thābit ibn Qurra and the direct impression of David King, who had not noticed any omissions or changes relative to the 'normal' *Almagest* when he inspected the manuscript in Jaipur. Also, the number of folios occupied by the five books in the Jaipur manuscript speaks against substantial abbreviations. Despite such inconsistencies, the conclusion that the Jaipur text was a copy of al-Shīrāzī's epitome remained unquestioned during the following years, and the Jaipur manuscript was no longer considered among the extant Arabic copies of the *Almagest*.[43]

In my doctoral dissertation, I proposed a new hypothesis which conforms to the details in King's report and which also resolves the above contradictions. Since I have shown above that al-Shīrāzī took the chapter beginnings in the opening part of his epitome from Thābit's *Almagest*, the conclusion that the Jaipur text is a copy of al-Shīrāzī is no longer cogent. At the same time, the recent certitude about the existence of a version of the *Almagest* by Thābit ibn Qurra alone gives new importance also to King's description. The reports from Jaipur in combination with the new evidence prove that the Jaipur *Almagest* derives from what has meanwhile been identified as Thābit's own version of the *Almagest*. They further indicate that the Jaipur manuscript contains an unshortened copy of the first five books of that version.[44] A clearer assessment, including a placement of the Jaipur fragment in the complex tradition of Thābit's *Almagest*, will hopefully be possible in the future as soon as sub-

[40] See Kunitzsch, 'A Hitherto Unknown', p. 32, note 6. cf. also the *Nachtrag* in Kunitzsch, *Der Sternkatalog*, vol. III, p. 200, last paragraph.

[41] An Arabic rather than Greek origin for the Dresden *Almagest* was suggested in 2001, by Lorch, *Thābit ibn Qurra*, pp. 356 f. The Tehran manuscript Majlis-Senate 1231 was introduced in 2015, at the Ptolemy conference in London, in connection with the present paper.

[42] See Kunitzsch, 'A Hitherto Unknown', p. 32.

[43] See, for example, Kunitzsch, 'A Hitherto Unknown', pp. 31 f.

[44] I proposed this in 2013, in Grupe, *The Latin Reception*, p. 128, note 152, at that time based only on the available publications and my recent discovery of a Thābit version of the *Almagest*. Meanwhile, my comparison of the epitome discussed under 2.1 with the Latin Dresden translation proved that the Jaipur *Almagest* is indeed a representative of the Thābit tradition. From a copy of the transcriptions in Paul Kunitzsch's private archive I could also see that the wording of the Jaipur *Almagest* corresponds to the Latin even where the reworked texts under 2.1 and 2.3 differ in some details. This confirms that the Jaipur text is an independent, closer copy of Thābit's *Almagest* than the other known witnesses.

stantial portions of the Jaipur text are accessible for comparison with the other witnesses. The manuscript was acquired for astronomical activities by Maharaja Jai Singh in the early eighteenth century. It is thus also the latest known evidence of an active reception of Thābit's text.

2.5. Quṭb al-Dīn Maḥmūd b. Masʿūd al-Shīrāzī, *Talkhīṣ al-Majisṭī*

From al-Shīrāzī, the assumed author of the epitome discussed under 2.1, stems also this other *Summary of the Almagest*, in Persian, which has been preserved in at least two manuscripts (MS Istanbul, Süleymaniye, Lala Ismail Efendi 288 and MS Tehran, Majlis, 600; in what follows reference will be made to the Tehran copy).[45] As in the epitome, also in the present *Talkhīṣ al-Majisṭī* al-Shīrāzī substantially shortened Ptolemy's arguments while including once again distinctive elements from Thābit's version.

The *Talkhīṣ al-Majisṭī* shows Thābit's insertion of Theon of Alexandria's proof of the first form of Menelaus' Theorem from its second form, and also Thābit's use of Theon's proof of the addition theorem of chords in place of Ptolemy's (cf. fols 5v–6r).[46] Al-Shīrāzī further determines the chord of one degree using a diagram with three, instead of only two, chords (cf. fol. 5v). This deviation from Ptolemy can again be found in several witnesses of Thābit's version, for example in the Latin Dresden translation (cf. MS Dresden, SLUB, Db. 87, fol. 9v), in a quotation from the *Almagest* in al-Nasawī's *Commentary* (cf. MS Istanbul, Topkapı, Ahmet III 3464, fol. 209r), in Ibn Sīnā's *Kitāb al-shifāʾ* (cf. MS Paris, BnF, ar. 2484, fol. 8r) and in the Tehran reworking of Thābit's *Almagest* (cf. MS Tehran, Majlis-Senate, 1231, fol. 9r).

It is clear from the epitome discussed under 2.1 that al-Shīrāzī had detailed knowledge about Thābit's *Almagest*. The *Talkhīṣ al-Majisṭī* further proves that al-Shīrāzī was acquainted also with some of Thābit's mathematical arguments that do not appear in the epitome. Since al-Ṭūsī could already be identified as one of the key witnesses of Thābit's *Almagest*, it is not surprising to find related indications also in the works of al-Shīrāzī. Having been a student and co-worker of al-Ṭūsī in Marāgha, al-Shīrāzī can be easily imagined to have had access to the same sources.

2.6. Athīr al-Dīn al-Abharī, *Kitāb fī ṣināʿa al-Majisṭī*

Al-Abharī's (d. *c.* 1264) *Commentary on the Almagest* is preserved in at least three manuscripts (MS Tehran, Majlis, 6195; MS Istanbul, Süleymaniye, Ayasofia 2583 bis, and MS Tehran, National Library, 20371; in what follows refer-

[45] Mª José Parra, 'Making a Catalogue' (see above, note 4).

[46] Both elements were described by al-Ṭūsī as features of Thābit's *Almagest*; cf. Grupe, 'The Thābit-Version', p. 151.

ence will be made to MS Tehran, National Library, 20371).[47] Aside from using Thābit's diagram with three chords instead of two when determining the sine of one degree (fol. 14v; cf. above, 2.5), al-Abharī also presents an extended discussion of the sector figure (fols 21r–22r) that is very similar to what is found in Thābit's *Almagest*. In this context, al-Abharī further includes an alternative proof of the Menelaus' Theorem (fols 22v–23r), which does not appear in Thābit's *Almagest* but which Thābit presented elsewhere, together with the additional arguments from his *Almagest*, in his comprehensive *Commentary on the Sector Figure*.[48]

Al-Abharī seems acquainted with Thābit's *Commentary on the Sector Figure*, as he expressly attributes the alternative proof to Thābit (fol. 22v:8, starting: طريق اخر لثابت بن قره في برهان الشكل القطاع). Surprisingly, though, al-Abharī attributes to Ptolemy the extensions to the sector figure which Thābit had inserted in his version of the *Almagest* (see fol. 22r:1, concluding the previous discussion of Thābit's arguments: هذا تقرير ما ذكر بطلميوس في الشكل القطاع), although al-Abharī would have found the same arguments also in Thābit's *Commentary on the Sector Figure*, whereas none of these arguments appears in the classical tradition of the *Almagest*.

An explanation could be that al-Abharī studied the *Almagest* based on Thābit's version. Unaware of Thābit's insertions to Ptolemy, or unable to identify their origin, al-Abharī accredited Ptolemy with the entire content of that particular *Almagest*.[49]

2.7. Further Influences

Borrowings from Thābit's *Almagest* can be found at various other places in the Arabic astronomical literature. In many cases, however, a direct reception of Thābit's text is difficult to determine. This is true especially due to the strong influence of astronomers who had been using Thābit's work. Elements and adaptations from Thābit in a particular treatise may have been taken directly from Thābit's *Almagest* or, alternatively, from an intermediary tradition. Some of Thābit's concepts may also have become commonly known in the field.

An early example is the work of the Persian astronomer Kūshyār ibn Labbān (fl. *c.* 1000 CE).[50] In Book 4 of his *Jāmiʿ Zīj* Kūshyār ibn Labbān pres-

[47] Mª José Parra, 'Making a Catalogue' (see above, note 4). A brief account of what is known about the life and work of al-Abharī is given in Sarıoğlu, 'Abharī'.

[48] An edition of Thābit's *Commentary on the Sector Figure* has been made by Lorch, *Thābit ibn Qurra*, pp. 42 ff. Thābit's alternative proof is found on pp. 62–72 of Lorch's edition.

[49] This has a parallel in the work of al-Nasawī, who also attributes to Ptolemy the extended discussion of the sector figure in Thābit's *Almagest*, cf. Lorch, *Thābit ibn Qurra*, p. 357.

[50] For a brief account of what is known about the life and work of Kūshyār ibn Labbān see Bagheri, 'Ibn Labbān, Kūshyār'.

ents a summary of the theorems from Chapter I, 10 of the *Almagest*.[51] Aside from presenting the theorems in a Euclidised form in the manner of Thābit's *Almagest*,[52] Kūshyār ibn Labbān also uses a very similar wording as Thābit, and he makes the same replacement of Ptolemy's proof of the addition theorem of chords with Theon's (cf. above 2.5).[53] When determining the chord of one degree, however, Kūshyār follows Ptolemy instead of using Thābit's modified diagram with three chords (cf. above 2.5).[54] While an influence of Thābit's *Almagest* is thus clearly recognisable, Kūshyār ibn Labbān neither mentions Thābit as his source nor does he follow Thābit's version literally or consistently.

Different is the situation concerning another summary of the theorems from *Alm.* I, which is extant on fols 55r–60v of MS Tehran, Majlis, 6417. The text includes an explicit reference to 'Thābit's version' (fol. 57v:1 ff.). However, in this regard, the author seems to depend fully on an identical reference in al-Ṭūsī's *Taḥrīr*.[55]

Direct knowledge of Thābit's *Almagest* is more probable again in the case of the Persian astronomer Niẓām al-Dīn Ḥasan al-Nīsābūrī (d. 1328/9).[56] In the early fourteenth century, al-Nīsābūrī composed a commentary on al-Ṭūsī's *Taḥrīr of the Almagest*, entitled *Tafsīr Taḥrīr al-Majisṭī*, which has survived in several copies (in what follows reference will be made to MS London, BL, Add 7476).[57] When writing the *Tafsīr*, al-Nīsābūrī necessarily came across al-Ṭūsī's repeated references to Thābit's *Almagest*. However, al-Ṭūsī's discussion of deviant contents in Thābit's version do not seem to have attracted much of al-Nīsābūrī's attention as a commentator. On the other hand, al-Nīsābūrī includes in his *Tafsīr* further elements from Thābit's *Almagest* which are not found in al-Ṭūsī's *Taḥrīr*, such as the diagram with three chords instead of two when determining the chord of one degree (fol. 30r), and throughout the *Tafsīr* al-Nīsābūrī applies a system similar to that of Thābit for referring to numbered propositions (cf., e.g., fols 3r:3f. and 3r:10 for references to Euclid's *Elements* and fol. 30r:5 f. for two cross-references to propositions in the *Almagest*). All this suggests a natural acquaintance with Thābit's *Almagest* on al-Nīsābūrī's part. However, it must be expected that many of Thābit's concepts as adopted by al-Nīsābūrī were also known by the time independently from their original context.

51 The book has been edited by Bagheri, *Az-Zīj al-Jāmiʿ*. Kūshyār's summary of Ptolemy's theorems is found on pp. 79–88 of the Arabic part of Bagheri's edition.

52 This has been identified as a characteristic of Thābit's *Almagest* in Grupe, 'The Thābit-Version', p. 149.

53 cf. Bagheri, *Az-Zīj al-Jāmiʿ*, p. 85 of the Arabic part.

54 cf. Bagheri, *Az-Zīj al-Jāmiʿ*, pp. 87 f. of the Arabic part.

55 cf. Grupe, 'The Thābit-Version', p. 151.

56 A brief account of Nīsābūrī's life and work is given in Morrison, 'Nīsābūrī'.

57 Mª José Parra, 'Making a Catalogue' (see above, note 4).

Similar observations can be made in a later commentary, written by a certain Kāfī Qā'inī, on an astronomical treatise by Khāzimī.[58] Kāfī Qā'inī's commentary is preserved in a manuscript in Mashhad, Holy Shrine, 7345, where it is dated to the early seventeenth century.[59] In connection with the determination of the chord of one degree, Kāfī Qā'inī uses Thābit's diagram in addition to Ptolemy's (fol. 24v); he also includes several of Thābit's arguments on the sector figure (fols 34r–36v). Since these elements are not found in Khāzimī's treatise (extant, e.g., in MS Mashhad, Holy Shrine, 12297), Kāfī Qā'inī must have used further sources. However, also in view of the late production of the text, the reworked appearance of Thābit's concepts in Kāfī Qā'inī's commentary leaves one in doubt as to whether the latter really had a copy of Thābit's *Almagest* in his hands.

3. Conclusions

The identified sources reveal that Thābit ibn Qurra's version of the *Almagest* was used by several of the most prominent Islamic astronomers. The work also continued to influence astronomical studies in the Islamic world at least until the eighteenth century. Not all the texts presented here prove a direct use of Thābit's text. Nevertheless, in most cases a good knowledge of the work is highly probable. Conversely, isolated and modified borrowings from Thābit, based on second-hand information, suggest that at least some of Thābit's ideas had become part of a common reservoir from which later Islamic authors were able to draw when composing their own treatises. With further investigation it is likely that we shall be able to link further texts to the tradition of Thābit's *Almagest*. In addition, many of Thābit's interventions to the original content of the *Almagest* consist in commenting insertions. It is therefore possible that some of today's speculations about unknown commentaries by Thābit ibn Qurra will find their answer in the existence of 'Thābit's version' of the *Almagest*.

Bibliography

Bagheri, Mohammad, 'Ibn Labbān, Kūshyār', in Thomas Hockey et al. (eds), *The Biographical Encyclopedia of Astronomers*, New York: Springer, 2007, pp. 560–61.

Bagheri, Mohammad, *Az-Zīj al-Jāmiʿ by Kūshyār ibn Labbān. Books I and IV. An Arabic Astronomical Handbook*, Frankfurt am Main: Institute for the History of Arabic-Islamic Science, 2009.

Carmody, Francis J., *The Astronomical Works of Thabit b. Qurra*, Berkeley: University of California Press, 1960.

[58] Khāzimī's work is briefly discussed by Sezgin, *GAS* VI, p. 92.

[59] Mª José Parra, 'Making a Catalogue' (see above, note 4). Digital copies of the manuscripts in Mashhad were acquired by the project *Ptolemaeus Arabus et Latinus*.

Grupe, Dirk, 'The "Thābit-Version" of Ptolemy's *Almagest* in MS Dresden Db.87', *Suhayl* 11 (2012), pp. 147–53.

Grupe, Dirk, *The Latin Reception of Arabic Astronomy and Cosmology in Mid-Twelfth-Century Antioch. The Liber Mamonis and the Dresden Almagest*, PhD dissertation, University of London, 2013.

King, David, 'A Handlist of Arabic and Persian Astronomical Manuscripts in the Maharaja Mansingh II Library in Jaipur', *Journal for the History of Arabic Science* 4 (1980), pp. 81–86.

Kunitzsch, Paul, *Der Almagest. Die Syntaxis mathematica des Claudius Ptolemäus in arabisch-lateinischer Überlieferung*, Wiesbaden: Harrassowitz, 1974.

Kunitzsch, Paul, *Ibn aṣ-Ṣalāḥ. Zur Kritik der Koordinatenüberlieferung im Sternkatalog des Almagest*, Göttingen: Vandenhoeck & Ruprecht, 1975.

Kunitzsch, Paul, *Claudius Ptolemäus. Der Sternkatalog des Almagest. Die arabisch-mittelalterliche Tradition*, 3 vols, Wiesbaden: Harrassowitz, 1986–1991.

Kunitzsch, Paul, 'A Hitherto Unknown Arabic Manuscript of the *Almagest*', *Zeitschrift für Geschichte der arabisch-islamischen Wissenschaften* 14 (2001), pp. 31–37.

Lorch, Richard, 'Greek-Arabic-Latin: The Transmission of Mathematical Texts in the Middle Ages', *Science in Context* 14 (2001), pp. 313–31.

Lorch, Richard, *Thābit ibn Qurra. On the Sector-Figure and Related Texts*, Frankfurt am Main: Institute for the History of Arabic-Islamic Science, 2001 [reprinted Augsburg: Rauner, 2008].

Morelon, Régis, *Thābit ibn Qurra. Œuvres d'astronomie*, Paris: Les Belles Lettres, 1987.

Morrison, Robert, 'Nīsābūrī', in Thomas Hockey et al. (eds), *The Biographical Encyclopedia of Astronomers*, New York: Springer, 2007, p. 837.

Pedersen, Olaf, *A Survey of the Almagest. With Annotation and New Commentary by Alexander Jones*, New York: Springer, 2011.

Ragep, F. Jamil, 'Shīrāzī', in Thomas Hockey et al. (eds), *The Biographical Encyclopedia of Astronomers*, New York: Springer, 2007, pp. 1054–55.

Riḍā Madwar, M., I. Aḥmad and I. Madkour, *Ibn Sīnā, Al-shifā'. Al-riyāḍiyyāt*, vol. IV: *'Ilm al-hay'a*, Cairo: Al-Hay'a al-miṣriyya al-'āmma lil-kitāb, 1980. Reprinted Tehran, 1984/5.

Sarıoğlu, Hüseyin, 'Abharī', in Thomas Hockey et al. (eds), *The Biographical Encyclopedia of Astronomers*, New York: Springer, 2007, pp. 7–8.

Sezgin, Fuat, *GAS* VI = *Geschichte des arabischen Schrifttums*, Vol. VI: *Astronomie bis ca. 430 H.*, Leiden: Brill, 1978.

Sidoli, Nathan, and Takanori Kusuba, 'Al-Harawī's Version of Menelaus' *Spherics*', *Suhayl* 13 (2014), pp. 149–212.

Toomer, Gerald. J., *Ptolemy's Almagest*, London: Duckworth, 1984 [2nd ed. Princeton: Princeton University Press, 1998].

Revamping Ptolemy's Proof for the Sphericity of the Heavens: Three Arabic Commentaries on *Almagest* I.3

Y. Tzvi LANGERMANN

1. Introduction

In this project I finally return to one of my first projects in the history of science, one that I now know to have been constructed on false premises. The project was to search commentaries on the *Almagest* in order to see if and how thinkers may have challenged head on the cosmological principles that Ptolemy enunciates at the beginning of his work — the six principles, including the spherical shape of the cosmos and the rotation of the heavens about a fixed earth — that form the core of the so-called Ptolemaic system. But I should have known better; I had already learned from Thomas Kuhn that the Ptolemaic system was undermined not by a frontal assault on its basic principles, but rather by the painstaking efforts of astronomers to solve various specific, technical issues which resisted solution; and when this simply could not be done, the path to alternative models was opened. I know that Kuhn and his philosophy are now a very elaborate field of study, whose intricacies and complexities are beyond my comprehension. Nonetheless, as a simple historian I retain the insights I gained from Kuhn, and in the course of decades of research I see them hold well for cultural contexts (Islamic and Jewish) that he did not study.

The second false premise concerns the nature of the commentary in medieval Islamicate civilization.[1] It was expressly not a vehicle for criticism, but rather for elucidating issues that the author had not explained well enough (in the view of the commentator). This conception of the genre is elaborated by Maimonides in the introduction to his commentary on Hippocrates' Aphorisms. Maimonides observes:

'If most of what the book contains is in error, then the later composition which exposes those confusions is called *radd* ["rejection", "retort"], rather than *sharḥ* [commentary]'.[2] Hence, a commentary would not be the place to present a rejection of Ptolemy's cosmological principles.

[1] Wisnovsky, 'The Nature and Scope', pp. 149–91. Jamil Ragep made excellent use of several commentaries to al-Ṭūsī's *Tadhkira* in his *Naṣīr al-Dīn al-Ṭūsī's Memoir*.

[2] I translate from the Judaeo-Arabic text in Kafih, *Iggerot*, p. 143.

Ptolemy's Science of the Stars in the Middle Ages, ed. by David Juste, Benno van Dalen, Dag Nikolaus Hasse and Charles Burnett, PALS 1 (Turnhout, 2020), pp. 159–180

10.1484/M.PALS-EB.5.120178

With some adjustments, though, my project may still yield some information of interest. To begin with, it seems that writers of commentaries did not view the literary categories to be as rigid as Maimonides describes them. Fakhr al-Dīn al-Rāzī sharply criticized Ibn Sīnā in his commentaries to *al-Ishārāt* and *al-Qānūn fī al-Ṭibb*.[3] In his commentary on the *Almagest*, Ibn al-Haytham says clearly that he will be both a commentator (*mufassir*) and a writer of a précis (*mulakhkhiṣ*).[4] In the lists of his writings that same work is called *Tahdhīb al-Majisṭī*, 'The Improvement of the Almagest', and *Sharḥ al-Majisṭī wa-talkhīṣuhu*, 'Commentary and Summary of the Almagest'.[5] Ibn al-Haytham refers more than once to the multifarious nature of his exposition. Even if commentaries (in the limited sense of the term) do not take issue with Ptolemy, they will critically engage other commentaries; Ibn al-Haytham will do this with al-Nayrīzī.

More importantly, as a member of the so-called 'academy' I need not be beholden to the medieval Arabic literary conception of the commentary. Instead, I will operate on the basis of a wider understanding of the genre, and I will include in my presentation works that in one way or another present the *Almagest* or parts of it, restating Ptolemy's remarks, reorganizing them, and also criticizing them.

In this paper I will confine myself to just one of the cosmological principles, the sphericity of the heavens, which Ptolemy presents in *Almagest* I.3 and offers arguments on its behalf. The chapter in the *Almagest* bears the title (in Toomer's translation, bracketed), 'That the heavens move like a sphere'. Toomer remarks that the chapter titles in Greek are all interpolations; Ptolemy did not use any chapter divisions at all.[6] Even so, the chapter title in Arabic does not reflect the Greek interpolation. In the Isḥāq-Thābit version of the Arabic, Ptolemy announces here two topics (I consulted MS Paris, BnF, arabe 2482): *Inna al-samāʾ kuriyyat*^{un} *wa-ḥarakatuhā ayḍan kuriyyat*^{un} 'That the heavens are spherical and that their motion is also spherical'.[7] On the other hand, the

[3] Langermann, 'Criticism of Authority'.

[4] The passage will be cited in full below. Concerning this commentary see Sezgin, *GAS* VI, p. 91, no. 19, and the extensive description in Sabra, 'One Ibn al-Haytham', pp. 33–39.

[5] Sabra, 'One Ibn al-Haytham', p. 33. For some reason, Sabra translates *tahdhīb* as 'commentary'.

[6] Toomer, *Ptolemy's Almagest*, p. 5. Kunitzsch, *Der Almagest*, pp. 130–35, discusses the naming of the sections of the book in Arabic (*maqāla*, *qawl*, *nawʿ*) and exhibits the incipits of some sections of Book I; however, he does not address the chapter titles that are found in Arabic. I know of only one study of the Hebrew versions, which were translated from the Arabic: Zonta, 'La tradizione Ebraica', pp. 325–50. Zonta finds that the translator Jacob Anatoli would consult the Latin version of Gerard of Cremona when tackling difficult passages.

[7] The Arabic text has not been edited; the Paris text that I utilized is Kunitzsch's manuscript *q*, a Maghrebi copy of the Isḥāq-Thābit version dated 618/1221 which he describes on pp. 42–43.

title in the Ḥajjāj version states a problem to be solved: *Kayfa yuʿlamu anna ḥarakatu al-samāʾi kuriyyat*^{un}?[8] 'How is it known that the motion of the heaven is spherical?' These differences in the headings already suggest different understandings as to what the chapter sets out to accomplish.

The present foray will be limited to three commentaries, in the wider sense of the term: that of Ibn al-Haytham, already mentioned; the work of Jābir ibn Aflaḥ, which should probably be considered to be both an epitome and a critique or 'correction' (*iṣlāḥ*); and the section on Ptolemy's 'cosmological principles' at the beginning of Abū Rayḥān al-Bīrūnī's *al-Qānūn al-Masʿūdī*. Inspection of additional commentaries, and especially the untouched Hebrew commentarial tradition, must be left for another occasion.

2. Ibn Al-Haytham (965–1040)

Ibn al-Haytham's commentary is preserved uniquely in MS Istanbul, Topkapi Library, Ahmet III 3329; the codex contains other commentaries as well. Ibn al-Haytham introduces his commentary with a relatively long essay in which he has some interesting things to say about the *Almagest*, the genre and tradition of the commentary, the way astronomy was studied in his own day, and the envisioned contribution of his work, which, as noted, is both a commentary and an epitome. Note in particular Ibn al-Haytham's criticism of the commentary (at present considered lost) of al-Nayrīzī;[9] to some degree his commentary was meant to be a better alternative to that of his predecessor. I present here a liberal citation from his preface:[10]

> [38v] I have found that the main intent of all who have commented upon this book has been to explicate the ways of computation and their ramifications, as well as noting other aspects that Ptolemy did not take notice of, without, however, shedding light on those concepts that are obscure to the beginner. Al-Nayrīzī did something like this, burdening his book by greatly increasing the examples of the computational methods, and placing in this his hope for glorification and honor for what he writes.
>
> Therefore, I have seen it fit in commenting on this book to give an account in which my main approach is to facilitate for students the understanding of those concepts (*maʿānī*) that are subtle. I add to it some commentary that appertains to the computation of the astronomical tables (*zīj/azyāj*), things that were neglected by Ptolemy. In omitting to mention them he was being concise, relying upon commendable minds to derive (*istikhrāj*) them. The principles which Ptolemy did record in his book are

[8] MS London, BL, Add. 7474, fol. 3a, online at <http://www.qdl.qa/en/archive/81055/vdc_100023514339.0x000011>, visited 25 May 2016.

[9] On the commentary of al-Nayrīzī see Sezgin, *GAS* VI, p. 192, no. 4.

[10] After having prepared my own translation of this passage, I found that Sabra had translated most of it in 'One Ibn al-Haytham', p. 35. My translation is much more literal and far less elegant than his, but I have decided to nonetheless present here my own version. My single disagreement with Sabra is discussed in the following note.

more than enough [literally: 'sufficient and enough'] for someone who possesses the barest talent to be able to derive them. In doing so I shall be explicating (*mufassir*) and giving a précis of (*mulakhkhiṣ*) the concept about which I intend to speak, basing [myself] on his words, that is, that which he recorded in the book [called] the *Almagest*. In this way, should the person seeking to know the *Almagest* hit upon the word for that concept, he may refer for commentary (*sharḥ*) and a précis to what I have recorded in this book of mine. Combining the two books [the *Almagest* and my commentary/précis], he will understand it; both the words and the concepts will become clear. Indeed, had I done this book the way the commentators to books do, which is to cite the word and follow it with an explication of the concept, there would have been no benefit in it.[11] For then I would have been transcribing a book which is famous and available. I will endeavor to be brief, confidently abandoning prolixity and lengthiness. Eloquence (*balāgha*) does not consist in expressing a few concepts with many words. Instead, eloquence is the clarification of many concepts with measured (*yasīr*) words, because excessive verbiage leads comprehensions astray, and terseness falls short of encompassing the concepts. Moreover, that which this commentary and précis records will be by way of proof, as well as solving (*ikhrāj*) for that which requires a solution by computation, concise rather than drawn out, shortening and not lengthening.

Had I instead done as al-Nayrīzī did, the commentary would have been long, indeed it would have become twice [the size] of Ptolemy's book. It would have made it more difficult than obtaining knowledge from [Ptolemy's] book directly. My sole purpose in what I am doing is make knowledge more accessible and practice easier. Nonetheless, I aim for [the same goal] as do most of those who set their minds upon knowledge of the *Almagest*, namely the knowledge by means of which one grasps the reasons underlying the operations (*aʿmāl*), which is the [true] subject for the person in quest of this art. Many people do it; they have introduced approximations[12] that facilitate the operations for the practitioners.

Recall that Ibn al-Haytham, even if his ultimate goal is to clarify Ptolemy's text, explicitly rejects the format of citation followed by commentary. In this respect, his book has the form of an epitome. In keeping with this plan, the chapter of interest ('On clarifying that the heavens are spherical', beginning on fol. 39v) is, indeed, a newly written chapter that takes as its starting and refer-

[11] Sabra here adds in brackets the word 'without', which significantly alters the meaning: 'If I had followed the practice of the commentators of books by quoting the words (without) following them with an explanation…' This seems to me to be too severe an intervention; moreover, do commentators simply cite from the book they are glossing without explaining? I believe that Ibn al-Haytham announces here his rejection of the usual form of the commentary, which consists in citing the text (*matan*, often copied in red ink and/or larger letters; Ibn al-Haytham eschews that term, using instead *alfāẓ*, 'words') followed by a commentary (*tafsīr*). Indeed, that form is not employed in his commentary.

[12] *Taqrībāt*; 'shortcuts' is a possible alternative translation.

ence points the corresponding section of the *Almagest*. Ibn al-Haytham offers a logical, well-organized exposition, drawing alternatively on astronomical, mathematical, optical and philosophical arguments, as the issue may require.

Ibn al-Haytham begins with a demonstration that the heavens must be a body and that this body must maintain a daily rotation about a pole without, however, producing any apparent variation in the sizes of the stars to the observer on earth. These considerations (and others, such as the accuracy of sundials that are constructed on the basis of a spherical model) lead already to the selection of the sphere. Next, a series of subtle philosophical arguments are mustered in order to show that the cosmos is the largest body in existence but also finite. As such, it will be bounded by a body that 'encompasses but is not encompassed'. Now it remains to show which geometrical shape will give the greatest volume for a given surface area. In fact, it has already been shown that the cosmos must be a sphere; we are retracing, or, to use an anachronism, we are checking this finding by showing that the sphere, in addition to meeting the observational requirements that it has already been shown to meet, will also answer this new demand, i.e., of providing the greatest volume for a given surface area. To do this Ibn al-Haytham returns to the cone and the cylinder, which had already been rejected on other grounds. He easily shows, on the basis of formulae for volumes that were available to him, that for the cone, sphere, and cylinder constructed on the same circle, the sphere will have the greatest volume.

On the other hand, Ibn al-Haytham totally ignores Ptolemy's arguments against the theory that the stars move in a straight line toward infinity, or that they are kindled and extinguished every day. Those ancient theories, which Ptolemy had to refute, had no currency in Islamic civilization, and Ibn al-Haytham felt that he could safely ignore them. More generally, Ptolemy's chapter, which has at times the appearance of a series of unconnected arguments, takes on a greater deal of coherence, organization, and relevance for the science of the period in Ibn al-Haytham's commentary.

Let us look a bit more closely at some highlights from Ibn al-Haytham's epitome of the chapter. He first proves that the heavens are a body. His argument is that the planets carry out different motions in the east-west and north-south directions simultaneously. This can be true only if they are joined to bodies that are joined to other bodies. His discussion begins as follows [fol. 39v]:

> Now, as for [the statement that] the heavens are spherical, it must have been preceded by the knowledge that the heavens exist as a body. This is known by means of what I shall describe.
>
> That is to say, sensation perceives the corporeal sensibilia through the intermediary of a certain thing. For example, sensation perceives the body possessing color through the intermediary of the color, and sensation perceives the body of the air,

> which does not possess color, through the intermediary of the propulsion of the sensible body from its place. Now it is apparent to sensation that the swift stars move with two motions, one to the west, the other to north and south, [both happening] at the same time; but it is not possible for a single body to perform two different motions simultaneously. Hence these two motions that belong to the stars must be due to that to which they [the stars] are joined. But bodies join one to another only by contact, only by body touching body. Hence it is clear to sensation, through the intermediary of transpositional motion, that the heavens exist as a body.

Ibn al-Haytham then notes that in their daily rotation, the stars all appear to move about a single point, tracing parallel circles that get larger the more one moves away from that point. This indicates that there is a pole about which the stars rotate; only three models can account for this. To be more specific, only three 'surfaces' (*basāʾiṭ*) should be considered: the cylinder, the double cone, and the sphere. Autolycus has already shown how the 'moving sphere' can account for the motions.[13] The other two options will not work; since the radii, or distances from the center of the solid to stars moving on different circles (which are parallel to the circle around the pole) are different, the distances of the heavenly bodies from the center, i.e. the earth, will vary and so also their apparent sizes; but no such variation is observed. Another proof comes from sundials, which would not work if the cosmos had a non-spherical shape. In these arguments, which I will pass over, Apollonius' *Conics* and Euclid's *Phaenomena* are cited.

Ibn al-Haytham (41v ff.) then presents extended, subtle philosophical arguments that 'reality' (*wujūd*) must be finite, and at its outer exterior there must be a body that encompasses but is not itself encompassed; this must be the greatest body. Starting then with the knowledge that the cosmos is the greatest body (in volume), we prove that it must be a sphere.

Four shapes are to be considered: the cylinder, the cone, the sphere, and the polyhedron (with planar faces). First the polyhedron is eliminated from the competition by means of some geometrical proofs. Ibn al-Haytham shows that the area of a circle whose perimeter is equal to that of a given polygon will be greater than that of the polygon. Consider circle K and polygon A whose perimeters are equal. Now inscribe circle W in the polygon. The radius of the inscribed circle will be less than that of circle K, because its perimeter is smaller. Now the area of the circle K is (radius K) × (half the perimeter), and that of the polygon is (radius W) × (half the perimeter). But the perimeters have been assumed to be equal; since (radius K)> (radius W), the area of circle K is greater than the area of the polygon.

[13] On the Arabic translations of Autolycus see Sezgin, *GAS* VI, pp. 73–74.

Here is the diagram in the manuscript and the proof displayed in simple modern notation:

Circle K Polygon A with circle W inscribed

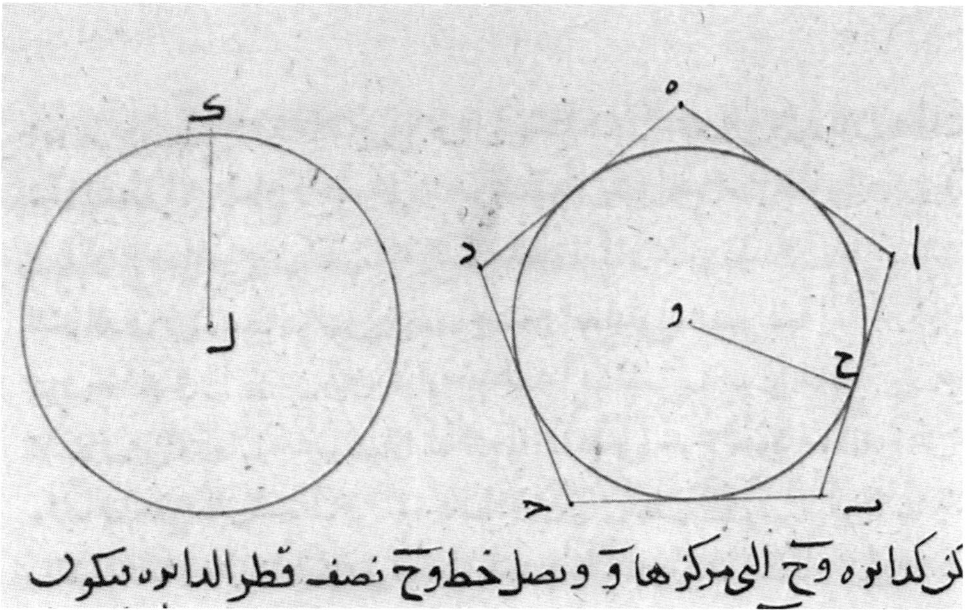

Area (circle K) = $\mathbf{r}_K \times (\text{perimeter}_K)/2$
Area (polygon A) = $\mathbf{r}_W \times (\text{perimeter}_A)/2$
$\text{perimeter}_K = \text{perimeter}_A$
$\mathbf{r}_K > \mathbf{r}_W$
$\therefore$ Area (circle K)> Area (polygon A)

Moving now to three dimensions, Ibn al-Haytham writes:

> This can also be proven by means of a geometrical demonstration. We take a sphere such as sphere K, and a polyhedron such as polyhedron ABCDE; the two have equal surfaces. We inscribe in this [polyhedral] body a sphere whose center is W [as in the example of the sphere and the pentagon]. The diameter of sphere K is greater than the diameter of sphere W. Since this is so, the same holds true for the surface areas of the solids as held true for the areas of the planar figures.

Ptolemy did not give a proof for his statement that the sphere has the greatest volume of any solid possessing the same surface area. However, Zenodorus did, and his proof is cited by Theon, who inserts the entire monograph on isoperimetry into his commentary to Book I of the *Almagest*; it is also found in the medieval tract known as *De isoperimetris*, which was published by Busard and studied further by Knorr, along with the text cited by Theon.[14] Knorr elegantly summarizes the proof as follows: Inscribe the sphere within the solid

[14] Busard, 'Der Traktat'; Knorr, *Textual Studies*, pp. 689–752.

(Euclid shows that this can be done). The sphere and solid will have the same radius, but the surface of the solid will be greater. Hence one must extend the radius of the sphere in order to have the sphere have the same surface as the solid. The volume of both objects is obtained by multiplying one-third of the radius by the surface; since the radius of the sphere is larger, its volume will also be larger than that of the solid.[15]

Theorem 6 in *De isoperimetris* proves the proposition for planar figures in the same way that Ibn al-Haytham does. Comparing the texts, however, I do not find that Ibn al-Haytham is citing a direct translation of the same text. Following Yushkevich, Busard, the editor of *De isoperimetris*, remarks that the Latin text was not based on an Arabic version.[16] The inclusion of this proof in the commentaries of Ibn al-Haytham and (as we shall see later) Jābir ibn Aflaḥ raises the possibility that the text, or portions of it, was available in Arabic.

Ibn al-Haytham continues [fol. 42v]:

> It has thus been shown that, since the heavens are the largest body, *prima facie* it must be the case that the heavens are a cylinder, a cone, or a sphere. But it is impossible that heavens be cylindrical or conical, because the first mover of the heavens, which, as Aristotle showed in the Metaphysics, has no magnitude (*'uẓm*) at all, moves them by means of an infinite power.[17] That which acts by means of an infinite power performs the utmost action within the realm of possibility (*mā huwa fī al-ghāya min unṣur al-imkān min al-af'āl*)...

Since the motion of the heavens is the swiftest of all motions, the body that carries out this motion must allow for the smoothest, most fluid and compliant motion. Superlatives such as these are included in the notion of the 'ultimate action in the realm of possibility'. This phase of the argument again rules out the cylinder and the cone — they have already been eliminated at least twice on other grounds — by the criterion of 'smoothness' (*salāsa*). Only a body that is round on all sides can meet this criterion.

'Smoothness' is mentioned by Ptolemy — that the motion of the stars must be 'the most unhampered and free of all motions' — before he states that the sphere will offer the greatest surface, as one of the additional considerations that lead to the sphericity of the heavens, the strongest being the 'revolution of the ever-visible stars' and their unvarying sizes and mutual distances in the course of their revolutions.[18] It seems that Ibn al-Haytham sees in Ptolemy's 'smoothness' of motion an allusion to Aristotle's description of the prime mover, and its imparting to the heavens the swiftest motion possible; this can only be the case if the heavens have the shape of a sphere, round on all sides.

[15] Knorr, *Textual Studies*, p. 716.

[16] Busard, 'Der Traktat', p. 62, with the reference to Juschkewitsch in note 2.

[17] Aristotle, *Metaphysics* XII, 1073a 6, states that the first mover has no magnitude (*megethos*).

[18] Toomer, *Ptolemy's Almagest*, p. 39.

Ibn al-Haytham is ready to sum up. He mentions again what seems to have been regarded as the strongest argument in favor of the sphere: any other shape would cause variation in the apparent sizes of the stars, something that is not observed. Finally, to those who claim that the heavens have the form of a polyhedron with elastic faces — this is how I must interpret the phrase '*shakl … dhū aḍlāʿ liṭāf*' — such that the sizes of the stars will not vary, Ibn al-Haytham responds: this theory demands that there be a vacuum beyond the heavens, to provide room for the 'angles' of the polyhedron to move. However, that has already been shown to be impossible [fol. 42v]. A few more pages contain some arguments from the science of optics, but I will stop here.

3. Jābir Ibn Aflaḥ (fl. *c.* 1100)

Jābir ibn Aflaḥ's book, which I will take up next, is one or more steps removed from a gloss on Ptolemy's book. The textual issues, and the manuscript traditions in three languages (Arabic — including Arabic in Hebrew characters, Hebrew, and Latin) have been studied, though perhaps not exhaustively. I have profited in particular from the publications of Richard Lorch and Josep Bellver.[19] In the present state of our knowledge, it emerges that the book (or books) is known under two titles, *Kitāb al-Hayʾa*, which would be appropriate for a basic textbook on astronomy, and *Iṣlāḥ al-Majisṭī*, 'Correction of the *Almagest*'. As the second title implies, Ibn Aflaḥ has some serious criticisms of Ptolemy. However, unlike Ibn al-Haytham's *Shukūk*, which is limited to a series of criticisms on specific points (and not limited to the *Almagest*, but addressing other writings of Ptolemy too), *Jābir's* book is a rather thorough reworking of the *Almagest*, both in terms of content and organization, which contains some criticisms as well.

Jābir's book has undergone some revision: perhaps on the part of the author, certainly on the part of Moses Maimonides, perhaps both. There are some substantive differences between the manuscripts, but just if and how these differences reflect the revisions remains unclear. Specifically, Richard Lorch has found the Arabic text that is labelled *Iṣlāḥ al-Majisṭī* on the front leaf of MS Berlin, SBPK, Landberg 132 (Ahlwardt 5652) differs from the version found in two Escorial manuscripts, árabe 910 and 930.[20] In any event, all versions of Jābir's book meet the wider definition of commentary that I set down at the beginning of this paper.

How does Jābir handle the question of the sphericity of the heavens? In Book I, which is dedicated to the geometry needed for the study of the *Almag-*

[19] Lorch, 'The Astronomy'; Bellver, 'The Role'. See also Bellver's article in this volume. I have inspected some of the manuscripts myself, without, however, carrying out a systematic study.

[20] Lorch, 'The Astronomy', p. 88; the historian Ibn al-Qifṭī mentions Maimonides' work on this text, see Lorch, 'The Astronomy', p. 89.

est, Jābir offers a proof for Ptolemy's unsubstantiated statement concerning the greatest equal-surface property of the sphere. Jābir limits himself to comparing the sphere with the five Platonic solids, declaring that 'there can be no regular solid other than them'. His proof is the same as that of *De isoperimetris*, which we have already met in the commentary of Ibn al-Haytham; however, that text, or rather, the different traditions within which that text was transmitted, extends the investigation to other solids as well, as does Pappus. Jābir apparently thought all of these extra proofs to be superfluous. Note further that the Hellenistic treatise on isoperimetry goes through the proof for each of the five solids; Jābir considers this too to be unnecessary.

Here is Jābir's proof, from Book I of his treatise, which I translate from one of the Hebrew versions (MS Paris, BnF, hébr. 1025, Jābir/Shmu'el of Marseilles), fol. 16a:

> After this has been shown, it will now be easily shown that the measure (scil. volume) of any sphere is greater than the measure of any regular solid whose surface area is equal to the surface area of that sphere.
>
> Let us take sphere AB whose surface area is equal to the surface area of regular solid C. I say that sphere AB is greater than solid C.
>
> Proof: Regular solid C must be one of the five solids mentioned by Euclid, because there can be no regular solid other than them. Imagine a figure similar to solid C so that sphere AB is [inscribed] in it. The surface area of this figure will be greater than the surface area of sphere AB, and so it will be greater than the surface area of solid C. Hence its height, which is the radius of sphere AB, is greater than the height of figure C. But multiplying the radius of sphere AB by one-third the surface area of sphere AB gives its volume; and multiplying the height of figure C by one-third of its surface area gives its volume. The surface area of the sphere has been assumed to be equal to the surface area of solid C. Therefore, sphere AB is greater than solid C. Q. E. D.

Jābir takes up his paraphrase of the *Almagest* in Book II of his work [MS Berlin, fol. 17r, bottom]. Following Ptolemy, he first discusses 'the order of the theorems' (*Almagest* 1.2), before proceeding to the six cosmological principles (*Almagest* I.3–8). According to Jābir the proper order is as follows:

> The first thing that we ought to examine in this book is the general issue of the earth as a whole relative to the heavens as a whole. After that we should try to learn the situation of the inclined orb and the inhabited places on earth, and the gradual difference of their horizons, which is due to the latitude. If we first learn the things that we have mentioned, the investigation into other issues will be easier.
>
> The second thing that we ought to seek to know is the motions of the sun and the moon, and all that they entail; until this knowledge has been attained, it is impossible to attain full knowledge of the stars. The final thing to be achieved according to what seems to be the [proper] order is the chapter on the stars, However, the chapter on the sphere of the fixed stars must come first... [ibid., fols 17r-v]

As we shall see, there seems to have developed a consensus of sorts among the Islamic astronomers that the discussion ought to begin with the most demonstrable of the principles, which is the earth and its shape. This provides a solid base for the other theorems, which are more speculative — even if their veracity is not called into question. However, as Jābir explains, Ptolemy takes up the topics in a different order:

> But the proposition that we ought to begin with would show that the heavens are spherical, and that their motion is spherical; that the shape of the earth in all of its parts is sensibly spherical; that its place is in the middle of the whole heaven, as its center; that it is, as far as distance and size are concerned, point-like relative to the fixed stars; and that it has no translational motion. Now let us first present a bit of the teaching concerning the proof of each one of the [items] that we have mentioned. [ibid.]

When he comes to discuss the spherical shape of the heavens, Jābir launches a virulent attack on Ptolemy's chain of reasoning leading to the inference, or conclusion, that the heavens are the greatest body in volume. Jābir has sensed something awkward in Ptolemy's text. He senses that the problem is not due to faulty expression alone; poor reasoning also must share in the blame. As we shall see presently, the problematics concerning which Jābir complains remain in the Arabic translation. He writes:

> Said the author: His [Ptolemy's] saying that, regarding different figures that have the same surface area,[21] that which has the most angles has the greatest volume, and, therefore, the circle must be the greatest among plane figures and the sphere among solids, is a statement that is as brief and general as can be, and as contrary [to fact] as can be. This is the first place where his insufficiency in the science of geometry can be seen. Let us now set out to interpret his statement, to explain what he intended, and then explain why it is contrary [to fact]. We say: what is to be understood from his statement — though the words do not convey this, the idea (*maʿnā*) that he is aiming at explaining does get it across — is [the following] ... [ibid.]

One can get a good idea of the problematic nature of Ptolemy's text from Toomer's translation, where we find not a few words added in brackets, most notably the word 'likewise', which indicates an inference:

> [S]imilarly, since of different shapes having an equal boundary those with more angles are greater [in area or volume], the circle is greater than [all other] surfaces, and the sphere greater than [all other] solids; [likewise] the heavens are greater than all other bodies.[22]

[21] The critical phrase 'that have the same surface area' is missing in the Berlin manuscript, fol. 18b, top line; it is found in the Hebrew translations, e.g. MS Paris, BnF, hébr. 1025, fol. 25v. It is hard to imagine that the omission in the Berlin manuscript — the one carrying the title, *Iṣlāḥ al-Majisṭī* — is anything but a scribal error.

[22] Toomer, *Ptolemy's Almagest*, p. 40.

Jābir launches into a lengthy diatribe, which I display below in Arabic, Hebrew, and English. The gist of his remarks seems to be that Ptolemy ought to have shown (as Jābir did in the first section of his book) that the sphere is greater in volume than any (regular) polyhedron having the same surface area. Instead, he talks about figures having 'more angles' and jumps from this to his conclusion about the heavens. Such reasoning is a *shinaʿā* — a word which, in mathematics, means 'absurdity'; but it also conveys an aesthetic or even moral judgement, namely 'abomination', 'something repulsive', and the like.[23]

Here are the texts: first Heiberg's Greek text, and the Arabic *Almagest*, then Jābir's text in Hebrew and Arabic. Once again, I have not scoured all of the manuscripts; the versions that I copy are good and may serve the purpose of this paper. The primary sources are followed by my translation of Jabir's critique.

From the edition in Heiberg, *Syntaxis mathematica*, vol. I, p. 13

> ... τῶν ἴσην περίμετρον ἐχόντων σχημάτον ἐπειδὴ μείζονά ἐστιν τὰ πολυγωνιώτερα, τῶν μὲν ἐπιπέδων ὁ κύκλος γίνεται μείζων, τῶν δὲ καὶ ὁ οὐρανὸσ τῶν ἄλλων σωμάτον.

From MS Paris, BnF, arabe 2482, fol. 3v

> ...وان الأشكال المختلفة التي احاطتها متساوية ما هو منها أكثر زوايا فهو أعظم قدرا ولذلك وجب أن تكون الدائرة أعظم السطوح والكرة أعظم المجسمات والسماء أعظم مما سواها من الأجسام

From Jābir's critique (Hebrew: Paris, BnF, hébr. 1025, fol. 25v-r)

> אמר המחבר אולם אמרו שהתמונות המתחלפות אשר הקיפם שוים מה שהיא מהם יותר רבת זויות הנה היא יותר גדולת שיעור ולכן התחיב שהעגול יותר גדול השטחים והכדור גדול שבמוגשמים הוא דבור בתכלית הכללות והקצור ובתכלית החלוף וזה הראשון שבמקומות שיראה בו קצורו בחכמת ההנדסה ונכוין עתה אל פירוש דבורו ונבאר מה שרצה ואחר זה נבאר חלופו בה.

> [25v]ונאמר שאשר יובן ממאמרו [נ"ב מדבורו] זה ואע"פ שלא תתנהו המליצה אבל יתנהו הענין אשר חתר לבארו שהוא מפני שהיה התכלית המכוון מן הגלגל שיכלול ושיחזיק מהגרמים השמימיים יותר מה שאפשר התחייב שיהיה תמונתו תמונה תתן לו זה והיא התמונה הכדורית לפי שהכדור גדול מכל אחת מהתמונות הרבות הזויות אשר הקפיהם שוים להקפת הכדור ולכן אמר והכדור יותר גדול המוגשמים והשמים גדולים ממה שזולתם מהגשמים

> וביאור זה הענין ר"ל שהכדור גדול התמונות הרבות הזויות אשר הקפיהם שוים להקף אותו הגדור קרוב הלקיחה כפי מה שזכרנו במאמר הראשון מזה הספר ודרך הוא אל ביאורו כפי מה שתתנהו מליצתו כשיבאר תחלה בתמונות הרבות הזויות השוות ההקפה כי מה שהוא מהם יותר רבת זויות הנה היא גדולת השיעור

[23] On the meanings of this word see Langermann, 'The Translation'.

וכשהתבאר זה העתיק המשפט אל העגול ואל הכדור ואם לא איך יצא אומרו ולכן יתחייב שהעגול גדול מכל השטחים והכדור גדול מכל המוגשמים ואלו ידע שבאור זה בתמונות קצתם עם קצת יותר קשה הרבה בביאורו בתמונות רבות הזויות והכדור לא יכוין אל זכרון התמונות קצתם עם קצת מאשר הכוונה המכוונת אמנם היא ענין הכדור עם תמונה רבות הזויות לא ענין תמונה עם תמונה

עוד שהוא יצא מכח דבורו בהעתיק המשפט מהתמונות אל העגול והכדור שהעגול מלא זויות שטחיות והכדור מלא זויות המוגשמות וזה מהגנות מה שלא יעלם

From Jābir's critique (Arabic: Berlin, SBPK, Landberg 132, fol. 18v):

قال جابر أما قوله أن الأشكال المختلفة ما هو منها أكثر زوايا فهو أعظم قدراً ولذلك وجب أن الدائرة أعظم السطوح والكرة أعظم المجسمات فكلام في غاية الإجمال والإيجاز وهو مع ذلك خلف في القول ولهذا أول موضع ظهر فيه مختلفة في علم الهندسة ولنقصد الآن الى شرح كلامه ونبين ما أراد وبعد ذلك نبين مختلفه فيه

فنقول إن الذي يفهم من كلامه هو إن كان لا يعطيه اللفظ لكن يعطيه المعنى الذي رام تبيينه أنه لما كانت الغايةِ المقصودة في الفلك الإحتواء وأن يسع من الأجرام أكثر ما يمكن وجب أن يكون شكله شكلاً يعطيه ذلك وهو الشكل الكري لأن الكرة أعظم من كل واحد من الأشكال الكثيرة الزوايا التي أحاطتها مساوية لإحاطة الكرة قريب المآخذ على ما ذكرناه في المقالة الأولى وتطرق هو الى تبيينه على ما يعطيه لفظه بأن نبين أولا في الأشكال الكثيرة الزوايا المساوية الإحاطة ان ما كان منها اكثر زوايا أعظم قدراً

فإذا استبان ذلك نقل الحكم الى الدائرة والكرة وإلا فكيف يخرج قوله ولذلك وجب أن الدائرة أعظم السطوح والكرة أعظم المجسمات ولو علم أن تبيين بعضها من الاشكال مع بعض أصعب بكثير من تبينه في الشكل والكرة لم يتعرض الى ذكر أحوال الأشكال بعضها مع بعض إذ الغرض المقصود إنما هو تبين أحوال الكرة مع شكل كثير الزوايا لا يتبين شكل مع شكل

ثم أنه يخرج من قوة كلامه في نقل الحكم من الأشكال إلى الدائره والكرة أن الدائرة مملوءة زوايا مسطحة والكرة مملوءة زوايا مجسمة وفي هذا شناعة ما لا يخفى

English translation:

Said the author: His [Ptolemy's] saying that, regarding different figures that have the same surface area, that which has the most angles has the greatest volume, and, therefore, the circle must be the greatest among plane figures and the sphere among solids, is a statement that is as brief and general as can be, and as contrary [to fact] as can be. This is the first place where his insufficiency in the science of geometry can be seen. Let us now set out to interpret his statement, to explain what he intended, and then explain why it is contrary [to fact].

We say: what is to be understood from his statement — though the manner of expression does not convey this, but the issue that he is aiming at explaining does convey it — is [the following]. Since the intended purpose of the orb is to encompass and hold within as many bodies as possible, it was necessary that its figure be a figure that will allow this, and that is the spherical figure. Indeed, the sphere is the greatest of all multiangular[24] figures that have the same surface area as the sphere.

[24] I do not use 'polygonal' since in English usage, 'polygon' is generally used to describe a two-dimensional figure.

> For this reason he said, 'and the sphere is the greatest of the solids, and the heavens are greater than other solids'.
>
> The explanation of this issue, i.e., that the sphere is the greatest of all multiangular figures that have the same surface area as the sphere is readily comprehensible, as we mentioned in the first book of this treatise. He approached its proof, as his manner of expression delivers it, by showing first that for multiangular figures having the same surface area, that with the most angles has the greatest volume. When this was shown, he transferred the theorem to the circle and the sphere. Otherwise, how can his statement, 'and therefore, the circle is the greatest of the planar [figures], and the sphere is the greatest of the solids', result?
>
> Had he known that showing [that] this [holds true] for figures [in relation] one to another is much more difficult than it is to show it for the sphere and the figure,[25] he would not have intended the issue of the [relation of multi-angular] figures one to another, but would rather [have limited himself] to the issue of the sphere and the multi-angular figures, but not [the issue of] figure to figure.
>
> Moreover, it emerges from the implications of his remarks about transferring the theorem from the figures to the circle and the sphere, that the circle is full of planar angles, and that the sphere is full of solid angles. The absurdity of this cannot be missed.

Richard Lorch explains Jābir's objection as follows: 'What Ptolemy said was, that of different figures of equal perimeter, those with more angles have greater capacity, and so the circle is the greatest of the plane figures — and similarly the sphere is the greatest of the solid figures. Jabir reasonably complains that this implies that circles and spheres are full of angles, and that it would be far easier to compare the sphere with a polyhedron of equal surface directly (as he does himself in book I)'. Lorch then remarks, 'But in its context in the *Almagest* the point is a small one: Ptolemy's slight clumsiness does not merit such a violent reaction'.[26]

Jābir, in sum, both proves Ptolemy right and rebuffs him. He provides (at least a partial) geometrical demonstration for the claim that the sphere has the greatest volume. In the next section he castigates Ptolemy's own schematic presentation of the argument that the shape of the heavens must be that of a sphere; Ptolemy's theorem that the heavens are spherical is correct, but he ought to have made a better case for this.

4. Al-Bīrūnī (973–1048)

Abū Rayḥān al-Bīrūnī did not write a commentary to the *Almagest*. Moreover, as George Saliba has pointed out, 'Bīrūnī does not appear to have been inter-

[25] Sic. Jābir means here any polyhedron other than the sphere.

[26] Lorch, 'The Astronomy', p. 96. The comparison that Lorch refers to in Book I is found, e.g., in MS Paris, BnF, hébr. 1025, fol. 19r-v.

ested in the genre of astronomical writing in which Ptolemaic planetary models were considered as describing both the apparent motion of the planets and the physical spheres responsible for the kinematic forces acting upon them'.[27] I agree that al-Bīrūnī evinced no interest in parametrized three-dimensional models for the heavens, of the type Ptolemy takes up in the *Planetary Hypotheses* and to which the so-called Maragha astronomers made their landmark contributions. However, this is not to say that he had no interest at all in the cosmological principles enunciated by Ptolemy at the beginning of the *Almagest*; far from it. In the second chapter of the first book of his comprehensive textbook, *al-Qānūn al-Masʿūdī*, he critically reviews, in order, each one of the six cosmological principles that are established in Book I of the *Almagest*. He calls these subchapters *mabāḥith*, 'inquiries' or 'research projects'. His remarks on the possibility of the earth's rotation were signaled a generation ago in a paper by Shlomo Pines.[28]

As we shall soon see, the question whether the cosmos has the shape of the sphere troubled him for decades. He claimed all along that he believed the heavens to be spherical, and that this could be proven. Nonetheless, I sense that he never freed himself from the doubt that closely related alternatives, such as the ovoid or lenticular shapes, would also meet the basic observational requirements. Aristotle briefly mentions these shapes, along with the cone, polyhedron, and cylinder in *De Caelo* II.4 (287a:16–22), but dismisses them on the grounds that in their revolution, those shapes would require more space than their volume; in other words, one would need empty space beyond the body of the cosmos. (Note that Ptolemy, to the best of my knowledge, pays no attention to *De Caelo* in *Almagest* I.) Aristotle's argument, however, is flawed; this was pointed out already by Alexander of Aphrodisias.[29] It seems that al-Bīrūnī would have rejected out of hand the cone and cylinder — models that were considered by Ptolemy, if only heuristically, and refuted several times by Ibn al-Haytham — because their boundaries include flat surfaces; on the other hand, an ovoid or lenticular rotating on its major axis ought to answer all the observational (but not the aesthetic or philosophical) requirements.

Al-Bīrūnī thought the ellipsoid to be a serious enough alternative that he included it among the queries he sent to his contemporary, the great philosopher — and himself author of a revision of sorts of the *Almagest* — Abū ʿAlī Ibn Sīnā (Avicenna). Dimitri Gutas dates their exchange to circa 1000 with considerable certainty.[30] Al-Bīrūnī inquires:

> The Sixth Question: He [Aristotle] said in Book II [of *De Caelo*] the ovoid and lenticular shapes would require a vacuum and empty space, whereas the sphere does

[27] Saliba, 'Bīrūnī', p. 274.

[28] Pines, 'La théorie', pp. 301–06.

[29] Pellegrin, 'The Argument', pp. 174–76.

[30] Gutas, *Avicenna*, pp. 99, 449–50.

not; but the matter is not as he stated. In fact, the ovoid is generated by the rotation of an ellipse about its major axis, and the lenticular is generated by rotation about its minor axis. As there is no difference with regard to the rotation around the axes by which they are generated,[31] nothing of what Aristotle mentions would occur. The essential attributes of the sphere alone would follow necessarily [for all three]. If the axis of rotation of the ovoid is its major axis and if the axis of rotation of the lenticular is its minor axis, they would rotate like the sphere; neither would require empty space [exterior to the figure]. This would be the case, however, if the axis of [rotation of] the ovoid were its minor axis and the axis of [rotation of] the lenticular were its major axis. Then what he [Aristotle] states would necessarily follow. Even so, the ovoid may rotate around its minor axis and the lenticular around its major axis, both moving consecutively (*taʿāqqub*[an])[32] without need of a vacuum. It would be like the motion of individual entities (*ashkhāṣ*) in the interior of the orb, which contains no vacuum according to the opinion of many people. And I am not saying this in the conviction that the sphere of the orb[33] is not spherical, but rather ovoid or lenticular. I have tried hard to refute that theory. However, I am bewildered by the reasons offered by the master of logic (*ṣāḥib al-manṭiq*).[34]

Observe how al-Bīrūnī is careful to state that he remains committed to the spherical model; nonetheless, he is stupefied by the poor reasoning in its defense on the part of the great logician, Aristotle.[35]

Ibn Sīnā in his reply notes that al-Bīrūnī's objection is sound and 'all commentators' are somewhat embarrassed by this passage. He reminds al-Bīrūnī of the remark of Themistius, that 'the teaching of the Philosopher should be interpreted in the best of ways'.[36] Apparently in keeping with this approach, Ibn Sīnā says that the rotation of an ellipsoid *could* require a vacuum (if the figure rotated on its minor axis), whereas this is never the case with the sphere. That meek remark is the only direct response to al-Bīrūnī's query. Nonetheless,

[31] All three figures — lenticular, ovoid, and of course the sphere as well — are generated by rotation about an axis in a manner that requires no space beyond the generated figure, contrary to Aristotle's claim.

[32] That is, moving one instant after the other, consecutively, without gaps.

[33] *Kūrat al-falak*; here *kūra* obviously means the all-encompassing body, whose sphericity may not be precisely 1.

[34] My translation from the text published by al-Yāfī, *Ḥiwār al-Bīrūnī*, p. 51.

[35] Note as well that al-Bīrūnī states that the absence of a vacuum within the all-encompassing sphere is the view of 'many people'; he doesn't recognize it as a hard-and-fast cosmological doctrine. Paul Hullmeine calls my attention to al-Bīrūnī's questions regarding the vacuum in the letters to Avicenna (the sixth question regarding physical problems, and also the ninth and tenth question on *De Caelo*), where he criticizes the Aristotelian (and Avicennean) arguments against the vacuum, but without really indicating his personal opinion.

[36] Themistius' commentary to *De Caelo* is listed in the curriculum of Abū Sahl al-Masīḥī, Ibn Sīnā's contemporary and teacher in medicine; see Gutas, *Avicenna*, p. 172. See also the following notes.

the latter does not follow up on this query in his second round of correspondence.

As is well-known, Themistius' commentary to *De Caelo* survives only in a Hebrew translation from the Arabic.[37] Ibn Sīnā's brief snippet (*'innahu yanbaghī an yuḥtamila qawlu al-faylasūfi 'alā aḥsan al-wujūhi*) may be the only witness to the Arabic version. It is not entirely clear that Ibn Sīnā is referring to a gloss on *De Caelo* 278a 21, where Aristotle employs the faulty reasoning mentioned above; given the ponderous character of Zeraḥya's Hebrew translation one cannot be sure. Nonetheless, I do believe that Ibn Sīnā is referring to Themistius' gloss on the passage in question. I will attempt to produce a coherent translation of the Hebrew:

> Said Themistius: it is proper that we take the statement as a universal notion. That is, he did not state it as an open-ended statement, but rather in the case that the positioning is in one way. For it is possible for those figures to move in rotation, depending on the positioning, without there being a vacuum beyond the heavens. However, with regard to the sphere, there will never be need for something beyond it at all.[38]

In free translation, Themistius tells us that Aristotle did not mean that the ellipsoid will necessarily require a void beyond the cosmos; that would happen only in a certain positioning, namely, for example, the lenticular rotating about its major axis. However, for the sphere, it is always true: the rotating sphere never requires space beyond the volume of the sphere. This is, in fact, the resolution that Ibn Sīnā suggests in his response.

Now let us turn to al-Bīrūnī's very comprehensive textbook on astronomy, *al-Qānūn al-Mas'ūdī*, which he completed in 1030, some thirty years after his correspondence with Ibn Sīnā.[39] His critical review of Ptolemy's arguments for a spherical cosmos in the first of these 'inquiries' leads him to the following conclusion:

> Since the circuits of the stars cannot be on a flat planar surface, they must take place on the surface of a solid that is not flat [i.e. not made up of straight flat faces]. And since its motion is rotational, it doubtlessly takes place about an axis. Its actual reality [the fact that this body exists in actual reality] requires that it be finite [since no infinite body can exist]; and that the end-points of the axis be the poles of the axis. The heavens, therefore, have two poles, one of them lying below in the south by the same magnitude that the other is above in the north.
>
> This body may be spherical, just as it may be ovoidal or lenticular, or cylindrical or conical or polyhedral. Ptolemy's proof from the unvarying magnitudes of the stars in

[37] The text was edited by Landauer, *Themistii*; on this text and edition see further Zonta, 'Hebraica veritas'.

[38] Landauer, *Themistii*, p. 67 (Hebrew), lines 32–35.

[39] Bosworth, 'Bīrūnī'.

> all directions and regions of the heavens does not rule out the shape's being polyhedral...[40]

Thus, even the polyhedron, all of whose faces are flat, must be considered. The explanation is obscure; our argument thus far 'only rules out the very motion and tracings that the bodies would trace by their means'. I take this to mean, that we need to know the shape of the body on which the observer stands when he makes his observations, in order to be sure that the heavens themselves are not a polyhedron. Presumably, then, an observer located on or in a polyhedron within an enclosing, rotating polyhedron, may detect the same motions as the observer on the earth does now. For this reason, as he states in the passage that follows, al-Bīrūnī defers his proof for the sphericity of the heavens until after the sphericity of the earth is first established: 'We can decisively rule out alternative shapes for the heavens, other than the spherical one, only between the second and third of these investigations [that is, at the end of the second and before the third]; so we will put it off to its proper place'. The second investigation concerns the shape of the earth.

Al-Bīrūnī thus follows Ptolemy's order of the theorems but defers reaching his conclusion until after the sphericity of the earth has been demonstrated. In his discussion of the second principle, which is the sphericity of the earth, al-Bīrūnī again follows Ptolemy but offers a much fuller explanation. He discusses, for example, what happens during a lunar eclipse, so that the reader can understand how the evidence drawn from its simultaneity at different locations on the earth bolsters the claim that the earth is spherical, or at least curved in the direction of longitude. Curvature in the direction of latitude can be shown from the variation in the number of ever-visible stars as we move north or south. Moving beyond the trivial cases (as one may call them) of points on the same parallel of longitude or latitude, he provides a comparison between Aden and Bulghar, which differ with respect to both coordinates. Since curvature has been shown to hold in both latitude and longitude, al-Bīrūnī concludes, the earth's surface must be spherical.[41]

Al-Bīrūnī then takes up a possible objection which he ascribes to a certain *mutakallim*.[42] Perhaps all that has been shown regarding the spherical shape of the earth is true only for the inhabited portions? After all, we have no data from the uninhabited areas. In reply, he argues that the sphericity of the entire earth can be established by observing the shapes of the shadows which the earth casts on the moon during an eclipse. He concludes that all of the above

[40] My translation, with explanatory glosses in brackets, from al-Bīrūnī, *al-Qānunu'l-Mas'ūdī*, vol. I, 29–30.

[41] al-Bīrūnī, *al-Qānunu'l-Mas'ūdī*, vol. I, top of p. 36.

[42] The fourteenth century Jewish savant Hayyim Israeli held much the same view; see my study, in Hebrew: 'The Making of the Firmament'.

'removes the doubt in the matter of the earth, establishing its roundness in all directions; she is therefore sensibly (*fī al-ḥiss*) a sphere'.[43] Then, by observing the culmination of stars (for longitude) and the elevation of the pole (for latitude), we conclude that the shape of the heaven conforms to that of the earth and is thus spherical.

The heavens have thus been proven to have the shape of a sphere. I submit, nonetheless, that al-Bīrūnī may have retained some lingering doubts as to whether the heavens had the shape of a perfect sphere. Indeed, he knew (as did all cosmologists) that the earth had bumps and grooves, in the form of mountains and valleys; but these were negligible relative to the size of the earth. And what about the heavens? Could they have a sphericity that is less than 1? From the point of view of astronomical observations, it would not matter much, if at all; but from the philosophical, or aesthetical (and perhaps theological) perspectives, it seems that a perfect sphere was required.

In my contribution to the Festschrift for A. I. Sabra I described the optical proof for the sphericity of the cosmos put forward by Aḥmad ibn ʿĪsā, which is based on the sphericity of the earth.[44] I suggested that the Arabic-writing thinkers reversed the order of Ptolemy's arguments, since they took the sphericity of the earth to be the more evident of the two principles. Pierre Pellegrin sees the sphericity of the earth to be the best of Aristotle's arguments for the spherical shape of the cosmos. However, he adds, Aristotle argued in the reverse direction — from the sphericity of the outermost sphere down to the earth because the higher regions are 'naturally prior'.[45] A consequence of this approach is that one can explain the minor divergences from the perfect sphericity in the shape of the earth as a degradation of the perfect sphericity of the heavens as one moves below the sphere of the moon.

I now would like to press forward with this same analysis. Reversing the order — arguing from the sphericity of the earth for the sphericity of the heavens — would mean not only an argument that moves from the naturally posterior, but also one that seeks to prove the perfect sphericity of the heavens from the acknowledged imperfect sphericity of the earth. The difficulties are clear enough, but nonetheless, this seems to be the direction chosen by astronomers working in Islamic cultures. The reason for this, as it looks to me now (but maybe not tomorrow), is that the sphericity of the earth is the more secure and certain datum that our observations can provide. In other words, empirical evidence is prior to metascientific (or metaphysical) considerations.

[43] al-Bīrūnī, *al-Qānunu'l-Masʿūdī*, vol. I, pp. 36–37. Paul Hullmeine suggests that al-Bīrūnī may be making the same argument in Chapter 26 of his book on India (Sachau, *Alberuni's India*, vol. I, p. 268). Though the chapter carries the title 'On the Sphericity of the Heaven and the Earth [...]', it offers arguments only for the sphericity of the earth.

[44] Langermann, 'Transcriptions', pp. 247–60.

[45] Pellegrin, 'The Argument', p. 178.

Nonetheless, the perfect sphericity of the heavens remains a philosophical or aesthetic imperative. From the point of view of observational astronomy, a nearly spherical ellipsoid rotating on its major axis, as the shape of the outermost envelope of the world, cannot be ruled out. I somehow suspect that al-Bīrūnī never freed himself from doubt on this matter.

5. Conclusions

The commentary provided astronomers the opportunity to engage head-on with the *Almagest*, to go through it passage by passage in the effort to penetrate Ptolemy's reasoning. Although formally speaking the purpose of the genre was to explicate and clarify rather than to criticize, alter, or reject, in practice a good deal of criticism and reform is to be found in the commentary tradition — especially when one accepts the wider definition of the genre that I have applied in this paper. In particular, this paper fleshes out a form of silent criticism and reform. By that I mean that Ptolemy's proofs for the sphericity of the heavens are not challenged directly. Instead, proofs that seem no longer to be relevant are simply passed over, and the arguments are effectively rearranged so that the sphericity of the heavens is shown to conform to the sphericity of the earth, because the earth's sphericity was held to be the more secure datum.

All the commentators studied here accepted Ptolemy's statement that the heavens have the shape of the sphere; there was general agreement as well that the strongest argument on behalf of the sphere was the unvarying sizes of the stars in their rotations as well as the rotations of the circumpolar stars. This is observational evidence which is intrinsically linked to the position of the observer. Hence, some were of the opinion that this proof depends on the earth's being spherical, which Ptolemy had shown only in the sections following that on the sphericity of the heavens. Therefore, in principle, Ptolemy's ordering of the theorems is mistaken; nonetheless, the Arabic expositions remain loyal to Ptolemy's order of presentation.

The Arabic commentators also felt the need to tidy up Ptolemy's presentation. Arcane alternatives that Ptolemy had refuted at some length could be safely ignored. Unsubstantiated statements, such as the isoperimetric property of the sphere, should be demonstrated. The only serious alternatives were other geometrical solids, especially those with at least some curved surfaces. However, no one actually argued on behalf of one of those alternatives, though, in my own opinion, al-Bīrūnī may not have been entirely convinced that the ellipsoids had to be rejected.

This paper has studied only a small sampling of commentaries written in Arabic to the Arabic *Almagest*. Many more remain to be studied; at least one compares the *Almagest* in the original Greek with both Hebrew and Arabic translations.[46] Further study of the commentary tradition is sure to enrich our understanding of the way this great astronomical textbook was read, sympa-

[46] Langermann, 'Science in the Jewish Communities', pp. 446–49.

thetically and critically, in a wide spectrum of cultures and varying historical contexts.

Bibliography

Bellver, José, 'The Role of Jābir b. Aflaḥ's *Iṣlāḥ al-Majisṭī* in the So Called « Andalusian Revolt against Ptolemaic Astronomy »' (in Spanish), *al-Qanṭara* 30 (2009), pp. 83–136.

al-Bīrūnī, Abū Rayḥān Muḥammad ibn Muḥammad, *al-Qānunu'l-Mas'ūdī (Canon Masudicus). An Encyclopaedia of Astronomical Sciences*, 3 vols, Hyderabad: Osmania Oriental Publications Bureau, 1954–1956.

Bosworth, C. Edmund, 'Bīrūnī, Abū Rayḥān i. Life', in Ehsan Yarshater (ed.), *Encyclopædia Iranica*, vol. IV, London/Boston: Routledge & Kegan Paul, 1989, pp. 274–76. Available online at http://www.iranicaonline.org/articles/biruni-abu-rayhan-i-life (accessed on 10 December 2018).

Busard, Hubertus L. L., 'Der Traktat *De isoperimetris*, der unmittelbar aus dem Griechischen ins Lateinische übersetzt worden ist', *Mediaeval Studies* 42 (1980), pp. 61–88.

Gutas, Dimitri, *Avicenna and the Aristotelian Tradition. Introduction to Reading Avicenna's Philosophical Works*, 2nd ed., Leiden: Brill, 2014.

Heiberg, Johan L., *Syntaxis mathematica*, 2 vols, Leipzig: Teubner, 1898-1903 [*Claudii Ptolemaei opera quae exstant omnia*, vol. I].

Kafih, Rabbi Joseph, *Rabbi Moshe ben Maimon (Maimonides). Iggerot (Letters)*, Jerusalem: Mossad haRav Kook, 1987.

Knorr, Wilbur, *Textual Studies in Ancient and Medieval Geometry*, Boston: Birkhäuser, 1989.

Kunitzsch, Paul, *Der Almagest. Die Syntaxis Mathematica des Claudius Ptolemäus in arabisch-lateinischer Überlieferung*, Wiesbaden: Harrassowitz, 1974.

Landauer, Samuel, *Themistii in libros Aristotelis De caelo Paraphrasis. Hebraice et latine*, Berlin: Reimer, 1902.

Langermann, Y. Tzvi, 'The Making of the Firmament. R. Hayyim Israeli, R. Isaac Israeli, and Maimonides', in Moshe Idel et al. (eds), *Shlomo Pines Jubilee Volume on the Occasion of his Eightieth Birthday* [*Jerusalem Studies in Jewish Thought* 7], Jerusalem: Magnes, 1988, vol. I, pp. 461–76.

Langermann, Y. Tzvi, 'Transcriptions of Arabic Treatises into the Hebrew Alphabet. An Underappreciated Mode of Transmission', in F. Jamil Ragep, Sally P. Ragep and Steven Livesey (eds), *Tradition, Transmission, Transformation*, Leiden: Brill, 1996, pp. 247–60.

Langermann, Y. Tzvi, 'Criticism of Authority in the Writings of Moses Maimonides and Fakhr al-Dīn al-Rāzī', *Early Science and Medicine* 7 (2002), pp. 255–75.

Langermann, Y. Tzvi, 'Science in the Jewish Communities of the Byzantine Cultural Orbit. New Perspectives', in Gad Freudenthal (ed.), *Science in Medieval Jewish Cultures*, Cambridge: Cambridge University Press, 2011, pp. 438–53.

Langermann, Y. Tzvi, Rabbi Yosef Qafih's Modern Medieval Translation of the *Guide*, in Josef Stern, James T. Robinson and Yonathan Shemesh (eds), *Maimonides'* Guide of the Perplexed *in Translation. A History from the Thirteenth Century to the Twentieth*, Chicago: University of Chicago Press, 2019, pp. 257–78.

Lorch, Richard, 'The Astronomy of Jābir ibn Aflaḥ', *Centaurus* 19 (1975), pp. 85–107.

Pellegrin, Pierre, 'The Argument for the Sphericity of the Universe in Aristotle's *De Caelo*: Astronomy and Physics', in Alan C. Bowen and Christian Wildberg (eds), *New Perspectives on Aristotle's De caelo*, Leiden: Brill, 2009, pp. 163–85.

Pines, Shlomo, 'La théorie de la rotation de la terre à l'époque d'Al-Bîrûnî', *Journal Asiatique* 244 (1956), pp. 301–05.

Ragep, F. Jamil, *Naṣīr al-Dīn al-Ṭūsī's Memoir on Astronomy (al-Tadhkira fi ʿilm al-hayʾa)*, 2 vols, New York: Springer, 1993.

Sabra, Abdelhamid I., 'One Ibn al-Haytham or Two? An Exercise in Reading the Bio-Bibliographical Sources', *Zeitschrift für Geschichte der Arabisch-Islamischen Wissenschaften* 12 (1998), pp. 1–50.

Sachau, Edward, *Alberuni's India. An Account of the Religion. Philosophy, Literature, Geography, Chronology, Astronomy, Customs, Laws and Astrology of India About A.D. 1030*, 2 vols, London: Kegan Paul, 1910.

Saliba, George, 'Bīrūnī, Abū Rayḥān iii. Mathematics and Astronomy', in Ehsan Yarshater (ed.), *Encyclopædia Iranica*, vol. IV, London/Boston: Routledge & Kegan Paul, 1989, pp. 277–79. Available online at http://www.iranicaonline.org/articles/biruni-abu-rayhan-iii (accessed on 10 December 2018).

Sezgin, Fuat, *GAS* VI = *Geschichte des arabischen Schrifttums*, vol. VI: *Astronomie bis ca. 430 H.*, Leiden: Brill, 1978.

Toomer, Gerald J., *Ptolemy's Almagest*, London: Duckworth, 1984 [2nd edition Princeton: Princeton University Press, 1998].

Wisnovsky, Robert, 'The Nature and Scope of Arabic Philosophical Commentary in Post-Classical (ca. 1100–1900 AD) Islamic Intellectual History: Some Preliminary Observations', *Bulletin of the Institute of Classical Studies* 47 Suppl. 83/2 (2004), pp. 149–91.

al-Yāfī, ʿAbd al-Karīm, *Ḥiwār al-Bīrūnī wa-Ibn Sīnā*, Damascus: Dār al-Fikr, 2002.

Zonta, Mauro, 'La tradizione Ebraica dell'Almagesto di Tolomeo', *Henoch* 15 (1993), pp. 325–50.

Zonta, Mauro, 'Hebraica veritas: Temistio, *Parafrasi* del *De caelo*. Tradizione e critica del testo', *Athenaeum* 82 (1994), pp. 403–28.

The Arabic Versions of Jābir b. Aflaḥ's *al-Kitāb fī l-Hayʾa*

José Bellver

Abū Muḥammad Jābir b. Aflaḥ, the Latin Geber, was an Andalusi mathematician and theoretical astronomer who probably flourished in early sixth/twelfth century Seville. He is the author of the *al-Kitāb fī l-Hayʾa*, or the *Book on Astronomy*, a reedition of Ptolemy's *Almagest*, which is now better known as *Iṣlāḥ al-Majisṭī*, or *Correction of the Almagest*. Jābir b. Aflaḥ's *al-Kitāb fī l-Hayʾa* was translated into Latin and Hebrew. To date, there are four known Arabic manuscripts in Arabic script[1] which transmit four different versions. It is not clear whether these versions were authored by Jābir b. Aflaḥ himself or by a later author. The aim of the present contribution is to discuss the authorship and chronological order of the different versions of Jābir b. Aflaḥ's *al-Kitāb fī l-Hayʾa* based on the earlier witnesses.[2] In order to fulfil this aim, I will survey the data we know about Jābir b. Aflaḥ, his main work, and its influence on the astronomy of the Islamicate world. I will also elaborate on the title under which this work was known to its contemporaries.

1. Jābir b. Aflaḥ

Little is known about Jābir b. Aflaḥ's life. Biobibliographical dictionaries, even with a specific interest in scientists, such as Ibn Abī Uṣaybiʿa's (d. 668/1270) *ʿUyūn al-anbāʾ*, remain silent. The facts and circumstances of his life provided by Lorch in his seminal work on Jābir b. Aflaḥ[3] are still valid today. Lorch placed Jābir b. Aflaḥ in the first half of the sixth/twelfth century based on references given by Ibn Rushd (the Latin Averroes, d. 595/1198) and Mūsā b. Maymūn (the Latin Maimonides, d. 601/1204). In his *Compendium of the Almagest*, Ibn Rushd stated that Ibn Aflaḥ lived in his own century, that is the sixth/twelfth century. More detailed is the reference on Jābir b. Aflaḥ by

[1] There are at least three extant Arabic manuscripts in Hebrew script of which two seem to be incomplete. See Lorch, 'The Manuscripts'.

[2] I am preparing the critical edition of Jābir b. Aflaḥ's *al-Kitāb fī l-Hayʾa* based on the extant Arabic manuscripts in Arabic script. This contribution is the result of this ongoing edition.

[3] Lorch, 'The Astronomy'.

Ptolemy's Science of the Stars in the Middle Ages, ed. by David Juste, Benno van Dalen, Dag Nikolaus Hasse and Charles Burnett, PALS 1 (Turnhout, 2020), pp. 181–199
© BREPOLS PUBLISHERS 10.1484/M.PALS-EB.5.120179

Mūsā b. Maymūn in his *Guide of the Perplexed* II.9, which I translate from the Arabic:[4]

> Then, there appeared groups of people from the later generations (*muta'akhkhirūn*) in al-Andalus who were very proficient in mathematics (*ta'ālīm*) and clarified, according to the principles laid down by Ptolemy, that Venus and Mercury were above the Sun. On this topic, Ibn Aflaḥ al-Ishbīlī, whose son I met, wrote a famous book. Thereupon the excellent philosopher Abū Bakr b. al-Ṣā'igh [i.e. Ibn Bājja], under the guidance of one of whose students I myself have studied his books, examined this question and exposed some ways of argumentation, which we copied, by which it is shown to be implausible that Venus and Mercury are above the sun. Nevertheless, what Abū Bakr mentioned is an argument showing its implausibility, not an argument proving its impossibility.

Therefore, since Ibn Bājja's argumentation seems an answer to Jābir b. Aflaḥ and bearing in mind that Ibn Bājja passed away in Ramaḍān 533/May 1139 when he was still middle-aged, Jābir b. Aflaḥ should have been active at least during the first third of the sixth/twelfth century. Along these lines, since the first steps in the education on the secular sciences included astrology, mathematics and astronomy,[5] Mūsā b. Maymūn probably met Jābir's son at a young age when he was still in al-Andalus before he left, early in his twenties, for Fez with his family around the year 554/1160. His remark that he met Jābir's son and not Jābir b. Aflaḥ himself may indicate that Jābir b. Aflaḥ had passed away, or at least was no longer teaching, by the beginning of the second half of the sixth/twelfth century. In addition, the reference in Ibn Rushd's *Compendium of the Almagest* also suggests that Jābir b. Aflaḥ was no longer alive by the time Ibn Rushd completed the *Compendium* in the period between 554/1159 and 557/1162.[6]

Jābir b. Aflaḥ's interest in the *Almagest* links him to the Sevillian intellectual circles around the prominent Abū 'Abd Allāh Mālik b. Yaḥyā b. Wuhayb (d. 525/1130–1), known in his lifetime as the Philosopher of the West. Mālik b. Wuhayb lived most of his life in Seville, although later in life he was called to Marrakesh where he was a distinguished jurist in the service of the Almoravids. He was proficient both in the transmitted (*naqlī*) and in the intellectual (*'aqlī*) sciences with interests ranging from philosophy and logic to astronomy and astrology. The historian 'Abd al-Wāḥid al-Marrākushī (d. 647/1250) reports that he saw copies of the *Almagest* and the *Centiloquium* in the hand

[4] cf. Ātāy, *Dalālat al-ḥā'irīn*, p. 293. For a different translation, see Pines, *The Guide of the Perplexed*, vol. I, pp. 268–69. The translation in Lorch, 'The Astronomy', p. 85, from Friedländer contains some inaccuracies.

[5] Kraemer, 'Moses Maimonides', p. 13.

[6] For the period of the composition of Ibn Rushd's *Compendium*, see Lay, 'L'Abrégé de l'*Almageste*', p. 25.

of Mālik b. Wuhayb.[7] Al-Marrākushī also points out that in that copy of the *Almagest* there were marginal notes in Ibn Wuhayb's hand indicating the sections that he had studied under the direction of a certain Ḥamd or Ḥamad al-Dhahabī of Cordoba.[8] Mālik b. Wuhayb and Jābir b. Aflaḥ were probably of the same generation and lived in Seville at roughly the same period.

2. The *al-Kitāb fī l-Hayʾa*

Jābir b. Aflaḥ is mostly known for his main work in nine books, his *al-Kitāb fī l-Hayʾa*, currently better known under the title *Iṣlāḥ al-Majisṭī*, in which he rewrote Ptolemy's *Almagest* to make it accessible to his contemporaries. Jābir b. Aflaḥ removed any practical contents, computations or tables from his revision of the *Almagest*, simplified its trigonometric proofs by resorting to the rule of four quantities, and introduced some corrections of a mathematical tenor, the most important being his criticism of the Ptolemaic order of the spheres. Contrary to Ptolemy, Jābir b. Aflaḥ suggested that the spheres of Mercury and Venus should be above that of the Sun, and not below. Jābir b. Aflaḥ's *al-Kitāb fī l-Hayʾa* was also remarkably famous because of its first book consisting of an introduction to plane and spherical trigonometry, which was extremely influential in Medieval Europe. Another distinctive feature in *al-Kitāb fī l-Hayʾa* was the inclusion of a new instrument in Book V, similar to the *torquetum*, that Jābir claimed to substitute all four measuring instruments included by Ptolemy in the *Almagest*.[9]

The title *Iṣlāḥ al-Majisṭī* was not used during his own lifetime and was not widely used before recent times. It is taken from the first recto folio of Ms. Berlin, SBPK, Lbg. 132 (Ahlwardt 5653), copied in Damascus in 626/1229, one of the hitherto known Arabic manuscripts in Arabic script of Jābir b. Aflaḥ's *al-Kitāb fī l-Hayʾa*, although this title is written in a different hand from that of the scribe. Jābir b. Aflaḥ's work was referred to either as *al-Kitāb fī l-Hayʾa*,[10] *Kitāb al-Hayʾa*,[11] or simply *Hayʾa Ibn Aflaḥ*,[12] that is the *Book on Astronomy*

[7] al-Hawārī, *Al-Muʿjib fī talkhīṣ*, p. 140.

[8] It is unlikely that this Ḥamd or Ḥamad al-Dhahabī can be identified with the son, also named Ḥamd or Ḥamad, of Abū ʿAbd Allāh Muḥammad b. Najjāḥ al-Dhahabī al-Qurṭubī (d. 532/1138), who led the burial prayers for his father, since his father Muḥammad b. Najjāḥ, born in 455/1063, had roughly the same age as Mālik b. Wuhayb, born in 453/1061, his would-be student. Nevertheless, since frequently the same names run in families, it is likely that this Ḥamd or Ḥamad al-Dhahabī from Cordoba would be an older relative of Muḥammad b. Najjāḥ al-Dhahabī. On Muḥammd b. Najjāḥ al-Dhahabī and his son, see al-Ḍabbī, *Bughyat al-multamis*, p. 133.

[9] On this instrument, see Lorch, 'The Astronomical Instruments', pp. 11–34.

[10] Ms. Escorial, RBMSL, ár. 930, 1r.

[11] Ms. Escorial, RBMSL, ár. 910, 1r.

[12] See Ṣāliḥānī al-Yasūʿī, *Taʾrīkh mukhtaṣar al-duwal*, p. 423; and Yāltaqāyā and Bīlga, *Kashf al-ẓunūn*, col. 2047. Ḥājjī Khalīfa knew only the title. He also had access to an anon-

by Jābir b. Aflaḥ. In this sense, entitling a comprehensive book on astronomy as the author's *hayʾa* was a general practice in sixth/twelfth century al-Andalus, such as for instance Ibn Bājja's *Kalām fī l-hayʾa* or al-Biṭrūjī's *al-Kitāb fī l-Hayʾa*.[13]

The title *Iṣlāḥ al-Majisṭī*, that only appears in the Berlin manuscript, is probably derived from the single use of the root *ṣlḥ* in Jābir b. Aflaḥ's introduction to his *al-Kitāb fī l-Hayʾa* where he lists some technical mistakes (*awhām*) that he believes are present in the *Almagest* and states that he will correct them in their corresponding places.[14] The reference reads *wa-qad aṣlaḥnā jamīʿ mā dhakarnā-hu mim-mā wahama fī-hi fī l-mawāḍiʿ allatī dhakarnā-hā min kitābi-nā hādhā*, that is 'we have corrected (*aṣlaḥnā*) all the mistakes that we have mentioned [previously in this introduction] in the corresponding places of this book of ours that we have [also] mentioned [previously]'.[15] In the central and eastern Islamicate world, where there were different abridgements of the *Almagest* available, Jābir b. Aflaḥ's *al-Kitāb fī l-Hayʾa* mostly caught the attention because of the mistakes that Jābir b. Aflaḥ claimed were present in the *Almagest* and the corrections he provided, in much the same way that this work may be appealing to a modern historian of science. But for Jābir b. Aflaḥ himself and his readers in the western Islamicate world, his *al-Kitāb fī l-Hayʾa* was first and foremost a corrected abridgement (*talkhīṣ*)[16] of the *Almagest* providing a comprehensive self-contained mathematical description of the celestial spheres — in short, a *hayʾa*. That the title *Iṣlāḥ al-Majisṭī* only appears in a single manuscript and there is no other later reference to it allows us to believe

ymous and untitled copy of Jābir b. Aflaḥ's *al-Kitāb fī l-Hayʾa*, since he provides elsewhere its incipit under a category of works entitled *Sharḥ al-Majisṭī*, that is *Commentary of the Almagest*, and attributes it to a scholar of a recent generation. See Yāltaqāyā and Bīlga, *Kashf al-ẓunūn*, col. 1595.

[13] The distinctive feature of books including the term *hayʾa* in their titles is that they cover the different celestial spheres either separately or as a system, regardless of whether their approach is fully mathematical, as in the case of Jābir b. Aflaḥ's *al-Kitāb fī l-Hayʾa* or Ṭūsī's *Tadhkira fī l-hayʾa*, or non-mathematical, which includes works proposing a new physically consistent astronomy such as al-Biṭrūjī's *al-Kitāb fī l-Hayʾa*, and merely introductory works, such as Jaghmīnī's *al-Mulakhkhaṣ fī al-hayʾa al-basīṭa*. Jābir b. Aflaḥ's *al-Kitāb fī l-Hayʾa* is a work of *hayʾa*, not because his criticism of the Ptolemaic order of spheres has cosmological implications, but because it is a comprehensive astronomy covering the different celestial spheres. In this sense, the foundational work of *hayʾa* is the *Almagest* itself. Accordingly, the *Almagest* is sometimes called in Arabic *al-Majisṭī fī l-hayʾa*: see, for instance, Cheikho, *Kitāb Ṭabaqāt al-umam*, p. 31. For the translation, see Salem and Kumar, *Science in the Medieval World*, p. 28.

[14] For a short description of Jābir b. Aflaḥ's criticisms, see Bellver, 'On Jābir b. Aflaḥ's Criticisms'.

[15] See Bellver, 'El lugar', p. 135.

[16] In Ms. Escorial, RBMSL, ár. 910, 1r, Jābir b. Aflaḥ's *al-Kitāb fī l-Hayʾa* is deemed to be an abridgement of the *Almagest*, a *talkhīṣ al-Majisṭī*. Jābir b. Aflaḥ also used the root *lkhṣ* in his introduction to his *al-Kitāb fī l-Hayʾa*. See Bellver, 'El lugar', p. 129.

that it was not widely used other than in the manuscript in which it appears. The only other reference conveying a similar idea, i.e. correcting the *Almagest*, as in the title *Iṣlāḥ al-Majisṭī* occurs in Ibn al-Qifṭī's *Taʾrīkh al-ḥukamāʾ*, where the author calls it *al-Istikmāl li-Jābir b. Aflaḥ fī l-hayʾa*. This title appears next to a reference to al-Muʾtaman b. Hūd's (d. 478/1085–6)[17] *al-Istikmāl* ('the Completion'). Thus, the title *al-Istikmāl li-Jābir b. Aflaḥ fī l-hayʾa* seems to copy the title of the work by Ibn Hūd and to apply it to Ibn Aflaḥ's *al-Kitāb fī l-Hayʾa*.[18] The title *Iṣlāḥ al-Majisṭī* is more distinctive than the commonly used *hayʾa*, but gives only a partial idea of the aim and intent of the author. Thus, I suggest referring to Jābir b. Aflaḥ's main work as *al-Kitāb fī l-Hayʾa*, the shortened form of the title *Kitāb al-Shaykh Abī Muḥammad Jābir b. Aflaḥ al-Ishbīlī fī l-hayʾa* occurring in the closest witness to the author, Ms. Escorial, RBMSL, ár. 930, as we shall see below.

Jābir b. Aflaḥ's *al-Kitāb fī l-Hayʾa* had a moderate success in the Islamicate world. During the sixth/twelfth century, the authors of the so-called Andalusi revolt against Ptolemaic astronomy[19] extensively used it as an introduction to the *Almagest*. They read it instead of the *Almagest* or in parallel to it.[20] Jābir b. Aflaḥ's *al-Kitāb fī l-Hayʾa* was also read in the central and eastern Islamicate world at least during the seventh/thirteenth century. Ibn al-Qifṭī reports one transmission to the central Islamicate world.[21] According to him, Joseph ben Jehuda (d. 623/1226) brought a copy with him from Ceuta to Fusṭāṭ. In Fusṭāṭ, under the direction of Mūsā b. Maymūn, he studied mathematics and astronomy, and there the two aimed at correcting (*hadhdhaba*) Jābir b. Aflaḥ's *al-Kitāb fī l-Hayʾa* since the original (*aṣl*) was in a disordered state (*takhlīṭ*).[22] In another place, Ibn al-Qifṭī reports that both scholars committed themselves to correcting (*iṣlāḥ*) and reediting (*taḥrīr*) it.[23] Joseph ben Jehuda remained for a short time in Fusṭāṭ and left for Aleppo in 583/1187 where he finally settled and devoted himself to medicine. It is very likely that he took his original copy of Jābir b. Aflaḥ's *al-Kitāb fī l-Hayʾa* with him to Syria along with the revised version.[24]

[17] For him, and particularly for the date of his death, see Shiḥāda and Zikār, *Taʾrīkh*, vol. IV, p. 209.

[18] Lippert, *Taʾrīkh al-ḥukamāʾ*, p. 319.

[19] Sabra, 'The Andalusian Revolt'.

[20] Lay, 'L'Abrégé de l'*Almageste*', p. 40; Bellver, 'El lugar', p. 112.

[21] Lippert, *Taʾrīkh al-ḥukamāʾ*, pp. 392–93.

[22] Lippert, *Taʾrīkh al-ḥukamāʾ*, p. 319.

[23] Lippert, *Taʾrīkh al-ḥukamāʾ*, p. 393. In this reference, the word *iṣlāḥ* in *wa-saʾala-hu iṣlāḥ Hayʾat Ibn Aflaḥ al-Andalusī*, i.e. 'and he asked him to correct the *Astronomy* of Ibn Aflaḥ al-Andalusī', should not be understood as part of a title *Iṣlāḥ al-Hayʾa*.

[24] The Berlin manuscript copied in Damascus in 626/1229 may stem from Joseph ben Jehuda's original manuscript.

During the seventh/thirteenth century, Jābir b. Aflaḥ's *al-Kitāb fī l-Hay'a* had some impact on the central and eastern Islamicate world. Quṭb al-Dīn al-Shīrāzī (d. 710/1311), who was a student of Naṣīr al-Dīn al-Ṭūsī (d. 672/1274) at Marāgha, wrote a short summary of Jābir b. Aflaḥ's *al-Kitāb fī l-Hay'a* under the title *Fawā'id min al-Kitāb al-mawsūm bi-l-Majisṭī li-Ibn Aflaḥ al-Maghribī* (*Useful notes from the Book referred to as the Almagest by Ibn Aflaḥ al-Maghribī*) – a work that seems to have been commissioned by al-Ṭūsī himself.[25] Al-Shīrāzī finished this abridgement during the second ten days of *Rabī' al-awwal* of the year 663 (30 December 1264 to 9 January 1265), probably during his stay at Marāgha under the direction of al-Ṭūsī or shortly thereafter during his trip with al-Ṭūsī to Khurāsān.[26] The impact of Jābir b. Aflaḥ's *al-Kitāb fī l-Hay'a* on al-Shīrāzī can also be traced in al-Shīrāzī's short opuscule entitled *Faṣl fī Kayfiyyat taḥṣīl al-zamān al-dawrī li-l-qamar 'alā mā dhakara Baṭlamiyūs fī awā'il al-maqāla al-rābi'a min al-Majisṭī* (*Section on how to obtain the lunar period according to what Ptolemy mentioned in Book IV of the Almagest*)[27] where he closely follows Jābir b. Aflaḥ's original method for finding the lunar period in anomaly, although he does not acknowledge his debt.[28] Another important witness to Jābir b. Aflaḥ's influence that still needs to be studied is the *Kitāb Mukhtaṣar fī 'ilm al-hay'a min Hay'at Kūshyār wa-min Hay'at Ibn Aflaḥ al-Ishbīlī* (*Abridgement on astronomy from the Astronomy of Kūshyār and the Astronomy of Ibn Aflaḥ al-Ishbīlī*) by Athīr al-Dīn al-Mufaḍḍal b. 'Umar al-Abharī (d. 663/1264).[29]

But despite the initial esteem with which Jābir b. Aflaḥ's *al-Kitāb fī l-Hay'a* was regarded, external factors prevented it from being extensively copied in the Islamicate world after the eighth/fourteenth century. The most important one is that in 644/1247, shortly after the reception of Jābir b. Aflaḥ's *al-Kitāb fī l-Hay'a* in the central Islamicate world, Naṣīr al-Dīn al-Ṭūsī completed his *Taḥrīr al-Majisṭī*, which was to replace the *Almagest* itself and any other of its abridgments in the central and eastern Islamicate world. While Jābir b. Aflaḥ's *al-Kitāb fī l-Hay'a* was the work of a mathematician mostly interested in the geometrical consistency of the *Almagest* and, more importantly, any practical element present in the *Almagest* was left aside, the *Taḥrīr al-Majisṭī* was the work of a mathematical astronomer that met the requirements of potential astronomers needing an introduction to both the theoretical and practical sides of the *Almagest*. Once the *Taḥrīr al-Majisṭī* entered the postclassical canon,

25 Ms. Oxford, Bodleian, Thurston 3, 75v–92v.

26 See the authorial colophon in Ms. Oxford, Bodleian, Thurston 3, 92v.

27 Ms. Dublin, Chester Beatty Library, Ar. 3637, 168v and 171r.

28 See Bellver, 'Jābir b. Aflaḥ on the Four-Eclipse Method'.

29 This work is extant in Ms. Istanbul, Süleymaniye Library, Carullah 1499, 11v–81r, copied in Cairo in 677/1279 shortly after Abharī's death. The title is given according to that in fol. 1r.

the *Almagest* itself and any other abridgment were hardly copied afterwards. In the western Islamicate world, however, the factor that prevented the success of Jābir b. Aflaḥ's *al-Kitāb fī l-Hayʾa* was simply the lack of interest in highly technical theoretical astronomy, so that the audience for the work rapidly decreased, and accordingly it was seldom copied afterwards.[30] However, beyond the fact that the *al-Kitāb fī l-Hayʾa* was not extensively copied, the impact of Jābir b. Aflaḥ's criticisms on later Arabic works, probably through works produced in Marāgha, has still to be studied in depth.

Jābir b. Aflaḥ's *al-Kitāb fī l-Hayʾa* was translated into Latin by Gerard of Cremona (d. 583/1187). The most common titles in Latin manuscripts are *Astronomia Gebri* and *Liber Geber super Almagesti*.[31] This translation was later printed by Petrus Apianus (d. 1552) with the title *Gebri libri IX de astronomia*.[32] As Lorch suggested,[33] it is likely that Gerard of Cremona translated Jābir b. Aflaḥ's *al-Kitāb fī l-Hayʾa* before his translation of the *Almagest*, since those sections in Jābir's *al-Kitāb fī l-Hayʾa* directly quoting the *Almagest* are different from his rendition of the *Almagest*. It is also reasonable to translate Jābir b. Aflaḥ's abridgement before the *Almagest* itself in order to get acquainted with its complexities. Thus, Jābir b. Aflaḥ's *al-Kitāb fī l-Hayʾa* may have been translated between *c.* 544/1150, when Gerard of Cremona arrived in Toledo, and 571/1175, the date when a certain master Thadeus the Hungarian copied Gerard of Cremona's translation of the *Almagest*.[34] The translation by Gerard of Cremona therefore provides an additional early witness to Jābir b. Aflaḥ's work.

As already summarized by Lorch, Jābir b. Aflaḥ's *al-Kitāb fī l-Hayʾa* was also translated twice into Hebrew. The first translation was carried out in 672/1274 by Moshe ibn Tibbon (fl. between 637/1240 and 682/1283). Some time later, it was also translated by Jacob ben Maḥir ibn Tibbon (d. 703/1304). In 735/1335, the latter translation was revised by Samuel ben Jehuda of Marseille (fl. 735/1335). Samuel ben Jehuda, along with his brother David, had tried to translate Jābir b. Aflaḥ's *al-Kitāb fī l-Hayʾa* from an Arabic manuscript found in Arles, but they had to return the manuscript after they had only copied an eighth of it. Samuel ben Jehuda later located an autograph of Jacob ben Maḥir's translation, but found it faulty by collating it with the Arabic manuscript they had previously used. Thus, he took pains to correct Jacob

[30] This is clearly shown by the almost complete absence of manuscripts on mathematical *hayʾa* in Maghrebi libraries compared to the more abundant presence in these libraries of works on *mīqāt*, instruments and tables.

[31] Lorch, 'The Astronomy', p. 91.

[32] *De Astronomia Gebri*.

[33] Lorch, 'The Astronomy', p. 91.

[34] See the colophon in Ms. Florence, BML, Plut. 89 sup. 45.

ben Maḥir's translation despite the troubles that a correction represented in comparison to a fresh translation.[35]

Both accounts, i.e. Joseph ben Jehuda and Mūsā b. Maymūn's revision in Fusṭāṭ, and Samuel ben Jehuda's revision of Jacob ben Maḥir's translation after comparing it with an Arabic manuscript, suggest that the Arabic edition of Jābir b. Aflaḥ's *al-Kitāb fī l-Hay'a* was somehow problematic. This is corroborated by differences found in the extant Arabic manuscripts and the Latin and Hebrew versions.

3. Witnesses

To date, there are four known Arabic manuscripts in Arabic script of Jābir b. Aflaḥ's *al-Kitāb fī l-Hay'a*, three of which are complete. Two manuscripts are located in the Escorial library. The first one is Ms. Escorial, RBMSL, ár. 910 (henceforth referred to as Ea),[36] with the title (fol. 1r) *Kitāb al-Hay'a li-Abī Muḥammad Jābir b. Aflaḥ al-Ishbīlī, wa-huwa talkhīṣ Kitāb al-Majisṭī, wa-hiya al-nuskhat al-muḥadhdhafa* (*The Book of Astronomy by Jābir b. Aflaḥ, which is an abridgement of the Almagest. This is the shortened version*). The second one is Ms. Escorial, RBMSL, ár. 930 (henceforth referred to as Eb),[37] with the title (fol. 1r) *al-Nuskhat al-kubrā min Kitāb al-Shaykh Abī Muḥammad Jābir b. Aflaḥ al-Ishbīlī fī l-Hay'a* (*The Long Version of the Book on Astronomy by the Shaykh Abū Muḥammad Jābir b. Aflaḥ al-Ishbīlī*). Both manuscripts are from the western Islamicate world and probably of Andalusi origin. They do not provide any date, but Derenbourg tentatively dates them to the eighth/fourteenth century; the dating of Ms. Escorial, RBMSL, ár. 930 will be discussed below. A third manuscript is the one already mentioned, located in Berlin and copied in Damascus in 626/1229 in a *naskh* hand, i.e. Ms. Berlin, SBPK, Lbg. 132 (henceforth referred to as B),[38] with the title (fol. 1r) *Iṣlāḥ al-Majisṭī li-Jābir b. Aflaḥ*. Since this is an eastern manuscript, the absence of a *nisba* in the author's name linking him to al-Andalus or the Maghrib is rather unusual. The last manuscript, located at the Parliament Library of Iran in Tehran, was identified in September 2015 by Mohammad Mozaffari, who, in a private communication, informed a number of scholars that Ms. Tehran, Ketāb-ḵāna-ye Majles-e šurā-ye Eslāmi (Parliament), 1440S (henceforth referred to as T) contained Jābir b. Aflaḥ's *al-Kitāb fī l-Hay'a*.[39] This manuscript, acephalous and incomplete, only covers the first six books. It lacks any scribal or authorial colophon. Thus, it can be inferred that this is the first volume of a two-vol-

[35] Lorch, 'The Astronomy', p. 93.

[36] For this manuscript, see Derenbourg and Renaud, *Les manuscrits arabes*, pp. 10–11.

[37] For this manuscript, see Derenbourg and Renaud, *Les manuscrits arabes*, p. 39.

[38] For this manuscript, see Ahlwardt, *Verzeichniss*, vol. V, p. 141 no. 5653.

[39] For this manuscript, see Dāneš-Pažowh and ʿElmi Anvāri, *Fehrest-e kotob-e ḵaṭṭi*, p. 255.

ume set. Lacking the scribal colophon, it is undated and does not provide any location, but Dānesh-Pažowh and Anvāri date it to the seventh/thirteenth century, what considering the type of paper and size of the script seems very likely. The manuscript contains two different titles: *Kitāb fī ʿIlm al-hayʾa* (*Book on the Science of Astronomy*) and *Šarḥ-e Majesṭi* (*Commentary on the Almagest*). This manuscript, in a single clear *naskh* hand, is of eastern origin, maybe Persian, although that cannot be firmly concluded.

All four manuscripts are different to a varying degree. They provide four different versions, so that a stemma cannot be clearly delineated. Most of Jābir b. Aflaḥ's work is the same in all four Arabic manuscripts with no remarkable variations, but Book I and some sections throughout the rest of the work are rewritten in three different versions –those transmitted by Ea, B and T. Even though B and T transmit different versions, the differences between B and T are less significant when both are compared to Ea. Thus Ea, on the one hand, and B and T, on the other, present the major differences.[40]

Eb is a mixed version that follows Ea in Book I, and B in the remaining books.[41] However there are some differences between Ea and Eb in Book I. Even though the contents of Book I in Ea and Eb are the same, they are arranged differently. In addition, in Ea the propositions defining poles and great circles are omitted, whereas they are present in Eb. Thus, Ea omits basic well-known contents that do not add much to the introductory Middle Books. And last, in Book I, even though Eb follows Ea despite minor differences, Eb gives in marginal glosses some proofs found in B.

T follows B closely, although it also presents some differences from B, namely: There are some demonstrations in B given in a summarized form in T.[42] Some marginal glosses in B, for instance containing proofs of converse propositions,[43] made their way into the body text of T. And finally, there are some newly rewritten sections in T that differ from Ea, Eb and B. Thus, even though T is very close to B, it should be considered a distinctive version on its own.

There is no evidence that the version in Ea (including Book I from Eb) was transmitted to the central and eastern Islamicate world, while the remaining versions, i.e. B (including Books II–IX from Eb) and T, clearly circulated in the east, since both were copied in the east, and the wording of al-Shīrāzī's

[40] For differences between Ea and B regarding Book I, see Lorch, 'The Astronomy', p. 88.

[41] See, for instance, the section edited in Bellver, 'Jābir b. Aflaḥ on the Four-Eclipse Method', where Eb follows B instead of Ea.

[42] See, for instance, T, 48v–49r compared to B, 23v–24r.

[43] cf. B, 6v and T, 11v.

Fawā'id shows that it was based on B, or maybe T.[44] On the other hand, at least versions Ea, Eb and B circulated in al-Andalus and the Maghrib.

Additionally, the fact that the translation by Gerard of Cremona was done at most some twenty-five to fifty years after the completion of Jābir b. Aflaḥ's *al-Kitāb fī l-Hay'a*,[45] provides a very early witness to Jābir b. Aflaḥ's work and its circulation in al-Andalus. Nevertheless, it inaugurates an independent Latin transmission which introduced variations particular to its tradition.

The Latin translation by Gerard is mostly based on the version transmitted in B. However, the Latin translation is not a straight rendition of B, since in Books II–IX the Latin translation transmits small sections found in Eb, but not present in B. These sections, few in number, are mostly borrowings from the *Almagest*.[46] Thus, for Books II–IX, it is more appropriate to say that the Latin translation follows Eb. Additionally, when T differs from B the Latin translation follows B (or Eb for Books II–IX).

Moreover, the Latin translation has some distinctive features. The sections on the Milky Way and the solid globe in *Almagest* VIII.2–4 are included in Gerard's translation as part of the main body text of Book VI,[47] whereas these sections are not part of the equivalent sections in the main body text of any of the Arabic versions transmitted in the Arabic manuscripts. However, these sections, as Parra has shown,[48] are placed in an appendix at the end of manuscript Eb and thus have an Arabic origin.[49] In any case, these sections in the appendix are literal borrowings from the Isḥāq-Thābit version of the *Almagest* and add nothing specific to Jābir b. Aflaḥ's *al-Kitāb fī l-Hay'a*.[50]

[44] The source of al-Abharī's *Kitāb Mukhtaṣar fī 'ilm al-hay'a min Hay'at Kūshyār wa-min Hay'at Ibn Aflaḥ al-Ishbīlī* is not so clear, since al-Abharī summarizes his sources, omits sections, and rearranges the order, mixing contents from both works. The first section of al-Abharī's work, devoted to principles and introductory contents, partly follows Book I of Jābir b. Aflaḥ's *al-Kitāb fī l-Hay'a*. The wording of this section is closer to B. Therefore, it is likely that al-Abharī used a version similar to B or T.

[45] It is worth pointing out that, at the time of the translation, Jābir b. Aflaḥ's son was probably still alive, since the translation was done at roughly the same time when Mūsā b. Maymūn met him. That means that at the time of the translation it was still possible to commission the obtention of a manuscript of the final version of the *al-Kitāb fī l-Hay'a* through contacts with the circles of Jābir b. Aflaḥ's son.

[46] For instance, the quotation in Book II (Eb, 19r-v; *De Astronomia Gebri*, 22) from *Almagest* I.3 (Toomer, *Ptolemy's Almagest*, p. 40) on the physical considerations on the sphericity of the earth taken from the Isḥāq-Thābit translation (Ms. Tunis, BN, 7116, 3r-v) is not present in B or T.

[47] *De Astronomia Gebri*, 95–102.

[48] Parra, 'A Previously Unnoticed Appendix', pp. 113–28.

[49] Eb, 142r–150v.

[50] In this point too, Gerard's translation of the *Almagest* is different from his translation of this section in Jābir b. Aflaḥ's *al-Kitāb fī l-Hay'a* borrowed from the *Almagest*. See *Almagesti*, 89r–91v compared to *De Astronomia Gebri*, pp. 95–102.

The second distinctive feature of the Latin translation concerns the instrument which Jābir b. Aflaḥ claimed to supersede all four measuring instruments included by Ptolemy in the *Almagest*. Like the extant manuscripts of the Arabic versions, the Latin translation contains only one instrument, but the Latin translation transmits a completely different instrument than the one transmitted in the extant Arabic versions. Since manuscript T was identified after the study by Lorch, it is worth pointing out that the instrument in T is the same as the one in the other Arabic manuscripts.[51] However, the instrument in the Latin translation must be of Arabic origin, as Lorch has shown, since it is also described in the Hebrew translation of Jacob ben Maḥir ibn Tibbon revised by Samuel ben Jehuda of Marseille.[52] Thus, the single instrument contained in the Latin translation is the only witness to an instrument of Arabic origin of which there is no other witness in the extant Arabic manuscripts.

As to the Hebrew translations, according to Lorch,[53] the translation by Moshe ibn Tibbon follows B, while the translation by Jacob ben Maḥir ibn Tibbon revised by Samuel ben Jehuda of Marseille follows the version in Ea, at least for Book I.

In short, we have four Arabic manuscripts in Arabic script transmitting three different versions (Ea, B and T) and an additional mixed version (Eb), plus a Latin rendition translated shortly after the original composition of the work. Thus, two questions follow. Which of these versions is the earliest? And are all of them by Jābir b. Aflaḥ?

4. Editions

Lorch, in his seminal work of 1975, suggested that the version in B was the first and unrevised version and Ea the revised one. He also suggested the possibility that the revised version, i.e. the one in Ea, was the one done by Mūsā b. Maymūn and Joseph ben Jehuda in Fusṭāṭ.[54] He based his suggestion on the fact that the version in Ea was more complete and less prolix, and thus more elegant than the version in B.[55] Whatever the case, B cannot be the revised version by Mūsā b. Maymūn and Joseph ben Jehuda done in Fusṭāṭ, since B is the one transmitted in the Latin translation by Gerard of Cremona who died in 583/1187, whereas the revision by Mūsā b. Maymūn together with Joseph ben Jehuda was done only shortly before Joseph ben Jehuda left for Aleppo during the same year. Since Derenbourg dated both Escorial manuscripts to the eighth/fourteenth century, Lorch considered possible that the version in Ea

[51] cf. T, 107v–111r.

[52] Lorch, 'The Astronomical Instruments', p. 31.

[53] Lorch, 'The Astronomy', p. 93.

[54] Lorch, 'The Astronomy', p. 89.

[55] Lorch, 'The Astronomy', p. 89.

would be the revision by Mūsā b. Maymūn and Joseph ben Jehuda. In a later work, Lorch hypothesized that both versions were authored by Jābir b. Aflaḥ, without ruling out the possibility that a second author worked out the second version.[56] In 2009, I pointed out that the version in B contains corrections of astronomical contents in Ea.[57] This would mean that, in fact, the revised version is B instead of Ea.

It is unlikely that the versions Ea and B were done by different authors. A revision of the contents of the *al-Kitāb fī l-Hayʾa* would presuppose a very advanced and self-confident reader, and it is unlikely that his name would not have been mentioned. It is rather implausible that an advanced reader with a thorough knowledge of the *Almagest* would have corrected in B such a minute mistake in Ea as the one present in the computation of the longitude of the apparent conjunction of the sun from the true conjunction in solar eclipses[58] and would have overlooked in B that Jābir b. Aflaḥ's bitter criticism regarding lunar eclipses had no textual basis in the *Almagest* and arose from a misunderstanding because of his faulty manuscript of the *Almagest*.[59] This can only be explained by both versions being worked out by the same author on the basis of the same faulty manuscript of the *Almagest*. Generally speaking, the manuscript tradition of Jābir b. Aflaḥ's *al-Kitāb fī l-Hayʾa* shows a great respect for the literality of its text. For instance, B, a manuscript copied in Damascus, contains a table of the western abjad numerals on fol. 1r, so whenever there are western abjad numerals in the text, they are left as they are. Quṭb al-Dīn al-Shīrāzī is cautiously respectful and introduces his observations by a tentative 'I believe' (*aẓunnu*) when he agrees with or cast his doubts on Jābir b. Aflaḥ's *al-Kitāb fī l-Hayʾa*. All these peripheral reasons point to the fact that unacknowledged revisions by others than Jābir b. Aflaḥ are rather unlikely.

But the decisive witness to the authorship and the order of the different versions of Jābir b. Aflaḥ's *al-Kitāb fī l-Hayʾa* can be found in Ms. Escorial, RBMSL, ár. 930. This manuscript contains the name of the scribe of the manuscript, which has previously been unnoticed. Contrary to the usual practice, the name of the scribe is not part of the colophon, but is present on the first recto folio below the title of this single-work manuscript. As already noted, the title on fol. 1r reads *al-Nuskhat al-kubrā min Kitāb al-Shaykh Abī Muḥammad Jābir b. Aflaḥ al-Ishbīlī fī l-hayʾa*. Then follows a statement containing the name of the scribe, which reads *khaṭṭa-hu li-nafsi-hi ʿAlī b. Aḥmad b. Mufarrij, raḥima-hu Allāh*, that is "ʿAlī b. Aḥmad b. Mufarrij, God have mercy on him, copied it for his personal use'. The scribe can be identified as the Andalusi

[56] Lorch, 'Jābir b. Aflaḥ and the Establishment', p. 34.

[57] Bellver, 'El lugar', p. 89 and n. 24.

[58] See B, 67v and Ea, 65v.

[59] See Ea, 63v–64r and B, 65v–66r. For a discussion of this criticism, see Bellver, 'Jābir b. Aflaḥ on Lunar Eclipses', pp. 47–91.

faqīh ʿAlī b. Aḥmad b. Mufarrij b. Ziyād al-Sayyārī who was active in 530/1136, when he also copied (*khaṭṭa*) Ibn Rushd al-Jadd's (d. 520/1126) *al-Bayān wa-l-taḥṣīl*.[60] Thus, Ibn Mufarrij was active during Jābir b. Aflaḥ's lifetime or shortly afterwards. However, it is very likely that this manuscript was copied when Jābir b. Aflaḥ was still alive, since the scribe refers to him in the title of the work as *shaykh*, which suggests that Jābir b. Aflaḥ was still an elderly person when he copied the manuscript. The formula *raḥima-hu Allāh* referring to Ibn Mufarrij is written on top of an erased one. Thus, it was changed upon Ibn Mufarrij's death, which suggests that the manuscript was owned by him until his death or that the new owner knew him. Therefore, Eb should be dated to the first half of the sixth/twelfth century, which makes it the closest witness to Jābir b. Aflaḥ's lifetime. Since Eb was copied when Jābir b. Aflaḥ was still alive, or less likely shortly after he passed away, the shortened title on the first recto folio of this manuscript, that is *al-Kitāb fī l-Hayʾa*, should be the one by which Jābir b. Aflaḥ's work was known in his own lifetime, and thus this is the title by which I have suggested above that this work should be known, instead of the rather spurious title *Iṣlāḥ al-Majisṭī*, with only one late occurrence.

However, this extraordinary manuscript is not only unique because it is the witness closest to the author, but also because it combines both versions of the *al-Kitāb fī l-Hayʾa*. As already noted, Eb contains the same version as in Ea for Book I (although rearranged), and the version in B for Books II–IX, with only minor differences. Thus, it is safe to assume that both versions were worked out by Jābir b. Aflaḥ himself. The only other possibility is that one of these versions was made by an independent Andalusi author alive during Jābir b. Aflaḥ's lifetime, but this can be ruled out, since such a close witness as Eb clearly attributes this work in both versions (the one similar to Ea transmitted in Book I of Eb, and the one similar to B transmitted in Books II–IX of Eb) to Jābir b. Aflaḥ himself. At most, one of these versions may have been done by a disciple under Jābir b. Aflaḥ's supervision. So, in all, it is safe to attribute both versions to Jābir b. Aflaḥ himself.

Thus, we should ask whether Eb is a mixed manuscript created from two different versions, or an in-between version. The possibility that this is an in-between version, i.e. that this manuscript was copied amid the rewriting of the first into the second version, can be excluded. The element that makes clear that this is a manuscript created from two already existing versions lays in the instrument. The description of the instrument in Ea is appended at the end of the manuscript. This means that in the first redaction of the version now transmitted in Ea there was no instrument. However, in the Introduction

[60] On ʿAlī b. Aḥmad b. Mufarrij, see Sharīfa and ʿAbbās, *al-Dhayl wa-l-takmila*, vol. V, p. 181, no. 355.

to the *al-Kitāb fī l-Hay'a* in Ea, Jābir b. Aflaḥ actually announces that he is going to provide an instrument to supersede Ptolemy's measuring instruments, although the section devoted to that instrument is found only in an appendix in Ea.[61] But in the Introduction present in Eb, there is no such reference to an instrument.[62] Thus, the source of both the Introduction and Book I as transmitted by Eb is earlier than the source of Ea. Since it does not make sense to first devise and provide the description of an instrument, later remove it and then reintroduce it, it should be concluded that Eb was created from two different source manuscripts of the two versions transmitted by B and an earlier version than Ea, and was not an in-between version. That also answers which of the two versions is the earliest and confirms that the version in Ea is earlier than the version in B.

Lorch suggested that Ea was the revised version since he adduced that the version in B was more prolix and less elegant than the version in Ea. But this is only so for Book I, since in the rest of the *al-Kitāb fī l-Hay'a* there are rewritten sections in B more succinct than in Ea.[63] A more prolix Book I, as in B, strongly suggests that Jābir b. Aflaḥ used it for teaching purposes. As a teaching text, his revision was aimed at making his exposition clearer for his students, although it would be less elegant than his first redaction. This would help to explain the apparent contradiction regarding why a more prolix and less elegant version in Book I has also undergone some corrections from an astronomical point of view in the remaining books.

Eb, or the source from which it was copied, was put together from two different versions. The more likely possibility is that the scribe first obtained a copy of an older version and continued from Book II onwards with a newer version obtained shortly thereafter. Accordingly, there are glosses in Book I in the hand of the scribe and introduced by *wa-fī nuskha ukhrā*, i.e. 'and in another copy', which are taken from the same version transmitted by B.[64] This probably makes Ibn Mufarrij the one who mixed both versions in this manuscript. However, the sections on the Milky Way and the solid globe in Eb are probably a later addition, although the scribe is the same as that of the main text. As singled out by Parra,[65] a marginal gloss on the last folios of Book VI[66]

[61] Ea, 3r.

[62] This reference to the instrument in the introduction is omitted from the equivalent section in Eb, 2v. See Bellver, 'El lugar', p. 128. In Eb, this omission affects the sentences where Jābir b. Aflaḥ introduces this instrument. Thus, in this case an omission by scribal error is not very likely.

[63] See, for instance, Ea, 29r-v and B, 29r.

[64] See, for instance, the marginal glosses in Eb, 8v.

[65] Parra, 'A Previously Unnoticed Appendix'.

[66] Eb, 94v. The hand of this marginal gloss pointing to the appended sections and the hand of the glosses in Book I seem the same. Even though the script of these marginal glosses

in Eb indicates the point in the main body text where various available copies included the appended sections. It is worth noting that the scribe pointed out that the appended sections were written on a quire of different colour, which suggests that they were appended at a significantly later stage.

In all, from the information in the manuscripts and the Latin version, we can attempt to draw a list of editions and revisions along with their witnesses. That is:

- First Edition – Eb Book I;
- Revised First Edition – Ea;
- Second Edition – Eb Books II–IX;
- Augmented Second Edition – Latin translation;
- Revised Second Edition – B;
- Third Edition – T.

The first edition of Jābir b. Aflaḥ's *al-Kitāb fī l-Hayʾa* is transmitted by Eb Book I and Ea. However, initially the first edition of Jābir b. Aflaḥ's *al-Kitāb fī l-Hayʾa* did not contain an instrument. Our only witness to the first edition in its initial version is Eb Book I. Thus, it is very likely that the arrangement of the trigonometrical propositions in Eb Book I was the earlier one.

The first revision of the first edition introduced the instrument and rearranged the contents of Book I. The witness to this first revision, and the major witness to the first edition in general, is the version transmitted in Ea.

The second edition is a partial rewriting of the first edition. Almost all of Book I, and at least sections of Books II, IV and VII are rewritten. There may also be small differences in wording that are difficult to be distinguished from scribal errors. The witness to this second edition in its initial version is Eb, Books II–IX, while B is a slightly revised version of the initial second edition. Since the Latin translation closely follows the second edition, as witnessed by Eb Books II–IX, a comparison of Book I in B and Book I in the Latin translation will indicate the degree by which Book I in B departs from the initial second edition of Book I, which up to now has no witness in Arabic.

B is a slightly revised version of the initial second edition witnessed by Eb Books II–IX. The revisions are very minor, and they mostly amount to omissions of short quotations of the *Almagest* that are found in Eb Books II–IX. However, the presence of these quotations is significant since they are also

---

is different from that of the main text, we can guess that they were written by the same scribe since the hand of this marginal gloss is expected to be the same as that of the appendix, which is the same of the main text. These differences in the script between glosses and body text most probably owe to the use of a different *qalam* for marginal glosses and main body text, and not to a different scribe.

found in Ea, and thus Eb Books II–IX should precede B. Additionally, since the Latin translation also transmits these short quotations from the *Almagest*, the Latin translation mostly follows the second edition witnessed by Eb Books II–IX rather than the slightly revised second edition transmitted by B. The main departure of the Latin translation from the second edition is the new instrument superseding the one in the Arabic manuscripts, already described by Lorch.

The Latin translation was done from an extended second edition, since in Book VI the Latin translation includes in the main text quotations from *Almagest* VIII.2–4 that are in an appendix in Eb, as Parra has already shown. But, as pointed out by the scribe of Eb, there were Arabic copies of the *al-Kitāb fī l-Hayʾa* which included these sections taken from *Almagest* VIII.2–4 in the main text. Thus, the second extended edition had an Arabic origin.

The third edition is witnessed by T. It rewrites some small sections from the revised second edition witnessed by B. Thus, T also omits the same quotations of the *Almagest* as B does. The rewritings in T compared to B are less substantial than the rewritings of the second edition compared to the first edition. Nevertheless, although quantitively not very important, since these are conscious rewritings affecting some proofs throughout Jābir b. Aflaḥ's *al-Kitāb fī l-Hayʾa*, I have listed T as a new edition. Since T is an acephalous and incomplete manuscript with no colophon and with no cover containing statements of ownership, purchase or reading, no information about the authorship of this reedition and the history of the manuscript can be inferred. Jābir b. Aflaḥ may have been the author of this third edition represented by T. But it also may be the revision authored by Joseph ben Jehuda and Mūsā b. Maymūn, or by any other later author. However, I favour the possibility that the author of the rewritings in this third edition is Jābir b. Aflaḥ himself, or a very close disciple under his supervision, since, except for Book I which serves as an independent work on its own,[67] I have found no important rewritings in T of sections already rewritten in B from Ea. This is particularly striking in Book II where rewritings from Ea into B and from B into T follow one another without ever affecting a section already rewritten. This suggests, although not conclusively, that the author behind the rewritings in B and in T is the same person, that is Jābir b. Aflaḥ. However, the fact that the Latin translation was not based on the, to our knowledge, last version of the *al-Kitāb fī l-Hayʾa*, i.e. T, may be a counterargument for Jābir b. Aflaḥ's authorship, since it casts a doubt on whether it was available in al-Andalus. In any case, it should be underlined that the differences between B and T are not very important.

[67] A few sentences of the discussion of the sine rule for spherical triangles are rewritten in T from B in a far less prolix way. See T, 19v–20r and B, 10r.

The Latin translation stems from a particularly complete version. It mostly follows an extended version of the second edition before a few minor quotations from the *Almagest* were skimmed from it. Thus, apparently the source for the Latin translation was either carefully selected or, as an extended version, was a popular one in al-Andalus in a time when Jābir b. Aflaḥ's son was probably still alive. It is doubtful that the instrument in the Latin translation, which is of Arabic origin, would have been designed by Jābir b. Aflaḥ. The short description of the instrument in the Latin translation of the Introduction is the same as in the Introduction of the Arabic manuscripts, although the actual description of the instrument in the Latin translation is different from the one in the Arabic manuscripts. And the punctilious character that Jābir b. Aflaḥ shows throughout his work does not fit with the lack of detail in the description of the instrument in the Latin translation. Since this instrument had an Arabic origin and since the translation was executed shortly after Jābir b. Aflaḥ's lifetime, the most consistent possibility is that the instrument witnessed by the Latin translation was designed by disciples of Jābir b. Aflaḥ in an attempt to overcome the unfeasibility[68] of Jābir b. Aflaḥ's instrument witnessed by the extant Arabic manuscripts. That also casts doubts on whether the third edition transmitted by T was authored by Jābir b. Aflaḥ, since the instrument witnessed by the Latin translation was part of the second edition and not the third.

The final consideration is that the version in Ea is referred to as *al-nuskhat al-muḥadhdhafa*, the shortened version, whereas the version in Eb is referred to as *al-nuskhat al-kubrā*, the long version. However, despite the fact that Ea and Eb Books II–IX transmit two different editions, there are no major differences in length between them. The main difference between the two editions are rewritings, not additional contents, except for the appended borrowings from the *Almagest* in Eb regarding the Milky Way and the solid globe. The only significant departure in the contents covered is in Book I where, as indicated above, Ea omits propositions aimed at defining the concepts of the pole and great circle. It is possible that by 'shortened version' (*al-nuskhat al-muḥadhdhafa*), the scribe referred to the omission of such basic well-known contents defining the concepts of the pole and great circle that did not add much to the introductory Middle Books. These omissions seem deliberate and do not preclude Eb Book I from being earlier than Ea.

In all, except for the version in the Tehran manuscript, whose authorship cannot be fully ascertained, Jābir b. Aflaḥ was the most likely author of the different editions of his *al-Kitāb fī l-Hayʾa*, currently better known as *Iṣlāḥ al-Majisṭī*. He kept rewriting his *al-Kitāb fī l-Hayʾa* over a rather long period. He first added an instrument, and then rewrote substantial parts of his work

[68] Lorch, 'The Astronomical Instruments', p. 31.

in order to polish it and to accommodate it to the needs of his students. A critical edition will further expand our understanding of how the composition of this work developed over time. Additionally, should a new manuscript of the *al-Kitāb fī l-Hayʾa* be identified, it is likely that it will modify the tentative list of editions and revisions presented here.

Bibliography

Ahlwardt, Wilhelm, *Verzeichniss der arabischen Handschriften der Königlichen Bibliothek zu Berlin*, 10 vols, Berlin: Asher, 1887–1899.

Almagesti = Gerard of Cremona (tr.), *Claudius Ptolemaeus. Almagesti*, Venice: Petrus Lichtenstein, 1515.

Ātāy, Ḥusayn (ed.), *Mūsā b. Maymūn. Dalālat al-ḥāʾirīn*, Cairo: Maktabat al-thaqāfa al-dīniyya, 1422/2002.

Bellver, José, 'Jābir b. Aflaḥ on the Four-Eclipse Method for Finding the Lunar Period in Anomaly', *Suhayl* 6 (2006), pp. 159–248.

Bellver, José, 'Jābir b. Aflaḥ on Lunar Eclipses', *Suhayl* 8 (2008), pp. 47–91.

Bellver, José, 'On Jābir b. Aflaḥ's Criticisms of Ptolemy's *Almagest*', in Emilia Calvo, Mercè Comes, Roser Puig and Mònica Rius (eds), *A Shared Legacy. Islamic Science East and West*, Barcelona: Publicacions i Edicions de la Universitat de Barcelona, 2008, pp. 181–89.

Bellver, José, 'El lugar del *Iṣlāḥ al-Maŷisṭī* de Ŷābir b. Aflaḥ en la llamada « Rebelión andalusí contra la astronomía ptolemaica »', *Al-Qanṭara* 30 (2009), pp. 83–136.

Cheikho, Louis (ed.), *Ṣāʿid b. Aḥmad al-Andalusī. Kitāb Ṭabaqāt al-umam*, Beyrouth: Imprimerie Catholique, 1912.

al-Ḍabbī, Aḥmad ibn Yaḥyá, *Bughyat al-multamis fī taʾrīkh rijāl ahl al-Andalus*, Cairo: Dār al-kātib al-ʿarabī, 1967.

Dāneš-Pažowh, M. T., and Behāʾ al-Din ʿElmi Anvāri, *Fehrest-e kotob-e ḵaṭṭi-e Ketāb-ḵāna-ye Majles-e šurā-ye Eslāmi (Senā-ye sābeq)*, Tehran: Entešārāt Ketāb-ḵāna-ye Majles-e šurā-ye Eslāmi, 1359sh (1980/1).

De Astronomia Gebri = Apianus, Petrus. Instrvmentvm primi mobilis. Accedunt ijs Gebri filii Affla Hispalensis, Astronomi uetustissimi pariter et peritissimi, libri IX de astronomia, ante aliquot secula Arabice scripti, et per Giriardum Cremonensem latinitate donati, nunc uero omnium primum in lucem editi, Nuremberg: Johannes Petreius, 1534.

Derenbourg, Hartwig, and Henri-Paul-Joseph Renaud, *Les manuscrits arabes de l'Escurial décrits d'après les notes de Hartwig Derenbourg revues et complétées par le Dr H.-P.-J. Renaud*, tome 2, fascicule 3, Paris: Geuthner, 1941.

al-Hawārī, Ṣalāḥ al-Dīn (ed.), *ʿAbd al-Wāḥid al-Marrākushī. Al-Muʿjib fī talkhīṣ akhbār al-Maghrib*, Beirut: al-Maktaba al-ʿaṣariyya, 2006.

Kraemer, Joel L., 'Moses Maimonides. An Intellectual Portrait', in Kenneth Seeskin (ed.), *The Cambridge Companion to Maimonides*, Cambridge: Cambridge University Press, 2005, pp. 10–57.

Lay, Juliane, 'L'Abrégé de l'*Almageste*. Un inédit d'Averroès en version hébraïque', *Arabic Sciences and Philosophy* 6 (1996), pp. 23–61.

Lippert, Julius (ed.), *Ibn al-Qifṭī. Taʾrīkh al-ḥukamāʾ*, Leipzig: Dieterich'sche Verlagsbuchhandlung, 1903.

Lorch, Richard P., 'The Astronomy of Jābir b. Aflaḥ', *Centaurus* 19 (1975), pp. 85–107 [reprinted in Id., *Arabic Mathematical Sciences. Instruments, Texts, Transmission*, Aldershot: Variorum, 1995, chapter VI].

Lorch, Richard P., 'The Astronomical Instruments of Jābir ibn Aflaḥ and the Torquetum', *Centaurus* 20 (1976), pp. 11–34.

Lorch, Richard P., 'The Manuscripts of Jābir b. Aflaḥ's Treatise', in Id., *Arabic Mathematical Sciences. Instruments, Texts, Transmission*, Aldershot: Variorum, 1995, chapter VII.

Lorch, Richard P., 'Jābir b. Aflaḥ and the Establishment of Trigonometry in the West', in Id., *Arabic Mathematical Sciences. Instruments, Texts, Transmission*, Aldershot: Variorum, 1995, chapter VIII.

Parra, Mª José, 'A Previously Unnoticed Appendix in the *Iṣlāḥ al-Majisṭī* of Jābir ibn Aflaḥ', *Suhayl* 15 (2016/17), pp. 113–28.

Pines, Shlomo, *Moses Maimonides. The Guide of the Perplexed*, 2 vols, Chicago-London: University of Chicago Press, 1963.

Sabra, Abdelhamid I., 'The Andalusian Revolt against Ptolemaic Astronomy. Averroes and al-Biṭrūjī', in Everett Mendelsohn (ed.), *Transformation and Tradition in the Sciences. Essays in Honor of I. Bernard Cohen*, Cambridge: Cambridge University Press, 1984, pp. 133–53.

Salem, Semaʿan I., and Alok Kumar (eds), *Ṣāʿid al-Andalusī. Science in the Medieval World. Book of the Categories of Nations*, Austin: University of Texas Press, 1991.

Ṣāliḥānī al-Yasūʿī, Anṭūn (ed.), *Ibn al-ʿIbrī [Bar Hebraeus], Ghrīghūriyūs Abī l-Faraj b. Ahrūn al-Ṭabīb al-Malaṭī. Taʾrīkh mukhtaṣar al-duwal*, Beirut: al-Maṭbaʿa al-Kāthūlīkiyya, 1890.

Sharīfa, M. b., and Iḥsān ʿAbbās (eds), *Ibn ʿAbd al-Malik al-Marrākushī. al-Dhayl wa-l-takmila*, 5 vols (I, IV, V, VI, VIII), Beirut: Dār al-thaqāfa, n.d.–1984.

Shiḥāda, Khalīl, and Suhayl Zikār (eds), *Ibn Khaldūn. Taʾrīkh*, 8 vols, Beirut: Dār al-fikr, 1431/2001.

Toomer, Gerald J., *Ptolemy's Almagest*, New York: Springer, 1984.

Yāltaqāyā, Muḥammad Sharaf al-Dīn, and Rifʿat Bīlga al-Kilīsī (eds), *Ḥājjī Khalīfa, Katip Çelebi. Kashf al-ẓunūn ʿan asāmī al-kutub wa-l-funūn*, 2 vols and suppl., Istanbul: Maṭābiʿ Wikālat al-maʿārif al-jalīla, 1941–1947.

The Astrological Computations Attributed to Ptolemy and Hermes in Medieval Arabic Sources

Josep CASULLERAS

1. Introduction

Many medieval Arabic sources attribute some of the most popular mathematical procedures and geometrical definitions applied to the practice of astrology either to Ptolemy (*c.* AD 150) or to the legendary Hermes. However, these attributions have little basis either in Ptolemy's astrological work[1] or in the writings related to the Hermetic tradition.[2] Focusing on this apparent disagreement between authors and attributions, in Section 1 we review the basic concepts of natal astrology and draw up a list of the computations that have been associated with either Ptolemy or Hermes. In Section 2, we will try to explain the meaning of these attributions with reference to some medieval authors who were concerned with this same question. Finally, we present some conclusions in Section 3.

2. Houses, rays and progressions. Methods and attributions

The main practices associated with natal astrology correspond to the three concepts of houses, rays and progressions, all of them taken from the ancient Greek tradition.

The astrological houses are the twelve divisions of the ecliptic around the local horizon, as represented in Figure 1. During one apparent daily revolution of the celestial sphere, any celestial body will pass through all twelve houses. Unlike zodiac signs, the houses vary depending upon the time and latitude for which they are calculated. Therefore, the operation of equalizing the astrological houses (in Arabic *taswiyat al-buyūt*) relates the positions of the celestial objects to our place and moment.[3]

[1] The *Tetrabiblos* or *Quadripartitum*, in Arabic *Kitāb al-arbaʿ*.

[2] On Hermes and his astrological works see Sezgin, *GAS* VII, pp. 50–58. See also van Bladel, 'Hermes'.

[3] See, for instance, Bouché-Leclercq, *L'astrologie grecque*, pp. 256–86; Casulleras and Hogendijk, 'Progressions', pp. 38–39, and the references given there.

Ptolemy's Science of the Stars in the Middle Ages, ed. by David Juste, Benno van Dalen, Dag Nikolaus Hasse and Charles Burnett, PALS 1 (Turnhout, 2020), pp. 201–221
10.1484/M.PALS-EB.5.120180

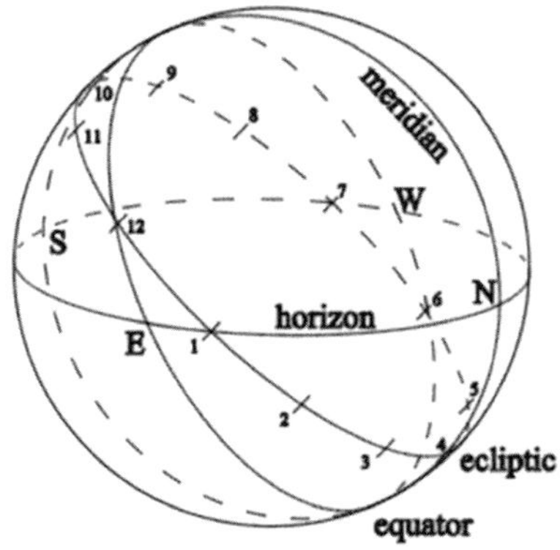

Figure 1. The cusps of the astrological houses

The doctrine of the planetary aspects or projection of rays (in Arabic *maṭraḥ al-shuʿāʿāt*) considers the astrological significance of certain angular distances (60°, 90°, 120° and 180°) defined between the objects on the celestial sphere. Figure 2 is a schematic representation of the different rays of a planet P on the ecliptic circle: the 'rays' PP_1, PP_2, PP_3 (PP_7, PP_6, PP_5) are called the left (right) sextile, quartile and trine rays respectively, and PP_4 is called the opposition.[4]

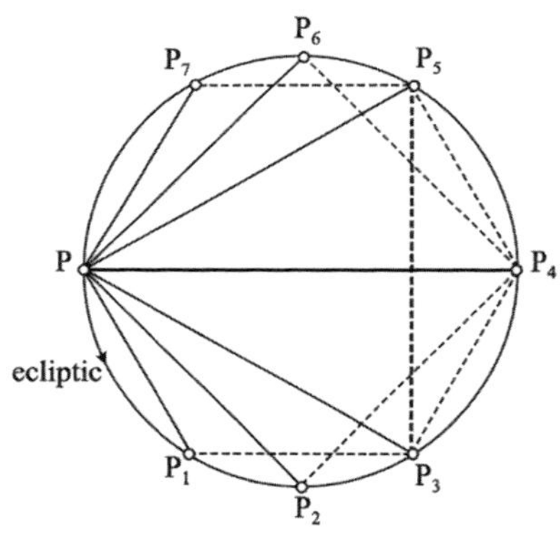

Figure 2. The planetary aspects or rays

The astrological theory of progressions (in Arabic called *tasyīr*) establishes a relationship between angular distances and periods of time as a basis for astrological predictions. A typical example of this practice is the attempt to find the moment of death, by giving a value of one year of life per degree of the angle between two significant objects selected in the celestial configuration at the moment of the individual's birth. One of these objects is thought of as emitting the life-force, and the other is seen as destroying life. In Figure 3,

[4] cf. Bouché-Leclercq, *L'astrologie grecque*, pp. 165–79; Casulleras and Hogendijk, 'Progressions', pp. 40–41.

the point *F* represents the destructive point that will reach the initial position of the emitting point *P* after rotation over *n* degrees around the celestial axis. According to this theory, the angle *n* corresponds to the arc of *tasyīr* and the individual will live *n* solar years.[5]

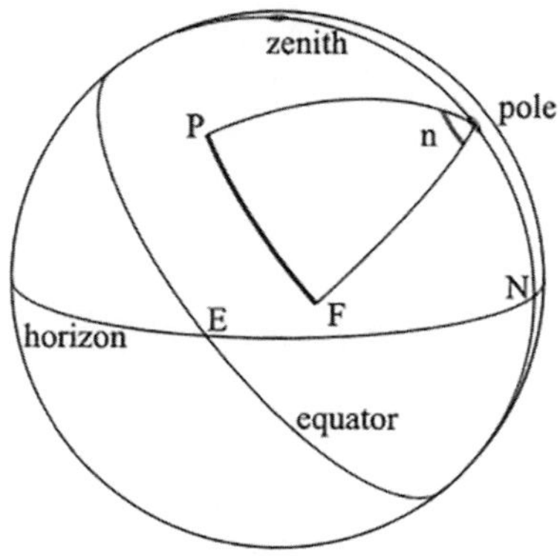

Figure 3. The system of progressions

A variety of geometrical approaches are used to calculate the houses, rays and progressions. For each one, the sources describe various methods of computation, which all produce different results. Fortunately, this feature of astrological technique has played an important role in the development of applied mathematics, and one major consequence of the research into the history of these methods is that we have well-established designations for all of them. John North first classified the medieval systems for the houses in 1986,[6] and Edward S. Kennedy extended North's classification in 1996.[7] Casulleras and Hogendijk published a classification of the methods for the rays and the progressions in 2012.[8] For the purpose of our discussion, we will focus only on the methods that have been attributed to either Ptolemy or Hermes in the Arabic sources.

2.1. Houses

2.1.1. Ptolemy: the Standard Method and the Hour Lines Method

The most popular method for the division of houses in the Middle Ages is the one that North called the Standard Method.[9] In this method, the houses are defined on the ecliptic by meridians crossing equal divisions of the equatorial

[5] cf. Bouché-Leclercq, *L'astrologie grecque*, pp. 411–22; Casulleras and Hogendijk, 'Progressions', pp. 41–43.

[6] North, *Horoscopes and History*.

[7] Kennedy, 'The Astrological Houses'.

[8] Casulleras and Hogendijk, 'Progressions'.

[9] cf. North, *Horoscopes and History*, p. 4.

arcs lying between the local meridian and the meridians that pass through the ascendent and the descendent points of the ecliptic. Figure 4 shows the cusps (i.e. beginning points) of the houses of the eastern hemisphere using this procedure.

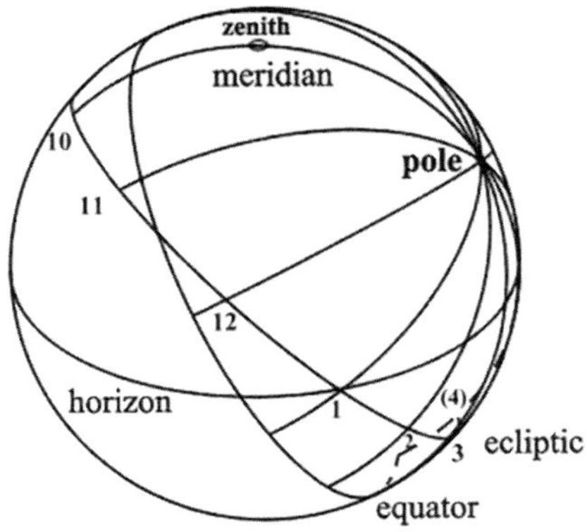

Figure 4. The Standard Method for the division of the houses

This system is pre-Islamic in origin, and it was called 'the well-known method' by the eleventh-century Iranian astronomer al-Bīrūnī.[10] In many Andalusī sources it is attributed to Ptolemy,[11] and it is usually implemented with an astrolabe, moving either the rule or the *spider* with the help of the inscribed hour lines, in order to find the beginnings of the houses according to the definition of the method.[12]

The other method for the houses which is sometimes attributed to Ptolemy is the Hour Lines Method.[13] In this method, the cusps of the houses are the intersections of the ecliptic with the lines of the even seasonal hours.

Modern astrologers attribute this system to Placidus, a seventeenth-century Perugian monk, but the Andalusī astronomer Ibn al-Samḥ (d. 1035)[14] said in his *Treatise on the Use of the Astrolabe* that, according to Ḥabash [al-Ḥāsib] (ninth century), this is Ptolemy's method.[15] This attribution was repeated by some other Andalusī authors. The procedure can be performed with any standard astrolabe plate which has lines for the seasonal hours.

[10] cf. North, *Horoscopes and History*, p. 6; al-Bīrūnī, *Al-Qānūn*, vol. III, pp. 1357–1359. On this author see, for example, Yano, 'Bīrūnī'.

[11] cf. Calvo, 'La résolution graphique', pp. 35–36; Casulleras, 'Mathematical Astrology', pp. 265–67.

[12] The two options are described in North, *Horoscopes and History*, p. 59.

[13] cf. North, *Horoscopes and History*, pp. 20–27; Kennedy, 'The Astrological Houses', p. 538; Casulleras, 'Mathematical Astrology', p. 265; Calvo, 'La résolution graphique', p. 36; Casulleras and Hogendijk, 'Progressions', pp. 83–85.

[14] On this author see Rius, 'Ibn al-Samḥ'.

[15] Viladrich, *El Kitāb al-ʿAmal*, pp. 66, 124.

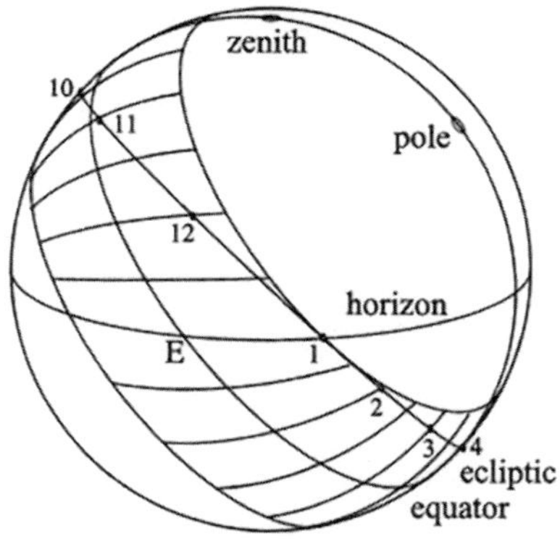

Figure 5. The Hour Lines Method for the houses

2.1.2. Hermes: the Prime Vertical Method and the Equatorial Method

The two methods for the houses that have been associated with Hermes are called the Prime Vertical Method[16] and the Equatorial Method[17] in the North-Kennedy classification. They define the houses as the intersections of the ecliptic with certain great circles on the celestial sphere passing through the North and South points of the local horizon, which are called position circles.

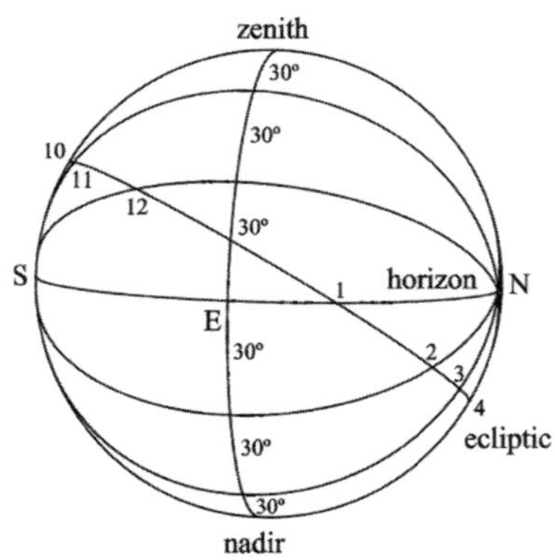

Figure 6. The Prime Vertical Method for the houses

For the first method (Figure 6), these position circles cross equal divisions of the prime vertical, which is the great circle passing through the local zenith and the East and West points on the local horizon. For the Equatorial Method, they cross the celestial equator (Figure 7).

[16] North, *Horoscopes and History*, pp. 32–33, 47; Kennedy, 'The Astrological Houses', pp. 541–43; Casulleras and Hogendijk, 'Progressions', pp. 82–83.

[17] North, *Horoscopes and History*, pp. 27–30, 47; Kennedy, 'The Astrological Houses', pp. 543; Casulleras and Hogendijk, 'Progressions', pp. 80–82.

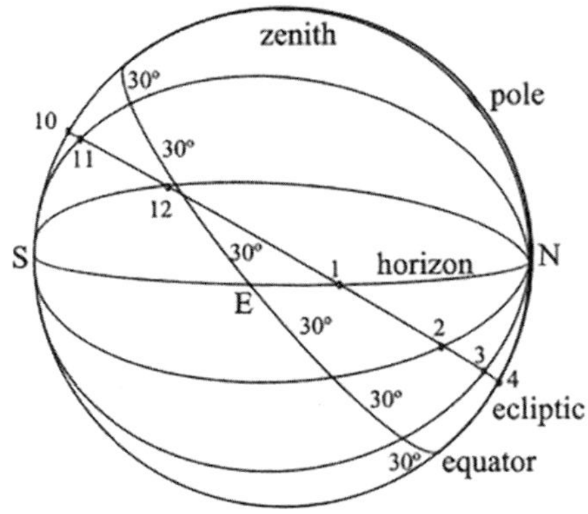

Figure 7. The Equatorial Method for the houses

The Prime Vertical Method is attributed to the thirteenth-century Italian author Campanus of Novara in the Latin West, to al-Bīrūnī in the medieval Islamicate East, and to the mythical Hermes in al-Andalus.

The Equatorial Method for the houses is not found in Eastern Arabic sources. In the Western Arabic area, it is attributed to Hermes and to the Andalusī mathematician and astronomer Ibn Muʿādh al-Jayyānī (d. 1093), who described the method for the first time in a treatise on the computations applied to the division of the houses and the projection of rays.[18] Modern astrologers call it the system of Regiomontanus, a fifteenth-century author who was seemingly the owner of a manuscript of the Latin translation by Gerard of Cremona of the canons to the astronomical tables of Ibn Muʿādh, in which the method is also described.[19]

2.2. Rays

In the case of rays, the name of Ptolemy is related to what we call the Single Hour Line Method in many Eastern and Western Arabic sources.[20] This method can be considered the standard procedure for computing the projections of the rays in the medieval Islamicate area. In it, the arcs defining the different rays are measured on the equator using the hour line that passes through the planet that casts its rays. In Figure 8, point L stands for the left sextile ray of a Planet P. If L is rotated around the celestial axis over an angle

[18] Study, edition and Spanish translation in Casulleras, *La astrología*.

[19] cf. North, *Horoscopes and History*, p. 35; North, 'The Alfonsine Books', p. 49; Hermelink, 'Tabulae Jahen', p. 109. The canons were printed in Nuremberg in 1549, after Regiomontanus' death (in 1476) and probably using his Latin manuscript (I thank Benno van Dalen for drawing my attention to this).

[20] cf. Kennedy and Krikorian-Preisler, 'The Astrological Doctrine', p. 5; Hogendijk, 'The Mathematical Structure', pp. 178–80; Casulleras and Hogendijk, 'Progressions', pp. 68–71.

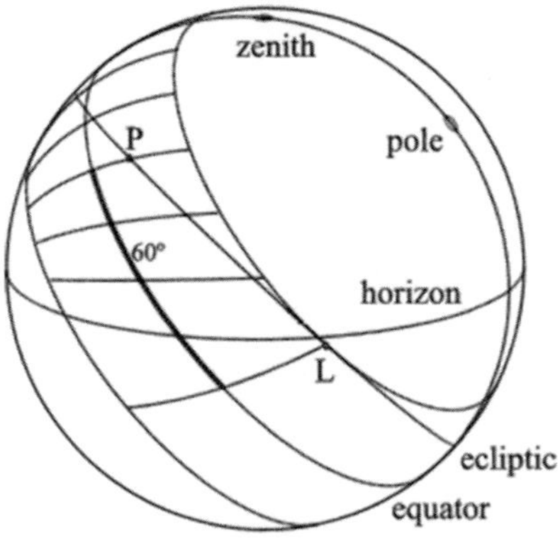

Figure 8. The Single Hour Line Method for the rays

of 60 degrees in the direction of the daily motion of the celestial sphere, its image after rotation is on the hour line through *P*.

Two methods for the projection of rays were associated with Hermes: the Single Position Semicircle Method (Figure 9) and The Four Position Circles Method (Figure 10).

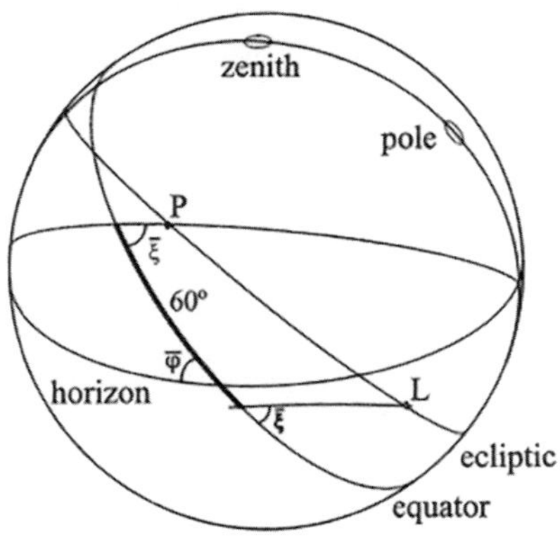

Figure 9. The Single Position Semicircle Method for the rays

In the first method, the arcs defining the different rays are measured on the equator using only one position semicircle. In Figure 9, the image of point *L* after rotation of 60 degrees in the direction of the daily motion of the celestial sphere is on the position semicircle through *P*. We call this method the Single Position Semicircle Method, because it involves only the position semicircle through *P*. The attribution of this procedure to Hermes is found in treatises on astronomical instruments from al-Andalus.[21]

[21] cf. Casulleras and Hogendijk, 'Progressions', pp. 67–68.

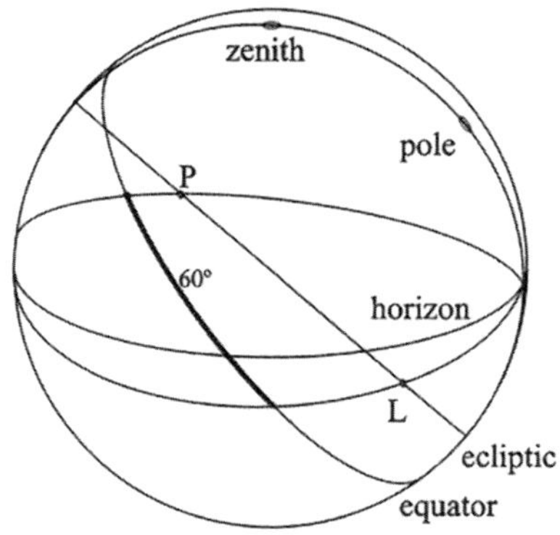

Figure 10. The Four Position Circles Method for the rays

The Four Position Circles Method is based on the same geometrical approach as the Equatorial Method for the houses and it is also attributed to Ibn Muʿādh al-Jayyānī, who provides the only known algorithm for this method. Figure 10 shows the construction of *L*, the left sextile ray of *P*. This method involves a total number of four position circles, and the different rays would correspond to the beginnings of the third, fourth, fifth, seventh, ninth, tenth and eleventh houses of a hypothetical horizon passing through *P* and the North and South points of the local horizon. All the sources mentioning this method come from the Western area.[22]

2.3. Progressions

Book III, Chapter 10, of Ptolemy's *Tetrabiblos*[23] contains references to most of the systems for progressions (*tasyīr*s) in connection to the problem of finding the length of an individual's life.

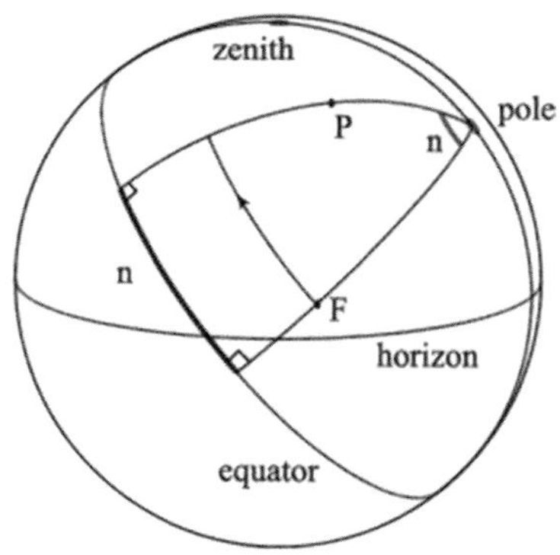

Figure 11. The Right Ascension Method for progressions

[22] cf. Casulleras and Hogendijk, 'Progressions', pp. 71–73; Casulleras, 'Mathematical Astrology', pp. 248–51.

[23] cf. Robbins, *Tetrabiblos*, pp. 286–305; Casulleras and Hogendijk, 'Progressions', p. 89; Hogendijk, 'Al-Bīrūnī', pp. 286–90.

In what we call the Right Ascension Method (Figure 11) the arc of *tasyīr* is measured on the equator, between the meridians through the two objects involved in the progression. According to Ptolemy, this procedure is correct only for objects in the meridian plane. Later sources repeat this reasoning[24] but do not attribute the method to either Ptolemy or Hermes.

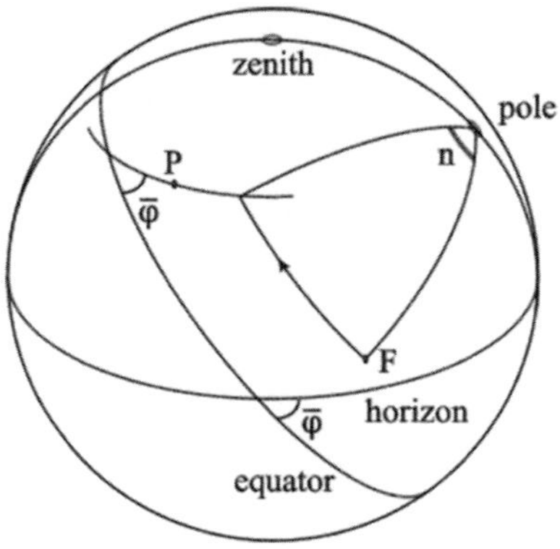

Figure 12. The Oblique Ascension Method for progressions

In the Oblique Ascension Method (Figure 12) the arc of *tasyīr* is determined by the difference in oblique ascension between the two objects. Ptolemy says that this is the usual system but that it is correct only if the emitting point is on the horizon. As in the case of the Right Ascension Method, later authors repeat the same principle,[25] without attributions to either Ptolemy or Hermes.

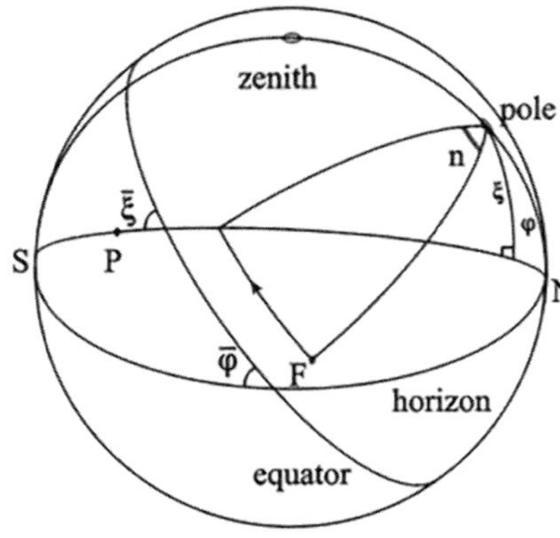

Figure 13. The Position Semicircle Method for progressions

In the Position Semicircle Method (Figure 13), the arc between the two significant objects is measured on the equator by means of the position semicircle

[24] cf. Casulleras and Hogendijk, 'Progressions', pp. 45–46.

[25] cf. Casulleras and Hogendijk, 'Progressions', p. 47.

through the emitting point. Ptolemy states the principle of this method in the *Tetrabiblos*, and he probably thought that this was the true system. However, the sources attribute this method to Hermes, not to Ptolemy.[26]

Finally, the method actually described by Ptolemy is the Hour Line Method, presented with numerical examples in the *Tetrabiblos* as an approximation to be used in the computations.[27] In this method (Figure 14), the arc between the two significant objects is measured on the equator by means of the hour line passing through the emitting point.

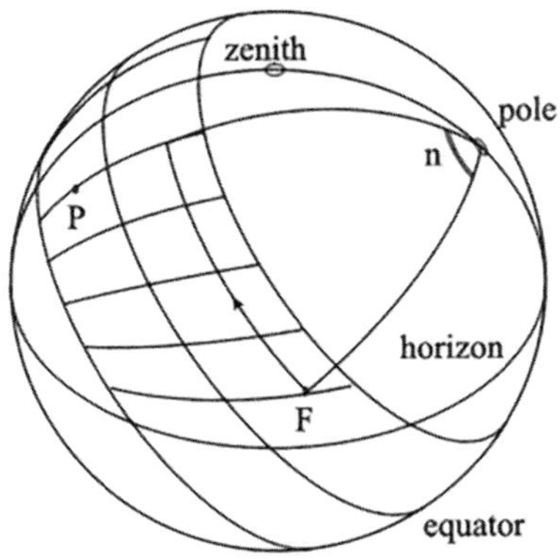

Figure 14. The Hour Line Method for progressions

2.4. Summary table

This table summarizes the attributions in Arabic sources of astrological methods to Ptolemy and Hermes:

	Ptolemy	Hermes
Houses	Standard Hour Lines	Equatorial Prime Vertical
Rays	Single Hour Line	Single Position Semicircle Four Position Circles
Progressions	Hour Line	Position Semicircle

3. Interpretations and opinions

The problem that emerges from this overview of the methods attributed either to Ptolemy or to Hermes is that these attributions do not seem to correspond to the preserved works that constitute the Hermetic and Ptolemaic traditions.

[26] Robbins, *Tetrabiblos*, pp. 290–91. See Casulleras and Hogendijk, 'Progressions', pp. 48–53.

[27] The numerical examples have been analysed in Hogendijk, 'Al-Bīrūnī', pp. 286–90; see also Casulleras and Hogendijk, 'Progressions', pp. 53–59.

We have just seen that, in his *Tetrabiblos*, Ptolemy states the principle of the Position Semicircle Method for the *tasyīr* and, instead of giving the corresponding calculation for this procedure, presents the Hour Line Method as a computational approximation. Therefore, this latter method was correctly attributed to Ptolemy, but no single procedure for the computation of the projection of rays or the division of houses is to be found in the *Tetrabiblos*.[28] The case of Hermes is even less clear, since the preserved works ascribed to him do not seem to justify any of the attributions mentioned here.

Following a suggestion made by al-Bīrūnī (see the next sub-section), Hogendijk comes to an interesting conclusion:[29] these ascriptions of methods do not refer to the actual authorship of a procedure but to the type of elements it uses. Thus, the procedures related to Ptolemy correspond to the ones that use hour lines, whereas the methods attributed to Hermes correspond to the ones using position circles or semicircles.[30] Another point worth making is that the attributions to Hermes are normally found in Western Arabic sources, and not before the time of Ibn al-Samḥ.[31]

In this section, we will look at the details of al-Bīrūnī's reasoning, assess how the issue of the attributions of methods is addressed in the thirteenth-century Alfonsine *Libro Segundo de las Armellas*, and examine the opinions of Ibn Muʿādh al-Jayyānī on the methods for the houses and the rays associated with Hermes and Ptolemy. Finally, we will compare these medieval points of view with a passage in the *Tetrabiblos* that may shed some light on the matter.

[28] This in spite of the fact Ibn al-Samḥ says that the Single Hour Line Method for the rays 'is mentioned by Ptolemy in the *Tetrabiblos*'. See Viladrich, *El Kitāb al-ʿAmal*, p. 149.

[29] In a first draft of Casulleras and Hogendijk, 'Progressions', cf. pp. 87–88.

[30] There are a few exceptions in the Islamicate area: for example, the Samarkand ruler Ulugh Beg (d. 1449) attributes the Single Position Semicircle Method for the rays to Ptolemy (Sédillot, *Prolégomènes*, p. 209; cf. Casulleras and Hogendijk, 'Progressions', pp. 68, 75); the eleventh-century Andalusī astronomer al-Istijjī relates the Oblique Ascension Method for the rays also to Ptolemy (Samsó and Berrani, 'World Astrology', pp. 303–04 and Samsó and Berrani, 'The Epistle', pp. 199, 200, 234; Casulleras and Hogendijk, 'Progressions', pp. 66); the Jewish mathematician Abraham Ibn ʿEzra (d. 1167) attributes the Hour Line Method for progressions to Hermes (but also to Ptolemy, among other authors, cf. Viladrich and Martí, 'Sobre el Libro', p. 91; Casulleras and Hogendijk, 'Progressions', p. 56: n. 48); the Maghribī astronomer Ibn ʿAzzūz (d. 1354) attributes to 'Ptolemy and Hermes' the Single Hour Line Method for the rays (Casulleras, 'Ibn ʿAzzūz', pp. 63–64, 81, 89; Casulleras and Hogendijk, 'Progressions', p. 70).

[31] One exception may be the case of the Christian astrologer Ibn Hibintā (fl. Baghdad, early tenth-century), who states that 'Hermes said in his book related to "The Latitude" that the trine, and sextile, and quartile (rays) are made in equal degrees', but this book of Hermes seems to be lost and the quotation is too imprecise to pinpoint the attribution to Hermes to a particular method. See Sezgin et al., *Ibn Hibintā,* vol. I, p. 293 and vol. II, p. 66; Kennedy and Krikorian-Preisler, 'The Astrological Doctrine', p. 13; Casulleras and Hogendijk, 'Progressions', p. 88.

3.1. Al-Bīrūnī

In Book XI, Chapter 4 (On the Projection of the Rays), Section 1 of his *al-Qānūn al-Masʿūdī*,[32] al-Bīrūnī describes the Single Hour Line Method for the rays, but he says that the computation is incorrectly attributed to Ptolemy, and that this attribution was made because the procedure is based on Ptolemy's method of progressions.[33] Elsewhere, in the chapter on progressions,[34] al-Bīrūnī says that this procedure for the rays is adapted from the procedure for progressions, that is to say, the Hour Line Method.[35] However, in another passage in the same chapter, al-Bīrūnī is more critical: he regards the Hour Line Method for the progressions as 'unsatisfactory from a theoretical point of view',[36] and he presents the Position Semicircle Method as 'the method which I preferred' and a way to 'get rid of the carelessness and approximation which are involved in it [i.e. in the Hour Line Method]'.[37]

3.2. The Alfonsine *Libro Segundo de las Armellas*

As for the Western sources, an interesting examination of the attributions of astrological methods to Ptolemy and Hermes is found in Chapter 52 of the Alfonsine *Libro Segundo de las Armellas*.[38] This text was written in the thirteenth century, probably by Rabiçag (Rabbī Isḥāq ibn Sīd), the chief scientific advisor of the Castilian king Alfonso the Wise, and it is preserved in medieval Spanish. The previous chapters deal with the application of an armillary sphere to the resolution of astrological problems but, in Chapter 52, the author explains that his source for the methods ('opinions') attributed to Hermes is a book by Ibn Muʿādh, that he *did not find* in the books of Ptolemy what is said to be Ptolemy's method about the projection of the rays and the *tasyīr*, and

[32] Spanish translation of this section in Casulleras, *La astrología*, pp. 303–11.

[33] *Fī l-ʿamal al-mansūb ilā Baṭlamiyūs* [...] *faʾinna al-marjaʿ fī-hi ilā al-ʿamal al-musnad ilā Baṭlamiyūs wa-ʾin lam yakun la-hu bal mustanbaṭan min raʾyi-hi fī l-tasyīr*, al-Bīrūnī, *Al-Qānūn*, vol. III, p. 1377, line 14 [...] p. 1378, lines 4–5; cf. Casulleras, *La astrología*, p. 303.

[34] Book XI, Chapter 5. English translation and study of the whole chapter in Hogendijk, 'Al-Bīrūnī'.

[35] *Liʾanna dhālika al-ʿamal muqtabas min ʿamal al-tasyīr*, al-Bīrūnī, *Al-Qānūn*, vol. III, p. 1394, line 3; Hogendijk, 'Al-Bīrūnī', p. 293. See Casulleras and Hogendijk, 'Progressions', p. 71.

[36] *Ghayr murḍī fī ṭarīq al-naẓar*, al-Bīrūnī, *Al-Qānūn*, vol. III, p. 1397, line 1; Hogendijk, 'Al-Bīrūnī', p. 297.

[37] *Al-ṭarīq alladhī āthartu-hu* [...] *wa-tajarrada ʿam-mā fī-hi min al-tasāhul wa-l-taqrīb*, al-Bīrūnī, *Al-Qānūn*, vol. III, p. 1397, lines 14 [...] lines 16–17; Hogendijk, 'Al-Bīrūnī', p. 298. See also Casulleras and Hogendijk, 'Progressions', p. 59.

[38] Edited in Rico y Sinobas, *Libros*, vol. II, p. 68. See Casulleras and Hogendijk, 'Progressions', n. 95 on pp. 88–89; Casulleras, *La astrología*, pp. 112–15; On the *Libros de las Armellas* see Samsó, *Las ciencias*, pp. 175–80; Comes, *Historia*, pp. 190–202.

that these attributions may derive from mistakes in the transmission. Finally, he also says that a correct understanding of the *Tetrabiblos* shows that the method ('opinion') of Ptolemy is closest to that of Ibn Muʿādh.

Thus, it seems that in the thirteenth century Rabiçag identified the systems of Hermes with Ibn Muʿādh, and he also said that the methods of Ibn Muʿādh and Ptolemy are almost the same.

This is a paraphrase of the text:

> On knowing how to perform the *tasyīr* following the opinion of Ibn Muʿādh [Aben-Mohat].
>
> This man called Ibn Muʿādh [Aben-Mohat] was a great scholar of geometry and astrology, and wrote a book dealing with the projection of the rays, and the *tasyīr*, and the equalization of the twelve houses, and he showed in it reasons and proofs very near the truth, and he also showed how those mentioned things [i.e. rays, *tasyīr* and houses] have to be performed. And he said about Hermes that he agreed with him. And from that book the projection of rays and the *tasyīr* according to the method of Hermes were taken. And they are those that we put here in this book under the name of Hermes. And later scholars in this science agreed with him on that. And I, being smaller than all of them, also agree with them. And if you read in that book, you will see the good things that it contains. And all that I said here to be Ptolemy's method about the projection of the rays and the *tasyīr*, you must know that I did not find it in the books of Ptolemy, but Abū Maʿshar [Abumassar] and Azarquiel [Abuçac el Zarquiel] said that it was so.[39] And it may well be that it was not Ptolemy's method. But someone attributed it to him, as often happens, due to transmission errors, and they change one name for another, even more considering the long time that has passed since Ptolemy's day, more than one thousand years ago. However, if you correctly understand Ptolemy's *Tetrabiblos* [*quarto partido*] you will understand that his opinion is nearer to what Ibn Muʿādh [Aben-Mohat] said than to any other's [opinion]. And were it not because it would be too long, I would show you the places where you could understand it. And I wanted to put here all the methods, in order for the book to be more complete and lest it seemed that I did not include them for laziness. And you should choose among those aforementioned methods the one that you find to be the best.

3.3. Ibn Muʿādh

As mentioned above, the eleventh-century Andalusī mathematician and astronomer Ibn Muʿādh al-Jayyānī addressed the question of the division of the houses and the projection of the rays in two works: his *Astronomical Tables* (the *Tabulae Jahen*) and a short monograph dealing with the mathematical aspects of these practices.

[39] A different version of this passage, in Rico y Sinobas, *Libros*, vol. I, pp. LI–LII, mentions al-Battānī [Albateni], Ibn al-Samḥ [Abulcacin Abnaçam], Azarquiel [Abuiz-hac-Azarquiel] and others ... [et algunos otros ...], but not Abumassar.

The stated intention of Ibn Muʿādh's treatise is to draw attention to some errors that have spread from Ptolemy's *Tetrabiblos*. For this purpose, he analyses the procedures that existed in his time for the rays and the houses, and proposes his own solutions. In this respect, his work represents not only a classification of these methods but also an investigation of how far they conform to the principles that may maintain these astrological practices within a consistent geometrical, astronomical and geographical whole. According to these principles, Ibn Muʿādh contends that the division of houses and the projection of rays share the same theoretical basis, and defends the use of the Equatorial Method only for the houses and the Four Position Circles Method for the rays.

In the course of his investigation, Ibn Muʿādh also discusses the question of the methods for the houses and the rays according to the methods attributed to Ptolemy and to Hermes. Ibn Muʿādh considers that the Standard Method for the houses, attributed to Ptolemy, developed out of a confusion when the division of houses was performed at latitudes other than the equator ($\varphi = 0°$), where all the methods are equivalent. He is probably referring to the fact that horoscopes of the world — a practice of the Hindu-Iranian astrological tradition — were cast for latitude 0° before the Greek tradition of nativities reached the Islamicate regions.[40]

The other method for the division of the houses which is of particular interest to Ibn Muʿādh is the one attributed to Hermes by Ibn al-Samḥ, that is, the Prime Vertical Method. In this case, Ibn Muʿādh quotes a passage from a lost *zīj* (a set of astronomical tables) by Ibn al-Samḥ (d. 1035) containing an erroneous algorithm for this method, and presents a correct solution. However, Ibn Muʿādh disapproves of the use of the prime vertical and does not find its attribution to Hermes convincing: he suspects that it was created by Ibn al-Samḥ's misunderstanding of Hermes' texts. There are good grounds for believing that Ibn Muʿādh is right, because the attribution to Hermes does not appear before Ibn al-Samḥ.[41]

Concerning the methods for the projection of rays, Ibn Muʿādh focuses in particular on the method attributed to Ptolemy and 'transmitted by Abū Maʿshar (787–886), among others',[42] which corresponds to what we called the Single Hour Line Method. To alert his readers to the errors in this method, Ibn Muʿādh uses several numerical demonstrations including an example of the application of the method for a geographical latitude of 49°: in that situation, if the initial point of Capricorn (i.e., at an ecliptic longitude of 270°) is on the eastern horizon, the computation according to the 'Method of Ptolemy' will give its left trine ray at the beginning of Cancer (i.e., at an ecliptic longitude of

[40] cf. Casulleras, *La astrología*, pp. 179–80, and the references mentioned there.

[41] cf. Casulleras, *La astrología*, pp. 185, 193, 256–57, 288–90.

[42] cf. Casulleras, *La astrología*, pp. 240 (Spanish) and 268 (Arabic). On Abū Maʿshar see Yamamoto, 'Abū Maʿshar'.

90°), that is, the point diametrically opposite the beginning of Capricorn, with the astrologically absurd consequence that the left trine ray and the opposite ray fall on the same place. After other numerical proofs of the inconsistencies of this procedure, Ibn Muʿādh concludes that the method ascribed to Ptolemy is clearly erroneous.[43]

It may be worth recalling that, in a different historical context, John North mentioned another mathematical flaw related to the use of hour lines in computations. In this case it was in connection with the Hour Lines Method for the division of the houses, when this procedure failed to cast a horoscope for Marie Peary (1893–1978), the daughter of the Arctic explorer, who was born at a latitude (φ = 77;44°) at which the method could not be applied.[44]

The Treatise of Ibn Muʿādh also raises the question of an intriguing connection between the mythical Hermes, the ninth-century philosopher al-Kindī,[45] and the Four Position Circles method for the projection of rays.

When dealing with the idea that a single geometrical approach should be applied in an analogous way to the division of houses and the projection of rays, Ibn Muʿādh says:

> What we consider correct — and we disregard any other procedure — consists of dividing the equatorial circle into equal parts, for it [i.e. the equator] is the origin and the direction of the movement; and of that the dividing circles, for all the climates, are drawn from the common points of the horizon and the meridian [i.e. according to the definition of the position circles]. This is what expresses our opinion and our intention, and Ibn al-Samḥ intended to do the same *in what he attributed to Hermes* about the projection of the rays [...] his specific instructions for the projection of the rays indicate that *the circle which is divided is just the equator.*[46]

In another passage, after describing his computation for the Method of the Four Position Circles, Ibn Muʿādh states that:

> Among the things which this procedure verifies is the fact that the two quartiles [i.e. the quartile rays] are diametrically opposite in it [i.e. in this system], and similarly the left trine and the right sextile, and similarly the left sextile and the right trine. *The same is true in the doctrine of Hermes.*[47]

These allusions indicate that Hermes' method for the projection of rays has position circles crossing divisions of the equator and that it retains the expected symmetries between the different rays (Figure 15). These two characteristics

[43] cf. Casulleras, *La astrología*, pp. 203–10, 241–43 (Spanish) and 270–73 (Arabic).

[44] cf. North, *Horoscopes and History*, p. 21.

[45] On this author see, for example, Cooper, 'Kindī' and Adamson, *al-Kindī.*

[46] Casulleras, *La astrología*, pp. 258–59 (Spanish), 292 (Arabic). Here, and in the following citations, the italics are mine.

[47] I use the edition and translation by Hogendijk, 'Applied Mathematics', pp. 102 (Arabic) and 105 (English).

conform perfectly to the Four Position Circles Method, and confirm the attribution to Hermes of this method in Western sources.[48]

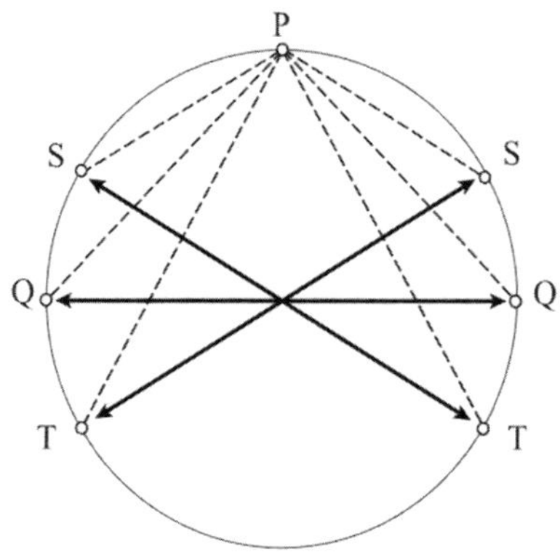

Figure 15. Symmetries between the rays

But Ibn Muʿādh goes further and finds grounds for basing the same analogy between the computations applied to the division of houses and the projection of rays on an astrological work by al-Kindī (d. *c.* 873), now lost.

In the relevant passage[49] Ibn Muʿādh states that:

> Among the things which strengthen our view and our doctrine in combining the equalization [i.e. determination] of the houses and the projections of the rays is the following: Al-Kindī [...] said in one of his writings on the rays: If the star [i.e., any celestial body] is on the horizon at the degree of the ascendent, the degree of the eleventh house and the degree of the third house are the sextiles [i.e., sextile rays] of it, and the degree of midheaven and lower culmination are the quartiles [i.e., quartile rays] of it, and the degree of the ninth house and the degree of the fifth house are the trines [i.e., trine rays] of it.

But, according to Ibn Muʿādh:

> *This does not occur in any way in the doctrine attributed to Ptolemy*, and we have demonstrated the absurdity of that doctrine. But this occurs if the equator is divided into equal parts, as we mentioned, and if circles are drawn to these division points from the common points of the horizon and the meridian. So those circles divide the ecliptic into its twelve houses. If any star is on any of those circles, its trines, quartiles and sextiles are at a distance of four, three and two houses.

[48] And, incidentally, they dismiss the use of the Single Position Semicircle Method, because this method does not retain the mentioned symmetries. See Casulleras, *La astrología*, p. 214; Casulleras and Hogendijk, 'Progressions', p. 79.

[49] Hogendijk, 'Applied Mathematics', pp. 100 (Arabic) and 102–03 (English).

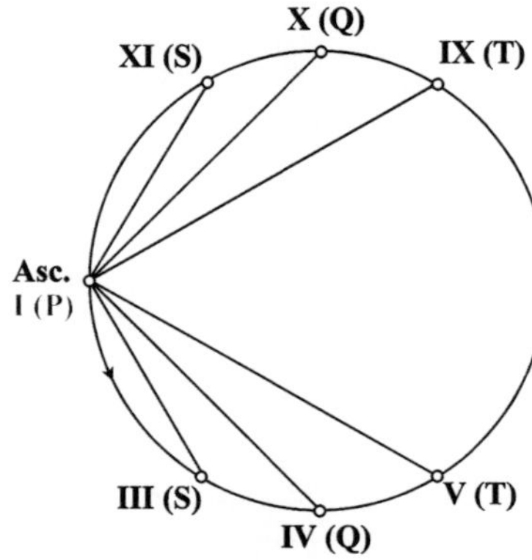

Figure 16. Rays and houses according to al-Kindī

Thus, there seems to be a possibility that the Four Position Circles Method for the projection of rays, attributed to Hermes in al-Andalus and defended by Ibn Muʿādh, was geometrically equivalent to the one used by al-Kindī, and clearly different from '*the doctrine attributed to Ptolemy*', that is, the Single Hour Line method for the projection of rays.

3.4. Ptolemy's *Tetrabiblos*

As we have seen, the *Tetrabiblos* does not indicate how the astrological rays and houses should be computed. There is, however, one relevant passage. In Book III, Chapter 10, when defining the astrological elements involved in the progressions, which are called the 'prorogative places', in connection with finding the length of life, Ptolemy establishes the following relationship between the houses and the rays:

> In the first place we must consider those places prorogative in which by all means the planet must be that is to receive the lordship of the prorogation; namely, the twelfth part of the zodiac surrounding the horoscope, from 5° above the actual horizon up to the 25° that remains, which is rising in succession to the horizon [i.e. the first house]; the part sextile dexter to these thirty degrees, called the House of the Good Daemon [11th house]; the part in quartile, the midheaven [10th house]; the part in trine, called the House of the God [9th house]; and the part opposite, the Occident [7th house].[50]

[50] Robbins, *Tetrabiblos*, pp. 272–73 (the ordinal numbers of the houses between square brackets are my interpretation). In footnote 2 Robbins lists the Greek names for the twelve houses and mentions that slightly different names are given in P. Mich. 149, col. ix, 13–19. On Papyrus Michigan 149, see Greenbaum, *The Daimon*, pp. 152–55 and the references mentioned there. The papyrus can be viewed at https://quod.lib.umich.edu/a/apis/x-1451/149R_A.TIF (accessed 12 December 2018).

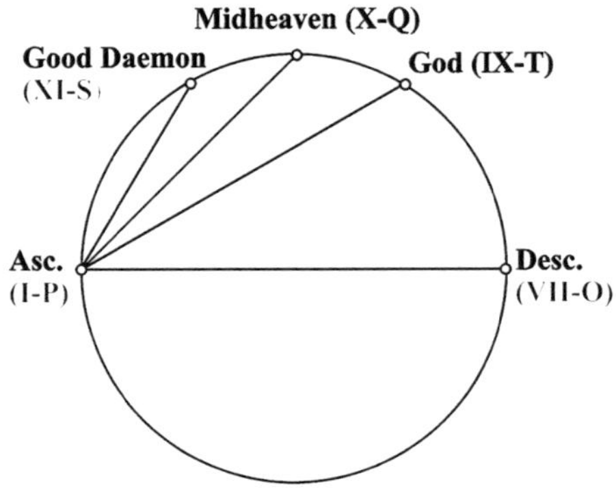

Figure 17. Houses and Aspects according to Ptolemy, Tetrabiblos, III, 10

The information contained in this passage is obviously insufficient to establish which method (if any) Ptolemy had in mind for the construction of the houses. The five-degree displacement from the ascendent is a convention that was sometimes associated with Ptolemy in the Middle Ages,[51] but what is relevant for our discussion here is that Ptolemy establishes the same relationship between rays and houses mentioned by Ibn Muʿādh in connection with al-Kindī and Hermes (compare Figures 16 and 17).[52] The fact that this relationship, as demonstrated by Ibn Muʿādh, 'does not occur in any way in the doctrine attributed to Ptolemy', is clear evidence that the method attributed to Ptolemy (that is the Single Hour Line Method) is not really his.

4. Conclusions

This sketch of the attributions of methods to Ptolemy and Hermes confirms the idea that they should not be regarded as bearing direct historical information regarding the transmission of these methods. Probably used to give prestige to astrological practices, the names of Ptolemy and Hermes are also ascribed to different geometrical approaches which are correctly understood only when compared to each other, and generally indicate that the methods used hour lines or position circles (or semicircles) respectively without any further considerations. Note, in this respect, that the principle of the Position Semicircle Method for progressions is found in Ptolemy's *Tetrabiblos*, so it should have been attributed to Ptolemy and not to Hermes.

Another conclusion that can be drawn is that perhaps the astrologers and the mathematicians had different understandings of the *Tetrabiblos*. A practi-

[51] cf. North, *Horoscopes and History*, p. 111 and the index on p. 225; Bouché-Leclercq, *L'astrologie*, p. 270 and n. 1.

[52] It may be a coincidence that Ibn Muʿādh, like Ptolemy, does not pay much attention to the houses below the horizon. See Casulleras, *La astrología*, pp. 175–77.

tioner of astrology may not have seen the geometrical differences between the use of hour lines (presented by Ptolemy as a computational approximation) and position circles, whereas the mathematicians probably did. This may be the reason why competent astronomers like al-Bīrūnī, Rabiçag or Ibn Muʿādh complain[53] about the usual attributions of methods and warn their readers about their inconsistencies; and all of them defend the use of position circles which, according to the principle expressed by Ptolemy in the *Tetrabiblos*, represent the true system for the *tasyīr*.

Bibliography

Adamson, Peter, *Al-Kindī*, Oxford: Oxford University Press, 2007.

al-Bīrūnī, Abū'l-Rayḥān, *Al-Qānūnu'l-Masʿūdī*, 3 vols, Hyderabad: Osmania Oriental Publications Bureau, 1954–1956.

BEA: Thomas Hockey et al. (eds), *The Biographical Encyclopedia of Astronomers*, New York: Springer, 2007.

Bouché-Leclercq, Auguste, *L'astrologie grecque*, Paris: Leroux, 1899 [reprinted Paris: Leroux, 1963].

Calvo, Emilia, 'La résolution graphique des questions astrologiques à al-Andalus', in *Histoire des Mathématiques Arabes. Actes du 3me Colloque Maghrébin sur l'Histoire des Mathématiques Arabes, Tipaza, 1–3 Décembre 1990*, Alger: Association Algérienne d'Histoire des Mathématiques, École Normale Supérieure, 1998, pp. 31–44.

Casulleras, Josep, 'Ibn ʿAzzūz al-Qusantīnī's Tables for Computing Planetary Aspects', *Suhayl* 7 (2007), pp. 47–114.

Casulleras, Josep, 'Mathematical Astrology in the Medieval Islamic West', *Zeitschrift für Geschichte der Arabisch-Islamischen Wissenschaften* 18 (2008/2009), pp. 241–68.

Casulleras, Josep, *La astrología de los matemáticos. La matemática aplicada a la astrología a través de la obra de Ibn Muʿāḏ de Jaén*, Barcelona: Universitat de Barcelona, 2010.

Casulleras, Josep, and Jan P. Hogendijk, 'Progressions, Rays and Houses in Medieval Islamic Astrology: A Mathematical Classification', *Suhayl* 11 (2012), pp. 33–102.

Comes, Mercè, *Historia de la esfera armilar. Su desarrollo en las diferentes culturas*, Madrid/Barcelona: Fundación Juanelo Turriano/Universitat de Barcelona, 2012.

Cooper, Glen M., 'Kindī: Abū Yūsuf Yaʿqūb ibn Isḥāq al-Kindī', in *BEA*, pp. 635–36.

Greenbaum, Dorian G., *The Daimon in Hellenistic Astrology: Origins and Influence*, Leiden: Brill, 2016.

Hermelink, Heinrich, 'Tabulae Jahen', *Archive for History of Exact Sciences* 2 (1963), pp. 108–12.

[53] Ibn Muʿādh is clearly disappointed by the astrologers' lack of mathematical knowledge. See Casulleras, *La astrología*, pp. 174, 240–43 (Spanish) and 268–73 (Arabic).

Hogendijk, Jan P., 'The Mathematical Structure of Two Islamic Astronomical Tables for "Casting the Rays"', *Centaurus* 32 (1989), pp. 171–202.

Hogendijk, Jan P., 'Applied Mathematics in Eleventh Century al-Andalus', *Centaurus* 47 (2005), pp. 87–114.

Hogendijk, Jan P., 'Al-Bīrūnī on the Computation of Primary Progression (*tasyīr*)', in Charles Burnett and Dorian G. Greenbaum (eds), *From Māshā'allāh to Kepler. Theory and Practice in Medieval and Renaissance Astrology*, Ceredigion: Sophia Center Press, 2015, pp. 279–308.

Kennedy, Edward S., 'The Astrological Houses as Defined by Medieval Islamic Astronomers', in Josep Casulleras and Julio Samsó (eds), *From Baghdad to Barcelona. Studies in the Islamic Exact Sciences in Honour of Prof. Juan Vernet*, Barcelona: Instituto Millás Vallicrosa de Historia de la Ciencia Arabe, 1996, vol. 2, pp. 535–78 [reprinted in Id., *Astronomy and Astrology in the Medieval Islamic World*, Aldershot: Variorum, 1998, ch. XIX].

Kennedy, Edward S., and Haiganoush Krikorian-Preisler, 'The Astrological Doctrine of Projecting the Rays', *Al-Abḥāth* 25 (1972), pp. 3–15 [reprinted in Edward S. Kennedy, Colleagues and Former Students, *Studies in the Islamic Exact Sciences*, Beirut: American University of Beirut, 1983, pp. 372–84].

North, John D., *Horoscopes and History*, London: Warburg Institute, 1986.

North, John D., 'The Alfonsine Books and Some Astrological Techniques', in Mercè Comes, Roser Puig and Julio Samsó (eds), *De Astronomia Alphonsi Regis. Proceedings of the Symposium on Alfonsine Astronomy Held at Berkeley (August 1985) Together with Other Papers on the Same Subject*, Barcelona: Instituto 'Millás Vallicrosa' de Historia de la Ciencia Arabe, 1987, pp. 43–50.

Robbins, Frank E., *Ptolemy. Tetrabiblos*, Cambridge (Mass): Harvard University Press, 1940.

Rico y Sinobas, Manuel, *Libros del Saber de Astronomía del Rey D. Alfonso X de Castilla*, 5 vols, Madrid: Aguado, 1863–1867 [reprinted in *Islamic Mathematics and Astronomy*, vols 109–12, Frankfurt: Institute for the History of Arabic-Islamic Sciences, 2002. Digital versions (accessed 28 May 2016) at http://www.galiciana.bibliotecadegalicia.xunta.es/ and http://bibliotecadigital.jcyl.es].

Rius, Mònica, 'Ibn al-Samḥ: Abū al-Qāsim Aṣbagh ibn Muḥammad ibn al-Samḥ al-Gharnāṭī', in *BEA*, p. 568.

Samsó, Julio, and Hamid Berrani, 'World Astrology in Eleventh-Century al-Andalus: the Epistle on *tasyīr* and the Projection of Rays by al-Istijjī', *Journal of Islamic Studies* 10 (1999), pp. 293–312 [reprinted in Id., *Astronomy and Astrology in al-Andalus and the Maghrib*, Ashgate: Variorum, 2007, ch. V].

Samsó, Julio, and Hamid Berrani, 'The Epistle on *tasyīr* and the Projection of Rays by Abū Marwān al-Istijī', *Suhayl* 5 (2005), pp. 163–242 [reprinted in Julio Samsó, *Astrometeorología y astrología medievales* [Col·lecció Homenatges 31], Barcelona: Universitat de Barcelona, 2008, ch. XIV].

Samsó, Julio, *Las ciencias de los antiguos en al-Andalus*, Almería, 2011. Second edition of the original published in 1992, with a comprehensive *addenda et corrigenda* by J. Samsó and M. Forcada.

Sédillot, Louis P. E. Amélie, *Prolégomènes des Tables astronomiques d'Oloug-Beg*, 2 vols, Paris: Typographie de Firmin Didot Frères, 1847–1853.

Sezgin, Fuat, *GAS* VII = *Geschichte des arabischen Schrifttums*, vol. VII: *Astrologie-Meteorologie und Verwandtes bis ca. 430 H*, Leiden: Brill, 1979.

Sezgin, Fuat et al., *Ibn Hibintā. The Complete Book on Astrology. Al-Mughnī fī aḥkām al-nujūm* (facsimile edition), 2 vols, Frankfurt: Institute for the History of Arabic-Islamic Sciences, 1987. Note that the text in vol. I, pp. 231–365, also appears in vol. II, pp. 4–138.

van Bladel, Kevin, 'Hermes and Hermetica', in Kate Fleet et al. (eds), *Encyclopaedia of Islam, THREE*. Consulted online.

Viladrich i Grau, Mercè, *El 'Kitāb al-'Amal bi-l-Asṭurlāb' (Llibre de l'ús de l'astrolabi). Estudi i Traducció*, Barcelona: Institut d'Estudis Catalans, 1986.

Viladrich, Mercè, and Ramón Martí, 'Sobre el *Libro dell Ataçir* de los *Libros del Saber de Astronomia* de Alfonso X El Sabio', in Juan Vernet (ed.), *Nuevos estudios sobre astronomía española en el siglo de Alfonso X*, Barcelona: Instituto de Filología, 1983, pp. 75–100.

Yamamoto, Keiji, 'Abū Ma'shar Ja'far ibn Muḥammad ibn 'Umar al-Balkhī', in *BEA*, p. 11.

Yano, Michio, 'Bīrūnī: Abū al-Rayḥān Muḥammad ibn Aḥmad al-Bīrūnī', in *BEA*, pp. 131–33.

III. The Latin Tradition

Glosses on the *Almagest* by Campanus of Novara and Others in Paris, Bibliothèque nationale de France, lat. 7256

Henry Zepeda

While studies of the Arabic and Latin translations of Ptolemaic texts are *desiderata* in the history of medieval science, analysis of the commentaries on these texts is also needed in order to understand Ptolemy's importance in the history of medieval astronomy and astrology. The actual text of Ptolemy's *Almagest* appears to have been more authoritative among Latin than Arabic astronomers. This is evidenced by the differing ratios of the number of surviving manuscripts containing the *Almagest* to the number of manuscripts containing commentaries. There are 73 manuscripts with the Latin translations of the *Almagest* written before the middle of the fifteenth century and 178 manuscripts containing commentaries written during the same period; however, among Arabic manuscripts, there are only 12 surviving witnesses of the *Almagest* and well over a hundred manuscripts containing just al-Ṭūsī's reworking of the *Almagest*.[1] Despite the relative prominence of the Latin translations of Ptolemy's own text, there were at least seventeen separate commentaries on the *Almagest* that were translated into or composed in Latin before the mid-fifteenth century. A few of these commentaries on the *Almagest* had wide circulation in medieval Europe (e.g., the translation of Thābit ibn Qurra's 'Things necessary before reading the *Almagest*', Gerard of Cremona's translation of Jābir ibn Aflaḥ's *Correction of the Almagest*, and the *Almagesti minor*); however, even these can only provide limited insight into the reception of Ptolemaic thought. While such commentaries provide knowledge of their authors' thinking, they and the *Almagest* are only influences on the thought of other medieval students of astronomy. They cannot reveal how their readers incorporated what they read into their own understandings of astronomy. The lesser known commentaries written in Latin[2] provide us with more representative evidence of what medie-

[1] For the numbers of Latin manuscripts, I have relied upon David Juste's work for his forthcoming manuscript catalogue of the Latin Ptolemaic corpus. For the Arabic, I relied upon María José Parra Pérez's work for the Arabic side of the Ptolemaeus Arabus et Latinus project and her article 'A List of Arabic Manuscripts'.

[2] These include the Erfurt Commentary I, the Vatican Commentary, Simon Bredon's Commentary, which I have discussed and partially edited in Zepeda, *The Medieval Latin Transmission*, pp. 184–251, 282–301, and 493–686.

Ptolemy's Science of the Stars in the Middle Ages, ed. by David Juste, Benno van Dalen, Dag Nikolaus Hasse and Charles Burnett, PALS 1 (Turnhout, 2020), pp. 225–244

10.1484/M.PALS-EB.5.120181

val students of Ptolemy thought — e.g., which parts of the *Almagest* they were interested in, why they read it, in what contexts they studied Ptolemy, how it was taught, and whether they agreed with his methods and results. For the same reasons, the examination of the glosses that medieval scholars added to manuscripts of the *Almagest* is critical for a comprehensive view of medieval Ptolemaic astronomy. The material is plentiful. Of the 45 manuscripts containing Gerard of Cremona's translation of the *Almagest* that I have been able to examine, 28 contain a large number of significant glosses, and scattered short glosses are found in some of the others. I offer here my examination of a sample of glosses to not only provide an example of what such glosses contain, but also to suggest some ways of approaching them.

For this first foray into medieval glosses on the *Almagest*, I have turned to Paris, Bibliothèque nationale de France, lat. 7256, a French manuscript from the mid-thirteenth century, that contains many marginal glosses, almost all of which are also found in a fifteenth-century manuscript, Vatican, Biblioteca Apostolica Vaticana, Barb. lat. 336, which is likely copied from BnF, lat. 7256.[3] Although in both manuscripts almost all of the notes are in the same hands as the main text, these glosses were composed by multiple scholars during the span of time from the late twelfth century to the mid-thirteenth century. Gerard of Cremona appears to have composed some of these glosses in these manuscripts himself. He had originally translated different parts of the *Almagest* from the Ḥajjāj and Isḥāq-Thābit versions of the *Almagest*, and in one group of manuscripts (Kunitzsch's **B**-class), the margins contain his alternate translations of some passages (i.e., where he originally translated from Ḥajjāj, he supplied translations from Isḥāq-Thābit, and *vice versa*).[4] While the number and location of all of Gerard's marginal additions have not been established, seven of the notes in BnF, lat. 7256 seem to be Gerard's, because they appear in the manuscripts of the **B**-class, including Florence, Biblioteca Medicea Laurenziana, Plut. 89 sup. 45, which was copied from an exemplar dated to 1175.[5]

[3] I examined the glosses in these manuscripts having to do with the Menelaus Theorem for my dissertation. See Zepeda, *The Medieval Latin Transmission*, pp. 130, 134–366, 142–60, and 398–431. Paris, BnF, lat. 7256 is described in Kunitzsch, *Der Sternkatalog*, vol. II, p. 13; and in *Catalogus codicum*, p. 331. Vatican, BAV, Barb. lat. 336 is described in Silverstein, *Medieval Latin Scientific Writings*, pp. 101–02.

[4] Kunitzsch, *Der Sternkatalog*, vol. II, pp. 2–3 and 7. These glosses by Gerard should be considered an integral part of his translation of the *Almagest* and should be included in the critical edition of the text. In Clagett, *Archimedes in the Middle Ages*, vol. I, pp. 231–33, it is established that Gerard wrote similar notes in his translation of the Banū-Mūsā's *Verba filiorum*.

[5] Florence, BML, Plut. 89 sup. 45, 183ra. As in this Florentine manuscript, these notes are found in the same hand as the main text of the *Almagest* in another early witness, Melbourne, SLV, RARES 091 P95A.

Other notes in BnF, lat. 7256 and Barb. lat. 336 were composed approximately 75 years later. The first of a pair of notes on *Almagest* III.9 provides the dates according to various epochs that correspond with 30 June, 1251 AD, and the second reports the value for the sun's position that the commentator calculated for that date using Ptolemy's tables.[6] Additionally, many notes in the manuscript are attributed to Campanus of Novara, who is best known for his version of Euclid's *Elements*. The works of Campanus, who died in 1296, that can be dated with any amount of certainty come from the 1250s and 1260s, so his glosses on the *Almagest* probably spring from those decades or shortly before or after.[7]

That Campanus is the author of the glosses attributed to him is not to be doubted.[8] One of the notes attributed to Campanus, in which he discusses the different cases of the Menelaus Theorem and how Ptolemy does not treat all of them, ends, 'Et ideo Thebit fecit tractatum unum qui intitulatur Thebit De figura sectore, in quo hec omnia probat. Ego etiam feci tractatum alterum de eodem planiorem ut puto et manifestiorem'.[9] And another note includes a justification, '... per librum de proportione et combinationibus proportionum quem composuimus'.[10] Campanus did indeed write works on the 'sector figure' (i.e. the Menelaus Theorem) in which he proves the various cases of the theorem, and on the compound ratio and its 'modes'.[11] Additionally, a table (unrelated to the *Almagest*) has the heading 'ad latitudinem 45 graduum', which agrees with the latitude of Novara.[12]

It appears that only the notes specifically attributed to Campanus are part of his set of glosses and that the other notes are by other commentators. A few of the glosses that are attributed to him are very similar to notes that do not

[6] BnF, lat. 7256, 34v.

[7] Benjamin and Toomer, *Campanus of Novara*, pp. 5, 9, and 13; this is supplemented by Toomer, 'Campanus of Novara'.

[8] Silverstein, *Medieval Latin Scientific Writings*, p. 102. Silverstein noticed that the *Almagest* in Barb. lat. 336 was 'heavily glosed [*sic*] in marg. (esp. through dictio 3a, fol. 72) with extensive passages from Campanus (many marked Campanus, Camp., or C)', but he did not determine that these notes are Campanus' commentary on the *Almagest* and not merely excerpts from Campanus' other works.

[9] BnF, lat. 7256, 10r.

[10] BnF, lat. 7256, 21r.

[11] An edition of Campanus' work on the sector figure is found in Lorch, *Thābit ibn Qurra*, pp. 436–42, and an edition of the work on compound ratio in Busard, 'Die Traktate *De Proportionibus*', pp. 213–22. While it has been unclear whether these works should be considered one work or two (Lorch, *Thābit ibn Qurra*, pp. 426–33), these notes show that Campanus himself considered them to be separate works. Nevertheless, Lorch does establish that the two works were often treated as a pair.

[12] BnF, lat. 7256, 103v.

bear his name; for example, most of the notes that provide outlines of the text do not contain an attribution to Campanus, but one does. Because these notes are all similar in style, it is tempting to take the single attribution as evidence that he wrote all of them; however, because the attribution to Campanus is not in Barb. lat. 336, its presence in the Parisian manuscript is probably due to scribal error, which would have been easy to make if the Parisian manuscript's exemplar ended the note with an abbreviated form of 'et cetera', which could have appeared very similar to the letter 'c' for Campanus.[13] Also, Campanus' name is attached to two notes that appear to include one of the numbered set of enunciations (to be discussed below), and again one might see this as evidence that he composed the entire set; however, this only occurs when the note includes the enunciation and additional commentary.[14] A simple explanation for these cases is that the numbered enunciations are not Campanus' and that the attribution to Campanus only applies to whatever is written before and after the enunciation. Additionally, even very small pieces of commentary are attributed to him. For example, there is one small added line in a figure, and it is marked with '·c·' for Campanus.[15] Similarly, an interlinear gloss that consists of only the two words 'puncto sumitur' is followed by 'Campanus'.[16] This suggests that Campanus or whoever added the attributions had a fastidious concern with marking each of his contributions, and that any gloss without his name was not composed by him. It is also very possible that Campanus wrote more glosses on the *Almagest* and that only the notes that a subsequent reader found interesting were transmitted.

The Parisian manuscript provides an example of how sets of notes were copied from one manuscript into the margins of another. Because this manuscript was made in the mid-thirteenth century, the marginal notes from Gerard of Cremona were clearly copied into it; moreover, it is likely that many or all of the glosses were copied from other manuscripts. We find evidence of this in five notes, including two from Campanus, two from the series of *divisiones textus*, and one proof that is not attributed to Campanus. In these the scribe omitted text (due to eye-skip in two of the cases) and then supplied the missing text alongside the note.[17] As stated above, the notes from BnF, lat. 7256 are also found in Barb. lat. 336. Also, Oxford, All Souls College, 95, and Oxford, New College, 281, contain notes derived from one in BnF, lat. 7256 and Barb. lat. 336, i.e. the note providing the dates corresponding to 30 June, 1251 AD

[13] BnF, lat. 7256, 2r; Barb. lat. 336, 3r.
[14] BnF, lat. 7256, 9v and 21r.
[15] BnF, lat. 7256, 21v.
[16] BnF, lat. 7256, 21v.
[17] BnF, lat. 7256, 4v, 10r, 10v, 11v, and 14r.

mentioned above.[18] There are other examples of notes being copied from manuscript to manuscript: Gerard's alternate translations were copied into the margins of many manuscripts along with the main text, as stated above, and another sets of glosses are found in Vatican, BAV, Vat. lat. 2057; Vatican, BAV, Barb. lat. 173; Cracow, Biblioteka Jagiellońska, 590, and Florence, Biblioteca Nazionale Centrale, Conv. Soppr. J.IV.20 (San Marco 182).[19] A consequence of this medieval practice of copying glosses is that historians must be cautious in using dates in marginalia to date manuscripts. Crossley writes about the note with the date in the All Souls College manuscript, '... it strongly suggests that the scribe was writing this text in the sixth month, i.e. June, 1250'.[20] However, this date can only suggest that the commentator who originally composed the note, wrote it at that time. It can only provide a *terminus post quem* for the manuscripts into which it was copied. In this case, it is seen for other reasons that the manuscript was written at the end of the thirteenth century, not 1250.[21]

In order to gain a rough overview of glosses, the *Ptolemaeus Arabus et Latinus* project is cataloguing marginal and interlinear notes in select chapters and manuscripts that are more than around ten words or that appear to be important for one reason or another. Although such a method of counting notes has its deficiencies — e.g. some notes consisting of only a few words are more important than longer ones, and it is not always clear what constitutes a single note, this is still an important first step in studying glosses that can indicate which sections of the text were of interest or were in need of further elucidation.

I made such a catalogue of the notes of BnF, lat. 7256 (using Barb. lat. 336 when notes were illegible in my reproductions of the Parisian manuscript). The text is glossed unevenly. Of the 144 chapters contained in the thirteen books of the *Almagest*, only 49 have glosses while 95 do not. Approximately 87% of the notes are written on the first three books of the *Almagest*. There are no glosses of more than approximately ten words in Books VII, VIII, X, or XI. However, one cannot make sweeping claims about which parts of the *Almagest* were of interest to medieval readers from this manuscript alone. It appears to be more normal for the glosses to be spread more consistently, and some

[18] For comparison, the beginning of the notes are: 'Creditur quod isti anni sibi conveniant ...' in All Souls College, 95, 27r; 'Isti anni omnes sibi conveniunt sine errore aliquo secundum quod invenitur discreta inquisitione ...' in New College, 281, 57v; 'Isti anni omnes sibi conveniunt secundum quod inveni discreta inquisitione ...' from BnF, lat. 7256, 34v and Barb. lat. 336, 70r.

[19] See Zepeda, *The Medieval Latin Transmission*, pp. 130 and 386–89.

[20] Crossley, 'Ptolemy's *Almagest*', p. 122.

[21] Watson, *A Descriptive Catalogue*, pp. 195–96.

manuscripts have very different distributions of notes than that which we find in BnF, lat. 7256. For example, Cracow, BJ, 589 has more glosses on Book VI than on Books I–III combined, and has many glosses on Books IX–XIII. Also, Paris, BnF, lat. 7258 has almost no glosses in Books I, II, XI, XII, and XIII, but many in the remaining books. Memmingen, Stadtbibliothek, 2° 2,33 provides an extreme example; it contains approximately 100 notes on the table of fixed stars and only three glosses of more than a few words on the entirety of the remainder of the *Almagest*. While BnF, lat. 7256's preponderance of notes in the early books may be a rarity among *Almagest* manuscripts, this bears a likeness to some of the stand-alone commentaries: the *Almagesti minor* only covers the first six books, the Erfurt Commentary treats only the first two, and Simon Bredon's commentary covers only the first three.

There are a variety of types of notes in this manuscript. Editorial comments concerning the establishment of the text include not only Gerard's alternate translations of passages, but also what appears to be a reading from another exemplar,[22] and a remark upon chapter division.[23] Many of the notes are summaries, which make the main thrust of a chapter or passage apparent quickly, and which sometimes serve as 'roadmarks' that allow readers to find relevant passages or to understand how the passages in chapters are related. There are also a number of ways in which notes provide explanations of Ptolemy's text; they express words, phrases, or longer passages in simpler wording; they elaborate on passages; they supply steps implicit in Ptolemy's calculations and proofs; and they provide citations to justify claims made by Ptolemy. References can be internal, as with 'per 25 huius',[24] or to other works 'per 1 sexti Euclidis',[25] 'per primam propositionem *De ysoperimetris*',[26] 'per quartam primi libri Milei',[27] 'per 21 primi Theodosius'.[28] Glosses also include lists of values, calculations for values that Ptolemy reports, and lists of places where unusual words appear. For the sake of brevity, I will discuss in more detail only a few types of notes: outlines, enunciations, figures, proofs, and corrections.

[22] BnF, lat. 7256, 27v. This note is not found in Florence, BML, Plut. 89 sup. 45 or Melbourne, SLV, RARES 091 P95A, and thus does not appear to be one of Gerard's notes.

[23] BnF, lat. 7256, 2r. The proposed chapter division matches that found in Toomer, *Ptolemy's Almagest*, p. 38.

[24] BnF, lat. 7256, 20r. As is clear here, internal references were not only to book and chapter, but also to the division of propositions as provided by the numbered enunciations.

[25] BnF, lat. 7256, 6r.

[26] BnF, lat. 7256, 2v. The enunciations of the cited propositions of this work are also given in the margin.

[27] BnF, lat. 7256, 13r.

[28] BnF, lat. 7256, 16r.

The *divisio textus* is one type of note that is found in BnF, lat. 7256. In the notes of this type, the commentator outlines the text and reports where parts of the text begin. Such divisions are made for passages of differing organizational strata: books, groups of chapters, chapters, and parts of chapters. Notes are given at the beginning of Books I and II, summarizing their main goals and dividing them into main sections (e.g., Book I is divided into sections on the form of the universe, chords, declinations, and right ascensions), most of which consist of multiple chapters. While some other chapters (I.3–4, 6, II.9) receive notes providing short outlines, the bulk of the outlining occurs in Book I, ch. 9–10, 12–4, and Book II, ch. 1–6.[29] These sections of the text are not merely divided into sections, but into layer after layer of subsection. At points the division reaches the twelfth level of subdivision. [See the table below.] Outlining the text in such detail would have been quite an undertaking. While the outline I have provided based upon these glosses is relatively brief, the notes generally express the relations between the sections of text not graphically, but with words only (there are some exceptions such as in Figure 1). The glossator starts to divide the second part of Book I in this way:

> Terminata parte prima principali huius primi libri in qua determinavit de forma universi. In hac secunda intendit demonstrare qualiter sciatur corda omnis arcus noti. Et dividitur in duobus. In prima ponit prohemium in quo utitur transitu, continuans dicta dicendis. In secunda prosequitur intentum, ibi 'Dividam igitur' etc. Hec secunda dividitur in 3. In prima ponit demonstrationes per quas cognoscitur omnis corda que subtenditur alicui arcui minori semicirculo non minori medietate partis. In secunda ponit modum compositionis tabularum cordarum arcuum. In tercia comparat demonstrationes ad tabulas. Secunda ibi 'Et quoniam necesse est nobis scire numerum partium' etc. Tercia ibi 'O quam bene' etc. Prima istarum dividitur in 3...[30]

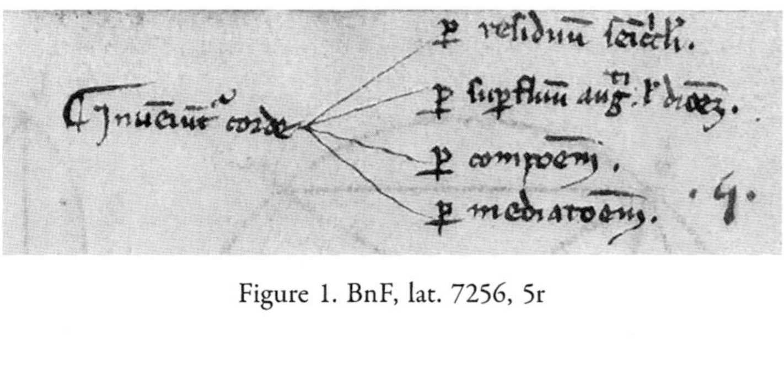

Figure 1. BnF, lat. 7256, 5r

[29] Notes including divisions of the text are found in BnF, lat. 7256, 1v–2v, 3v, 4v, 5v–6r, 9r–14v, 15v, and 19v.

[30] BnF, lat. 7256, 4v.

Division of Section on Chords according to Glosses of Paris, BnF, lat. 7256

Chords of arcs ('Summa vero principiorum . . .')
- Preface [84 words]
- Intended subject ('Dividam igitur . . .')
  - — Proofs through which chords of arcs are found
    - ▪ Hypotheses
      - ◦ 1st hypothesis [27 words]
      - ◦ 2nd hypothesis ('Et dividam dyametrum . . .') [76 words]
      - ◦ 3rd hypothesis ('Et assumemus numerum . . .') [42 words]
    - ▪ Proofs ('Sit itaque . . .')
      - ◦ Proofs through which chords are found directly from diameter
        - • Chords that are sides of inscribed figures
          - — How to find them
            - ▪ Side of decagon & pentagon
              - ◦ DZ is side of decagon, BZ side of pentagon
                - • Decagon [175 words]
                - • Pentagon ('Et similiter quoniam latus penthagoni . . .') [47 words]
              - ◦ Each of these is known ('Et quia dyametrum circuli . . .')
                - • Decagon [119 words]
                - • Pentagon ('Et etiam quoniam linea DZ est . . .') [105 words]
            - ▪ Side of square & triangle ('Iam ergo manifestum est . . . ') [103 words]
          - — Summary ('Iam ergo leviter novimus. . .') [9 words]
        - • Chords of their supplements ('Et declarabitur nobis quod cum . . .') [117 words]
      - ◦ Proofs through which chords are found indirectly from diameter ('Et declarabo in sequentibus . . .')
        - • Preface [24 words]
        - • Tract ('Sit itaque . . .')
          - — Chords known geometrically
            - ▪ Chords found by geometry
              - ◦ Preliminaries [219 words]
              - ◦ Propositions ('Et postquam hoc iam premisimus . . .')
                - • Chord of difference of arcs
                  - — Proof [142 words]
                  - — Application ('Declarabo etiam . . .') [36 words]
                - • Chord of half arc ('Quod si etiam arcus . . .')
                  - — Proof [263 words]
                  - — Application ('Per hoc ergo capitulum . . .') [89 words]
                - • Chord of sum of arcs ('Describam etiam circulum . . .') [191 words]
            - ▪ Construction of table ('Post hoc autem capitulum . . .') [80 words]
          - — Chord not known geometrically ('Quod si nos reperiremus . . .')
            - ▪ Chord of 1/2°
              - ◦ It cannot be found exactly [78 words]
              - ◦ How to find approximately ('Perscrutabor igitur . . .')
                - • Preface [53 words]
                - • Tract ('Et ad hoc premittam . . .')
                  - — Premiss [350 words]
                  - — Finding of chord ('Postquam affirmavimus . . .') [303 words]
            - ▪ How to complete tables ('Et per hoc complebitur . . .') [62 words]
    - ▪ Summary ('Autem sciatur . . .') [31 words]
  - — How to make table of arcs and chords ('Et quoniam necesse est. . .') [138 words]
  - — Proofs for tables ('O quam bene. . .') [77 words]

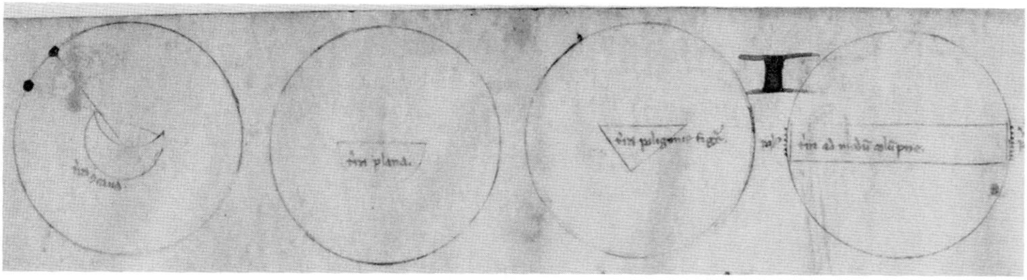

Figure 2. BnF, lat. 7256, 3r

The note continues in this vein. It and three other notes, consisting of over 700 words in total, divide the texts of I.9–10, which are made up of 3027 words. Keeping this ratio, it would require approximately 47,000 words just to outline the complete *Almagest*. Lecturers at universities often provided divisions of the text for their students, so it is possible that these notes arose from university lectures on the *Almagest*. Whether or not these outlines were given in lectures or were only written commentary, producing such detailed commentaries must reflect a medieval scholar's deep interest in the *Almagest*, especially the chapters on trigonometry and calculations of spherical astronomy found in *Almagest* I–II.

Another type of marginalia consists of the added geometrical figures and illustrations. Most of these are part of complete proofs added in the margins (to be discussed below). Of the other figures that appear in the glosses, some illustrate arguments presented in the text of the *Almagest*. For example, figures [Figure 2] are added in the margins to illustrate Ptolemy's argument by *reductio ad absurdum* that the earth is spherical. They depict the heavens surrounding the earth drawn as if it were concave, flat, polygonal, or cylindrical, and the first of these includes lines of sight drawn to a rising star to show that if the earth were concave, a star would appear first to people who lived further west.[31] Likewise, the manuscript includes figures to accompany Ptolemy's argument that the earth is in the middle of the universe.[32] Such illustrations were probably intended to aid the imagination of astronomy students, as were realistic depictions of instruments, such as the one described in *Almagest* V.12. [Figure 3] It is not a mere geometrical figure, but portrays the various parts of the instruments and even includes a disembodied arm holding a plumb in order to set up the instrument properly.[33] Besides accompanying or presenting

[31] These can barely be made out on my reproductions of BnF, lat. 7256, 3r, but they are clear on Barb. lat. 336, 4v.

[32] BnF, lat. 7256, 3r–3v. Some of these are difficult to make out in my reproductions, but they are all clear in Barb. lat. 336, 5r–5v.

[33] Florence, BML, Plut 89 sup. 45 and Melbourne, SLV, RARES 091 P95A do not include any depiction of this instrument. Another realistic image of a man using an instrument is found on BnF, lat. 7256, 56v.

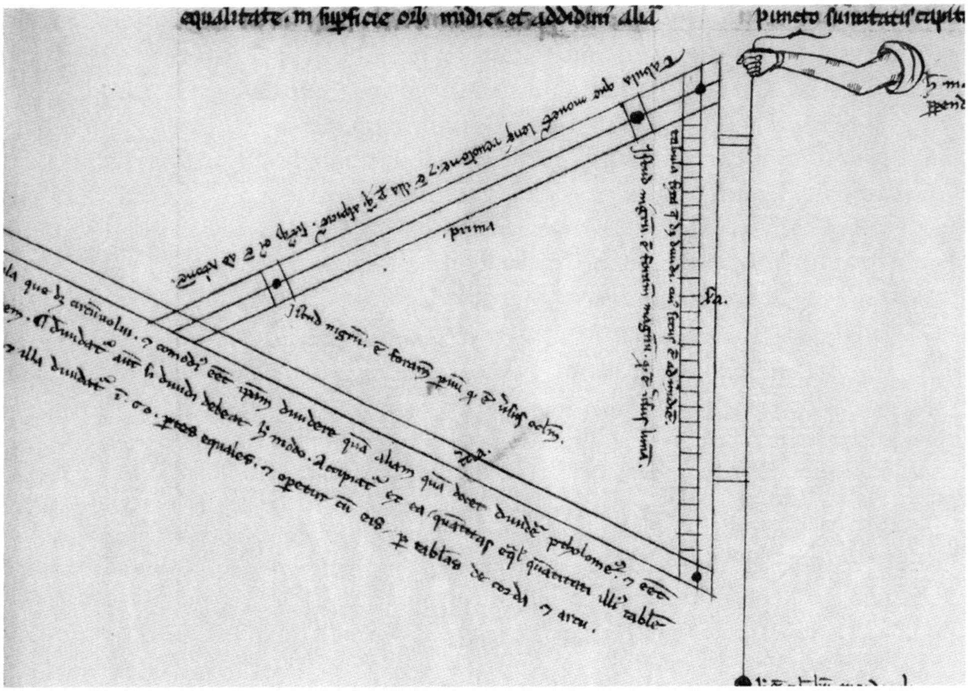

Figure 3. BnF, lat. 7256, 55r

ideas already present in the text in a visual manner, figures in the margins can suggest concepts or entire arguments that are not contained in the text at all. An excellent example is a figure [Figure 4] placed next to a note clarifying that while Ptolemy claims that the phenomena shows that the heavens are spherical, his arguments only show that the heavens move spherically and that the heavens could indeed be square-shaped. The figure indeed shows a square surrounding the earth, but the labels show that the text does not only represent the argument written in the note, but that there is also an implicit argument. The points labeled 'A' seem to identify the same star at equal intervals of time as it travels at a constant speed along a square path, not a circular one. The lines from the earth to the star show that angles would not be equal, and thus the star would appear to change its speed if its course were not circular.[34]

Besides adding new figures, commentators also clarified the connection between Ptolemy's text and his figures, which are in the margins in this manuscript. This was sometimes accomplished without the use of any words, but only with pairs of identical marks by the text and the corresponding figures.[35] Also, labels identifying the astronomical significance of different parts of

[34] BnF, lat. 7256, 2v.

[35] E.g. BnF, lat. 7256, 123v–124r.

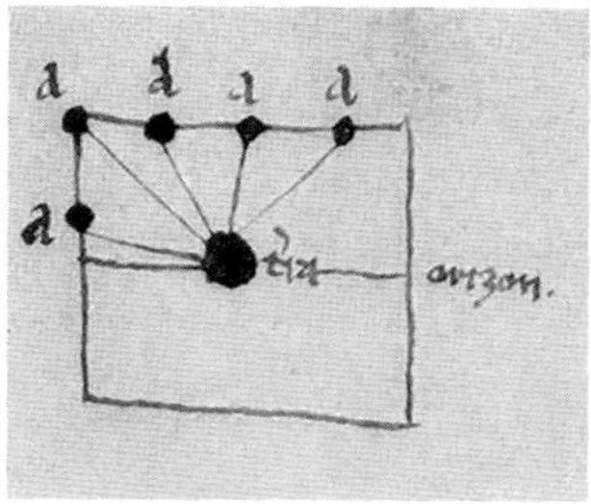

Figure 4. BnF, lat. 7256, 2v

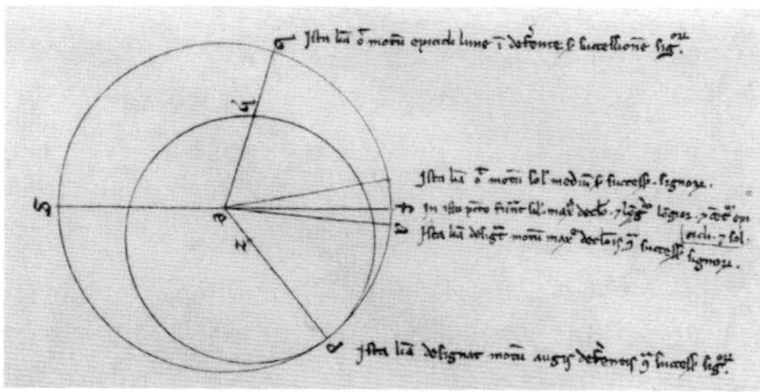

Figure 5. BnF, lat. 7256, 48v

geometrical figures were added. Some of these 'astronomical labels' may have been included in Gerard's translation, but it is clear that commentators added more.[36] By allowing the reader to focus on the figure and to not constantly move his gaze back and forth from text to figure, these labels would have made it easier for readers to grasp complex models. The figure for V.2, [Figure 5] which attempts to depict the motions of the lunar model in one static image, is able to be understood without constantly consulting the text because of its astronomical labels and its added lines representing the northernmost point of the moon's inclined circle, the position of the mean sun, and the starting position of all the motions.

Enunciations are a type of note that is found frequently in BnF, lat. 7256. Ptolemy intended many of his calculations to serve as examples that could be applied to other situations, but he often did not pose them in general terms. Likewise, Ptolemy's general proofs are interwoven into the surrounding text and are not given in as structured of a format as medieval scholars encountered in geometrical and arithmetical works such as Euclid's *Elements* or Jordanus de Nemore's *Arithmetica*. Commentators provided generalized statements in the margins by the beginning of many calculations and proofs. For example, in

[36] Several labels of this kind are found in early manuscripts including Florence, BML, Plut. 89 sup. 45 and Melbourne, SLV, RARES 091 P95A, and very infrequently in Paris, BnF, lat. 7254.

I.13 we find: '16. Dato puncto orbis signorum declinationem eius ab equinoctiali circulo invenire. Unde manifestum est quod si sinus arcus orbis signorum qui intercipitur inter equatorem et punctum datum ducatur in sinum maxime declinationis, et productum dividatur per sinum quarte, exibit sinus declinationis puncti dati'.[37] By providing the enunciation (i.e. the first sentence), the glossator gives a summary of the chapter to assist the reader in locating Ptolemy's treatment of declinations of the ecliptic. This sort of note also makes the universal applications more apparent and makes Ptolemy's astronomy match the ideals of a mathematical science closer. While there were competing ideas of what constituted *scientia*, most medieval scholars believed that it was derived deductively from certain principles and that it should concern universal truths. For a model of such a science, medieval scholars turned to Euclid's *Elements*, the books of which include lists of principles (definitions, postulates, and common notions), followed by proofs, each consisting of a general proposition and its demonstration that relies upon the principles and the earlier propositions.[38] The enunciations are a step in that direction. The corollaries that are often found with the enunciations (the second sentence in the example given above) provide rules for performing calculations. In fact, the important part of many of the proofs is not that a quantity is known geometrically, but that there is an algorithm that can be used to calculate its value from other given values.[39]

The enunciations included in the margins of BnF, lat. 7256 are found only in the first two books. They are numbered sequentially 1–17 for Book I and 1–39 for Book II. The enunciations and their corollaries are derived from the *Almagesti minor*, which was a summary of *Almagest* I–VI. In a few cases, enunciations directly from the *Almagesti minor* are given in addition to the set of enunciations with continuous numbering.[40] By fitting it into a structure and style that imitated the *Elements*, the author of this early thirteenth-century work 'Euclidized' Ptolemy's astronomical work. One of the many

[37] BnF, lat. 7256, 10r.

[38] For example, see Høyrup, 'Jordanus de Nemore' and Evans, 'Boethian and Euclidean'.

[39] This is clear from the passage in *Almagest* I.9 that corresponds to the first enunciation: '1. Data circuli dyametro ex ipsa latus exagoni, decagoni, pentagoni, quadrati, trigoni, equilaterorum circulo inscriptorum elicere' (BnF, lat. 7256, 5r). Ptolemy does not merely inscribe these equilateral polygons in the circle, as one might think from the enunciation alone, instead he proves the validity of procedures for finding their lengths in terms of the parts of the diameter (Toomer, *Ptolemy's Almagest*, pp. 48–50). In this case, no rule is provided in the corollary, probably because it would be extremely wordy.

[40] BnF, lat. 7256, 5v. The enunciations of *Almagesti minor* I.3–5 (but here numbered 4–6) are given at the bottom of the folio.

changes that the author of this work adopted to achieve this was to add these enunciations.[41]

Some of the enunciations are taken from the *Almagesti minor* with no changes or only trivial ones. For example, the enunciation '8. Maximam declinationem per instrumenti artifitium et considerationem reperire' is identical to one in the *Almagesti minor*, except it is the 15th proposition in that work.[42] Approximately ten other enunciations match ones in the *Almagesti minor*. Several others share common language. For example, compare the following corresponding enunciations (differences besides word order italicized):

> *Almagesti minor*: '25. Maxima declinatione nota angulum ex meridian*o* et *circulo* signorum aput punctum equinoctii proven*ientem notum esse oportet*. Unde patet quod *si* maxima*m* declinatione*m* add*as* super quartam vel ab ea *subtrahas*, *exibit* angulus quesitus'.
>
> BnF, lat. 7256, 20r: '*29*. Nota maxima declinatione angulum *qui* proven*it* ex *sectione* meridian*i* et *orbis* signorum aput *utrumlibet* punctum equinoctii *invenire*. Unde patet quod maxima declinatione add*ita* super quartam vel ab ea *diminuta*, *provenit* angulus quesitus'.

Since one change, such as the word 'sectione', added for clarity, requires changes in the endings of other words, the changes are even fewer than the simple marking suggests. Not all enunciations, however, correspond closely to those in the *Almagesti minor*. For example, the following enunciations from Book I convey the same meaning, but with scarcely any sign of the glosses' reliance upon the prior work (differences are italicized):

> *Almagesti minor*: '*3*. *Si* in semicirculo cord*e* arcuum inequalium *certe fuerint*, cord*a quoque arcus quo maior minorem superat erit nota*'.
>
> BnF, lat. 7256, 5v: '*4*. *Cognitus duabus* cord*is duorum* arcuum inequalium in semicirculo, cord*am superflui inter eas invenire*'.

There are no more similarities between these two enunciations than one would expect of any two passages written separately that convey the same meaning. The reason for the change in language is not apparent and may be due to the mere fact that the glossator wanted to try his hand at expressing the ideas of his source in his own wording. Other differences from the source are the reordering of some enunciations and a different division (i.e. sometimes multiple enunciations correspond to single enunciations from the *Almagesti minor*, and vice versa).

[41] For more on these themes and on the *Almagesti minor*, which has often been erroneously called the 'Almagestum parvum' by me and other scholars, see Zepeda, 'Euclidization in the *Almagestum parvum*', and Zepeda, *The First Latin Treatise*.

[42] BnF, lat. 7256, 9r.

Glosses	*Almagesti minor*	Reason for Change
Book I	Book I	
1–2	1	The glossator considers the corollary of *Almagesti minor* I.1 to be its own proposition.
3–7	2–6	
8	15	The *Almagesti minor* does not follow order of the *Almagest*, in order to emphasize mathematical proofs.
9–10	7–8	
10 (bis)	9	Two enunciations are numbered 10 in the glosses.
11–15	10–14	
16–17	16–17	
Book II	Book II	
1–2	2–3	The *Almagesti minor* does not follow the order of the *Almagest*, probably in order to first prove effect from cause.
3	1	
4	4	
5	—	This is mentioned in the text of *Almagesti minor* I.4, but is not given in the enunciation or as a separate enunciation because the mathematics is fundamentally the same.
6	5	
7–8	6	The glosses split one enunciation with two parts into two separate enunciations.
9–20	7–18	
21–22	19	The *Almagesti minor* does not follow the order of the *Almagest*.
23	20	'...'
24	19	'...'
25–35	21–31	
36	32–33	The glosses combine two enunciations because they are cases of the same proof.
37–39	34–36	

Enunciations for the astronomy of the *Almagest* are not only found in BnF, lat. 7256, Barb. lat. 336, and the 23 manuscripts of the *Almagesti minor* (among which is one *Almagest* manuscript, Paris, BnF, lat. 16200, that has the *Almagesti minor* written in the margins), but they also appear in several other *Almagest* manuscripts: Cracow, BJ, 589; Cracow, BJ, 619 (enunciations excerpted from the *Almagesti minor*); Erfurt, Universitätsbibliothek, Dep. Erf. CA 2° 375;

Florence, BML, Plut. 89 sup. 57; Oxford, New College, 281; and Melbourne, State Library of Victoria, RARES 091 P95A. Also, the Erfurt Commentary, Simon Bredon's commentary, Johannes Blanchinus' *Flores Almagesti*, and Peurbach and Regiomontanus' *Epitome Almagesti* contain enunciations.

Complete proofs are another class of notes that were written in *Almagest* manuscripts. There are 28 such notes in BnF, lat. 7256. The bulk of the proofs are on I.9 (5 notes), I.12 (5 notes), and I.13 (8 notes), and the remaining proofs are scattered throughout the *Almagest*: 2 notes in I.14, 2 in II.11, and 1 each in I.5, II.7, II.10, V.15, XII.1, and XIII.3. Much of the imbalance in the dispersal of these glosses is due to Campanus's interest in establishing the proportion theory that is needed to prove and utilize the Menelaus Theorem. Of the 15 proofs in notes on *Almagest* I.12–14, 13 are the work of Campanus, and 8 of these involve proportion theory;[43] however, Campanus only wrote 3 of the proofs in the other chapters (II.11 and XIII.3). Among the notes that were not written by Campanus, there is a clear focus upon *Almagest* I.9, in which Ptolemy gives the theory behind his table of arcs and chords (5 of the 12 non-Campanus notes are on I.9). These notes with proofs were added for a variety of reasons. Most of these proofs justify parts of Ptolemy's arguments and proofs. As stated above, Campanus' many proofs regarding proportions give the theoretical background for proving the Menelaus Theorem and performing calculations with it. Another example is a proof that supplies part of Ptolemy's argument in *Almagest* I.5, showing the impossibility of the earth standing under the equator but not on the axis that runs between the north and south poles; the glossator proves the general statement, 'Quod omnium duorum gnomonum equalium illuminatorum ab aliquo corpore luminoso inequaliter tamen ab eo distantium, magis distantis maior est umbra'.[44]

Of the remaining notes with proofs, one supplements one of Ptolemy's proofs: after Ptolemy proves a property of inscribed quadrilaterals, a note includes the proof for the special case when the quadrilateral is a square.[45] Another proof in the margins provides a geometrical demonstration that is more restrictive than one given in the text by Ptolemy. In *Almagest* XII.1, Ptolemy gives and uses a proof for a theorem from Apollonius that applies for both the situation in which a certain line is equal to a second or the situation in which it is greater than the second; however, Ptolemy applies it only for the latter of these two cases. A commentator reproves the lemma in a slightly different form so that it only applies to the case used by Ptolemy.[46]

Some notes provide alternates proofs, i.e. they demonstrate things proved by Ptolemy in other ways. While it is conceivable that some alternate proofs could

[43] BnF, lat. 7256, 9v, 10v, 11v.

[44] BnF, lat. 7256, 3v.

[45] BnF, lat. 7256, 5v.

[46] BnF, lat. 7256, 138r.

be included in the glosses merely because the commentator was compiling different methods, the alternate proofs in the Parisian manuscript are at least to some degree criticisms of Ptolemy. For example, two of Jābir ibn Aflaḥ's trigonometrical proofs are included, one for finding the chord of the half of an arc of which the chord is known, and the other for finding the chord of the sum of two arcs with known chords. The commentator explains that the first proof leads to more accurate calculations than Ptolemy's corresponding proof, and that the second is both simpler and easier than Ptolemy's. The commentator recreated Ptolemy's calculations for the chords of 1°30′ and 45′ according to the methods of calculation that can be derived from Ptolemy's half-arc proof, and he arrived at the values $1^{p}34'14''$ and 47′7″; however, Ptolemy writes that the chords are $1^{p}34'15''$ and 47′8″. The commentator found that calculating with Jābir's method leads to the values that Ptolemy reports in the *Almagest*, and from this he concluded that Ptolemy used the same method as Jābir and did not find the values of these chords in the way that he had laid out in the *Almagest*.[47] Another proof that tacitly critiques Ptolemy is found in a note on *Almagest* II.7. Because Ptolemy's second way of finding oblique ascensions does not easily reveal the corollary taken from the *Almagesti minor*, a commentator provides another proof of it that remains on the general level and makes the corollary manifest.[48] While not as obvious of a correction as the previous example, the inclusion of this alternate proof suggests that Ptolemy's calculation did not provide what the commentator wanted, i.e. the justification for a simple, general rule for performing calculations. Similarly, because Ptolemy finds the angle formed by the ecliptic and the horizon for any point of the ecliptic in a way that does not reveal the corollary found in the margins, Campanus provides a proof that shows this more clearly.[49]

A large correction that is found in the Parisian manuscript is a complete second table of chords and arcs reworked by Campanus. Although Ptolemy's table of arcs and chords is in *Almagest* I.11, this corrected table is placed after *Almagest* V.4. This table, which is entitled 'Tabule cordarum et arcuum secundum quod eas verificavit Magister Campanus Novariensis', is based on the table found in Gerard's translation, but approximately 16% of the values of the chords have been changed.[50] Presumably Campanus calculated the value of the chord of each of the 359 arcs of the table. In marginal notes, a commentator — it is unclear whether Campanus or another glossator — states that he calculated the values of the chords for 5°30′ and 6° according to both Ptolemy's

[47] BnF, lat. 7256, 5v. Several modern scholars have addressed the discrepancies between Ptolemy's proofs in *Almagest* I.9 and the values that he gives for chords in that chapter and in his table of chords. For an overview see Van Brummelen, *Mathematical Tables*, pp. 46–73.

[48] BnF, lat. 7256, 17r.

[49] BnF, lat. 7256, 21r.

[50] BnF, lat. 7256, 101v–103r.

methods and Jābir's alternatives, which had been given earlier in the glosses, and that while the values according to Jābir's methods are more accurate, he did not replace the values in the table because doing so would make the increase of the column of thirtieths inconsistent.[51]

As a final topic, there are two notes, in which commentators relate the mathematics of the *Almagest* to other problems that do not involve astronomy. In one note, a commentator provides a discussion of how one could determine how long it takes one mean conjunction of the sun and moon to the next. This is followed by a more complex question concerning a tower whose base is of a certain size that is filled from water from a stream at one rate as some of the water is drained from the tower into a moat of a certain size at another rate. The note asks how long it would take to fill the moat and how tall the tower must be so that it is filled at the same time the moat is filled. This question is followed by the solution to the problem, marked with 'Respondeo'. That this problem is more complicated than the one concerning the sun and moon shows that it is not intended as an explanation of the astronomy. Rather, the astronomical question was seen as an opportunity to practice skills of problem solving.[52] The content and the format suggest the possibility that this note was the result of a disputation held after a lecture on the *Almagest*. A similar case of the astronomy leading to this type of problem is found in the margins by *Almagest* XIII. There Campanus proves that if two quantities are known and the ratio of the remainders are known when one unknown quantity has been subtracted from each of the two known quantities, then the size of the remainders and of the subtracted quantity can be found. After proving this generally, Campanus proposes a problem concerning a *paterfamilias* who hires two men and their sons to complete two jobs. He then shows how the problem is solved through the algorithm whose validity he had proved. Again, this shows the astronomy serving as a springboard to discuss other mathematics of the sort found in algebraic or abbaco texts.[53]

From this brief exposition of some of the features of the glosses found in BnF, lat. 7256, and which are copied into Barb. lat. 336, it is clear that there is much to gain from an examination of the glosses in the *Almagest* manuscripts. Although difficult to decipher and understand, *Almagest* glosses provide a unique opportunity to observe the practice of the science of the stars during the Middle Ages. Therefore, one of the goals of the *Ptolemaeus Arabus et Latinus* project is to provide a survey of glosses on Ptolemy's astronomical and astrological corpus. The first book-length study devoted to this effort will be a cat-

[51] BnF, lat. 7256, 101v.

[52] BnF, lat. 7256, 103v. I also consulted Barb. lat. 336, 201v because very little of the text was legible in my reproductions of the Parisian manuscript.

[53] BnF, lat. 7256, 153v.

alogue of glosses from all manuscripts of Gerard's translation of the *Almagest* that is being produced by Stefan Georges. In ten selected chapters, each note over 10 words will be given an entry that includes a transcription for shorter notes or its incipit and explicit for lengthy notes, which manuscripts have it, and any names or dates mentioned in it. This catalogue will allow researchers to identify sets of notes that occur in more than one manuscript, and to easily determine which manuscripts have notes that merit closer examination. The project will also attempt to identify glossators and to date their comments. Further transcriptions of a select number of representative or innovative glosses may also be made. The survey of glosses will be an invaluable resource for other historians of medieval astronomy, especially when paired with the manuscript reproductions that will be available on the project's website or with the digital collection of all the manuscripts of the Ptolemaic corpus, which will be available to those collaborating closely with the project. Researchers interested in a specific topic, e.g. the comparative sizes of the earth, moon, and sun, can easily find which manuscripts may contain relevant notes and then turn to them in the manuscripts in a matter of seconds. The project website may also offer an opportunity to experiment with ways of displaying transcriptions of glosses. While the spatial arrangement on the folio, which is especially important for glosses, is usually lost in the editing process, digital media allow us to present transcriptions of glosses that are searchable but yet retain a clear, visual connection to the commented text. In the coming years (and perhaps at the following Ptolemy conferences), we shall hopefully see many fruits of the catalogues and transcriptions of glosses, as a clearer picture of medieval astronomy emerges.

Bibliography

Manuscripts

Cracow, Biblioteka Jagiellońska, 589

Cracow, Biblioteka Jagiellońska, 590

Cracow, Biblioteka Jagiellońska, 619

Erfurt, Universitätsbibliothek, Dep. Erf. CA 2° 375

Florence, Biblioteca Nazionale Centrale, Conv. Soppr. J.IV.20 (San Marco 182)

Florence, Biblioteca Medicea Laurenziana, Plut. 89 sup. 45

Florence, Biblioteca Medicea Laurenziana, Plut. 89 sup. 57

Melbourne, State Library of Victoria, RARES 091 P95A

Memmingen, Stadtbibliothek, 2° 2,33

Oxford, All Souls College, 95

Oxford, New College, 281

Paris, Bibliothèque nationale de France, lat. 7254

Paris, Bibliothèque nationale de France, lat. 7256

Paris, Bibliothèque nationale de France, lat. 7258

Paris, Bibliothèque nationale de France, lat. 16200

Vatican, Biblioteca Apostolica Vaticana, Barb. lat. 173

Vatican, Biblioteca Apostolica Vaticana, Barb. lat. 336

Vatican, Biblioteca Apostolica Vaticana, Vat. lat. 2057

Secondary Literature

Benjamin, Francis S., and Gerald J. Toomer, *Campanus of Novara and Medieval Planetary Theory. Theorica Planetarum*, Madison: University of Wisconsin Press, 1971.

Busard, Hubertus L. L., 'Die Traktate *De Proportionibus* von Jordanus Nemorarius und Campanus', *Centaurus* 15 (1971), pp. 193–227.

Catalogus codicum manuscriptorum Bibliothecae Regiae, Pars Tertia, Tomus IV, Paris: Typographia regia, 1744.

Clagett, Marshall, *Archimedes in the Middle Ages*, vol. I: *The Arabo-Latin Tradition*, Madison: University of Wisconsin Press, 1964.

Crossley, John N., 'Ptolemy's *Almagest*. Its Dates and the Dating of Oxford, All Souls College, ms 95', *La Trobe Journal* 81 (2008), pp. 118–26.

Evans, Gillian R., 'Boethian and Euclidean Axiomatic Method in the Theology of the Later Twelfth Century. The *Regulae Theologicae* and the *De Arte Fidei Catholicae*', *Archives Internationales d'Histoire des Sciences* 30 (1980), pp. 36–52.

Høyrup, Jens, 'Jordanus de Nemore. A Case Study on Thirteenth-Century Mathematical Innovation and Failure in Cultural Context', in Jens Høyrup (ed.), *Measure, Number, and Weight. Studies in Mathematics and Culture*, Albany: SUNY Press, 1994, pp. 173–201.

Kunitzsch, Paul, *Der Sternkatalog des Almagest. Die arabisch-mittelalterliche Tradition*, vol. II: *Die lateinische Übersetzung Gerhards von Cremona*, Wiesbaden: Otto Harrassowitz, 1990.

Lorch, Richard, *Thābit ibn Qurra. On the Sector-Figure and Related Texts*, Augsburg: Rauner, 2008.

Parra, Maria José, 'A List of Arabic Manuscripts of Naṣīr al-Dīn al-Ṭūsī's *Taḥrīr al-Majisṭī*', *Suhayl* 16-17 (2018-2019), pp. 251-322.

Silverstein, Theodore, *Medieval Latin Scientific Writings in the Barberini Collection. A Provisional Catalogue*, Chicago: University of Chicago Press, 1957.

Toomer, Gerald J., 'Campanus of Novara', in Charles C. Gillispie, Frederic Lawrence Holmes, Noretta Koertge and Thomas Gale (eds), *Complete Dictionary of Scientific Biography*, vol. XV, Detroit: Charles Scribner's Sons, 2008, p. 70.

Toomer, Gerald J., *Ptolemy's Almagest*, Princeton: Princeton University Press, 1998.

Van Brummelen, Glen, *Mathematical Tables in Ptolemy's Almagest*, PhD dissertation, Simon Fraser University, 1993.

Watson, Andrew, *A Descriptive Catalogue of the Medieval Manuscripts of All Souls College Oxford*, Oxford: Oxford University Press, 1997.

Zepeda, Henry, 'Euclidization in the *Almagestum parvum*', *Early Science and Medicine* 20 (2015), pp. 48–76.

Zepeda, Henry, *The Medieval Latin Transmission of the Menelaus Theorem*, PhD dissertation, University of Oklahoma, 2013.

Zepeda, Henry, *The First Latin Treatise on Ptolemy's Astronomy: The Almagest Minor (c. 1200)*, Turnhout: Brepols, 2018.

A Discussion on Ptolemy's Authority: Henry Bate's Prologue to His Translation of Ibn Ezra's *Book of the World*

Carlos STEEL

1. Henry Bate translator of Abraham Ibn Ezra

The philosopher-astronomer Henry Bate of Mechelen (Malines) (1246–*c.* 1310) played a crucial role in the dissemination of the astrological works of the twelfth-century Jewish scholar Ibn Ezra in the Latin world.[1] Some time around 1270, during his studies in Paris, Henry may have been introduced to Abraham Ibn Ezra's work in a circle of scholars interested in astrology, who had contacts with Jewish scholars. Once back in his native town, Bate decided to start working on a translation of the available Hebrew corpus of Ibn Ezra. In 1273, he had some astrological treatises of Ibn Ezra translated into French in his house in Mechelen. As we learn from the colophon of one of the translations (which are preserved in ms. Paris, BnF, fr. 24276 and fr. 1351), the translation was a joint venture of a Jewish scholar, named Hagins, who translated from the Hebrew, with a certain Obert de Montdidier, who edited and wrote down the French version. The Paris manuscript contains four treatises of Ibn Ezra, but we may suppose that more texts had been translated at that time in Bate's house. In fact, in the *Nativitas*, which Bate composed in 1280, one finds references and quotations not only from these four treatises, but also from other works of Ibn Ezra, though Bate would 'publish' Latin translations of these texts much later, when he was residing in Orvieto in 1292. We must therefore suppose that Bate, at the time he composed his *Nativitas*, had already translations of much more work of Ibn Ezra than the four now preserved in the Parisian manuscript. Although Bate quotes from these treatises in Latin, they probably existed at that time only in a draft French version with some occasional glosses in Latin. When Bate had more time later, he would make Latin versions of these translations.

In this contribution, I intend to study Bate's most succesfull translation of Ibn Ezra, that of the *Book of the World* (*Sefer ha-'Olam*). This book, of

[1] On life and work of Henry Bate see Steel et al., *The Astrological Biography*. I refer to the edition in this work as *Nativitas*. On the role of Bate in the dissemination of Ibn Ezra's works, see Sela, 'The Ibn Ezra-Henry Bate Astrological Connection'.

Ptolemy's Science of the Stars in the Middle Ages, ed. by David Juste, Benno van Dalen, Dag Nikolaus Hasse and Charles Burnett, PALS 1 (Turnhout, 2020), pp. 245–281

10.1484/M.PALS-EB.5.120182

which there are two Hebrew versions, deals with the prognostication of the influences of the conjunctions of the planets (in particular the so-called great conjunctions) on political and historical events. It concludes with a section on weather forecasting. Scholars are fortunate to have a critical edition of the two versions of this Book with an English translation and abundant annotations by Shlomo Sela: his edition will be the main reference throughout this contribution.[2] Henry Bate translated the first version of the *Book of the World*. His translation is dated and located in a colophon, which has been transmitted in one family of the manuscript tradition:

> Explicit liber de mundo vel seculo, completus die Lune post festum beati Luce, hora diei quasi decima, anno Domini 1281, inceptus in Leodio, perfectus in Machlinia, translatus a magistro Henrico Bate de Hebreo in Latinum.[3]
>
> Thus ends the Book of the World, i.e. the Age, translated from Hebrew into Latin by master Henricus Bate, started in Liège, completed in Mechelen on Monday after the feast of saint Luke, at about the tenth hour, in the year of the Lord 1281.

The exact time references are characteristic of Bate: we find them in the colophons of his other translations. The feast of St Luke is on the 16th of October, which was a Thursday in 1281. The translation was thus finished on the 20th of October of that year. It is implausible that Bate himself made this translation from the Hebrew. Although he may have learned some elements of Hebrew during the intensive translation weeks in Mechelen, he would never have known the language adequately to make this translation without help. Moreover, in the years 1280–1281, Bate was extremely busy advancing his career, defending himself against accusations, traveling to Paris and writing his *Nativitas*. This period certainly did not leave him much time to work on *De mundo*. He may have made his Latin translation starting from a French translation produced by the Hagins-Montdidier team in Mechelen some years earlier. In the *Nativitas*, which was composed at the same time as the *De mundo* translation, there are three references to Ibn Ezra's book.[4] Interestingly, Bate never refers to the treatise with the title *De mundo*, as found in the colophon. He calls the works 'Liber coniunctionum' or 'Liber revolutionum annorum mundi'. That there are only a few references to this treatise in the *Nativitas*

[2] Sela, *Abraham Ibn Ezra. The Book of the World*. As I do not know Hebrew, I am fully dependent on Sela's translation. I thank Shlomo Sela for his valuable comments and his answers to my multiple questions.

[3] Surprisingly this full colophon is lacking in the oldest and most important manuscript, Paris, BnF, n.a.l. 3091 (P), which only has a short version: 'Explicit liber Auenesre de mundo translatus de Hebreo in Latinum a magistro Henrico Bate anno domini 1281'. For an explanation of the different forms of the colophon see my edition of the complete text of Bate's translation in Steel, 'Henry Bate's Translation', pp. 233-234.

[4] See *Nativitas*, 233–35 ('in Libro Revolutionum annorum mundi'), 380–82 ('in libro Coniunctionum') and 2049–2051 ('in Libro Coniunctionum').

is understandable. After all, Ibn Ezra's book is about events on a world scale, not about personal biography. But why, then, did Bate choose to make a translation of exactly that work, which was of not much use to him in composing the *Nativitas*? Probably because at that time he was interested in the astrological explanation of historical events. Two years before, in 1278, he had already translated from Hebrew a treatise of Ibn Ezra, which seems to be the third version of *Sefer ha-'Olam*, together with two treatises attributed to Alkindi 'De iudiciis revolutionum annorum mundi'.[5] According to the colophon, that translation was made at the request of Johannes van Milanen, who was alderman in Mechelen.[6] It is possible that Bate made the translation of *De mundo*, which deals with similar issues, for the same patron.

In the present contribution, I will not discuss Bate's translation of the whole treatise, but only focus on the preface of the translator and his additions in the translation. This preface is in many ways a remarkable document.[7] We shall see how the young Henry, notwithstanding his admiration for Ibn Ezra, whose works he translated and helped to disseminate in the Latin world, is very critical about Ibn Ezra' attitude towards Albumasar and Ptolemy. What at first seems to be a polemical text becomes in fact an interesting manifesto on method in astronomy. In what follows, I will present an analysis of this text. In an appendix, I offer an edition of the Latin text of this preface and the additions of Bate, together with a translation.[8]

2. Ibn Ezra's misrepresentation of Albumasar

Henry starts his preface by expressing his indignation and his unbelief at Ibn Ezra's criticism of Albumasar:

> When we started working on the translation of Ibn Ezra's treatise *On the conjunctions of the planets and the revolutions of the years of the world*, we were shocked at the opening of this work, as we did not understand for what reason [the author] neglected to pay respect to the prince of the astrologers, Albumasar:[9] why had he not at least interpreted the words of such a great philosopher in a more charitable way? For it seems to be a want of judgement to say what he dared to say here, that one

[5] For an edition and study of this newly discovered text see Sela et al., *A Newly discovered Treatise*.

[6] See on this translation Steel et al., *The Astrological Biography*, pp. 49–50.

[7] The preface and the additions attracted the attention of many scholars. See Grant, *Nicole Oresme*, pp. 111–16 and 168 (notes) with reference to previous discussions in Thorndike's *A History of Magic* and Duhem's *Le système du monde*. Unfortunately, they draw conclusions from a problematic text edition, as will be shown below. Lemay analysed the preface in his edition of Albumasar; see Lemay, *Abū Maʿšar*, vol. VII, pp. 63–70.

[8] I shall refer to this edition as *Praef.* or *Add.* with line numbers.

[9] As I am discussing a Latin text, I use throughout the Latinized name Albumasar for Abū Maʿshar, also in the translations I adopted from Sela's edition.

> should not give assent to what Albumasar says in his *Book of Conjunctions*, since he makes his judgments according to the mean motions (*Praef.* 2–8).

One finds indeed a harsh criticism of Albumasar in the opening section of *the Book of the World*, which I now quote in Sela's translation:

> If you come across Albumasar's *Book on the Conjunctions of the Planets* you would neither like it nor trust it, because he relies on the mean motion for the planetary conjunctions. No scholar concurs with him, because the truth is that the conjunctions should be reckoned with respect to the zodiac. [...] Rather, the correct approach is to rely on the astronomical tables of the scientists of every generation who rely on experience (§ 1:1–4, tr. Sela).

As Sela notices, Ibn Ezra refers here to a passage in Albumasar's work *On the Great Conjunctions*, where the astronomer uses the 'mean motion' of the planets to calculate the period between two successive conjunctions of Saturn and Jupiter.[10] That Ibn Ezra is here so negative about the 'prince of astrologers' is indeed surprising, for, as Sela observes, Ibn Ezra usually draws heavily on Albumasar's introductions to astrology and even extracts literally from his works. Henry, however, is irritated by Ibn Ezra's negative comments on Albumasar. As he says, even if certain passages in Albumasar may raise problems, given the way they are formulated, one should first try to interpret them 'in a more charitable way': 'quin saltem tanti uerba philosophi in partem interpretatus fuisset meliorem'. The expression 'in meliorem partem interpretari' is often used by patristic and medieval authors, whenever a sentence of an authority (above all a biblical text) has an ambiguous meaning. The right interpreter will always attempt to interpret it 'in the best sense'.[11] This is Bate's own attitude towards authorities. Moreover, whenever he notices an opposition between Ibn Ezra and Albumasar, he will not hesitate to defend the latter.[12] He will thus use his preface to his translation of Ibn Ezra to write an extensive refutation of the author's criticism of Albumasar.

Henry opens his refutation with an argument *ad hominem*. How could Ibn Ezra criticise Albumasar on this issue, when even his own master, 'Abraham called the prince, whom Ibn Ezra respects so much', used the same method of calculation from 'mean motions' in the fifth section of the treatise *On the*

[10] See Sela, *Abraham Ibn Ezra. The Book of the World*, ad I § 1:1–2, p. 102.

[11] See e.g. Thomas Aquinas, *Summa theologiae*, II–II, 60, 4, especially the contra argument, based on the Glossa ordinaria: 'Dubia in meliorem partem sunt interpretanda'. I owe this reference to Guy Guldentops.

[12] Bate's preference for Albumasar over Ibn Ezra is evident from his commentary on Albumasar's *De magnis coniunctionibus* partially preserved in extracts by Pierre d'Ailly. I also noticed that Bate in translating Ibn Ezra's *Ṭeʿamim* I (*Liber rationum*) adds some comments in which he expresses his preference for Albumasar (on which Ibn Ezra depends): see ms. Leipzig, UB, 1466 f. 61v: 'Dicit translator: Si autem veritatem confitendum est multo melius assignat Albumasar huius rationem in suo maiori Introductorio' and again 'Inquit translator quod hic insufficienter dictum est, satis completum est per Albumasar'.

Redemption of Israel? Abraham the Prince, to whom Bate refers, is Abraham Bar Ḥiyya (*c.* 1065–*c.* 1136), an astronomer who was known as Abraham ha-Naśi' (Abraham the Prince). Bate refers to the fifth chapter of his *Megillat hamegalleh* (*Scroll of the revealer*), which was known in Latin as *Liber de redemptione Israel*.[13] Bate quotes the same work further on in this preface, and he also refers to it in his *Nativitas*.[14] He must have had a Latin translation of this interesting astrological history, which is no longer extant.[15] That Bate calls Abraham the Prince the venerated master of Ibn Ezra poses many problems, which have been resolved in a recent article by Shlomo Sela.[16] In fact, Abraham Bar Ḥiyya was not at all Ibn Ezra's master, and the latter did not respect him much.

However, even if Henry Bate had known that he had made a false claim, he would have felt justified to point to a contradiction, not, then, between Ibn Ezra and his 'master', but in Ibn Ezra himself. In fact, notwithstanding his rejection of Albumasar's use of the 'mean motions', Ibn Ezra relies on that method in the calculation of the intervals between the great conjunctions, as Bate explains:

> Moreover, Ibn Ezra himself noticeably establishes the number <of years> of the conjunctions according to the mean motions, just as Albumasar did. For the changes of the triplicities do not always happen after 240 or 260 years according to the true motions, but sometimes faster, sometimes slower, as is clear from experience. For what reason, then, can Ibn Ezra criticize both the famous Albumasar and himself? (*Praef.* 18–23).

In fact, when making a calculation of the years that must have passed in the shift from one triplicity to another to reach another conjunction of Jupiter and Saturn, Ibn Ezra remarks:

> They [Saturnus and Jupiter] proceed in this manner until 240 or 260 years have passed, so that they conjoin in the houses of the <same> triplicity 12 or 13 times (§ 9,1, tr. Sela).

[13] cf. Poznanski and Guttmann, *Abraham Bar Ḥiyya,* p. 117.16-19.

[14] See *Praef.* 100–03; *Nativitas*, 2354–2358: 'Iupiter et Saturnus in eodem gradu coniunctionis sunt secundum medios motus ipsorum prout testatur Hispanus Abraham cognomine Princeps in suo tractatu *Coniunctionum*'. One finds also long quotations of the Megilat ha-Megaleh in Pierre d'Ailly's *Elucidarium*. The extracts start on f. 128v: 'Abraham Iudeus dictus Avenezre in quodam tractatu de magnis coniunctionibus' and go as far as f. 133v. Pierre refers in this work three times to Henricus Bate and his lost commentary on the Great conjunctions of Albumasar. Most probably Pierre found the extracts in Bate's commentary. The translation he uses is certainly different from the translation of Theodoricus de Northem (see n. 15).

[15] There exists a Latin translation of the text preserved in three fifteenth-century manuscripts (see Steel et al., *The Astrological Biography*, p. 82, no. 14). According to the colophon this translation was made from a French translation by the Dominican Theodoricus de Northem, a baccalarius theologie. See Federici Vescovini, 'Una versione latina', pp. 6–7. According to father E. Panella (referred to by Federici Vescovini) Theoderic of Northeim was working around 1300.

[16] Sela, 'The Ibn Ezra-Henry Bate Astrological Connection', pp. 175–79.

Ibn Ezra would have certainly admitted that astronomical calculations are not possible without applying mean motions, and he would never have criticized Albumasar for that reason. What he criticized in Albumasar is the fact that he only relied on mean motions in making astrological judgments. Bate insists, however, that Ibn Ezra should have been more charitable in his reading of Albumasar. For even when Albumasar relies upon the times of the 'mean motions', he does so in order to find what the 'true motions' are. With this practice, he follows the scientific method which requires both in investigation as in the exposition of a doctrine, to start from mean motions.[17] Hence, one should not deduce from Albumasar's statements that he really thought that astrological judgements should be based upon 'mean motions'.

> Even though Albumasar sets times of the mean conjunctions and changes of the triplicities, he does so that the positions and times of the true [conjunctions and changes] may be more suitably investigated; this is what the order of both invention and exposition requires; hence one cannot conclude therefrom with syllogistic force that the author thought that judgments of the conjunctions should be related to the mean motions, unless one may try to impute some madness to the author (*Praef.* 14–8).

3. The discussion of Ptolemy's authority

Next, Bate discusses a second point where he disagrees with Ibn Ezra, namely the way in which he manipulates the authority of Ptolemy.

> What the same Ibn Ezra attempts to establish as an assertion of Ptolemy, deserves certainly to be examined and discussed, I think, namely his claim that it was not possible, either for Ptolemy or for his predecessors and his successors, to find the ascendant degree in the hour of the entrance of the Sun into Aries because of the incertitude of the observations, which comes both from an error in the preparation of the instruments and from the different judgment regarding the length of the year because of the discordant observations of the experimental masters (*Praef.* 31–39).

This is indeed what Ibn Ezra makes Ptolemy say:

> Now I will give you another explanation. Ptolemy said: The scientists of our generation boast that they can find the sign of the ascendant in any city at the revolution of the year, which is the moment when the Sun enters Aries. But I say that I cannot do so and that those who preceded me did not know how, nor will those who come after me (§ 12,1–3, tr. Sela).

[17] 'prout expostulat doctrine ordo ac inuentionis' (*Praef.* 16). The distinction between 'inuentio' and 'doctrina' (based on Aristotle's *Posterior Analytics*) is often found in scholastic authors: 'doctrina' stands for the systematic exposition of a given science from its first principles to its last conclusions, 'inventio' for the investigation of a certain matter which may lead to new insights. The 'ordo doctrinae' and 'ordo inuentionis' may be different, not however in this case (where always the astronomer starts from mean motions, both in investigation and in exposition).

And after having developed what he considers to be Ptolemy's arguments — the imprecision of astronomical instruments (see § 13,3–6) and the difficulty to determine exactly the length of the year (see § 14–17) –, Ibn Ezra concludes:

> So now you realize that no man can know the sign of the ascendant at the revolution of the year. This is why Ptolemy said, [...] that we should always observe the moment of the luminaries' conjunction or opposition, whichever occurs last before the Sun enters Aries, for we can be precise about this without approximation, in any place we wish, and from it we can know all the judgments of the world (§ 18,1–2, tr. Sela).

In these texts Ibn Ezra seems to be sceptical about the possibility of knowing with precision the time of the entrance of the Sun into Aries (i.e. the vernal equinox), and he finds in Ptolemy a similar doubt. Therefore, he advocates with Ptolemy another method for making prognostications. One should rely upon the time of the syzygies of the luminaries that immediately preceded that event: for this can be known with precision.[18]

Bate is not at all convinced by this conclusion, and certainly not by Ibn Ezra's claim that this is the view of Ptolemy himself. First, he expresses with some sarcasm his surprise that Ibn Ezra claims to establish this unacceptable conclusion by relying upon arguments taken from Ptolemy's *Tetrabiblos*. In fact, Bate notices that Ibn Ezra declares elsewhere that the *Tetrabiblos* has no value at all. How, then, could he rely on a text whose authority he rejects?

> What is first of all astonishing here is the fact that he [Ibn Ezra] grounds his argument upon something about which the same Ibn Ezra affirms the following in his *Book of Reasons*, in the first section, first chapter, when speaking about Ptolemy's *Tetrabiblos*, from which also the argument mentioned before was taken: 'But I, Abraham, the author, say that this book was not composed by Ptolemy, because there are many arguments in that have no weight when compared to science and experience'. Likewise in the *Book on Nativities* in the chapter on the fifth house: 'I warn you not to rely somehow upon the arguments of this book, because it has no

[18] As David Juste informed me, already al-Battānī defended the view that it is preferable to take into account the syzygies preceding the entrance of the Sun into Aries and that this was also Ptolemy's doctrine. However, most Arabic astrologers thought that the horoscope of the year should be based on the vernal equinox. See on al-Battānī, Kennedy et al., 'Al-Battānī's Astrological History', in particular the quotation on p. 19: '[...] since we have looked into what was said about it by most of its practitioners in our time, and we found them to have searched for this knowledge and sought its lore through year-transfers, the beginnings of which are the entry of the Sun into the sign of Aries at the times of conjunctions, by casting the horoscopes of such times, and similar things of this sort, for the use of which there is no justification, nor are there any principles to go by. They did not take into account eclipses accompanying conjunctions in their places in the ecliptic, as well as the participation of the planets in these configurations, since that (the occurrence of eclipses) is among the best indicators regarding variations and the beginnings of changes. Since Ptolemy, with all his preeminence in this art, paid no heed to anything other than it concerning events coming to pass in this world, we deem it well to adopt his doctrine'.

value' and in the second chapter of the *Book of Reasons*: 'I give you a general rule: anything that Ptolemy says about the orbs is correct and no one surpasses him; but his astrological judgments do not befit his science' (*Praef.* 37–48).

As we can learn from this passage, at the time he wrote his preface, Bate was already well acquainted with Ibn Ezra's works. He knows indeed the main passages in Ibn Ezra's œuvre wherein the author expresses his negative judgment about the scientific value of the *Tetrabiblos*, and even his doubts about the authenticity of this work. He first refers to a passage in the *Book of Reasons* (*Te'amim*) (I § 1 5:5, tr. Sela, *Abraham Ibn Ezra. The Book of Reasons*, p. 35): 'But I, Abraham, the author, say that this book was not written by Ptolemy, because there are many things in it that have in them nothing of rational thought or experience, as I shall explain in the *Book of Nativities*'. Interestingly, this self-reference of Ibn Ezra made Bate look up the *Book of Nativities*, where he found the following aspersion of Ptolemy's authority: 'Ptolemy said in the *Tetrabiblos* that regarding children we should always observe the tenth and eleventh mundane houses. All those who came after him, including Māshā'allāh, laugh at him; and they are right. I have mentioned this so that you will not rely on everything written in that book [i.e., the *Tetrabiblos*], because it has no substance' (III, V, 4, 3, tr. Sela, *Abraham Ibn Ezra on Nativities*, p. 145). Finally Bate quotes again from the *Book of Reasons*: 'Now I give you a general rule: anything that Ptolemy says about the orbs is correct and no one surpasses him; but his astrological decrees and judgments do not befit his wisdom'. (I § 2 18:1–2, tr. Sela, *Abraham Ibn Ezra. The Book of Reasons*, p. 59). Ibn Ezra clearly opposes the Ptolemy of the *Almagest*, whom he admires, and the Ptolemy of the *Tetrabiblos*. In making this opposition Ibn Ezra follows an Arabic tradition in which scholars express their mistrust of the author of the *Tetrabiblos*.[19] A notorious early example of this hesitant attitude towards Ptolemy is Albumasar's introduction to Book IV of the *Introductorium maius*. According to Albumasar, some scholars attributed the *Tetrabiblos* to the same Ptolemy who also composed the *Almagest*; others, however, ascribed it to another author named Ptolemy. He leaves the question open, but notices that the author of the *Tetrabiblos* is much less reliable in his exposition of the nature of the stars.[20] The Arabic commentator on the *Tetrabiblos*, Ali ibn Riḍwān (Haly Abenrudian), deals with these doubts in his introduction and develops a lengthy argument to defend the attribution of the *Tetrabiblos* to the same author who also composed the *Almagest*.[21]

[19] For Ibn Ezra's attitude towards Ptolemy, see Sela, *Abraham Ibn Ezra and the Rise*, pp. 240–56.

[20] *Introductorium maius* (tr. Hermann of Carinthia), IV.1, ed. Lemay, *Abū Ma'šar*, vol. VIII, p. 56.19–23.

[21] See *Liber Quadripartiti Ptholomei*, ed. Venice, Bonetus Locatellus, 1493, fol. 2va: 'de nomine compositoris dico quod fuit Ptolomeus Pheludianus ille qui fecit Almagesti'. In what

Henry Bate does not defend the authority of the *Tetrabiblos* against Ibn Ezra. He is more concerned with the question of where, in Ptolemy, Ibn Ezra could have found the argument he attributes to the astronomer, namely that it is impossible to determine the time of the entrance of the Sun into Aries (i.e. the moment of the vernal equinox) and that one should therefore rely upon the time of a conjunction or opposition of the luminaries preceding that event. If, however, one cannot rely on the *Tetrabiblos*, as Ibn Ezra says, where else, Bate asks, can this conclusion be found in Ptolemy? It is certainly not found in the *Almagest*, where one reads rather the contrary. Where then? It seems that, after all, even Ibn Ezra has to rely upon the *Tetrabiblos* to make his claim:

> if it should be found somewhere, it has to be in the *Tetrabiblos*, in the 11th chapter of the second book, which deals with the beginning of the year (*Praef.* 50–51).

It is this chapter, Bate believes, that Ibn Ezra has in mind when he attributes the above claim to Ptolemy. However, as Bate intends to show, Ibn Ezra interprets this text perversely. To counter this interpretation, Bate first wants to establish the text of that chapter with the greatest precision. He quotes the full text of Ptolemy in three different translations. The first translation, which comes from an anonymous scholar, circulated around 1250–1260 in Paris and was used by Roger Bacon and by the author of the *Speculum astronomiae*.[22] The second is the well-known translation of Plato of Tivoli.[23] The third is the translation made directly from the Greek by William of Moerbeke, from whom Bate had received a personal copy.[24] Apart from Bate, no other medieval author ever quoted this translation. As one can see, Bate exploits his extraordinary erudition to present Ptolemy's text in its purest state without corruptions.[25] He places Moerbeke's translation in the last position because, whenever

follows, Haly gives the main arguments in favor of the attribution to Ptolemy. On this text see Boudet, 'Ptolémée dans l'occident médiéval', in particular pp. 200–04.

[22] Ptolemaeus, *Quadripartitum*, II.10 transl. anonyma (PAL A.2.4), cf. Vatican, BAV, Vat. lat. 4075, f. 28v28-40. On this anonymous translation see Juste, 'Ptolemy, Quadripartitum'.

[23] Ptolemaeus, *Quadripartitum*, II.10 transl. Platonis Tiburni, ed. Ven. 1484, f. C 5va.

[24] Ptolemaeus, *Iudicialia* (=*Quadripartitum*), II.11, 688-693, ed. Vuillemin-Diem and Steel. On Bate's use of this translation see Vuillemin-Diem and Steel, *Ptolemy's Tetrabiblos*, pp. 39–44.

[25] When writing his preface, Bate did not yet know the new translation made by Aegidius de Thebaldis (after 1257). One of the first scholars in Paris to use the new translation was Pierre de Limoges, who owned a personal copy of the text (Paris, BnF, lat. 16653). In Aegidius' translation the quoted text reads as follows: 'Et quod teneo melius et proximius esse opinioni naturali in facto anni est operari per hec quatuor principia et inspicere ad coniunctiones et ad oppositiones Solis et Lune que sunt proximiores istis principiis ante hec tempora, et in ea propria que erit eclypsis; ita quod sciamus a principio hore qua ingreditur Sol Arietem qualiter erit status ueris, et a principio quando Cancrum ingreditur quomodo erit estas, et a principio quando Capricornum ingreditur quomodo erit hiems; ita quod qualitates uniuerales temporum et suorum statuum erunt propter Solem'.

possible, he prefers to use translations made immediately from the Greek, since he believes that they are less corrupt than translations through Arabic intermediaries.[26] Moerbeke's translation is undoubtedly the most accurate and the closest to the Greek, but because of its literality almost unintelligible. Plato of Tivoli's translation, though made from the Arabic and less literal, renders Ptolemy's argument rather well. The anonymous translation, however, is a disaster: it only makes sense when compared to Plato's version. In one section, however, both translations from the Arabic lack an element of Ptolemy's text.[27] Whereas Moerbeke has 'coniugationes Solis et Lune *coniunctionales aut pleniluniares*' to render 'ἡλίου καὶ σελήνης συζυγίας συνοδικὰς καὶ πανσεληνιακάς', i.e. 'the sizygies of the Sun and Moon at new or full Moon', Plato of Tivoli translates 'coniunctionum et preuentionum Solis et Lune' and the anonymous 'eorundem conuentum aut oppositionem'. The addition 'at new or full Moon' is absent in the Arabic text, maybe because it was thought to be redundant.

As we cannot rely here on a Latin translation to follow Bate's argument, I insert Robbins's excellent translation from the Greek, though making it somewhat more literal:

> It seems more appropriate and natural to me, however, to employ the four starting-points for investigations which deal with the year, *observing the syzygies of the Sun and Moon at new or full Moon which most nearly precede them*, and among these in particular the conjunctions at which eclipses take place, so that from the starting-point in Aries we may conjecture what the spring will be like, from that in Cancer the summer, from that in Libra the autumn, and from that in Capricorn the winter. For the Sun creates the general qualities and conditions of the seasons.[28]

After having quoted the *Tetrabiblos* chapter in three versions, Bate expresses his exasperation at Ibn Ezra's misinterpretation. Unless one manipulates this text ('nisi littere uiolentia fiat'), Bate says, it is clear that Ibn Ezra's view cannot be founded upon Ptolemy. In fact, one cannot deduce from Ptolemy's text, as a general rule, that in predictions one should take, not the ascendant at the exact hour of its occurrence, but the syzygy preceding it. What Ptolemy has to say in this chapter, is about how to predict the *general qualities* of the seasons of the coming year. It is in that perspective that he recommends that one should not only know the entrance of the Sun into the cardines (equinoxes and solstices), but also the syzygies of the Sun and the Moon preceding its entrance.

[26] cf. *Nativitas*, 2719–2720: 'Alia uero translatio que de greco melius habet hoc modo' and *De diebus creticis periodorumque causis*, c. 9 (ed. Dell'Anna, *Dies critici*, vol. II, pp. 111–12): 'oportet igitur litteram que de Arabico intelligi per translationem que de Greco in qua magis confidendum est cautius'.

[27] The same omission is found in Aegidius' translation.

[28] *Tetrabiblos*, II.11, tr. Robbins, *Ptolemy*, pp. 197–99.

> From what is said one cannot draw the conclusion Ibn Ezra made, unless one manipulates the text, but rather that, in order to know the disposition of the year and of its parts [i.e. the seasons], one has not only to know when the Sun enters the vernal equinox and the other points on the tropic; but, in addition to these, it is also appropriate to observe the conjunctions of the Sun and the Moon at new and full Moon (*Praef.* 83–7).

To confirm his position, Bate refers to Albumasar who says in his *Book on Conjunctions* that, when we have to make predictions of general conditions, which affect mankind as a whole, such as pestilence, 'it is necessary to have *two* ascendants, namely that of the conjunction or opposition preceding the entrance of the Sun in the mobile vernal point, and also that of the revolution in which the Sun enters that mobile vernal point'.[29] But when one has to make predictions regarding humidity or dryness of the coming year, Bate says, it is enough to examine the syzygies of the Sun and of the Moon preceding the seasons, as particular climatological changes follow from different relations of the Moon to the Sun. The case is different for cosmic changes that have a more pervasive and far-reaching effect: they depend on a more permanent cause in the celestial constellations; here the conditions for making predictions are more strict. And with a final blow to Ibn Ezra, Bate refers again to 'Abraham the prince, whom Ibn Ezra calls his teacher'. He too requires strict conditions, including the determination of the vernal equinox, for the prediction of general events. Thus, in the fifth chapter of the *Redemption of Israel*, where he talks about general events, such as the change of reigns, battles, famine, drought, low and high [prices] of grain, he says: 'all this we shall know through the revolution of the conjunction of Saturn and Jupiter, that is, when the Sun enters Aries'.[30]

[29] Bate refers to 'in primo Coniunctionum, differentia prima, et in octavo etiam, differentia prima'. See Albumasar, *De magnis coniunctionibus*, VIII,1,9: 'Qualitas autem rerum communium comprehendentium genus, ut pestes et bubones et fertilitas et siccitas et pluvie, scitur illud quidem ex ascendentibus inceptionum universalium que erunt ante equidistantiam luminaris signo mobili vernali, in hora equidistantie, ex parte Lune duorum locorum, et gradus coniunctionis et impletionis, in annis coniunctionum et aliis' (ed. Yamamoto and Burnett, *Abū Maʿšar on Historical Astrology*, pp. 296–98). Yamamoto and Burnett translate the Arabic text as follows: 'As for the question of the things of general kind like an epidemic, a plague, fertility, barrenness, and rain, this is known from the horoscopes of the universal beginnings occurring before the parallelism of the great luminary with the spring tropic, and at the time of its parallelism, and from the Moon in its two positions, i.e. the position of the conjunction (New Moon) and that of the opposition (Full Moon) in the year of the conjunctions or other <years>'. Bate's reference to the first book is less evident. See however I,1,28, pp. 16–17: 'veluti significatio signo profectionis ab ascendente coniunctionis que fit in Ariete super res universales atque generales ut sunt diluvia, terremotus et pestilentie et his similia'. The reference to book VIII is lacking in the oldest and best copy of the text (P). It was probably added later by Bate.

[30] cf. Poznanski and Guttmann, *Abraham Bar Ḥiyya*, pp. 117, 16–19.

Bate may be right in his claim that Ibn Ezra manipulates somehow Ptolemy's text to make him say what he, Ibn Ezra, wants to say. Even his modern editor, Sela, must admit that, although Ptolemy expresses a similar view in a number of places in the *Tetrabiblos*, 'it is not in the precise form as it is presented by Ibn Ezra'. Sela refers in his annotations to the same chapter (II.11) quoted by Bate as Ibn Ezra's possible source, but adds three other passages where one may find a similar view (II.1; II.12, and III.2).[31] In particular the last text which deals with 'the degree of the horoscopic point' is illuminating. I quote it in Robbins' translation:

> Difficulty often arises with regard to the first and most important fact, that is, the fraction of the hour of the birth; for in general only observation by means of horoscopic astrolabes at the time of birth can for scientific observers give the minute of the hour, while practically all other horoscopic instruments [...] are frequently capable of error [...]. It would therefore be necessary that an account first be given how one might discover [...] the degree of the zodiac which should be rising, *given the degree of the known hour nearest to the event* [...]. We must, then, take the syzygy most recently preceding the birth, whether it be a new moon or a full moon.[32]

This text does not as such deal with the question discussed by Ibn Ezra, namely how to establish the moment of the vernal equinox, but with another issue, how to establish the ascendant degree at the moment of birth, which is difficult to determine empirically. Nevertheless, it is relevant to the discussion for another reason. In fact, in both questions, we are confronted with the *difficulty of establishing empirically with instruments* a particular astronomical position. Ptolemy recommends an alternative method to establish the ascendant degree of the nativity: 'take the syzygy most recently preceding the birth, whether it be a new moon or a full moon'. In an analogous way, one can use a similar alternative (i.e. turning to the preceding syzygies) when it is difficult to establish the vernal equinox. However, even in this text Ptolemy does not give up the possibility of determining the ascendant at the very hour, he only says that it is often very difficult to know it with precision and, therefore, recommends another procedure.

4. How to deal with the incertitude of observations

Having shown that Ibn Ezra could not invoke Ptolemy's authority for his claim that one cannot determine the exact moment of the entrance of the Sun into Aries, Bate next discusses Ibn Ezra's main *arguments* for his conclusion, which he had summarized in this way: 'the incertitude of the observations, which

[31] See Sela, *Abraham Ibn Ezra. The Book of the World*, pp. 114–15 and comment p. 108 (ad I § 12): 'This ostensible quotation from Ptolemy... has no clear basis in the Tetrabiblos'.

[32] *Tetrabiblos*, III.2, tr. Robbins, *Ptolemy*, pp. 229–31.

comes both from an error in the preparation of the instruments and from the different judgment regarding the length of the year because of the discordant observations of the experimental masters' (*Praef.* 27–30).

Ibn Ezra discusses the problem of the instruments in § 13: 3–6:

> The instruments used to determine the Sun's altitude at noon, if they are very accurate, can provide a result in minutes but not seconds. The Banū Shākir [brother] said that they made three instruments that were graduated in minutes and with which they could also measure ten seconds. But when they measured the Sun's altitude at noon there was a discrepancy among them of two minutes, which stems from the imprecision of the instruments. There will be an error if we cast a shadow on the earth, too, either because the surface of the earth is not straight, or because the stake is not straight, or because it does not stand upright (tr. Sela).

Bate only briefly refers to the problem of instruments in the preface, but he returns to the question at the end of his life, in the last part of his *Speculum divinorum*, where he refers to the same example of the Banū Shākir brothers:

> With regard to sight it is quite clear that it can be frequently be changed and deceived in its state, as also Alhazen declares in many places.[33] An absolutely accurate fabrication of precise instruments and their exact installation, executed with indubitable verification so that no error can occur from it, is a so difficult and nearly impossible operation that only a few could believe it except for experts who have frequently tried it out. Hence Abraham the Jew says in his book *On Tabulation* that *he found two brothers Bensechit, who had composed two instruments,* i.e. *astrolabes, with a diameter of 9 spans. They divided the degrees of the quadrant to measure the sun's altitude in minutes, and the minutes in five. And when these two men measured the Sun's altitude, when the Sun was entering the head of Aries, a difference of 2 minutes was found between the two instruments.*[34] Likewise, in our time, two technical specialists ('artificiosi viri'), making use of two very large skilfully produced and ingeniously verified quadrants, measured in Paris the maximal elevation of the Sun at noon on the summer solstice: the result found by one was 64°42'; while the other 64°45'; and so there was a difference of 3 minutes.[35]

As one can see, Bate was as much aware as Ibn Ezra of the difficulty, the virtual impossibility of making absolutely precise observations. Nevertheless, this was not for him a sufficient reason to give up all attempts to come closer to the

[33] See Alhazen, *Optica*, III.2 (ed. Basel, 1572, p. 75); III.7 (ed. p. 102); VI.2–9 (ed. p. 188); VII.7 (ed. p. 230).

[34] Bate does not take the example of the two brothers from *De mundo*, but from the parallel text *De rationibus tabularum*, ed. Millás Vallicrosa, *El libro de los fundamentos*, pp. 81, 7–12. See on these two parallel texts, Sela, *Abraham Ibn Ezra. The Book of the World*, p. 110.

[35] *Speculum divinorum* XXII.17, lines 31–40 (ed. Steel and Guldentops, *Henricus Bate*, p. 337, translation Guldentops). This passage is quoted and commented by Giovanni Pico della Mirandola, see Garin, *Pico*, vol. II, pp. 322–24.

truth, or to opt for an alternative method to circumvent the difficulty, as Ibn Ezra seems to suggest. This is clear in the way Bate summarizes the discussion among astronomers on the length of the solar year, which he read in *the Book of the World* § 13:7–9, 17:1 and 17:9–11. He agrees with Ibn Ezra that the masters who made observations came to different results in their calculations: some put the motion of the Sun too fast, others too slow. Although they seem to turn in circles, now putting a time below than above the true time, nevertheless

> they could in the end, through a proportional division of the differences, investigate the truth or what is so close to the truth that it excludes an error that may harm (*Praef.* 107–109).

Bate confirms his optimistic conclusion with examples he found in Ibn Ezra's own discussion of the problem (in § 17–18). The motions Ptolemy proposed are faster than those that Hipparchus had proposed and yet 'fall short of the due velocity'. Al-Battānī posited a solar motion that is too fast, yet closer to the truth than that of Ptolemy. Al-Ṣūfī and Ibn Ezra posited motions that are somewhat slower than those of al-Battānī and yet they too fall short from the due velocity.[36] Finally, Bate refers to his own efforts in correcting and improving the astronomical tables, starting from his observations in Mechelen:[37]

> Finally, in our own times, by saving appropriately the observations of Ptolemy, al-Battānī as well as our own, we posited a motion [with a velocity] of almost intermediate proportion between al-Battānī and Ibn Ezra, so that this way we may come closer to the middle, where the truth lies (*Praef.* 117–120).[38]

To be sure, with this method of proportional division, one may not 'reach the indivisible truth'. However, as Bate believes, we may come closer to it, and that is enough, given our human possibilities. Even if our calculations are imperfect, there is no reason to despair about the possibility of astronomical science. Bate refers to what Ptolemy says in his introduction to the *Tetrabiblos*: 'Even if prognostication be not entirely infallible, at least its possibilities have appeared worthy of utmost zeal'.[39] And he associates with him Albumasar who, in the first chapter of the *Introductorium maius* says that 'an error of a number of minutes, or — rarely — even a whole degree, does not harm a lot'.[40] Moreover,

[36] This argument on the value of the precession of the equinoxes (central to the problem of the length of the solar year) finds an exact parallel in *Speculum divinorum*, XXII.5, lines 146–53 (ed. Steel and Guldentops, *Henricus Bate*, p. 283).

[37] On the *Tabulae Mechlinienses*, see Nothaft, 'Henry Bate's *Tabule Machlinenses*'.

[38] For a more exact calculation of the different views on the length of the tropical year, see Nothaft, 'Criticism of Trepidation'.

[39] *Tetrabiblos*, I.3, tr. Robbins, *Ptolemy*, p. 31 (translation Robbins modified at the end).

[40] *Albumasar*, *Introductorium maius* (tr. Hermann of Carinthia), I.4, ed. Lemay, *Abū Maʿšar*, vol. VIII, p. 19.647–48: 'punctorum seu gradus etiam integri error et raro nec multum impedit'.

Bate adds, if an error originates from poor instruments, 'it can sufficiently be avoided through the skilfulness of an ingenious man and through frequent observations' (*Praef.* 125–7).

Bate keeps this optimistic belief in the possibility of an astronomical science starting from empirical data until the end of his life, as one can judge from an addition to the *Speculum divinorum* (dated after 31 January 1310), in which he comments on the information he had received on observations of a solar eclipse in Paris.[41] He notices how 'difficult, even impossible it is to indubitably and precisely grasp through some observation' the truth of this celestial phenomenon. Nevertheless, he concludes with some scholarly pride:

> Therefore, [our own tables of Mechelen, which have now been corrected for the third and last time (...)] should not at all be despised. For by means of these [tables], which are adjusted to the observations done previously by Ptolemy and by us later on, and which agree with truthful experience, it is possible to find the positions of the planets and their conjunctions, as well as the revolutions of the year and its seasons: I mean those conjunctions at least that are foremost worthwhile or of which one needs certainty.[42]

As we have seen, Bate defended a similar position already in his preface to *De mundo*, written forty years earlier. His belief in the possibility of establishing from empirical data the moment of the yearly vernal equinox made him reject Ibn Ezra's proposal, that we should rather rely upon the hour of the conjunction or opposition of the Moon with the Sun preceding that event. Granted that this is the right method to proceed, he says, it would nevertheless be impossible unless one knew beforehand the true motion of the Moon. However, it is impossible to know what occurs with the Moon (conjunctions, eclipses) without knowing the true place of the Sun. All astrologers, he says, Ptolemy, Geber, al-Battānī, and all masters of celestial observation, knew that it was impossible to obtain the true positions and times of eclipses unless the motion of the Sun was verified.

Nevertheless, Bate admits that, though this may be the ideal procedure in a scientific investigation — first determine the position of the sun, then the position of the moon in relation to it, then cast the horoscope of that time

[41] *Speculum divinorum*, XXII, additio ad capitulum 18, ed. Steel and Guldentops, *Henricus Bate*, pp. 346–48. See on this discussion of the solar eclipse Nothaft, 'Henry Bate's *Tabule Machlinenses*', pp. 276–78.

[42] 'Demum neque spernendae sunt omnino nostrae praefatae tabulae [i.e. Machlinenses tertio iam et ultimo correctae]; per ipsas enim concordatas utique considerationibus a Ptolemaeo prius observatis et a nobis posterius experientiaeque veraci convenientes inveniri possunt loca planetarum et eorum coniunctiones, anni quoque simul et quartarum eius revolutiones; coniunctiones autem inquam illae saltem de quibus principaliter operae pretium aut necesse est aliqualem habere certitudinem' (*Speculum divinorum*, XXII, additio ad capitulum 18, lines 44 and 59–63, ed. Steel and Guldentops, *Henricus Bate*, pp. 347–48).

–, this will not always be possible. If, then, for various reasons (such as lack of sufficient instruments), it is not possible to determine the positions of the luminaries with great precision ('ad unguem seu precise'), one should worry about [the rates of] the motions that have been found.[43] Maybe an appropriate [rate] was not found or there may be no trust that what was found is sufficient. If, then, an error arises in calculations, it will manifest itself more quickly regarding conjunctions and oppositions of the Moon with the Sun than regarding the entrance of the Sun into Aries. The reason is, as Bate explains, that the motion of the Moon is faster, which makes it possible to detect easier errors in calculation regarding the Moon's conjunctions and oppositions, whereas the passage of the Sun is much slower: it takes a long time before one can notice errors in calculation regarding its position. This may have been the reason, Bate thinks, that induced Ibn Ezra to take the lunar time as root of his calculations and not the solar time. However, there may be even another explanation for this preference, Bate suggests. After all, Ibn Ezra is a Jew and all people love their own folks and ways of life. Moreover, a calendar is not just a system of measurement of time; it is always linked to a religious interpretation of human life in time. Bate suggests that Ibn Ezra, as a Jew, could have been disposed to prefer the lunar calendar over the solar calendar. In fact, according to the Jewish calendar, the year begins at the conjunction of the Moon.[44]

> Besides, the love and inclination he had for his own Jewish sect, which begins the year at the conjunction of the Moon, could perhaps have contributed to that (*Praef.* 148–150).

This is a surprising *ad hominem* argument coming at the end of a scholarly discussion, but it is not unfriendly.

Bate admits that the method advocated by Ibn Ezra may be easier, but it is wrong in its principles. A perspicacious astronomer, Bate says, should not be content 'with rough estimation of an incompetent and indolent mind'. As we can learn from Albumasar, 'before we can make skilfully astrological judgments, we must beforehand have some scientific certitude about the celestial motions. For that reason, it is necessary to lay first the fundament regarding the Sun'. Even if we have difficulty in knowing the true motions, we can still rely on our mathematical models and calculations to approach it. And again Bate confronts Ibn Ezra with statements he makes in his other works. He refers to his treatise *De rationibus tabularum*, where Ibn Ezra set the study of motion of the Sun before the study of the motion of the Moon.[45] This reference is

[43] 'de inuentis sit cura motibus'. I thank the anonymous referee for his comment on this passage.

[44] Regarding Ibn Ezra's attitude towards the Jewish calendar see Sela, *Abraham Ibn Ezra and the Rise*, pp. 273–88.

[45] The treatise has been published by Millás Vallicrosa, *El libro de los fundamentos*.

another example of Bate's astonishing knowledge of the works of Ibn Ezra at the time he was writing this preface.

To conclude my analysis of this remarkable preface of a young astronomer full of self-confidence I can do nothing better than offer him the last words:

> To make a general conclusion, it is more connatural and suitable to the achievement of science and its further perfection to look not only at the conjunction of the luminaries or their opposition in order to know the state and being of the world, but to consider also skilfully, with all precaution, the entrance of the Sun into Aries and, if required, into the other tropical points, as in their conjunction are rooted the great events of the world, as is clear from the statements of the scholars (*Praef.* 161–7).

5. Additions in the translation

Bate not only wrote the long preface to his translation but also added his own comments at two places in the translation.

The first addition comes after I § 2–6. Bate expresses here his frustration with and incomprehension of Ibn Ezra's calculation of the 120 possible conjunctions of the seven planets. He inserts the following comment:

> The translator says: This is Ibn Ezra's argument as it is found in Hebrew,[46] but it seems to us that either the text has been truncated in the exemplar or, given that the text is sound and well written, that the doctrine he transmitted is too confused and not skilful enough.[47]

More important, however, is a second longer addition in which Bate again takes some distance vis-à-vis the astronomer he promoted through his translations. This comment follows on the following passage in Ibn Ezra (§ 24: 3–4 and 7–8, tr. Sela):[48]

> If someone argues that every cycle of seventy-five years is the same as the previous ones, because the planets and those that share their power are the same, the answer is as follows. Know that it is impossible according to proportion[49] that, when there is an ascendant with a proportional relationship <of the planets> to it, that the proportion of one to another will always be uniform and the same, even were the world to last forever. [...] Therefore, it is not possible for the nativity of one person to be the same as that of another. For the orb never remains in the same pattern, and at every moment there emerges a new proportion,[50] whose like has never existed and never will; and the mathematicians know that.

[46] 'Arabo' in manuscripts of the β group.

[47] See on this addition and its context Clagett, *Nicole Oresme*, pp. 445–47.

[48] I adapted Sela's translation to make it correspond more to the Latin translation of Bate. In particular, I introduced the term 'proportion' because it may explain Bate's critical reaction.

[49] 'secundum viam proportionis' (Bate) 'arithmetically' (tr. Sela).

[50] 'proportio' (Bate), 'pattern' (Sela).

Ibn Ezra seems to reject in this section the hypothesis that there is a perfect cyclical regularity in celestial motions. As he argues, every planet at any time has so many diverse moving relationships to the other planets and to the fixed stars that there will never be found one and the same proportion between them after a long period of time, 'even were the world to last for ever'. Bate adds a long digression in his translation to explain what Ibn Ezra may have meant. As he says, there is a radical difference between mathematical calculations, which may go on in infinity, and astronomical calculations:

> Although the multiplication of a number can increase to infinity, the revolutions of the celestial bodies are necessarily finite in species, as has been demonstrated with certitude in another part of philosophy. It is necessary, therefore, that similar [celestial] configurations should sometimes return, even though the [interval of] the time of such revolutions is incomprehensible to us because of the enormity of these intervals (*Add.* 175–80).

To defend Ibn Ezra, Bate offers a charitable interpretation. Ibn Ezra did not really reject the circular regularity of the celestial motions, he only wanted to say that we could never calculate exactly when a certain proportionate relation between planets and stars would return given the infinite possibilities of combination. However, to admit the difficulty of calculation of the exact return of a celestial configuration does not yield the conclusion that the celestial motions would go on in infinity without ever returning to a certain configuration:

> One must not suppose, however, that, because of the manifold diversity of the motions of celestial bodies, they may not come together nor be coordinated, as is the case with incommensurable lines, which, in the tenth book of the *Elements*, Euclid calls irrationals or surds, because of their incapacity to communicate with one another. For, as the Philosopher testifies in the twelfth book of the Metaphysics, 'all things are ordered together'[*Met.* XII 9, 1075a18-19]; and the Commentator says about this that 'all the actions of celestial bodies in their communication with one another are in the organization of the world as the action of freemen in the organization of a house'[Averroes, *In Met.* XII c.52, 338 BC]. For it is evident even to someone considering these matters a little, that if there must be a communication between some things, this communication must be more excellent in divine things. Therefore, it is absurd to think that the motions of the superior [i.e., celestial] bodies are irrational, or surd. This is what Pythagoras and other ancients wanted to discover through the music of the world; and Plato says similar things on this matter in the *Timaeus* and elsewhere, as does Calcidius together with innumerable other philosophers.[51] (*Add.* 180–93)

This addition deals with an issue that will be often discussed in the fourteenth century, the question whether the motions of the different luminaries in their reciprocal relations are commensurable or not. Best known is the position of

[51] cf. Plato, *Timaeus* 36E-37A; Calcidius, *In Tim.* c.95 (ed. p. 147, 26-148, 9); Simplicius, *In De Caelo* II 9 (p. 469,1-32); Bate, *Speculum* XXII, c.23 (p. 367, 106-198).

Nicole Oresme, who wrote different treatises in which he argued for the irrationality of the celestial motions. As is clear from the addition quoted above, Bate radically opposes such a view. All celestial motions have rational proportions which can in principle be perfectly calculated even if they surpass the capacities of the human intellect.

It is worth examining in what sense Bate's addition had an influence upon the later debate through the diffusion of his translation of *De mundo*. Let me just give two examples of reactions to his comments.

The first and the most important example is a marginal note, which entered the text in a group of manuscripts of *De mundo*, probably related to the University of Paris. An unknown scholar added at the sentence beginning with 'Non est autem' (*Add.* 180) the following comment: 'Non est autem, etc.: nescio quare hic translator deturpauit pergamentum ponendo se in textu et ostendendo se scire mathematicam': 'I do not know why this translator has defiled the parchment by putting himself in the text and showing that he knows mathematics'. Because this note is also present in the edition of 1507, it attracted the attention of scholars who wrote on the late-medieval discussion on the incommensurability of the celestial motions. Pierre Duhem is the first to offer a survey of this debate.[52] He begins by presenting the views of Ibn Ezra in the above-quoted section from *De mundo*, in which the Jewish astronomer seems to admit the incommensurability of celestial motions. Duhem then turns to Henry Bate, who criticized this view in his additional note. Duhem gives a full translation of the note. However, he is puzzled by the invective 'nescio quare hic translator deturpauit pergamentum ponendo se in textu et ostendendo se scire mathematicam'. As he did not know that this sentence does not belong to the original version, but had been added, rather late, in one subgroup of manuscripts, Duhem thought it was Henry Bate's own reaction against the Jew who had made the translation of Ibn Ezra, and who, in Bate's view, got it all wrong. 'Ce passage [i.e from Ibn Ezra in Hagins' translation] provoque, de la part d'Henri Bate, cette brutale observation'.[53] Influenced by his reading of Duhem, Lynn Thorndike writes: 'Henry Bate represents the idea [i.e. of the incommensurability] as an innovation of the translator from Hebrew into French'.[54] The same error is made by Edward Grant in the introduction to his edition of Nicole Oresme's *Tractatus de commensurabilitate vel incommensurabilitate motuum celi*. In his survey of the discussion of the incommensurability before Oresme, Grant also deals with Henry Bate.[55]

[52] See Duhem, *Le système du monde*, vol. VIII, pp. 443–51 (esp. 445–47 on Ibn Ezra and Bate).

[53] See Duhem, *Le système du monde*, vol. IV, p. 28; cf. vol. VIII, p. 446, n. 1: 'Bate veut sans doute parler du Juif qui avait traduit en flamand (*sic*!) l'hébreu de Aven Ezra'.

[54] See Thorndike, *A History of Magic*, vol. III, p. 406.

[55] Grant, *Nicole Oresme*, pp. 111–16 and 164–66.

He begins with the following surprising statement: 'In 1281, Henry Bate of Malines criticized an anonymous translator of Abraham ibn Ezra's Book on the World [...]'. Grant believes that Bate is criticising the author of a 'translation [in French?] that may have served subsequently as the basis for Bate's own translation of the same treatise into Latin'. It seems that Grant understands the 'inquit translator' not as referring to Bate himself (for it is a third-person statement) but to the anonymous who made the translation from the Hebrew. Bate, then, would only interfere in the note starting from 'Nescio quare hic translator deturpavit...' up to the end ('let us return to the text'). According to Grant, the interpretation of the anonymous translator, who had tried to interpret charitably Ibn Ezra's statement, 'aroused the wrath of Henri Bate' as is clear from his reaction in what follows 'I do not know why the translator has defiled the parchment...'.

All this confusion and concatenation of errors could have been avoided if the scholars had known that the 'harsh reaction' did not come from Bate himself, and was not addressed by him to the 'translator' — he was himself the translator! —, but had been added by a scholar reading Bate in the late fourteenth or fifteenth century. This scholar (maybe it be Pietro d'Abano?) was angry at Bate's critique of Ibn Ezra.

The second reaction, this time a positive one, is found in a fifteenth century manuscript Paris, BnF, lat. 7438, fol. 273r.[56] After a revised version of Nicole Oresme's treatise *De commensuratione motuum celi* follows a long postscript that is written by the same scribe who had copied (composed?) the revised vision. 'When I was writing this text [i.e. *De commensuratione*] I remembered what Abraham Ibn Ezra says in his Book on the World about the fardar that return according to circularity every 75 years. This is what he says: 'If someone argues [...] and the mathematicians know that'. [Ibn Ezra § 24 3–8]. He does not say more, but the translator of this work from Arabic[57] into Latin, Henry Bate, a man great in the quadrivium, says the following [...]'. Then follows the complete text of Bate's additional note, 'Quamquam multiplicatio — philosophis infinitis'. [*Add.* 256–73]. The scribe-scholar then concludes: 'He does not say more about that issue. But I believe that he deals more with that question in his *Speculum divinorum* whose incipit is "Bonorum honorabilium preclariorem partem eligentes, etc."' Interestingly, the scribe of the postscriptum copies Bate's *Additio* without the invective 'nescio quare hic translator'. This should not surprise us, as it was not in the exemplar where he read the translation of

[56] This post-script was first noticed by Wallerand, *Henri Bate*, p. 16, n. 17. One finds a full transcription of the post-scriptum in Grant, *Nicole Oresme*, pp. 164–65 (with plate 8: fol. 273r).

[57] 'translated from the Arabic': this may be a slip of the pen of the copyist, or he may have found it in his exemplar of *De mundo* (the error is found in the first addition in many manuscripts).

De mundo, which belonged to another subfamily of manuscripts.[58] But, here again, Grant comes with a complicated explanation, which makes no sense.[59]

Interestingly, the scholar refers to Bate's *Speculum divinorum*, where a more extensive discussion of the problem of incommensurability might be found. In fact, some thirty years after *De mundo*, in the composition of the 22nd part of his *Speculum divinorum*, Bate returned to the question of the celestial harmony he had touched upon in his additional note, when he was still a young scholar. The last part of chapter 23 is devoted to the question of the harmony of the celestial motions. Here, Bate quotes long extracts from Plato's *Timaeus*, from Calcidius' commentary on the *Timaeus*, and from an author he did not yet know when he made the additional comment to *De mundo*, Simplicius in his commentary on Aristole's *De Caelo* II, dealing with Pythagoras' doctrine on the celestial harmony.[60] He does not tackle directly the issue of the incommensurability of the celestial motion, though it is evident from his argument that he rejects it. Moreover, he quotes long extracts from Aristotle's *Metaphysics* XII and Averroes' commentary to which he also referred in the additional note of *De mundo*.[61]

6. Conclusion

As the digression in the translation of *De mundo* shows, Bate is convinced that celestial configurations, even a thousand years from now, can in principle be calculated with precision, since they result from motions of limited bodies occurring in limited sections of time and space. Even if such a calculation is difficult, there is no reason to abandon the attempt to come closer to the truth. As we have seen, Bate advocates in his Preface a method of approximating the truth through calculations, tested with always new observations. Even if all data cannot always be obtained as precisely as we would wish, this is not a reason for despair. With such an approximate method a real science of nature, and in particular astronomy, is possible. In his *Speculum divinorum* Bate talks about his frustrations with the then-dominant Aristotelian 'logical mode' of science, which attempted to explain individuals through universal specific concepts expressing some apparent similarity. He opposes this Aristotelian mode of science to what he calls a 'real science', in which one examines individual things

[58] The texts of Ibn Ezra and Bate quoted by the anonymous scholar have particular variant readings of the α family.

[59] 'The few lines that were omitted [they were not omitted, but not yet added C. S.] explain why the author of the postscript mistakenly attributed the Latin passage quoted in this note to Henry Bate rather than to the anonymous translator whom Bate was to criticize' (Grant, *Nicole Oresme*, p. 114, n. 83).

[60] See Steel and Guldentops, *Henricus Bate*, pp. 367–70 (lines 106–208).

[61] See Steel and Guldentops, *Henricus Bate*, pp. 452–53.

thoroughly 'usque ad minima'.[62] Such a real science is difficult and almost impossible to obtain because the power of human intellect is not capable of comprehending all individual things and events. However, Bate thinks, we should not despair of acquiring some knowledge of real individual things, even if they exist only contingently and for a limited time. He refers to astronomy as an example of such a real science. 'For is not a scientific knowledge possible about a particular eclipse at a given moment of time (in the past or future) and about other particular things, as the coming of the Antichrist?'[63]

As is well known, it took a long time before a mathematical approach to the study of natural phenomena was accepted.[64] No doubt, medieval scholars were convinced that mathematical proportions were expressed in the creation of the world, and in particular of the celestial spheres, and they liked to refer to what is said in the *Book of Wisdom* (X, 20) 'omnia in mensura, et numero et pondere disposuisti'. However, they were aware that to know the exact proportions of the world order is beyond the capacities of human reason. Starting from sense perception, one can never obtain empirical data that could satisfy the exactitude of mathematical analysis. As Nicole Oresme noticed, an imperceptible error in measurement, even less than a thousandth part, would have an effect on the calculations. How, then, could one in such conditions be able to know the exact proportion ('punctualem proportionem') between celestial magnitudes and motions? And Nicole Oresme refers to al-Battānī who, with reference to Ptolemy's authority, said that 'it is not possible for anyone to understand the truth exactly' ('veritatem ad unguem comprehendere'). Nicole Oresme concludes: 'I will not vainly presume to solve this problem by mathematical demonstration' ('non ergo vane presumam mathematica demonstratione terminare predictum problema').[65] In fact, as Luca Bianchi argued against Alexandre Koyré, to make modern science possible, it was necessary to abandon this ideal of an impossible exactitude. 'The mathematical physics was only possible when innumerable accidental variations were set aside, if one does not take into account not essential deviations in measurement, if one renounces a complete correspondence between the result of demonstrations and the result of measurement, between

[62] *Speculum divinorum*, VI.13, lines 100–02 (ed. Van de Vyver and Steel, *Henricus Bate*, p. 53): 'physice quidem et exquisite perscrutando funditus usque ad minima, licet ad huiuscemodi nulla descendat expresse scientia ab aliquo philosopho tradita'. See on this search for a real science of nature my contribution 'Nature as Object of Science'.

[63] *Speculum divinorum*, VI.10, lines 104–06 (ed. Van de Vyver and Steel, *Henricus Bate*, p. 37): 'Numquid et de eclipsi determinata certo tempore, seu praeterito seu futuro, potest haberi scientia, similiter et de antichristo et consimilibus particularibus?'

[64] See on the deadlock of medieval science, Bianchi, 'L'impossibile exactitude'. The presentation of Bianchi's views in my conclusion comes from my 'Nature as an Object of Science'.

[65] *De commensuratione motuum caeli*, III, lines 14–20 and 36–37 (ed. Grant, *Nicole Oresme*, pp. 284–87).

theoretical values and observational values'.[66] As Bianchi said, the principal reason why the late-medieval philosophers did not succeed in their efforts to apply the new mathematical method was 'their incapacity or refusal to cover the inevitable variance between mathematical certitude and empirical data'. To make progress possible in natural science, we have to admit an approximate calculus of natural processes. That is what we see happen with Galileo, Bianchi explains: he passionately tried to invent always more sophisticated systems of measurement; he repeated with obstinacy his experiments (which were very primitive and far from mathematically exact); he collected data and he reflected on the evaluation of errors of estimation. In so doing, Galileo diverted attention away from the ideal of total (but impossible) precision, towards a non-ideal level of precision which is nonetheless indispensable for solving the problems under consideration and which is accessible via the available instruments. One may see in his scientific experiments a transition from the world of the impossible divine precision to *a universe of approximation* which is accessible to us, and in that sense, more human.

I by no means wish to diminish the genius of Galileo and the novelty of his scientific approach; it seems to me, however, that medieval astrologers had already taken that attitude. They knew that it was impossible to know with precision the true motions of heaven. However, this did not prevent them from developing a method of approximation to the truth. As Bate formulates it nicely at the end of his Preface: 'This is the way of proceeding in things so sublime granted to human smallness, thanks to which we have been left with the possibility to investigate at the end appropriately the truth'.[67]

Appendix

Bate's Preface and Additions to *De Mundo*

The Latin text that follows is taken from my edition in Steel, 'Henry Bate's Translation'.

[66] Luca Bianchi, 'L'impossibile exactitude', pp. 190–91.

[67] I am greatly indebted to David Juste for corrections and comments on the first version of this text. I express my gratitude to Dr Philipp Nothaft (All Souls, Oxford) who read carefully the last version of my contribution and helped me with his competence in medieval astrology to better understand what was at stake in the debate between Ibn Ezra and Henry Bate.

INCIPIT LIBER AVENESRE DE MUNDO VEL SECULO

Tractatus ***Auenesre*** DE PLANETARVM CONIVNCTIONIBVS ET ANNORVM REVOLVTIONIBVS MVNDANORVM translationem aggressuri in uestibulo quidem sermonis obstupuimus, ignorantes quo animo principi astrologorum **Albumasar** deferre neglexerit, quin saltem tanti uerba philosophi in partem interpretatus fuisset meliorem. Indiscretionis namque uisum est hoc fore ut hic dicere sit ausus quod acquiescendum non est dictis **Albumasar** in LIBRO CONIVNCTIONVM eo quod iudicet secundum motus medios, qui se gloriatur illius discipulum qui significationibus coniunctionum in triplicitatibus iudicia commiscet expresse coniunctionum mediarum, prout apparet ex uerbis **Abrahe principis** in 5 particula LIBRI REDEMPTIONIS ISRAEL, quem quidem **Abraham** iste cognominatus **Auenesre** magistrum suum profitetur, ut patet in LIBRIS NATIVITATVM et RATIONVM plerisque locis.

Adhuc quamquam ponat **Albumasar** tempora coniunctionum mediarum et mutationum triplicitatum, ut sic uerarum loca et tempora conuenientius inuestigentur, prout expostulat doctrine ordo ac inuentionis, non tamen ex hoc concludi potest uirtute sillogistica quod actor intellexerit coniunctionum iudicia ad medios motus fore referenda nisi rabiem actori imponere quis conetur. Insuper ipsemet **Auenesre** uisus est numerum ponere coniunctionum secundum cursus medios, quemadmodum **Albumasar**. Non enim mutationes triplicitatum semper fiunt in 240 annis aut 260 secundum ueros motus, sed aliquando citius, aliquando tardius, sicut apparere potest experienti. Qua igitur ratione famosum Albumasar et seipsum potest reprehendere Auenesre?

At uero pertractatione dignum uidetur necnon et discussione hoc quod idem **Auenesre** sub assertione **Ptolomei** confirmare nititur, uidelicet quod non fuit **Ptholomeo** possibile neque precedentibus ipsum neque sequentibus ut in hora introitus Solis in Arietem gradum ascendentis inuenirent propter considerationum incertitudinem, tum ex errore preparationis instrumentorum prouenientem, tum ex diuerso quantitatis anni iudicio ob discordantes magistrorum probationum obseruationes. Hiis itaque de causis probare nititur auctoritate **Ptholomei** intercedente quod non est possibile in anni reuolutione gradum ascendentis inueniri. Quocirca concludit ulterius sustentandum esse super gradum ascendentis in hora coniunctionis aut preuentionis luminarium utra earum immediate precedat ingressum Solis in Arietem. Hoc enim documentum ait esse **Ptholomei** atque anni principium quod absque uariatione conuenienter potest certificari et secundum ipsum iudicia propalari.

TRANSLATION

When we started working on the translation of Ibn Ezra's treatise *On the Conjunctions of the Planets and the Revolutions of the Years of the World*, we were shocked at the opening of this work, as we did not understand for what reason [the author] neglected to pay respect to the prince of the astrologers, Albumasar: why had he not at least interpreted the words of such a great philosopher in a more charitable sense? For it seems to be a want of judgment to say what he dared to say here, that one should not give assent to what Albumasar says in his *Book on Conjunctions*, since he makes his judgments according to the mean motions. Yet [Ibn Ezra] himself boasts of being the disciple of someone who obviously brings in judgments [based on] mean conjunctions when dealing with the significance of the conjunctions in the triplicities, as is evident from the words of Abraham, the prince, in the fifth section of the treatise *On the Redemption of Israel*, whom this Abraham named Ibn Ezra acknowledges as his master, as is evident in the *Book of Nativities* and the *Book of Reasons* in many places.

Besides, even though Albumasar sets times of the mean conjunctions and changes of the triplicities, he does so that the places and times of the true [conjunctions and changes] may be more suitably investigated; this is what the order of both invention and exposition requires; hence one cannot conclude therefrom with syllogistic force that the author thought that judgments of the conjunctions should be related to the mean motions, unless one may try to impute some madness to the author. Moreover, Ibn Ezra himself noticeably establishes the number <of years> of the conjunctions according to the mean motions, just as Albumasar did. For the changes of the triplicities do not always happen after 240 or 260 years according to the true motions, but sometimes faster, sometimes slower, as is clear from experience. For what reason, then, can Ibn Ezra criticize both the famous Albumasar and himself?

What the same Ibn Ezra attempts to establish as an assertion of Ptolemy, deserves certainly to be examined and discussed, I think, namely his claim that it was not possible, either for Ptolemy or for his predecessors and his successors, to find the ascendant degree in the hour of the entrance of the Sun into Aries because of the incertitude of the observations, which comes both from an error in the preparation of the instruments and from the different judgment regarding the length of the year because of the discordant observations of the experimental masters. For all these reasons, he attempts to demonstrate with the authority of Ptolemy that it is impossible to find the degree of the ascendant in the revolution of the year. Therefore, he concludes further that one should rely upon the degree of the ascendant at the time of the conjunction or the precession of the luminaries to see which of them precedes immediately the entrance of the Sun into Aries. For, as he says, this is the teaching of Ptolemy and this is the beginning of the year that can be suitably certified without variation and judgments can be issued according to it.

Sed non uidentur hec dicta sufficientiam omnino continere. Primum enim quod hic admiratione dignum apparet est illud quod radicem dicti sui super illud fundauit de quo in LIBRO RATIONVM prima particula capitulo primo idem **Auenesre** loquens de QVADRIPARTITO **Ptholomei** ex quo tractus est ille sermo pretactus, sic inquit: *Et ego Abraham compilator dico quod hunc librum non compilauit Ptholomeus. Nam in eo sunt multi sermones friuoli secundum scientie contraponderationem et experientie.* Item in LIBRO NATIVITATVM capitulo domus quinte: *Et ego premunio te quod non sustenteris aliquatenus super sermones illius libri. Non enim est in ipso ualor aliquis.* Item in LIBRO RATIONVM capitulo secundo: *unam itaque generalitatem tibi dico quod omnes sermones quos inuenies a Ptholomeo ubi de circulis loquitur ueri sunt, et non ab ipso alii magis. Iudicia uero sua scientie non conueniunt.*

Constat autem quod in ALMAGESTI non inuenitur illud dictum, sed eius contrarium. Igitur si usquam inueniri debeat, hoc esse deberet in QVADRIPARTITO, capitulo quidem 11° secundi libri, ubi tractatur de initio anni. Illud autem quod ibi scriptum est de hac materia, est huiusmodi secundum unam translationem: *Quia ergo nil magis conueniens nil naturali rationi propinquius hac obseruatione deprehendi potest, hiis quatuor punctis anni principia relinquuntur; nos itaque quid singulis accidat anni temporibus prescire uolentes in eisdem principiis Solis et Lune conuentum aut oppositionem, utrum scilicet istorum Solis ingressum in illius quarte principium preueniat considerare oportet. Si itaque eorundem conuentum aut oppositionem comitetur eclipsis, efficacior erit significatio. Quod si antequam Sol Arietis principium ingrediatur, contingat in totum uer et uernali qualitate, significatio illa dilatabitur in Cancro estatem, in Libra autumpnum, sed in Capricorno hyemem totam occupabit. Temporum namque alternatio et generalis eorum proprietas omnisque eorum status causam a Sole specialiter assumunt.*

Alia quidem translatio sic habet: *Et quod magis conueniens naturalique rationi propius in obseruatione rerum anni deprehendimus, est istorum quatuor principiorum obseruatio, necnon coniunctionum et preuentionum Solis et Lune, que predicta tempora precedentes prope ipsa fuerint, maxime autem in quibus fuerint eclipses, ita ut per principium quod ex Solis existentia in Ariete deprehenditur uernalem qualitatem cognoscimus et per initium ex eiusdem existentia in Cancro deprehendimus qualitatem estiualem, per principium autem quod ex ipsius in Libram ingressu cognoscitur qualitatem autumpnalem. Initium uero quod per eiusdem introitum in Capricornum accipitur, hyemalem qualitatem demonstrat, propterea quod temporum generales qualitates et eorum omnes modi non sunt nisi per Solem.*

But what is said does not at all contain sufficient proof. What is first of all astonishing here is the fact that he [Ibn Ezra] grounds his argument upon something about which the same Ibn Ezra in his *Book of Reasons*, in the first section, first chapter, when speaking about Ptolemy's *Tetrabiblos*, from which also the argument mentioned before was taken, says the following: 'But I, Abraham, the compilator (author), say that this book was not composed by Ptolemy, because there are in it many arguments that have no weight when compared to science and experience'. Likewise in the *Book on Nativities* in the chapter on the fifth house: 'I warn you not to rely somehow upon the arguments of this book, because it has no value'; and in the second chapter of the *Book of Reasons*: 'I give you a general rule: anything that Ptolemy says about the orbs is correct and no one surpasses him; but his astrological judgments do not befit his science'.

However, in the *Almagest*, one can certainly not find what is said [here about Ptolemy] , but rather the contrary. Therefore, if it should be found somewhere, it has to be in the *Tetrabiblos*, in the 11th chapter of the second book, which deals with the beginning of the year. What is written there on this topic, is as follows according to one translation: 'Since nothing more convenient can be found, nothing more proximate to natural reason than this observation, there remain the four starting points of the year [indicated by] these points. Therefore, if we want to know beforehand what will happen to each period of the year, we must consider the conjunction or opposition of the Sun and the Moon at these points, namely, which of them precedes the entrance of the Sun in the starting point of that quarter season. If an eclipse accompanies their conjunction or opposition, the signification will be more effective. Indeed, if it is the case that before the Sun enters the starting-point of Aries, generally spring happens with its quality, this meaning will expand in Cancer into summer, in Libra into autumn, but in Capricorn it will occupy the whole winter. For the alternation of the seasons and their general qualities and all their conditions are caused in particular by the Sun'.

In another translation it reads as follows: 'What we take as more appropriate and closer to natural reason in an observation dealing with the year, is an observation of these four starting-points, and also of the conjunctions and oppositions of the Sun and the Moon, which precede the aforementioned times and come close to them, in particular of [the conjunctions] at which eclipses take place. Thus, through the starting-point that is found in the Sun's being in Aries we may know what the spring will be like, through the starting-point of its being in Cancer we may know what the summer, through the starting-point of its entrance in Libra we may know what autumn. The starting-point of its entrance in Capricorn indicates what the winter will be like. For the general qualities and conditions of the seasons only exist because of the Sun'.

Tertia uero translatio, que scilicet immediate de Greco interpretata, hec tenet: *Conuenientius autem mihi uidetur et magis naturale ad annuales considerationes quatuor principiis uti, obseruando propinquissime sibi prius factas coniugationes Solis et Lune coniunctionales aut pleniluniares, et harum maxime rursum eclipticas, ut a principio quidem quod est penes Arietem uer quale sit, consideremus, ab eo autem quod est penes Cancrum estatem, ab eo uero quod est penes Chelas autumpnum, ab eo autem quod est penes Capricornum hyemem. Vniuersales quidem enim temporum qualitates et consistentias Sol facit.*

Ex hiis ergo, nisi littere uiolentia fiat, non potest elici illud dictum **Auenesre**, verum hoc potius quod ad sciendum anni dispositionem et partium eius non solum sufficit cognitionem habere de introitu Solis in punctum uernalis equinoctii et in reliqua tropica, sed cum hiis conuenientius est obseruare coniugationes Solis et Lune coniunctionales et pleniluniares, secundum etiam quod uult **Albumasar** in primo CONIVNCTIONVM, differentia prima, et in octavo etiam, differentia prima, in qua dicit quod ad *sciendum qualitatem rerum comprehendentium genus, ut pestes*, necesse est habere duo ascendentia, coniunctionis scilicet aut oppositionis, que precedit introitum Solis in punctum mobile uernale et cum hoc reuolutionis in qua ingreditur Sol ipsum punctum mobile uernale. Humiditas autem anni seu ariditas ex dispositionibus coniunctionum seu preuentionum luminarium perpendi debet, secundum quod etiam attestantur reliqui sapientes, et rationabiliter quidem cum aure mutatio particularis motum Solis et Lune sibi inuicem comparatum precipue comitetur. Reliqua uero magis permansiua mundi accidentia permanentiorem causam necessario habent imitari. Et hanc quidem rationabilem uiam nititur **Albumasar** in suis iudiciis obseruare.

Insuper et **Abraham princeps** quem **Auenesre** magistrum suum profitetur in 5° REDEMPTIONIS ISRAEL loquens de mutatione regnorum, de preliis, de fame et siccitate, leuitate et grauitate bladi sic ait: *et hoc totum sciemus per reuolutionem coniunctionis Saturni et Iouis idest Sole intrante in Arietem, etc.*

Sed redeamus ad probationem qua dictum quod **Ptholomeo** imponit, probare conatur **Auenesre**. Dicamus ergo quod magistri probationum diuersi diuersis usi considerationibus, nunc motum ponentes tardiorem debito, nunc uelociorem, sic quidem circueundo tandem per proportionalem differentie diuisionem ipsam potuerunt ueritatem inuestigare aut quod ueritati tam propinquum sit ut errorem qui obesse possit excludat. Hic enim modus procedendi in rebus tam sublimibus humane concessus paruitati, quo mediante nobis relinquitur

Finally, there is a third translation, made directly from the Greek: 'It seems more appropriate and natural to me, however, to employ the four starting-points for investigations which deal with the year, observing the conjunctions of the Sun and Moon at new and full Moon which most nearly precede them, and among these in particular the conjunctions at which eclipses take place, so that from the starting-point in Aries we may conjecture what the spring will be like, from that in Cancer the summer, from that in Libra the autumn, and from that in Capricorn the winter. For the Sun creates the general qualities and conditions of the seasons'.[1]

From what is said one cannot draw the conclusion Ibn Ezra draws, unless one manipulates the text, but rather that, in order to know the disposition of the year and of its parts, one has not only to know when the Sun enters the vernal equinox and the other points on the tropic; but, in addition to these, it is also appropriate to observe the conjunctions of the Sun and the Moon at new and full Moon. This is also what Albumasar affirms in the first book of the *Conjunctions*, first difference, and also in the eighth book, first difference, where he says that in order to know the qualities of general conditions, such as pestilence, it is necessary to have two ascendants, namely that of the conjunction or opposition, preceding the entrance of the Sun in the mobile vernal point, and also that of the revolution in which the Sun enters that mobile vernal point. The humidity of the year or its dryness must be judged according to the dispositions of the conjunctions or precessions of the luminaries, as also the other scholars confirm, and this is reasonable since a particular change of the air goes together with the motion of the Sun and of the Moon in comparison to one another. The other more permanent accidental qualities of the world are necessarily consequent upon a more permanent cause. This, then, is the rational way Albumasar tries to keep in his judgments.

Moreover, even Abraham the prince, whom Ibn Ezra calls his teacher, says in the fifth chapter of the *Redemption of Israel*, where he talks about the change of reigns, about battles, about famine, drought, low and high [prices] of grain: 'all this we shall know through the revolution of the conjunction of Saturn and Jupiter, that is, when the Sun enters Aries'.

But let us return at the argumentation by which Ibn Ezra attempts to demonstrate what he imposed upon Ptolemy. Let us say, then, that the masters of experiment, using each different observations, have sometimes posited a motion slower than required, sometimes one faster. Going around that way they could in the end, through a proportional division of the differences, investigate the truth or what is so close to the truth that it excludes an error that may harm.

[1] This is a slightly modified version of Robbins' translation.

facultas non incongrue ueritatem finaliter indagare, quemadmodum perpendere possumus in exemplis. **Ptholomeus** enim uelociores motus posuit quam **Abrachus**, motus tamen **Ptholomei** a debita uelocitate defecerunt. **Albategni** uero motum posuit nimis uelocem propinquiorem tamen ueritati ex parte quam **Ptholomeus. Alzophi** quidem et **Abraham Auenesre** motus posuerunt aliquantulum tardiores **Albategni** ac deficientes paululum a uelocitate debita. Denique nostris temporibus obseruationes **Ptholomei Albategni** necnon et nostras satis conuenienter saluantes motum posuimus proportionis fere medie inter **Albategni** et **Auenesre**, ut sic ad medium in quo consistit ueritas propinquius pertingamus. Et quamuis ad ipsum indiuisibile ueritatis non ueniamus, dummodo prope accedamus, nostre debet sufficere possibilitati. Ait enim **Albumasar** in primo INTRODVCTORII quod *punctorum seu gradus etiam integri error et raro nec multum impedit.* Vnde in primo QVADRIPARTITI: *estimo autem de ipsa pronosticatione etsi non ex toto sit sine errore tamen quod possibile est de ipsa maximo studio dignum uidetur.* Verum error qui ex instrumentis consurgere dicitur per artificiositatem hominis ingeniosi satis caueri potest necnon et per frequentiam considerationum. Demum uero propter ultimam **Auenesre** conclusionem minus sufficienter subillatam aduertendum est quod non est possibile scire horam coniunctionis seu oppositionis Lune cum Sole nisi presciatur motus Lune. Dicit autem **Ptholomeus** tertia dictione ALMAGESTI capitulo primo quod non est possibile scire aliquid eorum que contingunt in Luna ante scientiam Solis et eorum que in ipso contingunt. Vnde in omnibus eclipsium considerationibus tam **Ptholomeus** quam **Ieber** et **Albategni** cum ceteris magistris probationum ad accipiendum uera eclipsium loca et tempora necesse habuerunt motum Solis uerificatum presupponere, secundum quod in demonstrationibus ipsorum perpendi potest manifeste. Et hoc etiam cuilibet palam esse potest etiam parum consideranti. Etenim non est uera Lune coniunctio aut oppositio cum Sole nisi cum Luna fuerit in eodem puncto cum Sole uel in eius opposito. Quod si locum Solis uerum ignoremus, quomodo sciri potest, quando sibi Luna secundum uertitatem coniungi debeat aut opponi.

Sane licet secundum ordinem inquisitionis demonstratiue necessarium sit hoc ita se habere, si tamen non fit ad unguem seu precise, de inuentis sit cura motibus aut forsan conueniens desit inuentio aut inuentionis fiducia sufficientis, error inde proueniens citius seipsum prodit manifestando in coniunctionibus luminarium et oppositionibus propter uelocem Lune motum quam error contingens circa introitum Solis in Arietem propter motum tardiorem. Manifestior

This is the way of proceeding in things so sublime granted to human smallness, thanks to which we have been left with the possibility to investigate at the end appropriately the truth, as we can consider in the [following] examples. Ptolemy for instance posited motions that are faster than [those of] Hipparchus; yet Ptolemy's motions fall short of the due velocity. Al-Battānī posited a motion that is too fast, yet closer to the truth [than that of] Ptolemy. Al-Sufi and Abraham Ibn Ezra posited motions that are somewhat slower than those of Al-Battānī and falling short a little from the due velocity. Finally, in our own times, saving appropriately the observations of Ptolemy and Al-Battānī and our own, we posited a motion [with a velocity] of almost intermediate proportion between Al-Battānī and Ibn Ezra, in order to come closer the middle in which consists the truth. And although we did not reach the indivisible truth, if only we could come closer to it, this must be sufficient to our possibilities. For Albumasar says in the first <chapter> of his *Introduction* that 'an error of a number of minutes or rarely even a whole degree does not harm a lot'. Hence, in the first book of the *Tetrabiblos* it is said: 'But, I think, just as with prognostication, even if it be not entirely infallible, at least its possibilities have appeared worthy of the highest zeal'. Indeed, an error that is said to originate from the instruments, can sufficiently be avoided through the skilfulness of an ingenious man and through frequent observations. Finally, because of Ibn Ezra's last conclusion, which has been drawn in an unsatisfactory manner, one should notice that it is not possible to know the hour of the conjunction or opposition of the Moon with the Sun unless one knows beforehand the motion of the Moon. Now, Ptolemy says in the third section of the *Almagest*, chapter one, that 'it is not possible to know what occurs in the Moon without knowing the Sun and what occurs in the Sun'. Hence, in in all observations on eclipses, Ptolemy, Geber and al-Battānī, as well as other masters of experiment, thought that it was necessary for obtaining the true places and times of the eclipses, to have the motion of the Sun verified, as one can clearly notice in their demonstrations. And this can also become evident to anyone who considers this even a little. In fact, there is no true conjunction or opposition of the Moon with the Sun unless the Moon is at the same point as the Sun or in opposition to it. Hence, if we do not know the true place of the Sun, how could we know when the Moon is due to be in conjunction or opposition to [the Sun]?

Yet, although, according to the order of a demonstrative investigation, it is necessary that this must be the procedure, if it is not possible to do it precisely and exactly, one should worry about [the rates] of the motions that have been found; or maybe an appropriate [rate] was not found or sufficient trust in what is found may be lacking. An error that arises thereof quickly betrays itself, when it is manifest in the conjunctions and oppositions of the luminaries because of the fast motion of the Moon, more so than an error regarding the entrance of

itaque perceptibilitas transitus Lune per puncta coniunctionis et oppositionis eius cum Sole mouit **Auenesre** ut illud tempus poneret pro radice. Adhuc et amoris inclinatio quam habuit ad sectam suam Iudaicam, que annum initiat a coniunctione Lune, ad idem potuit cooperari forsan. Verum propter rationes supratactas non sufficere debet hoc astronomo perspicaci qui rationabiliori uie magis inniti debet quam inhertis ingenii et pigritantis estimationi grosse. Idcirco tutius est rationique magis consentaneum sapientibus quoque ceteris magis concordius sententie principis astrologorum **Albumasar** confidentius adherere. Postquam enim de motibus celestium aliquam certitudinem scientificam, antequam artificiose secundum astrologica iudicandum sit iudicia, presupponere debeamus, necessarium est de Sole primum subicere fundamentum. Quapropter etiam **Abraham** ipse in tractatu suo DE MOTIBVS ET OPERE TABVLARVM SVPER PISAS scientiam de motu Solis ante scientiam que de Luna preordinauit.

Vt igitur ad omne dicatur, connaturale est magis et conuenientius scientie complemento atque perfectioni eius ampliori non solum ad coniunctionem luminarium seu oppositionem eorum aspicere pro statu et esse mundi cognoscendo, verum etiam omni cautela mediante introitum Solis in Arietem et, si necesse fuerit, in reliqua puncta tropica artificiosius considerare, cum in coniunctione magnorum mundi accidentium non modica radix sit, secundum quod patet ex dictis sapientum. Nunc autem tempus est ut ad id quod intendimus accedamus, non pretermittentes que ab **Auenesre** rationabiliter dicta sunt et bene. Sermones igitur eius prout melius poterimus interpretemur.

ADDITIONES

1. **Inquit translator**: hic est itaque sermo Auenesre secundum quod iacet in Ebraico, sed uisum est nobis aut truncatam fuisse litteram in exemplari aut saluis bene dictis eius doctrinam nimis confusam tradidisse et minus artificiosam

2. **Dicit translator**: Quamquam multiplicatio numeri possit crescere in infinitum, reuolutiones tamen corporum celestium finite sunt secundum speciem necessario, quemadmodum in alia parte phie demonstrari habet cum certitudine. Quapropter necessarium est consimiles interdum redire constellationes, licet incomprehensibile sit a nobis tempus huiusmodi reuolutionum propter intervallorum immensitatem. Et hoc forsan est quod hic innuit actor iste. Non est autem opinandum quod propter multiplicem diuersitatem motuum corporum celestium possibile sit ipsos in reuolutionibus quibuslibet non conuenire seu communicare, quemadmodum est de lineis incommunicantibus quas in decimo

the Sun into Aries because of its slower motion. The fact that the passage of the Moon through the points of conjunction and opposition with the Sun is more manifest induced Ibn Ezra to take that time as root [of his calculation]. Besides, the love and inclination he had for his own Jewish sect, which begins the year at the conjunction of the Moon, could perhaps have contributed to that. However, for the aforementioned reasons, this [procedure] should not be sufficient for a perspicacious astronomer, who has to rely more on the rational way than on a rough estimation of an incompetent and indolent mind. Therefore, it is safer and more in agreement with reason and also more concordant with the other scholars to adhere faithfully to the view of the prince of astrologers, Albumasar. In fact, before we can make expertly astrological judgments, we must beforehand have some scientific certitude about the celestial motions. For that reason, it is necessary to lay first the foundation with regard to the Sun. Therefore, Abraham himself, in his treatise *On the Motions and the Use of [Astronomical] Tables for Pisa* set the study of motion of the Sun before that of the Moon.

To make a general conclusion, it is more connatural and suitable to the achievement of science and its further perfection to look not only at the conjunction of the luminaries or their opposition in order to know the state and being of the world, but to consider also skilfully, with all precaution, the entrance of the Sun into Aries and, if required, into the other tropical points, as in their conjunction the great events of the world are rooted, as is clear from the sentences of the scholars. Now it is time to proceed to what we intend to do, without leaving out what has been well said and reasonably by Ibn Ezra. Let us then translate his arguments, as well as we can.

Additions

(1) The translator says: this is Ibn Ezra's argument as it is found in the Hebrew, but it seem to us that either the text has been truncated in the exemplar or, given that the text is sound and well, that the doctrine he transmitted is too confused and not skilful enough.

(2) The translator says: Although the multiplication of a number could increase to infinity, the revolutions of the celestial bodies are necessarily finite in species, as has been demonstrated with certitude in another part of philosophy. It is necessary, therefore, that similar constellations should at some time return, even though the [period of] time of such revolutions is incomprehensible to us because of the immensity of these intervals. Perhaps this is what the author [i.e., Ibn Ezra] means here. It must not be thought, however, that, because of the manifold diversity of the motions of celestial bodies, it would be possible for them not to meet or to communicate, whatever the revolutions may be, as is the case with incommensurable lines, which, in the tenth book of the *Ele-*

ELEMENTORVM ***Euclides*** uocat irrationales siue eu surdas propter impotentiam communicandi. *Omnia namque coordinata sunt,* ut testatur ***Philosophus*** 12° METAPHYSICE, super quo dicit ***Commentator*** quod omnes *actiones corporum celestium in communicatione eorum ad inuicem in constitutione mundi sunt sicut actio liberorum in constitutione domus.* Palam autem est etiam modicum consideranti circa hoc quod, si inter aliqua debet esse ordo siue communicatio, excellentius esse debet in diuinis. Quare absurdum esset opinari motus corporum superiorum irrationales esse siue surdos, et hoc est quod Pitagoras et alii antiqui per musicam mundanam innuere uoluerunt. De qua similiter **Plato** in THYMEO et alii necnon et **Calchidius** cum aliis philosophis infinitis.

ments, Euclid calls irrationals, or surds, because of their incapacity to communicate. For, as the Philosopher declares in the twelfth book of the *Metaphysics,* 'all things are ordered together'; and the Commentator says about this that 'all the actions of celestial bodies in their communication with one another are in the organization of the world as the action of freemen in the organization of a house'. For it is evident even to someone considering these matters a little, that if there must be a communication between some things, this communication must be more excellent in divine things. Therefore, it would be absurd to think that the motions of the superior bodies are irrational, or surd. This is what Pythagoras and other ancients meant to intimate through the music of the world; and on this matter, Plato says similar things in the *Timaeus* and elsewhere, as does Calcidius along with innumerable other philosophers. But let us return to the text.[2]

Bibliography

Bianchi, Luca, 'L'impossibile exactitude. Science and 'calculationes' au XIVe siècle', in Luca Bianchi and Eugenio Randi (eds), *Vérités dissonantes. Aristote à la fin du Moyen Age*, Fribourg/Paris: Saint-Paul, 1993, pp. 153–94.

Boudet, Jean-Patrice, 'Ptolémée dans l'occident médiéval. Roi, savant et philosophe', *Micrologus* 21 (2013), pp. 193–217.

Clagett, Marshall, *Nicole Oresme and the Medieval Geometry of Qualities and Motions*, Madison: University of Wisconsin Press, 1968.

Dell'Anna, Giuseppe, *Dies critici. La teoria della ciclicita delle patologie nel XIV secolo*, 2 vols, Galatina: Congedo, 1999.

Duhem, Pierre, *Le système du monde. Histoire des doctrines cosmologiques de Platon à Copernic*, 10 vols, Paris: Hermann, 1913–1959.

Federici Vescovini, Graziella, 'Una versione latina medievale dell'opera escatologica di Abramo bar Hijja (Savosarda) "Megillat ha-Megalleh": il "Liber de redemptione Israhel"', in Michele Ciliberto and Cesare Vasoli (eds), *Filosofia e cultura. Per Eugenio Garin*, Roma: Editori Riuniti, 1991, vol. I, pp. 5–37.

Garin, Eugenio, *Pico della Mirandola. Disputationes adversus astrologiam divinatricem*, 2 vols, Firenze: Vallecchi, 1952.

Grant, Edward, *Nicole Oresme and the Kinematics of Circular Motion. Tractatus de commensurabilitate vel incommensurabilitate motuum celi*, Madison: University of Winconsin Press, 1971.

Juste, David, 'Ptolemy, Quadripartitum (tr. anonymous, before *c.* 1250)' (update: 27.09.2017), *Ptolemaeus Arabus et Latinus. Works*, URL= http://ptolemaeus.badw.de/work/34.

[2] For the translation of this addition, I modified Grant's translation, which was based on the 1507 edition of De mundo.

Kennedy, Edward S., Benno van Dalen, George Saliba, and Julio Samsó, 'Al-Battānī's Astrological History of the Prophet and the Early Caliphate', *Suhayl* 9 (2009–2010), pp. 13–148.

Lemay, Richard, *Abū Maʿšar al-Balḫī [Albumasar]: Liber introductorii maioris ad scientiam judiciorum astrorum*, 9 vols, Napoli: Istituto orientale di Napoli, 1995–1996.

Millás Vallicrosa, José Maria, *El libro de los fundamentos de la Tablas astronómicas de R. Abraham ibn 'Ezra. Edición crítica, con introducción y notas*, Madrid/Barcelona: Consejo Superior de Investigaciones Científicas, Instituto Arias Montano, 1947.

Nothaft, C. Philipp E., 'Criticism of Trepidation Models and Advocacy of Uniform Precession in Medieval Latin Astronomy', *Archive for History of Exact Sciences* 71 (2017), pp. 211–44.

Nothaft, C. Philipp E., 'Henry Bate's *Tabule Machlinenses*. The Earliest Astronomical Tables by a Latin Author', *Annals of Science* 75 (2018), pp. 275–303.

Poznanski, Adolf, and Julia Guttmann, *Abraham Bar Ḥiyya, Sefer Megillat ha-megalleh*, Jerusalem: Mekize Nirdamim, 1968.

Robbins, Frank E., *Ptolemy. Tetrabiblos*, Cambridge (Mass.) [Cambridge (Mass.): Harvard University Press, 1940.

Sela, Shlomo, *Abraham Ibn Ezra and the Rise of Medieval Hebrew Science*, Leiden/Boston: Brill, 2003.

Sela, Shlomo, *Abraham Ibn Ezra. The Book of Reasons. A Parallel Hebrew-English Critical Edition of the Two Versions of the Text*, Leiden/Boston: Brill, 2007.

Sela, Shlomo, *Abraham Ibn Ezra. The Book of the World. A Parallel Hebrew-English Critical Edition of the Two Versions of the Text*, Leiden/Boston: Brill, 2010.

Sela, Shlomo, *Abraham Ibn Ezra on Nativities and Continuous Horoscopy. A Parallel Hebrew-English Critical Edition of the Book of Nativities and the Book of Revolution*, Leiden/Boston: Brill, 2013.

Sela, Shlomo, 'The Ibn Ezra-Henry Bate Astrological Connection and the Three Abrahams', *Mediterranea. International Journal on the Transfer of Knowledge* 2 (2017), pp. 163–86.

Sela, Shlomo, Carlos Steel, C. Philipp E. Nothaft, David Juste and Charles Burnett, 'A Newly Discovered Treatise by Abraham Ibn Ezra and Two Treatises Attributed to al-Kindī in a Latin Translation by Henry Bate', *Mediterranea. International Journal on the Transfer of Knowledge* 5 (2020), pp. 193-305.

Steel, Carlos, and Guy Guldentops, *Henricus Bate. Speculum divinorum et quorundam naturaliam. Parts XX–XXIII. On the Heavens, the Divine Movers, and the First Intellect*, Leuven: University Press, 1996.

Steel, Carlos, 'Nature as Object of Science. On the Medieval Contribution to a Science of Nature', in Chumaru Koyama (ed.), *The Concept of Nature. Some Approaches East and West*, Leiden: Brill, 2000, pp. 125–52.

Steel, Carlos, 'Henry Bate's Translation of Ibn Ezra's Treatise *The Book of the World*', *Quaestio* 19 (2019), pp. 227-278 [part of *Stars, Kingdoms, Beliefs, and Masses.*

Political Astrology in the Mediterranean Area from Middle Ages to the Renaissance. Le stelle, i regni, le credenze e le masse. L'astrologia politica nel Mediterraneo tra Medioevo e Rinascimento, eds M. Benedetto, P. Arfé, P. Porro].

Steel, Carlos, Steven Vanden Broecke, David Juste and Shlomo Sela, *The Astrological Autobiography of a Medieval Philosopher. Henry Bate's Nativitas (1280–81)*, Leuven: Leuven University Press, 2018.

Thorndike, Lynn, *A History of Magic and Experimental Science*, vol. III, New York: Columbia University Press, 1934.

Vuillemin-Diem, Gudrun, and Carlos Steel, *Ptolemy's Tetrabiblos in the Translation of William of Moerbeke. Claudii Ptolemaei Liber Iudicialium*, Leuven: University Press, 2015.

Van de Vyver, Emiel, and Carlos Steel, *Henricus Bate. Speculum divinorum et quorundam naturalium. Parts VI–VII. On the Unity of Intellect. On the Platonic Doctrine of the Ideas*, Leuven: Leuven University Press, 1993.

Yamamoto, Keiji, and Charles Burnett, *Abū Maʿšar on Historical Astrology. The Book of Religions and Dynasties (On the Great Conjunctions)*, 2 vols, Leiden: Brill, 2000.

Wallerand, Gaston, *Henri Bate de Malines. Speculum divinorum et quorundam naturalium*, Louvain: Institut superieur de philosophie de l'université, 1931.

The Medieval Latin Versions of Pseudo-Ptolemy's *Centiloquium*: A Survey*

Jean-Patrice Boudet

The main Greek version of the pseudo-Ptolemaic *Καρπός* (*Kitāb al-thamara* in Arabic, *Liber fructus* or *Centiloquium* in Latin) has been edited by Emilie Boer in the Teubner Collection in 1952 and again in 1961 with some corrections.[1] This version contains the hundred astrological propositions or aphorisms alone, without any commentary and any transliteration from Arabic. It comes very probably from a Greek archetype and is preserved in *c.* 50 manuscripts of which none, nevertheless, is anterior to the fourteenth century.

We do not know exactly how and when the Greek archetype of this Byzantine version may have been translated into Arabic. A commentary to the *Kitāb al-thamara* is ascribed by al-Bīrūnī (*c.* 1040) to Abū'l Abbās al-Iṣbahānī (tenth century?),[2] but it seems to be lost, and the main Arabic version was drawn up by Abū Jaʿfar Aḥmad ibn Yūsuf, an Egyptian physician, mathematician and astrologer who wrote his commentary after 300H/912–13 AD.[3] According to Richard Lemay, the text and the commentary were written together by Aḥmad ibn Yūsuf and the commentary constituted the true substance of the text. As Maria Mavroudi has shown in her paper, Lemay's hypothesis is invalidated. Aḥmad ibn Yūsuf was probably inspired, indirectly or directly, by a Greek, a Syriac[4] or an Arabic version bereft of a commentary. But Aḥmad may be considered, in a way, as the real 'inventor' of the *Centiloquium* from a Medieval Latin point of view, even if his identity has been generally unknown to European scholars. As a matter of fact, most of the Latin translations of this text derive from his Arabic version.

* Many thanks to David Juste for his numerous and very helpful remarks on the Latin translations of the *Centiloquium*, to Maria Mavroudi for her remarks and corrections of Greek and Arabic words in this article, and to Charles Burnett for all his suggestions and corrections.

[1] Boer, Καρπός.

[2] Sezgin, *Geschichte*, vol. VII, p. 167; Haddad et al., 'Al-Bīrūnī on Astrological Lots', pp. 30–31 and 48.

[3] *c.* 922 according to Lemay, 'Origin and Success'. The date 300 in the Muslim calendar or 912–13 in the Christian calendar corresponds to a comet studied by Aḥmad in his commentary on *kalimah/verbum* 100. See Boudet, 'Les comètes'.

[4] See Nau, 'Un fragment syriaque'.

Ptolemy's Science of the Stars in the Middle Ages, ed. by David Juste, Benno van Dalen, Dag Nikolaus Hasse and Charles Burnett, PALS 1 (Turnhout, 2020), pp. 283–304

10.1484/M.PALS-EB.5.120183

The *Centiloquium* has attracted much attention from Latin scholars from the beginning of the twelfth century and six different translations may be ascribed to that century. Those six translations are the following:

1. Adelard of Bath's translation (*c.* 1116–1120?), incipit 'Doctrina stellarum'. He translates only the first 39 *verba* (propositions) without their commentary. Two independent manuscript copies of this translation survive: London, British Library, Sloane 2030, fols 87r–87v (this part of the codex is an author's manuscript, probably autograph, from the first half of the twelfth century), and Lyon, Bibliothèque Municipale, 328, fols 70r–74r (dated 1395).[5] Sometime during the twelfth or thirteenth century, Adelard's partial translation was incorporated into the tradition of the principal translation made by Plato of Tivoli, together with a third version, incipit 'Mundanorum mutatio'. This triple conflated text or threefold version (R. Lemay called it 'la version agglomérée du *Centiloquium*') seems to survive in more than 15 manuscripts, of which several good copies may be used for the edition of Adelard's translation.
2. Plato of Tivoli's version, finished in Barcelona in 1136, incipit 'Iam scripsi tibi, Iesure'. The majority of surviving manuscripts of the *Centiloquium* carry this version (*c.* 100 copies).[6] Aḥmad ibn Yūsuf's commentary (*expositio*) follows each *verbum* and is attributed by Plato to a certain 'Hali', which may be a confusion with ʿAlī ibn Aḥmad al-ʿImrānī (d. 955) — author of a treatise on astrological elections translated also in Barcelona in 1133 by the Jewish scholar Savasorda (Abraham Bar Ḥiyya)[7] — or with ʿAlī ibn Riḍwān (d. 1068), an Egyptian physician and astrologer who composed a famous commentary on Ptolemy's *Tetrabiblos* (*Kitāb al-arbaʿa* in Arabic, *Quadripartitum* in Latin), translated into Latin in the thirteenth century, but most probably already known and used by Plato in his translation of the *Quadripartitum* (1138).[8]

[5] See Burnett, 'Catalogue. The Writings of Adelard of Bath', pp. 166 and 183–84; Lemay, *Le Kitāb at-Tamara*, vol. I, pp. 86–88. The Latin manuscripts of the *Centiloquium* are in the process of being described by David Juste, see https://ptolemaeus.badw.de/manuscripts.

[6] Lemay, *Le Kitāb at-Tamara*, vol. I, pp. 88–92. For Plato of Tivoli, see Samsó, 'El procés de la transmissió'.

[7] See Millás Vallicrosa, *Las traducciones orientales*, pp. 328–39; Carmody, *Arabic Astronomical and Astrological Sciences*, pp. 137–39; and the 'New Carmody' catalogue prepared by Charles Burnett and David Juste, whom I warmly thank for the information they were kind enough to give me about Plato of Tivoli's and Abraham Bar Hiyya's translations.

[8] Carmody, *Arabic Astronomical and Astrological Sciences*, pp. 155–56.

3. An anonymous version with incipit 'Mundanorum mutatio' mentioned above, in which Richard Lemay thought it was possible to recognize the influence of Hermann of Carinthia or someone from his circle.[9] This version, of which Dag Nikolaus Hasse showed most recently that it was, at least partly, attributable to John of Seville,[10] survives in *c.* 50 manuscripts. But each proposition of *Mundanorum* is in fact given in two versions, one being a fresh translation from Arabic ('*Mundanorum 1*') and the other offering a more elegantly Latinized version, using and quoting Plato of Tivoli's translation ('*Mundanorum 2*').[11]
4. A version by Hugo Sanctelliensis, a contemporary of Plato and Hermann.[12] Just like most of Hugo's translations, this one is addressed to Michael, bishop of Tarrazona (1119–1151). On account of Hugo's connection to Michael during the later years of his office and of Hugo's more scrupulous rendering of the Arabic in comparison with his other translations, R. Lemay tended to consider it later than the other versions (*c.* 1138–1151?), especially Plato's translation, which Hugo seems to have known. But this new translation was very difficult to use from a technical point of view and it survives in only two manuscripts: Madrid, Biblioteca Nacional, 10009, fols 85ra-105vb (thirteenth century), and Naples, Biblioteca Nazionale, VIII. D. 4, fols 3r–30v (fifteenth century).
5. An anonymous version with incipit 'Iam premisi libros' seems also to date from the twelfth century. R. Lemay called this version 'Abugafarus' because he thought that it was the only one that gives the true name of the commentator in the transliterated form 'Abugafarus' (for Abū Jaʿfar), sometimes distorted into 'Bugafarus' or 'Bugufarus', but in fact it is also the case in at least three *codices* of Plato of Tivoli's version. This version seems to partly survive in eight manuscripts but it is also known by some indirect witnesses. Because of its 'coarse' Latin resembling to the style of John of Seville and Gerard of Cremona, R. Lemay tended to ascribe this translation to one or the other, or both (considering Gerard of Cremona' 'strategy of revision' of older translations from Arabic), but D. Hasse has

[9] Lemay, *Le Kitāb at-Tamara*, vol. I, pp. 111–17.

[10] Hasse, 'Stylistic Evidence', esp. pp. 27–30.

[11] See Lemay, *Le Kitāb at-Tamara*, vol. I, pp. 111–17. See also Boudet, 'Les comètes', pp. 206–09, for a primitive version of 'Mundanorum' containing many transliterations from Arabic, being rather close to that of Adelard de Bath.

[12] Lemay, *Le Kitāb at-Tamara*, vol. I, pp. 117–19. For Hugo Sanctelliensis' translations, see Haskins, *Studies*, pp. 67–81; Burnett and Pingree, *The Liber Aristotilis*; Martínez Gázquez, *The Attitude of the Medieval Latin Translators*, pp. 51–61. For the identity of Hugo, see now Santoyo, 'El Normano Hugo de Cintheaux', pp. 341–57.

recently shown that it is probably not attributable to either and appears to be a revision of Plato of Tivoli's version.[13]

6. A sixth Latin translation has been discovered by David Juste in MS Vatican, BAV, Vat. lat. 5714, fols 105r–112v, a Northern Italian codex which can probably be dated to before 1229. Unfortunately, the beginning and the end of this copy are lost: it begins in the middle of the 6th *sermo* and ends just after the beginning of the 88th *sermo*. Recently, this version has also been identified by D. Juste in two fragmentary folia found in the binding of a printed book in the Diocesan Library of Vác, Hungary, where we find the preface of the text and fragments of *verba* 1, 3–4, 24–26 and 28–31.[14] However, fols 1r and 2v are faded and almost entirely illegible. This means that, for the purpose of textual comparison, we will have to choose propositions 7 to 39, if we want to compare with Adelard of Bath's version, and propositions 7 to 87, if we want to compare with the five other versions. I have chosen two very important *verba*: 8 and 51.

Another Latin version (but it is not an original translation) was made by William of Aragon much later, *c.* 1300 (and, in any case, before 1330, which corresponds to the *terminus ante quem* of one of the *codices*). It is a sort of copy or paraphrase of previous translations (mainly Plato of Tivoli's), with a new title for each *verbum* and a personal *glosa*.[15] At least four manuscripts survive.

The Greek version of the eighth proposition, edited by Emilie Boer, has been translated by James Hershel Holden as follows:

> 8. The sagacious mind helps the heavenly effects, just as the best farmer helps nature through plowing and clearing [the fields].[16]

Here is a translation of Aḥmad ibn Yūsuf's Arabic version:

> Proposition (*kalimah*) 8. Ptolemy says: the wise soul (*al-nafs al-ḥikmīyah*) [of the astrologer] aids the action of the celestial sphere just like the sower aids the natural forces through plowing and clearing.

[13] Lemay, *Le Kitāb at-Tamara*, vol. I, pp. 108–11; Hasse, 'Stylistic Evidence', pp. 27–30. For Gerard of Cremona's translations, see Lemay, 'Gerard of Cremona'; Pizzamiglio, *Gerardo da Cremona*; Burnett, 'The Coherence'; Id., 'The Strategy of Revision'.

[14] Vác, Egyházmegyei Könyvtár, 708.012/Fragm. 2 (see http://ptolemaeus.badw.de/ms/591). The beginning of this version seems to have been inspired by Plato of Tivoli's translation but afterwards it becomes original and was probably used as a draft by Hugo Sanctelliensis. See *infra*.

[15] On this scholar who flourished around 1330 and wrote a very interesting treatise of oniromancy, see particularly Pack, '*De pronosticatione sompniorum*'; Weijers, *Le travail intellectuel*, pp. 101–03; Val Naval, *Estudio, edición crítica*.

[16] Holden, *Five Medieval Astrologers*, p. 72. Greek text in Boer, Καρπός, p. 39.

> Commentary. By 'wise soul', Ptolemy means the soul knowing the celestial forces and their application to each natural individual [i.e. each individual belonging to the natural world]. In fact, this soul, when it senses that a benefit will come to a certain individual, moves that individual to be well disposed to that benefit, so that it will be available to him [in advance] and will become manifest in him, but this argument was sufficiently explained in our commentary to the fifth proposition.[17]

The text of the fifth proposition says that 'the astrologer will be able to prevent many (bad) effects of the stars when he knows the nature of their influence on him (the client), and sometimes he will then prepare [his client] for action before it occurs, so that he is able to withstand it'.[18] Aḥmad ibn Yūsuf's commentary on the fifth proposition gives as an example a physician and astrologer who is able to adjust the balance of the temperament of a person, and, being aware of his nativity and of the indications of the stars associated to a disease of the nature of Mars, he can try to modify his balance, free his complexion from this bad influence and get rid of the disease.

Here are now the six Latin translations of the twelfth century:

Verbum 8

1. Adelard of Bath. MSS *S* = London, British Library, Sloane 2030, fol. 87v (twelfth century); *L* = Lyon, Bibliothèque municipale, 328 fol. 69 (dated 1395). Collation with some manuscripts of the 'threefold version' (Adelard + Plato + 'Mundanorum'): *P* = Paris, BnF, n.a.l. 3091 (end of the thirteenth century); *V* = Vatican, BAV, Reg. lat. 1452 (beginning of the fourteenth century).

> VIII. Anima sapiens actum[19] stellarum adiuvat sicut[20] seminator potens (reading Arabic ألزرّاع ألقوي /'al-zarrā'' for 'al-qawīy') adiuvat naturalia cum aratione et mundatione.

Adelard does not seem to know Aḥmad ibn Yūsuf's commentary. His literal translation is quite good and the comparison between 'anima sapiens' (of the astrologer) and 'seminator potens' (of the powerful sower) is elegant and correct.

2. Plato of Tivoli (1136). MSS: *F* = Florence, Biblioteca Riccardiana, 163, fol. 4r (thirteenth century); *B* = Basel, Universitätsbibliothek, F. III. 25, fol. 45ra (thirteenth century); *C* = Cambridge, University Library, Ii 3.3, fol. 221va-221vb (*c.* 1299); *M* = Madrid, Biblioteca Nacional, 10015, fol. 20va (dated 1251).

[17] Martorello and Bezza (eds), *Aḥmad ibn Yusūf ibn al-Dāya*, pp. 70–71. The Arabic text of the proposition says: ألحكيمة يعين على ألفعل ألفلكيّ كما يعين ألزرّاع ألقوى ألطبيعية بألحرث و ألتنقية. I follow the Italian translation of Martorello and Bezza.

[18] ibid., pp. 62–63.

[19] actum *SL*] actus *PV*.

[20] sicut *om*. *S*.

Verbum 8. Anima sapiens ita adiuvabit opus stellarum quemadmodum seminator fortitudines naturales.

Expositio. Sapiens anima est illa[21] que scit illud quod de fortitudine celi[22] diximus, et eius adiutorium est quando aliquod bonum alicui homini eventurum cognoverit, ei res suas sic aptare precipiat ut illud bonum venturum maius ac melius deveniat[23] quam eveniret nisi eum sic premuniret. Et iam locuti sumus de hoc sufficienter in quinto capitulo.

Plato of Tivoli's translation of the proposition is not literal but it is quite good from a stylistic point of view; his translation 'fortitudines naturales' is very close to the Arabic phrase القوى الطبيعية but he omits the following words ('through plowing and clearing'). The translation of the commentary is totally literal and does not pose any problem.

3. 'Mundanorum'. MSS: *A* = Paris, BnF, lat. 16204, p. 556b (thirteenth century); *B* = Paris, BnF, lat. 16204, p. 543b (thirteenth century, without commentary).

Verbum 8.[24] Anima sapiens potest adiuvare celestem operationem quemadmodum seminans virtutem naturalem[25] per cultum et purgationem. Dixit Ptholomeus: Anima sapiens, etc. Anima sapiens ita adiuvat opus stellarum quemadmodum seminator fortitudines naturales.[26]

Expositio. Sapiens est anima illa que scit id quod de fortitudinibus celi prediximus. Et eius adiutorium est quando aliquod bonum alicui hominum eventurum[27] cognoverit ei res suas sic aptare precipiat ut id bonum venturum maius et melius eveniat quam eveniret nisi eum premuniret[28]. Et iam locuti sumus de hoc sufficienter in 5°[29] capitulo.

As we can see, the 'Mundanorum 1' translation of the *verbum* tries to improve on Plato's version with the addition 'per cultum et purgationem', which is an accurate translation of the Arabic. But 'Mundanorum 2' only quotes Plato's version and the translation of the commentary is almost a copy of Plato's.

4. Hugo Sanctelliensis. MSS: *A* = Madrid, Biblioteca Nacional, 10009, fol. 87va (thirteenth century); *B* = Naples, Biblioteca Nazionale, VIII. D. 4 (fifteenth century).

21 Sapiens est illa anima *BC*.
22 de fortitudine celi] *om. F* – de fortitudinibus celi *M* – de celi fortitudinibus *B*.
23 eveniat *BM*.
24 9 [*sic*] *A*.
25 naturalem *om. AB*.
26 Dixit Ptholomeus... naturales *om. B*.
27 eventurum] eventurus *A*.
28 premuniret] premineret *A*.
29 5°] 3 [*sic*] *A*.

Verbum 8. Anima sapiens et discreta circularem actionem iuvat, quemadmodum agricola et purgando nature subvenit.

[Expositio] Per hoc quod dicit: anima scilicet[30] sapiens, eam vult intelligere que circuli vires comprehendit arte[31] atque potenciam eiusque processum atque ordinem in individuis naturalibus attendit. Si itaque huiusmodi anima aliquid prosperitatis alicui individuo futurum perpendat, ipsum individuum ut illud ipsum bonum suscipiat motu debito compellet. Hoc autem taliter collectum quin appareat nullatenus poterit occultari. Verum 5[32] sermonis exposicio que ad huius loci evidenciam sufficere debent satis ut arbitror comprehendit.

With Hugo Sanctelliensis' translation, the contrast with the three first versions is total. If we except the word 'anima sapiens' and 'quemadmodum', perhaps inspired by Plato of Tivoli, almost all the vocabulary is different and affected. The sense of 'discreta circularem actionem' is not clear (like 'circuli vires' in the commentary). But in the commentary, Hugo is sometimes more precise than the others translators and he seems to have been the first to render the full sense of the Arabic الطبيعة للأشخاص / *al-ṭabīʿa lil-ashkhāṣ* ('nature in individuals', the Arabic word for 'individual' applies to any class of individuals (as opposed to species and genera) whether human, animal, vegetable or mineral.). Significantly, he adds one phrase, 'Hoc autem taliter collectum quin appareat nullatenus poterit occultari' ('This is collected in such a way that it cannot be occultated'), showing that the astrologer is *par excellence* the person able to discover the occult part of the future of the individual.

5. MS Vatican, BAV, vat. lat. 5714, fol. 105rb (thirteenth century).

Sermo octavus. 8. Dicit Ptolomeus: Anima sapiens et discreta efficacie circulari auxiliatur, vel ut brachium robustum auxiliatur nature, inquam, et aliis operibus.

Explanatio. Per hoc dicit: Anima sapiens, illa vult intelligere que circulari vires dignoscit; etiam enim effectum in individuis naturalibus. Hec igitur anima, cum cognoverit aliquid astrorum alicui individuo provenire, ipsum ita preparabit ut illam bonam stellam suscipiat. Cum autem ita fuerit, multum inde colliget et maxime in eo apparebit. In quinto quidem sermone tantum quod ad huius loci evidentiam suf[f]icere debet.

This newly discovered translation has some typical words in common with Hugo Sanctelliensis' version: 'discreta', 'circulari' or 'circuli vires', 'individuis naturalibus', 'suscipiat', 'sermo'. This is certainly no accident. The translation by 'ut brachium robustum auxiliatur nature' can be explained by the confusion of two very similar-looking letters in the Arabic.[33] The end of the commentary is strange and shows a lack of understanding. I think, therefore, that this ver-

[30] scilicet *om. B.*

[31] arte] ante *B.*

[32] 5] quinti *B.*

[33] Reading ذراع ('dhirāʿ', 'forearm') for زرّاع ('zarrāʿ', 'sower').

sion has not been diffused because it was rather bad and that it may have been used as a kind of first draft by Hugo Sanctelliensis to make his own and much better translation. Thus it seems to be also a twelfth century translation, older than Hugo's final version.

6. 'Iam premisi'. MS Basel, Universitätsbibliothek, F. III. 33, fol. 2r (thirteenth century, basic manuscript).

[Verbum 8] Anima sapiens adiuvat celestem effectum sicut adiuvat seminans virtutem naturalem cum aratione et purgatione.

[Expositio] Animam sapientem intelligit que novit virtutes celestes et comprehendit singulares naturas. Talis anima cum accidet quod aliqua fortuna consequatur aliquod singularium movet illud ad bonam fortunam et aptabit ita quod meliorabitur bonum fortune per premunitionem. Et iam locuti fuimus sufficienter super hoc in 5to capitulo quod sufficit hoc loco.

The 'Iam premisi' version is clearly an improvement of Plato of Tivoli's version, perhaps also partly based on the 'Mundanorum' version. The translation of the *verbum* is perfect. The translation of the commentary is very clear, although less literal and accurate with the introduction of the notion of chance ('accidet'); the translation of *sa῾āda* as 'fortuna' and 'bona fortuna', on the other hand is an accurate representation of the Arabic.

So we have at least six different Arabic-Latin translations of the *Centiloquium* dating from the twelfth century. But there were probably more than six translations at the end of the thirteenth century. In his *Nativitas*, or *Liber super inquisitione et verificatione nativitatis incerte*, an astrological autobiography written in 1280, Henry Bate of Malines quotes five different translations of *verbum* 38, of which four may be identified with Plato of Tivoli's, Adelard of Bath's and 'Mundanorum' 1 and 2. But Henry Bate also refers to a Greek-Latin translation:

Et Ptolemeus in *Centilogio*, 38 verbo, quod 'si Mercurius fuerit in duobus signis Martis, dabit fortitudinem perfidie et stultitie. Et fortior duobus locis est Aries'.[34] Et hoc quidem secundum unam translationem. Alia vero translatio sic habet: 'et si fuerit Mercurius in domibus Martis, dabit ei acuitatem ingenii in astutia et maxime in Ariete'.[35] Item illa translatio: 'cum vero fuerit in signo Martis super primitias et

[34] That is Plato of Tivoli's version, *Iam scripsi*: 'Verbum 38. Cum fuerit Mercurius in aliqua domorum Saturni et ipse fortis in esse suo dat bonitatem intelligentie medullitus in rebus [*FM in marg.*: scilicet in radicalibus rebus et in principiis artium [alio in principiis *M*]. Et si fuerit in duobus signis Martis, dabit fortitudinem perfidie et stultitie [*M in marg.*: vel in alio bonitatem subite eloquentie et dehonestationis], et fortior duobus locis est Aries'.

[35] That is 'Mundanorum 1'. The 'Iam premisi' version is different: 'Cum fuerit Mercurius in aliqua domorum Saturni et ipse fortis in esse suo, dat bonitatem intelligencie medullitus in rebus, et si fuerit in duobus signis Martis dabit fortitudinem perfidie et astucie et forcior duo-

fortunam, potentior autem horum duorum locorum est Aries'.[36] Et iterum alia translatio: 'et cum in uno signorum Martis fuerit significat hominem qui cito respondet interrogationi, et Aries melius est Scorpione'.[37] Translatio denique de Greco autem talis est: 'in domo autem Martis dabit facilem linguam et maxime in Ariete'. Huic vero sententie ultime translationis concordant commentatores. Unde prima translatio impropria est.[38]

Henry Bate quotes earlier in his *Nativitas* a 'Liber arboris' which is the title of the *Centiloquium* given in some manuscripts (e.g. Boulogne-sur-Mer, Bibliothèque Municipale, 198; Venice, Museo Civico Correr, cod. Cic. 617; Erfurt, Universitätsbibliothek, Dep. Erf. CA 4° 361), and he also knows the real name of its commentator, Aḥmad ibn Yūsuf ibn Ibrāhīm.[39] He quotes later the eight *verbum* of the *Centiloquium* in an unknown version: This translation from the Greek has not been found in the manuscripts, but Bate quotes from it on several occasions, including in the following one from the eight *verbum*:

> Nam ut dicit Ptolemeus in *Centiloquio*: 'Anima sapiens cooperatur celesti effectui, sicut optimus agricola cooperatur nature per arationem seu purgationem'.[40]

A totally different translation from the Arabic of proposition 8 is preserved in the thirteenth-century Latin version of the *Ġāyat al-ḥakīm* / *Picatrix*, IV, iv, 48, with a selection of nine other *verba*:

> Spiritus operantis effectus celestes adiuvat sicut messes naturales, videlicet arare et cultu terrarum adiuvantur messes.[41]

bus locis est Aries'. Bate's quotation does not correspond either with Sanctelliensis', nor with the version of MS Vatican, BAV, Vat. lat. 5714.

[36] This is very close to Adelard's translation: 'Cum fuerit Mercurius in signo Saturni potensque in seipso, dabit fetui bonitatem cogitationis in radicibus. Cum vero fuerit in signo Martis, supra primitias et fortunam. Potentior autem horum duorum locorum Aries'.

[37] Here is the 'Mundanorum 2' version: '[...] Et cum in uno signorum Martis fuerit, significat hominem qui cito respondet interrogationi. Et Aries est melius Scorpione'.

[38] *Nativitas* 1065–1078, ed. Steel et al., *The Astrological Autobiography*, pp. 169–70. I thank the authors for giving me access to Bate's text before publication.

[39] '[...] ut testatur Avenezre in *Initio sapientie*, cui etiam concordat tota cohors astrologorum. Nichil autem confert planeta aut parum nisi in nativitate promiserit, ut habetur in *Libro arboris* 78° verbo, quod et manifestius affirmant eius expositores Haly et Abuiafar Hamet filius Ioseph filii Abrahe' (*Nativitas* 1025–1030, ed. ibid., p. 168).

[40] *Nativitas* 2709–2711. According to Carlos Steel, this translation exhibits the style of William of Moerbeke, see Steel et al., *Astrological Autobiography*, p. 85.

[41] Pingree, *Picatrix*, p. 193. See Ritter, *Pseudo-Mağrīṭī*, p. 324. The Andalusian author of the *Ġāyat*, Maslama al-Qurtubī, is one of the very first Arabic scholars who used and quoted Aḥmad ibn Yūsuf's version of the *Kitāb al-thamara*. For this author, see Fierro, 'Bātinism in Al-Andalus'; de Callataÿ, 'Magia en al-Andalus'.

This is not the place to examine in detail William of Aragon's fourteenth century version of the same *verbum*, since, as we have seen, his *textus* and *commentum* are copied mainly from Plato of Tivoli's translation. But his *glosa* is original and rather interesting for its explanations inspired by natural philosophy and its reference to the famous historico-mythical example of Nectanabo and the birth of his son, Alexander the Great:

> De significatione contracta ex stellis per animam sapientem super res futuras. 9um [*sic*].
>
> Textus. Anima sapiens ita ai[u]vabit opus stellarum quemadmodum seminator fortitudines naturales.
>
> Commentum. Sapiens est illa anima que scit illo [*sic*] quod diximus de fortitudinibus celi, et eius adiutorium est quando aliquod bonum alicui homini eventurum cognoverit, ei res suas sic aptare precipiat ut illud bonum venturum maius ac melius eveniat quam eveniret nisi sic eum premuniret. Et iam locuti sumus de hoc sufficienter in quinto capitulo.
>
> Glosa. Postquam docuit Ptholomeus per quid astrologus habet iudicare quia per bonam et prudentem acceptionem et applicationem complexionis et moris hic docet modum per quem [*sic*] ex quadam proportionali actione. Unde dicit quod anima sapiens, supple in hiis que dicta sunt, scilicet de complexione stellarum ex moribus et de eisdem a parte receptoris, talis anima sapiens potest iuvare opus vel virtutem stellarum sicut seminator virtutes naturales. Seminator duo considerat, scilicet naturam loci et seminis proprietatem, et naturam loci duppliciter, scilicet quod sit convenientis nutrimenti nec arida nec aquosa sed pinguis et temperata; secundo quod sit calida vel frigida secundum exigentiam seminis. Videmus enim quod semen vult calidam et temperata[m] regionem et plantari vult in aqua et sic de aliis proportionalibus diversificamus eis loca ut bonum eorum [earum *ms.*] sit maius post nativitatem; ante nativitatem etiam eligemus materiam in qua melius recipiatur materia quam in alia, sicut fecit Neptanebus de Alexandro.[42]

Let us now examine proposition 51, a very important proposition for historians of astrology because it shows quite clearly the difference, not to say the contradiction, between the authentic Ptolemy of the *Tetrabiblos/Kitāb al-arbaʿa/Quadripartitum*, and the Pseudo-Ptolemy of the *Centiloquium*. In the *Tetrabiblos*, III, 2, *Of the Degree of the Horoscopic Point*, Ptolemy gives a procedure to find the real ascendant of the native and rectify his nativity when the time of its birth is not precisely known. This procedure is in three steps: 1. 'Take the syzygy most recently preceding the birth, whether it be a new Moon or a full Moon'; 2. See what planet is the ruler of the degree of conjunction or opposition of the two luminaries; 3. Take the position of the ruler to be one of the cardines of the natal horoscope (mid-heaven, ascendant, etc.).[43] This method was known to Aḥmad ibn Yūsuf, who speaks about it in the 34th proposition

[42] MS Paris, BnF, lat. 7480 (fourteenth century), fols 23v–24v.

[43] Robbins, *Ptolemy. Tetrabiblos*, pp. 230–35.

of the *Kitāb al-thamara* and who says in his commentary that this doctrine of Ptolemy concerning the *namūḏār* ('indicator') is a valuable technique for the determination of the nativity of human beings, unlike that of animals.[44] But this method was already challenged in Antiquity. Vettius Valens, an astrologer of the second century AD, in the third book of his *Anthologies*, says: 'Assume the position of the Moon at the conception to be the same as the position of the ascendant at the time of the delivery, and from this you can know whether the conception was at new or full Moon'.[45] Vettius Valens, in his Book I, 21, also has a whole chapter on this subject,[46] which is perhaps the indirect source, via Abraham Ibn Ezra, of an appendix to the Latin *Centiloquium*, *verbum* 51, called from the thirteenth century *Trutina Hermetis*.[47] But this appendix seems to be lost in Arabic and I will focus only here on this *verbum*, not its appendix.

The Greek version of the 51st proposition of the pseudo-Ptolemaic *Καρπός* was probably inspired by Vettius Valens, *Anthologies*, III,10:

> 51. Where the Moon is at the time of birth, that sign will be at the ascendant at conception; and where it is at conception, it will be in the ascendant at birth.[48]

Here is now Aḥmad ibn Yūsuf's Arabic version:

> Proposition 51. Ptolemy said: The place of the Moon at the time of birth is the ascendant degree of the sphere at the time of the fall of the sperm (*suqūṭ al-nutfah*) and the place of the Moon at the time of the fall of the sperm is the ascendant degree at birth.
>
> Commentary. Scholars in the things of nature agree that the period of gestation of babies in maternal wombs is diverse and is not the same for all. In the case of an average period (of gestation), the Moon and the ascendant degree perform equal revolutions (*adwār*), and the Moon is then found in the ascendant degree of the nativity. A longer gestation corresponds to an additional duration compared to the average gestation, and after having completed full revolutions, the ascendant performs one revolution (*dawra*) less and the Moon is found to have already transited the ascendant degree. A shorter gestation corresponds to a period of time shorter than the

[44] Martello and Bezza, *Aḥmad ibn Yusūf ibn al-Dāya*, pp. 114–17.

[45] Pingree, *Vettius Valens*, p. 144.

[46] Bara, *Vettius Valens d'Antioche*, pp. 214–21.

[47] See Sela, *Abraham Ibn Ezra on Nativities*, pp. 41–45, 88–99, 220–21, 443–49; Id., 'Abraham Ibn Ezra's Role'; Boudet, 'Naissance et conception'. In the beginning of his *Nativitas*, Henry Bate refers to it. See *Nativitas* 81–91, ed. Steel et al., *The Astrological Autobiography*, pp. 129–30: 'Hiis vero prelibatis ad artem inveniendi cum precisione nativitatis dicte gradum ascendentis ex hora estimationis veritati propinqua accessimus in hunc modum. Dimisso annimodar Ptolemei ab auctoritatibus improbato, licet veritati ex parte consonum per nos aliquando sit inventum ad trutinam Hermetis, qui et Enoch, confugimus iuxta consilium Abrahe cognomine Principis et Ptolemei in *Centilogio* 51 propositione, et *Albumasar in Sadan*, nec non et reliquorum et precipue Abrahe Avenesre in suo *Libro de nativitate* corrigentis ibidem et verificantis pretactam trutinam sive annimodar Hermetis'. See now Sela, 'Calculating Birth'.

[48] Holden, *Five Medieval Astrologers*, p. 78. Greek text in Boer, Καρπός, pp. 48–49.

average period, its decrease being inferior by one revolution and, in this case, the Moon is decreased compared to the ascendant degree.

Astrologers have already abundantly spoken in their books about the gestation period of babies because it was very necessary for their judgments. For, it shows us the ascendant at the time of the fall of the sperm corresponds to the beginning of the life of the native, whereby one discerns the temperament of his body, the appearance of his limbs and many things that are completed before it comes out. There were [astrologers] already before Ptolemy, who painstakingly sought to know the time of the fall of the sperm. But Ptolemy showed us briefly and in a very simple way that the place of the Moon at the time of the nativity is the ascendent at the time of conception, and the ascendant of the nativity is the place of the Moon at the time of conception, because the Moon returns to its place after having performed equal revolutions. Similarly, the ascendant, compared to the average gestation, increases and decreases by increasing and decreasing.[49]

So Aḥmad ibn Yūsuf refers explicitly to an 'average', 'longer', and 'shorter' periods of gestation defined by the revolutions of the Moon, but without specifying their respective lengths in days or months.[50] This does not facilitate the full understanding of his commentary and contributes to explain the variety of the Latin translations of the twelfth century; none of them rendered, for example, the Arabic word 'dawr' or 'dawra' by 'revolutio'. Here are the five Latin translations of Aḥmad's version:

1. Plato of Tivoli. MSS: *F* = Florence, Biblioteca Riccardiana, 163, fol. 14r–14v; *B* = Basel, Universitätsbibliothek, F. III. 25, fols 47vb-48ra; *C* = Cambridge, University Library, Ii 3.3, fol. 229va-229vb; *M* = Madrid, Biblioteca Nacional, 10015, fol. 22va-22vb.

Verbum 51. Dixit Ptholomeus: Locus Lune[51] in nativitate est ipse gradus ascendens de circulo hora casus spermatis, et locus Lune hora casus spermatis est gradus ascendens hora nativitatis.

Expositio. In hoc concordati sunt physici quod more natorum in uteris matrum sunt[52] diverse et non sunt eedem[53] in omnibus. Ex illis igitur est mora media, et est cum Luna[54] et gradus ascendens perfecerunt[55] in tempore more[56] orbes equales et

[49] Martorello and Bezza, *Aḥmad ibn Yusūf ibn al-Dāya*, pp. 144–47.

[50] In the Latin appendix to aphorism 51, called 'Trutina Hermetis' (cf. *supra*, n. 47), the average duration of gestation ('mora media') amounts to 273 days corresponding to nine complete revolutions of the Moon, the longer gestation ('mora maior') to 288 days corresponding to 9.5 revolutions and the shorter gestation ('mora minor') to 258 days, corresponding to 8.5 revolutions.

[51] Lune] Luna *F.*

[52] sunt] sint *FM.*

[53] sunt eedem] sint heedem *F* – sint eedem *M.*

[54] nedia *add. M.*

[55] perfecerunt] perfecerint *FM.*

[56] *F in marg.*: 'scilicet natorum in ventre matrum'.

invenietur tunc Luna in gradu ascendente nativitatis. Maior vero mora est tempus auctum super tempus more medie post orbes perfectos minus orbe[57] et invenietur Luna per hoc ultra[58] gradum ascendentem. Minor mora est subtractio de tempore medio et subtractio illa minor est orbe, et invenitur Luna per hoc subtracta de gradu ascendente.

Iamque commemoraverunt astrologi in libris suis sermonibus prolixis moras nativitatum quia erant sibi valde necessarie in[59] iudiciis. Ille enim ostendunt nobis locum ascendentis in casu seminis[60] et illud est principium inceptionis nati ex quo discernimus[61] complexionem corporis eius et modum membrorum et multa que accidunt illi antequam egrediatur a matrice. Illi autem qui fuerunt ante Ptholomeum pro[62] inquisitione casus spermatis passi sunt laborem nimium. Patefecit autem nobis Ptholomeus breviter et[63] via levi quod locus Lune in tempore nativitatis est ascendens casus spermatis, et locus Lune in tempore casus spermatis est gradus ascendentis nativitatis,[64] quia[65] Luna et ascendens revertuntur[66] ad locum suum orbibus equalibus in mora media, et augetur et minuitur in maiori et minori mora.

Plato of Tivoli's translation here appears to be very reliable and congruent with its Arabic model, even if the matter is not an easy one. The choice to translate by 'orbis' the Arabic word *dawr/dawra* turns out to be sensible to denote the orbits of the Moon.

2. 'Mundanorum'. MS *A* = Paris, BnF, lat. 16204, p. 569a–569b

Verbum 51. Locus Lune qui est in nativitate ipse est gradus ascendens in hora casus spermatis. Et locus Lune in hora casus spermatis est gradus ascendens in hora nativitatis. Dixit Ptholomeus: Locus Lune, etc. Gradus in quo est Luna in nativitate alicuius est gradus ascendens in hora qua infusum est semen in matrice a patre. Et gradus in quo fuerit Luna hora qua infuditur sperma est ascendens in nativitate.

Expositio. Omnes homines periti in naturali scientia sciunt quod tempora quibus in matrum ventribus morantur infantes non sunt equalia omnibus, immo quidam faciunt mediocrem moram in quorum nativitatibus invenitur Luna in ascendente quia Luna perfectos circulos peragravit. In maiori autem mora addetur pars mediocri postquam perfecti sunt circuli, et minor minuetur a mediocri et minuetur quod Luna non perveniet in nativitate ad gradum ascendentis.

[57] orbe *om. C.*

[58] *F in marg.*: 'id est ultra gradum in quo est Luna hora casus spermatis qui erit gradus ascendens in hora nativitatis'. *B in marg.*: 'id est ultra gradum ascendentis, id est ultra gradum in quo est Luna hora casus spermatis qui erit gradus ascendens in hora nativitatis'.

[59] dandis *add. M.*

[60] seminis] spermatis *C.*

[61] discernimus] decernimus *C.*

[62] pro] in *C.*

[63] et] ex *C.*

[64] est ascendens casus spermatis... ascendentis in nativitate *om. M.*

[65] quia] quod quia *F.*

[66] revertuntur] revertitur *M.*

> Antiqui autem astrologi de spermatis casu multa dixerunt. Volebant enim scire quamdiu infans in utero matris morabatur, et qua hora semen infundebatur quia hoc eis necessarium erat. Erat enim initium generationis infantis, et inde poterat sciri complexio corporis sui et qualitates membrorum. Illi autem qui ante Ptholomeum fuerunt in extrahendo horam conceptionis valde laboraverunt. Ptholomeus autem leviter nos docuit quando dixit quoniam gradus Lune in nativitate erat ascendens quando infundebatur semen et in ascendente nativitatis erat iam hora infusionis seminis. Luna enim ad locum suum revertitur quando equales circuli sui perfecti sunt. Et ascendens similiter. Et addes vel minues quando debes addere vel minuere.

As is usually the case in the translation of each *verbum*, the text of 'Mundanorum' offers two versions for each proposition, i.e. a new translation from the Arabic ('Mundanorum 1') and Plato of Tivoli's translation ('Mundanorum 2'). The commentary consists of a new translation from the Arabic, more elegant than the previous versions.

3. 'Iam premisi'. MS Basel, Universitätsbibliothek, F. III. 33, fol. 4v.

> [Verbum 51.] Locus Lune in nativitate ipse est gradus ascendens de circulo hora casus spermatis, et locus Lune in hora qua ceciderit sperma est gradus ascendens in hora nativitatis.
>
> [Expositio.] Phisici noverunt quod more natorum in utero matrum sunt diverse nec sunt eedem in omnibus. Ex illis igitur est media mora et est cum Luna et gradus ascendens perfecerunt in tempore more orbes equales, et invenietur tunc Luna in gradu ascendentis nativitatis. Maior vero mora est tempus auctum super tempus more medie post orbes perfectos minus uno orbe, et invenietur per hoc Luna ultra gradum ascendentis. Minor mora est subtractio de tempore medio et illa subtractio minor est orbe, id est infra gradum in quo est Luna in tempore casus spermatis qui erit gradus ascendentis hora nativitatis et invenietur per hoc Luna subtracta de gradu ascendentis.
>
> Iam commemoraverunt ast<r>ologi in libris suis sermonibus prolixis moras nati quia erant valde sibi necessarie in iudiciis. Ille enim ostendunt nobis locum ascendentis in casu spermatis et id principium inceptionis nati ex quo decernimus principium complexionis eius et modum membrorum et multa que accidunt ei antequam egrediatur a matrice. Illi autem qui fuerunt ante Ptholomeum pro inquisitione casus spermatis passi sunt laborem nimium et patefecit nobis breviter et via levi quod locus Lune in tempore nativitatis est ascendens in tempore casus spermatis, et locus Lune in tempore casus spermatis est ascendens gradus nativitatis, quia Luna et ascendens revertuntur ad locum suum orbibus equalibus in mora media, et augebitur et minuetur in maiori et minori mora.

This translation is closer to that of Plato of Tivoli, with some improvements. Here is, by contrast, the very personal version of Hugo Sanctelliensis:

4. Hugo Sanctelliensis. MSS: *A* = Madrid. Biblioteca Nacional, 10009; *B* = Naples, Biblioteca Nazionale, VIII. D. 4.

Verbum 51. Locus Lune in nativitate est ipsius orientis in circulo gradus sub conceptione,[67] et quem sub conceptione possidet est ipsius[68] natalis orientis gradus.

[Expositio.] Astrologorum omnium et qui nature[69] secreta rimantur generalis est sentencia varios[70] esse status fetuum in materno utero existentium. Est enim quidam medius, dum[71] videlicet Luna et orientalis[72] gradus equalem debiti status perficiunt circuitionem. Unde Luna sub[73] huiusmodi nativitatis in orientis gradu repperiri necesse est. Maior quidem status est qui circuitione perfecta supra medium, minus quam circulum augmentat. Sub eo enim Luna orientalem pertransivit gradum. Minor rursum status citra medium minus una circuitione repperitur, necdum Luna ad orientalem gradum poterit pervenire.

Astrologi quidem in suis voluminibus et satis prolixo sermone nascentium status ad iudicia summe afferunt necessarios. Hec enim ex eorum documento[74] contrahimus quatinus conceptionis gradu comprehenso quod est origo universe generationis atque principium inde complexionem[75] atque membrorum formam et alia multa antequam de materno procedat utero plenissime decet[76] intueri. Verum qui ante Ptholomeum fuerunt astrologi ad conceptionis[77] horam dispendio longe difficili conabantur accedere de quo idem actor Ptholomeus compendiosum hoc in loco[78] subgerit[79] edictum. Inquid enim locum Lune sub hora nativitatis oriens esse conceptionis, et oriens natalis lunarem locum esse sub conceptione. Luna enim sub equali circuitione ad eumdem revertitur locum. Nec aliter status medius, idem etiam secundum augmentum et diminutionem addit et minuit.

We may note the emphasis at the beginning of the 'expositio' on the 'secreta nature', which shows a common interest for the secrets of nature with Hugo's translation of the 8th *verbum*. In spite of its elegance, the neo-classical and rather difficult Latin of Hugo Sanctelliensis clearly affected the diffusion of his translation, especially for technical reasons: instead of 'ascendens', we have here 'oriens'. But it is not the only problem and we can see that the copyist of the Naples manuscript does not seem to have understood the text at all. Some lexical features of this version are also found in the strange version discovered

67 conceptione] conclusione *B*.
68 ipsius] ipse *B*.
69 nature] inter *B*.
70 varios] id est varios *B*.
71 medius, dum] medius qui dum *B*.
72 et orientalis] est in orientali *B*.
73 Luna sub] Lunam in *B*.
74 documento] nocumento *B*.
75 complexionem] compressionem *B*.
76 decet] licet *B*.
77 conceptionis] contemptionis *B*.
78 hoc in loco] in hoc loco *B*.
79 subgerit] sugerit *B*.

by David Juste, including the word 'circuitio' for *dawr/dawra* (instead of 'orbis' in Plato of Tivoli and 'Iam premisi', and 'circulus' in 'Mundanorum'):

5. MS Vatican, BAV, Vat. lat. 5714, fol. 109rb-109va.

> Sermo quinquagesimus primus, 51. Dicit Ptolomeus: Lune locus in nativitatibus est pars surgens[80] in circulo in hora casus spermatis et locus Lune in hora casus spermatis est pars[81] ascendens in nativitate.
>
> Explanatio. Convenerunt sapientes et docti in natura dicentes quod status nati in utero matris diversificatur et non est idem in omnibus. Et eorum medius status est cum Luna et gradus ascendentis complent suas circu<i>tiones equalificatas et invenietur Luna[82] in hora ipsa in parte nativitatis ascendente. Status quoque maior est tempus maior isto statu medio post circu<i>tiones completas et inveniunt Lunam iam transactam gradum ascendentis. Status quidem minori minor? est tempus minus tempore medio et deest eius minus una circuitione et inveniunt Lunam in eo non pertigente[m] gradum ascendentis.
>
> Astrologi vero tractatibus suis dixerunt cum sermo[ne] suo prolixo statum nati in iudiciis maxime esse neccessarium. Ostendunt[83] namque nobis ascendente[m] manente[m] in hora casus spermatis et est prima inceptio nati per quam[84] commixtionem sui corporis dignoscant, et suorum membrorum figuram, multaque in eo complentur antequam de utero matris sue procedat. Illi qui antequam Ptolomeus esset extiterunt in dignoscenda hora casus spermatis maxime laborabant. Quorum postea nobis apperuit Ptolomeus brevi sermone et via plana, quoniam locus Lune in hora partus est ascendens in hora casus spermatis, ascendens vero nati est locus Lune in hora casus spermatis Luna namque ad locum suum revertitur et equatur suis circuitionibus equalibus. Similiter etiam ascendens in statu medio et addit et minuit in maiore et minore.

This rather over-literal translation ('pars surgens' or 'pars ascendens', retaining the Arabic term 'juz', whose primary meaning is 'part') instead of 'gradus ascendens'; 'inveniretur Lunam' instead of 'invenietur Luna') seems to be independent from the others, and in spite of the common technical word 'circuitio', connections with Hugo Sanctelliensis' version are less obvious for this *verbum* than for the eighth, as seen above. Anyway, it confirms that we have at least six different Arabic-Latin translations of the text of the *Centiloquium* dating from the twelfth century, and five different versions of Aḥmad ibn Yūsuf's commentary.

In the fourteenth century, William of Aragon copied Plato of Tivoli's version of this *verbum* 51 but added a new title for the proposition ('De significa-

[80] paras surgens [*sic*] *ms.*

[81] paras [*sic*] *ms.*

[82] Lunam [*sic*] *ms.*

[83] Ostendit *ms.*

[84] eam *ms.*

tione contracta ex loco Lune in nativitate ut sanitur [?] locus Lune hora casus spermatis et ascendens nativitatis') and his personal *glosa*.[85]

In all these versions, the *Centiloquium* enjoyed a wide circulation in Medieval Europe, seeing that Richard Lemay and David Juste have recorded more than 168 different manuscripts of the Latin versions from the Arabic, which puts it in second place in the David Juste's 'Top 50 of the most popular astrological texts in Latin', just after Alcabitius' *Liber introductorius*, preserved in more than 210 manuscripts.[86] The *Liber introductorius* was used as a basic manual for teaching astrology in medieval universities and other indications show that the *Centiloquium* owed a large part of its success to the fact that it held a certain place in the fourteenth and fifteenth centuries in the teaching of *astronomia* (e.g. astronomy-astrology) belonging to the *quadrivium*, notably in Italian (especially Bologna) and other European universities — including that of Paris.[87] As it also includes more than twenty propositions clearly destined for medical use, one can imagine the interest of physicians in this text. At a time when most astrologers were first and foremost medical practitioners (although the reciprocal was not true), the *Centiloquium* may be considered as a key for the study of the relationship between the two university disciplines, *astronomia* and medicine, and a detailed study of its manuscript tradition will be very useful for a clearer understanding of the controversial subject held by astrology in medical theory and practice at the end of the Middle Ages.

The influence of this medieval corpus was increased by printing. Plato of Tivoli's version was edited at least three times in Venice: in 1484 by Erhard Ratdolt, in 1493 by Bonetto Locatello, and in 1519 by the heirs of Ottaviano Scoto.[88]

More generally, the fascination astrologers held for the *Centiloquium* peaked at the end of the Middle Ages and it also aroused the interest of the humanists of the Quattrocento. This is why new commentaries of the *Centiloquium* appeared during this period: by Reimbotus of Castro († 1390), physician to Emperor Charles IV;[89] by the physician and astrologer Conrad Heingarter,

[85] MS Paris, BnF, lat. 7480, fols 86r–88v: 'Glosa. Postquam reducit nobis ad memoriam Ptholomeus ea que in omnibus iudiciis tanquam radices sunt accipienda in omnibus mutationibus huius mundi. In hoc capitulo docet nos dirigere nativitates hominum ne acceptio figure celestis circuli fiat fallax, unde iudex veniat in errore. [...]'.

[86] Burnett et al., *Al-Qabīsī*. See Juste, 'The Impact of Arabic Sources', p. 177.

[87] On this topic, R. Lemay's article, 'The Teaching of Astronomy', has to be corrected. See Boudet, *Entre science et nigromance*, pp. 283–95; Id., 'Un colliege de astrologie et medicine'.

[88] These editions reproduce the threefold version (Adelard + Plato + 'Mundanorum') of the *verbum primum* only. The conflated manuscript tradition, on the contrary, sustained the triple parallel presentation for *verba* 1 to 39 and the double one for aphorisms 40 to 100.

[89] *Glossa super Centiloquium*, MS Vaticano, BAV, Pal. lat. 1380, fols 65–80v (copied *c.* 1350–1366 at Bologna or Paris). See Thorndike, 'Pre-Copernician Astronomical Activity'; Schuba, *Die Quadriviums-Handschriften*, pp. 111, 113.

addressed to Jean II, Duke of Bourbon († 1488),[90] by Lorenzo Bonincontri in or before 1477,[91] and by several scholars at the University of Cracow in the second half of the fifteenth century.[92] And two new translations of the text itself appeared in the fifteenth century: in 1456 by the Byzantine humanist George of Trebizond, translated from the Greek version in Italy for the King Alfonso V of Aragon;[93] and in 1477 by the poet and humanist Giovianni Pontano, which was also translated from Greek.[94] Both were copied and printed many times until the middle of the sixteenth century.

From a qualitative point of view, the influence of the *Centiloquium* was crucial. It helped to construct the Church's doctrinal norm concerning astrology, a norm elaborated in the thirteenth century which could be qualified as quasi-consensual until the end of the Middle Ages. Aphorisms 5 and 8 of this text ascribed to Ptolemy played a major role in this process. The *Centiloquium* was therefore at the heart of the debate on the doctrinal validity of astrology in the Latin West from the middle of the thirteenth century. The sentence 'Vir sapiens dominabitur astris' ('the wise man will dominate the stars'), attributed to Ptolemy and inspired by the *verba* 5 and 8 of the *Centiloquium*, became, in the end of the Middle Ages, a leitmotiv of criticism against astrology. But at the same time, it was largely used as an adage by astrologers themselves, more than happy to show that they had a determinist but anti-fatalist conception of astral causality.[95] On this matter, aphorism 51, connected with aphorisms 5 and 8, is an excellent illustration of the utility of astrology and astrologers. If it is correct, one can chose the very best moment for the conception and for the birth of a child. Is it really reliable? I am very sceptical on this point. But I am not a medieval astrologer…

[90] MS Paris, BnF, lat. 7432, fols 134v–156r (see Juste, *Les manuscrits astrologiques latins*, pp. 152–53). For Conrad Heingarter, see in particular Préaud, *Les méthodes de travail*; Id., *Les astrologues*, pp. 71–100 and 240–42.

[91] See Rinaldi, 'L'inedita *Expositio*', and the entry made by D. Juste, 'Laurentius Bonincontrius'.

[92] MSS Cracow, BJ, 2703, fols 169r–175r (anon., *c.* 1492); Cracow, BJ 1857, pp. 71–128 (by Andreas Grzymala of Poznan), etc. See Markowski, *Astronomica et astrologica*, pp. 31–32, 302, 309. See also the *Circulum pro exitu geniture ab utero iuxta verbum Ptolomei 51 rectificare*, that may be ascribed to Albertus de Brudzewo: https://ptolemaeus.badw.de/work/115.

[93] See the classic studies of Monfasani, *George of Trebizond*, and *Collectanea Trapezuntiana*, pp. 97–100, 689–95 and 750–51. For this translation and Pontano's, see Rinaldi, 'Pontano, Traepezunzio'.

[94] Michele Rinaldi is about to publish Pontano's translation of the *Centiloquium* in a collection supported by the *Ptolemaeus Arabus et Latinus* project.

[95] See especially Boudet, 'Ptolémée dans l'Occident médiéval'; Id., 'Astrology between Rational Science'.

Bibliography

Bara, Joëlle-Frédérique, *Vettius Valens d'Antioche, Anthologies, livre 1. Etablissement, tradition et commentaire*, Leiden/Boston: Brill, 1989.

Boer, Emilie, Καρπός. *Pseudo-Ptolemaei Fructus sive Centiloquium*, Leipzig: Teubner, 1952 (new edition Leipzig: Teubner, 1961) [*Claudii Ptolemaei Opera quae extant omnia*, vol. III.2].

Boudet, Jean-Patrice, *Entre science et nigromance. Astrologie, divination et magie dans l'Occident médiéval, XIIe-XVe siècle*, Paris 2006.

Boudet, Jean-Patrice, 'Un colliege d'astrologie et medicine? Charles V, Gervais Chrétien et les manuscrits scientifiques du Collège de Maître Gervais', *Médiévales* 52 (2007), pp. 15–38. [Published in English in *Studies in History and Philosophy of Biological and Biomedical Sciences* 41 (2010), pp. 99–108].

Boudet, Jean-Patrice, 'Ptolémée dans l'Occident médiéval : roi, savant et philosophe', *Micrologus* 21 (2013), pp. 193–217.

Boudet, Jean-Patrice, 'Astrology between Rational Science and Divine Inspiration. The Pseudo-Ptolemy's *Centiloquium*', in Stefano Rapisarda and Erik Niblaeus (eds), *Dialogues among Books in Medieval Western Magic and Divination*, Firenze: SISMEL, Edizioni del Galluzzo, 2014, pp. 47–73.

Boudet, Jean-Patrice, 'Les comètes dans le *Centiloquium* et le *De cometis* du pseudo-Ptolémée', *Micrologus* 24 (2016), pp. 195–226.

Boudet, Jean-Patrice, 'Naissance et conception. Autour de la proposition 51 du *Centiloquium* du pseudo-Ptolémée', in *De l'homme, de la nature et du monde, Mélanges d'histoire des sciences médiévales offerts à Danielle Jacquart*, Genève, 2019, pp. 185–198.

Burnett, Charles, 'Catalogue. The Writings of Adelard of Bath and Closely Associated Works, together with the Manuscripts in which they Occur', in Charles Burnett, *Adelard of Bath. An English Scientist and Arabist of the Early Twelfth Century*, London: Warburg Institute, 1987, pp. 163–96.

Burnett, Charles, and David Pingree, *The Liber Aristotilis of Hugo of Santalla*, London: Warburg Institute, 1997.

Burnett, Charles, 'The Coherence of the Arabic-Latin Translation Program in Toledo in the Twelfth Century', *Science in Context* 14 (2001), pp. 249–88 [reprinted in Id., *Arabic into Latin in the Middle Ages. The Translators and their Intellectual and Social Context*, Ashgate: Variorum, 2009, article VII].

Burnett, Charles, 'The Strategy of Revision in the Arabic-Latin Translations from Toledo. The Case of Abū Ma'shar's *On the Great Conjunctions*', in Jacqueline Hamesse (ed.), *Les traducteurs au travail. Leurs manuscrits et leurs méthodes*, Turnhout: Brepols, 2002, pp. 51–113.

Burnett, Charles, Keiji Yamamoto, and Michio Yano, *Al-Qabīṣī (Alcabitius), The Introduction of Astrology. Editions of the Arabic and Latin Texts and an English Translation*, London/Torino: Warburg Institute, 2004.

Callataÿ, Godefroid de, 'Magia en al-Andalus. Rasa'il ijwān al-Ṣafa', Rutbat al-ḥakim y Gāyat al-ḥakim (Picatrix)', *Al-Qantara* 34 (2013), pp. 297–344.

Carmody, Francis J., *Arabic Astronomical and Astrological Sciences in Latin Translation. A Critical Bibliography*, Berkeley/Los Angeles: University of California Press, 1956.

Fierro, Maribel, 'Bātinism in Al-Andalus. Maslama b. Qāsim al-Qurtubī (d. 353/964), Author of the *Rutbat al-hakīm* and *Ghāyat-al-hakīm* (Picatrix)', *Studia Islamica* 84 (1996), pp. 87–112.

Haddad, Fuat I., David Pingree, and Edward S. Kennedy, 'Al-Bīrūnī on Astrological Lots', *Zeitschrift für Geschichte der Arabisch-Islamischen Wissenschaften* 1 (1984), pp. 9–54.

Haskins, Charles Homer, *Studies in the History of Mediaeval Science*, Cambridge: Harvard University Press, 1927 (2nd ed.).

Hasse, Dag Nikolaus, 'Stylistic Evidence for Identifying John of Seville with the Translator of Some Twelfth-Century Astrological and Astronomical Texts from Arabic into Latin in the Iberian Peninsula', in Charles Burnett and Pedro Mantas-España (eds), *Ex Oriente Lux. Translating Words, Scripts and Styles in Medieval Mediterranean Society*, Córdoba-London: Córdoba University Press, 2016, pp. 19–43.

Holden, James H., *Five Medieval Astrologers. An Astrological Miscellany*, Tempe: American Federation of Astrologers, 2008.

Juste, *Les manuscrits astrologiques latins conservés à la Bibliothèque nationale de France*, Paris: CNRS Editions, 2015.

Juste, David, 'The Impact of Arabic Sources on European Astrology. Facts and Numbers', *Micrologus* 24 (2016), pp. 173–94.

Juste, David, 'Laurentius Bonincontrius, *Commentum super Centiloquio Ptholomei*' (update: 19.07.2017), *Ptolemaeus Arabus et Latinus. Works,* URL = http://ptolemaeus.badw.de/work/113.

Lemay, Richard, 'The Teaching of Astronomy in Medieval Universities, Principally at Paris in the Fourteenth Century', *Manuscripta* 20 (1976), pp. 197–217.

Lemay, Richard, 'Origin and Success of the Kitāb Thamara of Abū Jaʿfar Aḥmad ibn Yūsuf ibn Ibrāhīm from the Tenth to the Seventeenth Century in the World of Islam and the Latin West', in Ahmad Y. al Hassan, Ghada Karmi and Nizar Namnum (eds), *Proceedings of the First International Symposium for the History of Arabic Science* (*Aleppo, April 5–12, 1976*), Aleppo: Institute for the History of Arabic Science, 1978, vol. II, pp. 91–107.

Lemay, Richard, 'Gerard of Cremona', in *Dictionary of Scientific Biography*, vol. XV, New York: Scribner's Sons, 1978, pp. 173–92.

Lemay, Richard, *Le Kitāb at-Tamara* (*Liber fructus, Centiloquium*) *d'Abū Ja'far Aḥmad ibn Yūsuf* [*Ps.-Ptolémée*], New York, 1999 (unpublished).

Markowski, Mieczysław, *Astronomica et astrologica Cracoviensia ante annum 1550*, Firenze: Olschki, 1990.

Martínez Gázquez, José, *The Attitude of the Medieval Latin Translators Toward the Arabic Sciences*, Firenze: SISMEL, Edizioni del Galluzzo, 2016.

Martorello, Franco, and Giuseppe Bezza, *Aḥmad ibn Yusūf ibn al-Dāya, Commento al Centiloquio tolemaico*, Milano/Udine: Mimesis, 2013.

Millás Vallicrosa, José María, *Las traducciones orientales en los manuscritos de la Biblioteca Catedral de Toledo*, Madrid: Consejo Superior de Investigaciones Científicas, Instituto Arias Montano, 1942.

Monfasani, John, *George of Trebizond. A Biography and a Study of His Rhetoric and Logic*, Leiden: Brill, 1976.

Monfasani, John, *Collectanea Trapezuntiana. Texts, Documents, and Bibliographies of George of Trebizond*, Binghamton, N. Y.: Medieval and Renaissance Texts and Studies, 1984.

Nau, François, 'Un fragment syriaque de l'ouvrage astrologique de Claude Ptolémée intitulé le *Livre du fruit*', *Revue de l'Orient chrétien* 28 (1931–1932), pp. 197–202.

Pack, Roger Ambrose, 'De pronosticatione sompniorum libellus Guillelmo de Aragonia adscriptus', *Archives d'Histoire Doctrinale et Littéraire du Moyen Âge* 41 (1966), pp. 237–93.

Pingree, David, *Picatrix. The Latin Version of the Ghāyat Al-Ḥakīm*, London: Warburg Institute, 1986.

Pingree, David, *Vettius Valens, Anthologiae, III*, Leipzig: Teubner, 1986. [English transl. by Mark T. Riley, available online at http://www.csus.edu/indiv/r/rileymt/ (updated 2011)].

Pizzamiglio, Pierluigi, *Gerardo da Cremona*, Cremona: Biblioteca statale e liberia civica, 1992.

Préaud, Maxime, *Les méthodes de travail d'un astrologue du XVe siècle. Conrad Heingarter*, thèse de l'École des chartes, Paris, 1969.

Préaud, Maxime, *Les astrologues à la fin du Moyen Âge*, Paris: J. C. Lattès, 1984.

Rinaldi, Michele, 'Pontano, Traepezunzio ed il Graecus Interpres del *Centiloquio* pseudotolemaico', *Atti della Accademia Pontaniana* 48 (1999), pp. 126–71.

Rinaldi, Michele, 'L'inedita *Expositio super Centiloquio Ptholomei* in Lorenzo Bonincontri', *MHNH* 15 (2015), pp. 271–80.

Ritter, Hellmut, *Pseudo-Mağrīṭī, Das Ziel des Weisen*, Leipzig/Berlin: Teubner, 1933.

Robbins, Frank E., *Ptolemy. Tetrabiblos*, Cambridge (Mass.): Harvard University Press, 1940.

Samsó, Julio, 'El procés de la transmissió científica al nord-est de la península Ibèrica al segle XII. Els textos llatins', in Juan Vernet and Ramon Parés (eds), *La ciència en la història dels països Catalans*, Valencia: Universitat de Valencia, 2004, vol. I, pp. 270–86.

Santoyo, Julio-César, 'El Normano Hugo de Cintheaux (*Hugo Sanctelliensis*), traductor en Tarazona (CA. 1145)', in Constance Carta, Sarah Finci and Dora Mancheva (eds), *Antes se agotan la mano y la pluma que su historia. Magis deficit manus et calamus quam eius hystoria. Homenaje a Carlos Alvar*, San Millán de la Cogolla: Cilengua, 2016, vol. I, pp. 341–57.

Schuba, Ludwig, *Die Quadriviums-Handschriften der Codices Palatini Latini in der Vatikanischen Bibliothek*, Wiesbaden: Dr Ludwig Reichert, 1992.

Sela, Shlomo, *Abraham Ibn Ezra on Nativities and Continuous Horoscopy. A Parallel Hebrew-English Critical Edition of the Book of Nativities and the Book of Revolution*, Leiden/Boston: Brill, 2013.

Sela, Shlomo, 'Calculating Birth: Abraham Ibn Ezra's Role in the Creation and Diffusion of the *Trutina Hermetis*', in *Pregnancy and Childbirth in the Premodern World*, ed. Constanza Gislon Dopfel, Alessandra Foscati and Charles Burnett, Turnhout: Brepols, 2019, pp. 79-106.

Sezgin, Fuat, *Geschichte des arabischen Schrifttums*, vol. VII: *Astrologie – Meteorologie und Verwandtes bis ca. 430 H.*, Leiden: Brill, 1979.

Steel, Carlos, Steven Vanden Broecke, David Juste, and Shlomo Sela, *The Astrological Autobiography of a Medieval Philosopher. Henry Bate's Nativitas (1280–81)*, Leuven: Leuven University Press, 2018.

Thorndike, Lynn, 'Pre-Copernician Astronomical Activity', *Proceedings of the American Philosophocal Society* 94 (1950), pp. 321–26.

Val Naval, Paula, *Estudio, edición crítica, traducción y comentario de la* Summa supra phisonomiam *de Guillermo de Aragón* (*c. 1300*), PhD dissertation, University of Zaragoza, 2006.

Weijers, Olga, *Le travail intellectuel à la Faculté des arts de Paris. Textes et maîtres* (*ca. 1200–1500*), fasc. III, Turnhout: Brepols, 1998.

Regiomontanus *versus* George of Trebizond on Planetary Order, Distances, and Orbs (*Almagest* 9.1)*

Michael H. Shank

It is a remarkable but largely ignored fact that the generation before Copernicus witnessed a protracted controversy about the understanding of Ptolemy's *Almagest*. During the last nine months of 1451, the rhetorician George of Trebizond completed two astronomical works in Rome. Using a Greek manuscript from Cardinal Bessarion's library, he made a new Latin translation of the *Almagest*, which he also commented.[1] Cast as a defense of Ptolemy against his critics, the 300-folio *Commentaria ad Magnam Compositionem* generated strong reactions from Bessarion. During the next two decades, the cardinal not only rebuked its contents, but also commissioned new works to replace it and expressed outrage at the uses to which George put it.[2] Bessarion's search for an alternative exposition of Ptolemy stimulated two other major works. The best-known is the *Epitome of the Almagest* that he commissioned from the Vienna-trained astronomer George Peuerbach (1423–61) and that the latter's colleague Johannes Regiomontanus (1436–76) completed by *c.* 1462. Regiomontanus also wrote a long attack on George's commentary, the *Defensio Theonis contra Georgium Trapezuntium*, which remains mostly unexplored. The primary focus of this paper is the *Defensio*'s striking critique of George's treatment of the order of the planets while commenting on *Almagest* 9.1.

Behind the question of planetary order lies the more fundamental one: what is the *ratio* of the cosmos, its organizing principle? Here the criteria and computations of mathematical astronomy contribute data as well as food for thought. Thus, an order based on the classification of mean motions seems very promising from the fixed stars to Mars, but it offers no help in distinguishing the relative distances of the Sun, Venus, and Mercury, which all share the Sun's mean motion. For astronomers who cared about the founda-

* For their critical readings and suggestions, I thank Bernard Goldstein, Dag Nikolaus Hasse, Nicholas Jacobson (also for several diagrams), Richard L. Kremer, Ronald L. Numbers, Noel M. Swerdlow, Scott Trigg, and Carol J. Troyer-Shank.

[1] For background, see Monfasani, *George of Trebizond*, pp. 71, 73–75; Monfasani, *Collectanea Trapezuntiana*, pp. 671–87, esp. 672.

[2] Monfasani, *Collectanea Trapezuntiana*, pp. 672, 676.

Ptolemy's Science of the Stars in the Middle Ages, ed. by David Juste, Benno van Dalen, Dag Nikolaus Hasse and Charles Burnett, PALS 1 (Turnhout, 2020), pp. 305–386

10.1484/M.PALS-EB.5.120184

tions of their subject, the order of these three bodies was a genuine unsolved problem.[3]

This wrangling about planetary order in the most acrimonious astronomical controversy before Copernicus's birth amply justifies the inquiry that follows. In addition, recent debates about the emergence of Copernican heliocentrism against the unfolding of fifteenth-sixteenth century astronomy and astrology have heightened interest in the historical puzzle of planetary order.[4] This chapter adds substantive new content and context to these discussions. In particular, it shows Regiomontanus challenging such staples of traditional astronomy as the project of computing cosmic distances, the order of the inferior planets, and the principle of uniform motion.

1. Introducing the controversy and its background

George's translation of and commentary on the *Almagest* were not disinterested. He had hoped they would generate new patronage from Pope Nicholas V, for whom he had successfully completed translations of non-mathematical Greek works. To assist George, Cardinal Bessarion had not only lent him one of his Greek *Almagest* manuscripts, but also advised him to use as a guide Theon of Alexandria's commentary on Ptolemy's book. To Bessarion's dismay, however, George attacked Theon in his *Commentaria*. George's opponents would later accuse him of plagiarizing the parts of Theon that he liked, and of incompetently criticizing those that he did not understand.[5] To evaluate the *Almagest* commentary, Pope Nicholas V consulted the Augustinian canon Jacobus Cremonensis (Jacopo di San Cassiano), a capable mathematician and translator of Archimedes, who returned the manuscript bristling with little sheets of paper marking its errors. When the pope rejected the *Commentaria*, George blamed that outcome on the collusion of Bessarion with Giovanni

[3] Simplicius's *Commentary on De caelo* quoted Geminus to emphasize that the order of the celestial bodies was a central problem of astronomy (2.2) and later devoted attention to it (2.10). See Duhem, *To Save the Phenomena*, pp. 10–11; Bowen, *Simplicius on the Planets*, pp. 97–110.

[4] Swerdlow, 'The Derivation and First Draft', esp. pp. 425–26; Goldstein, 'Copernicus and the Origin'; Westman, *The Copernican Question*, ch. 1–4; Swerdlow, 'Copernicus and Astrology'; Shank, 'Made to Order' (*Isis* 105 (2014), pp. 167–76), curiously followed immediately by Westman's reply (pp. 177–84) and Shank's rejoinder (pp. 185–87).

[5] This was a standard charge in the circle of Bessarion. The cardinal's protégé, Archbishop Niccolò Perotti, wrote a *Refutatio deliramentorum Georgii Trapezuntii cretensis* that accuses George of having deceived Pope Nicholas V and Mehmed II by dedicating to them his plagiarisms of Theon (Mohler, *Kardinal Bessarion*, vol. III, p. 366, lines 33–37; pp. 368–69, lines 31–42 and 1–9).

Tortelli, a scholar of Greek close to the papal curia and eventually the first Vatican librarian.[6]

In the next two decades, George of Trebizond and Cardinal Bessarion feuded with increasing bitterness. Their fight about the merits of the philosophies of Aristotle and Plato is well known,[7] but the place of the *Almagest* in the tensions between them is often overlooked even though this controversy had profound consequences for the history of astronomy. It was during Cardinal Bessarion's long diplomatic visit to Vienna in 1460–1461 that he had asked the Viennese university master Peuerbach to write a summary of the *Almagest*. His goal was a competent alternative to George of Trebizond's unsatisfactory commentary of 1451, the only Latin work that purported to explain the entire *Almagest*. Peuerbach was only half-way through the project when he died in April 1461, but he had laid the groundwork of the *Epitome*. The burden of completing it fell on his young collaborator Regiomontanus (1436–1476). When Bessarion returned to Italy in September 1461, Regiomontanus left the university to join the cardinal's retinue. By *c.* 1462, he had completed the second half (and presumably also edited the first half) of the *Epitome*. The two earliest manuscripts of the work belonged to Bessarion (they are still in his library, now in the Biblioteca Nazionale Marciana, Venice).[8] First printed in 1496, the *Epitome* has been called the 'finest text book of Ptolemaic astronomy ever written'.[9]

Regiomontanus's preface to Bessarion succinctly notes the link between the *Epitome* and the controversy with George of Trebizond.[10] For too long, however, these remarks have seemed a trivial detail with little bearing on the book's content. At first glance, this is not surprising. Indeed, like the anonymous thirteenth-century *Almagestum parvum*, Peuerbach had begun to recast the *Epitome of the Almagest* in a Euclidean vein, with spare geometrical propositions followed by proofs.[11] Regiomontanus completed it in this format, which by convention discourages digression and suppresses affect, leaving timeless proofs untouched by the cares of the world. Whatever its merits, the sobriety of the

[6] d'Alessandro and Napolitani, *Archimede latino*, pp. 70–72; Monfasani, *George of Trebizond*, pp. 106–09.

[7] Hankins, *Plato in the Renaissance*, vol. I, pp. 161–263; Kraye, 'The Philosophy of the Italian Renaissance', esp. 30–33.

[8] Codd. Marc. Lat. 328 and 329 (colloc. 1760 and 1843, respectively); see Labowsky, *Bessarion's Library*, pp. 211, 232, 450.

[9] Swerdlow, 'The Derivation and First Draft', pp. 425–26. The accolade predates Pedersen's *A Survey of the Almagest*, Neugebauer's *A History of Ancient Mathematical Astronomy*, and the revised edition of Pedersen by Alexander Jones.

[10] Schmeidler, *Joannis Regiomontani opera collectanea*, pp. 59–61 (*Epytoma*, fols a2r–a3r).

[11] Swerdlow, 'The Derivation and First Draft', p. 425; and Zepeda, 'Euclidization in the *Almagestum parvum*', esp. 74–75.

Epitome has effectively obscured the sharply polemical context from which it sprang.

1.1. Presenting the *Defensio*

One important key to reading the *Epitome* in context and to understanding Regiomontanus's tacit concerns soon after he finished it is his *Defensio Theonis contra Georgium Trapezuntium*. Its survival in one complete manuscript (and a fragment) was not what Regiomontanus intended. Ca. 1474, he had listed the *Defensio* among the more than two dozen editions that he planned for his press. His death in 1476 left that promise unfulfilled. The lack of circulation of the *Defensio* since the late fifteenth century in no way detracts from either its interest or its significance, however.[12]

Although the *Defensio* touches every book of the *Almagest*, its treatment of the material is unsystematic. Unlike the *Epitome*, it focuses combatively only on the errors in George's commentary, thus giving Regiomontanus free rein to express his anger and reveal his assumptions. Since the *Defensio* was still unfinished when Regiomontanus moved to Hungary sometime between 1465 and 1467, he left Bessarion a partial copy that was long considered lost, but David Juste found some parts of it in 2017.[13] The only complete copy of the *Defensio* so far is Regiomontanus's autograph (manuscript IV-1–935 of the Archive of the Russian Academy of Sciences in St Petersburg, Russia).[14] Its nearly 300 folios are often messy, peppered with deletions, addenda, marginalia, and stylistic corrections. Since the text ends well into book 13, it seems to be complete. When Regiomontanus announced his intention to publish it, the *Defensio* was in principle ready for *his* press, if not any press. Typesetting the book from this manuscript would have been frustrating, but not impossible: since the author was also the boss, his intentions were both accessible and enforcible.[15]

[12] See Kremer, 'Text to Trophy'; Zinner, *Leben und Wirken*, pp. 262–64.

[13] Although listed in the inventories of Bessarion's library for 1474 and 1543, it eventually became separated from most of the cardinal's other books (now in the Marciana); Labowski, *Bessarion's Library and the Biblioteca Marciana*, pp. 238, 324. In Escorial, RBSL, d.II.5, Juste discovered excerpts of the *Defensio* in Regiomontanus's hand, which probably belong to this early version; see Juste, 'MS Escorial, Real Biblioteca del Monasterio de San Lorenzo, d.II.5'.

[14] Prior to Bessarion's bequest of his library to Venice (the beginnings of the Biblioteca Nazionale Marciana), the 1468/69 inventory listed 'Corruptio Theonis per Trapeziuntium [*sic*] et tres quinterniones Ioannis contra eum in papiro, non ligatus', clearly a partial copy of the *Defensio*. Labowsky, *Bessarion's Library*, p. 238. Many astronomical manuscripts in this library have yet to be studied carefully.

[15] The positions of some marginal corrections are not always obvious. In other instances, Regiomontanus had not settled on a final choice among one or more synonyms. Such clarifications and decisions could all have been made in-house during typesetting.

High-resolution images of the manuscript and a working transcription are now accessible electronically. Thanks to the efforts of Rich Kremer, with the support of the Russian Academy of Sciences and the U. S. National Science Foundation, the high-resolution autograph is available on the *Defensio Theonis* website at Dartmouth College (http://regio.dartmouth.edu, the source for all citations of the *Defensio* below). My accompanying preliminary transcription, originally more than 900 typescript pages, is a work in progress from which I continue to purge errors.

George of Trebizond's *Commentaria*, the *Epitome*, and the *Defensio* all deserve critical editions, to say nothing of translations and comprehensive study. Together, they constitute hundreds of pages of critical (and uncritical) thinking about Ptolemy's *Almagest*. They would deserve attention if they had been written in any era, arguably much more so in the generation before Copernicus. Current generalizations about fifteenth-century Latin astronomy should be treated with much caution since they take almost no account of these contested explanations, interpretations, and criticisms of *the* fundamental work of ancient mathematical astronomy between 1450 and 1475.

In its day, the astronomical feud between Bessarion and Trebizond was so well known that it outlived the principals, not least because of its geopolitical repercussions. Bessarion discovered in 1467 that George had written secret letters dedicating his *Almagest* commentary to Mehmed II, the conqueror of Constantinople.[16] The two opponents died in 1472, but they had proxies. George's son Andreas offered new dedications of the *Commentaria* to Pope Sixtus IV, whereas Regiomontanus was planning to publish the *Defensio*. After Regiomontanus's own death in 1476, the acrimony of the famous controversy spawned the rumor (now discredited) that George of Trebizond's sons had poisoned the astronomer.[17] The content of the controversy is, however, only beginning to emerge. Whereas George's *Commentaria* was reasonably accessible (some 12 manuscripts survive), the *Defensio* must have had a very limited readership indeed. Parts of it came into the hands of Regiomontanus's acquaintance Hartmann Schedel, who copied the dedication and opening excerpts.[18] Regiomontanus's unique complete manuscript, which he had finished in Nuremberg, passed to his associate Bernard Walther with his *Nachlass*. Walther reportedly guarded this material closely until his death (1504). After more than two centuries in a secret closet, the manuscript was rediscovered, privately purchased, and given to Czar Alexander I in 1805. From St Petersburg, the manuscript traveled to Moscow, then back to the Pulkovo Observatory near St Petersburg.[19]

[16] Monfasani, *George of Trebizond*, pp. 185–94; Shank, 'The *Almagest*, Politics, and Apocalypticism' esp. 57–63; and the literature cited therein.

[17] Zinner, *Leben und Wirken*, pp. 238–39.

[18] Monfasani, *Collectanea Trapezuntiana*, p. 671.

[19] Zinner, *Leben und Wirken*, p. 264; Kremer, 'Text to Trophy'.

Its existence and whereabouts were finally confirmed in 1958. The manuscript is now preserved in the Archive of the Russian Academy of Sciences, St Petersburg branch.

The significance of the *Defensio* lies in the window it unexpectedly opens on what Regiomontanus thought, assumed, and loathed. In addition to his specific comments on the *Almagest* itself, his polemic brings out issues that he rarely or never addresses in his formal writings. To date, these include unguarded evaluations of Ptolemy's program as he understood it, and scattered natural philosophical comments and assumptions, including his conception of the astronomer's role.[20] His surprising views about planetary order, cosmic dimensions, and astronomical modeling appear below.

Importantly, the same controversy and antagonisms underlie both the *Epitome* and the *Defensio*, as Regiomontanus effectively became the cardinal's attack dog in matters astronomical.[21] Thus the *Defensio*, written during the last decade of his life, articulates some of the more enduring assumptions that tacitly guided the writing of the *Epitome*, completed by 1462. Even at this early stage of research, George's commentary (to which Regiomontanus had access while finishing the *Epitome*) and the arguments of the *Defensio* can help to explain puzzles in the *Epitome*, including oddities of both omission and commission.[22] It would, however, be a mistake to congeal Regiomontanus's evolving views into one systematic whole. Indeed, the analysis below documents a momentous shift, certainly for polemical purposes but perhaps seriously, away from the axiom of uniform circular motion.

This astronomical controversy between George of Trebizond and Cardinal Bessarion was but one facet of drawn-out political, philosophical, theological disputes between the two men. At Bessarion's behest, Regiomontanus's attack on George in the *Defensio* eventually served to undermine George of Trebizond before King Matthias Corvinus of Hungary, another potential patron to whom George sought to dedicate his commentary. No doubt with Bessarion's blessing, Regiomontanus moved to Hungary and dedicated the *Defensio* to the

[20] Shank, 'Regiomontanus on Ptolemy'; Shank, 'Regiomontanus as a Physical Astronomer'.

[21] The attacks are overwhelmingly directed at the *Commentaria*, occasionally addressing George's translation of the *Almagest* (e.g., *Defensio*, fol. 245r-v).

[22] One example is the *Epitome*'s handling of the beginning of *Almagest* 12, the foundation of Copernicus's mathematical bridge to heliocentrism. In *Epitome* 12.1–2, Regiomontanus quietly proves the equivalence of the eccentric and epicyclic models for the second anomaly for both the superior and the inferior planets, without noting that Ptolemy had declared it impossible for the inferior planets (*Almagest* 12.1; Swerdlow, 'The Derivation and First Draft', pp. 471–75). In the *Defensio*, however, Regiomontanus claimed that the equivalence for all five planets represented Ptolemy's own view, and berated George of Trebizond's correct summary of Ptolemy's error. See Shank, 'Regiomontanus as a Physical Astronomer', esp. pp. 336–41.

Hungarian king long before he had finished it.[23] When he left Hungary for Nuremberg in 1471, however, the work was still not complete. Before he had finished his autograph in Nuremberg in the mid-1470s, his opponent and his patron both died. By then, Regiomontanus had taken full ownership of the attack and planned to publish it himself.

This chapter highlights but one short passage from the *Defensio*. After surveying Ptolemy's discussion of planetary order in *Almagest* 9.1, it focuses on the surprising responses of George of Trebizond and Regiomontanus to that text.

1.2. Planetary order in *Almagest*, book 9, ch. 1

Although several works offer excellent summaries of *Almagest* 9.1,[24] I quote the passage in full to emphasize its narrow scope, in sharp contrast to that of George of Trebizond's remarks and Regiomontanus's subsequent critique. Most of the *Almagest*'s rare references to planetary spheres appear in the opening section of book 9. Referring to their order, Ptolemy states:

> ...almost all the foremost astronomers agree that all the spheres are closer to the earth than that of the fixed stars, and farther from the earth than that of the moon, and that those of the three [outer planets] are farther from the earth than those of the other [two] and the sun, Saturn's being greatest, Jupiter's the next in order towards the earth, and Mars' below that. But concerning the spheres of Venus and Mercury, we see that they are placed below the Sun's by the more ancient astronomers, but by some of their successors these too are placed above [the Sun's], for the reason that the Sun has never been obscured by them [Venus and Mercury] either. To us, however, such a criterion seems to have an element of uncertainty, since it is possible that some planets might indeed be below the Sun, but nevertheless not always be in one of the planes through the Sun and our viewpoint, but in another [plane], and hence might not be seen passing in front of it, just as in the case of the Moon, when it passes below [the Sun] at conjunction, no obscuration results in most cases.[25]

Ptolemy reports both a consensus about the extreme spheres (those of the fixed stars and the Moon), and a near-consensus about the locations of Saturn, Jupiter, and Mars immediately below the fixed stars and above the Sun (henceforth the 'superior' planets). He orders the spheres above the Sun explicitly by size (which the distances measure), and implicitly by velocity (why else make the sphere of Saturn the greatest after that of the fixed stars?).

For the spheres of Venus and Mercury, he reports no consensus about their order: the older astronomers placed them below the Sun (no justification is

[23] References to King Mathias appear in books 3 and 6 (138v, 139r–v, 194r, 240r, 301r) in particular. My transcription of the draft dedication is presently labeled Preface (fols 37r–39r; http://regio.dartmouth.edu/diplomatic/00.html).

[24] The fundamental work remains Swerdlow, *Ptolemy's Theory of the Distances*; summarized in Van Helden, *Measuring the Universe*, chapters 3 and 4.

[25] Toomer, *Ptolemy's Almagest*, p. 419.

given), whereas some later astronomers placed them above the Sun, citing the failure to observe what we now call transits (a planet's apparent passage across the solar disk).[26] Ptolemy dismisses this absence of evidence as unpersuasive, however. Just as solar eclipses do not occur at every New Moon, the much smaller Mercury and Venus may pass undetected below or above the Sun (the focus of latitude theory in *Almagest* 13).[27]

After rejecting the absence of evidence for transits as a reason for placing the spheres of Venus and Mercury above the Sun, Ptolemy notes the absence of any planetary parallax, the sole means for determining distances with certainty, and falls back on the plausibility of the traditional order, which he does not specify:

> And since there is no other way, either, to make progress in our knowledge of this matter, since none of the stars has a noticeable parallax (which is the only phenomenon from which the distances can be derived), the order assumed by the older [astronomers] appears the more plausible. For, by putting the Sun in the middle, it is more in accordance with the nature [of the bodies] in thus separating those which reach all possible distances from the Sun and those which do not do so, but always move in its vicinity; provided only that it does not remove the latter close enough to the earth that there can result a parallax of any size.[28]

Building on the (unproven) near-consensus about the location of Saturn, Jupiter, and Mars above the Sun (which he treats as an assumption), Ptolemy advances a plausibility argument for placing the Sun's sphere 'in the middle' (i.e., 3 planetary spheres above it, 3 below). In this position, the Sun's sphere separates the spheres of the 3 planets observed in opposition from those of the 2 that never are (no farther from the Sun than roughly 28° and 47°, respectively). Ptolemy here does not specifically order Venus and Mercury, implicitly following the 'older' astronomers (Mercury above the Moon, then Venus). Ptolemy qualifies this order as *physikōteron*, 'more in accordance with the nature [of the bodies]' (as Toomer expands it) or generally 'more in accordance with nature'.[29] Although its parallax proves that the Moon is the closest to the Earth, it obviously is seen in opposition, like the 3 superior planets. In short, the criterion for placing the Sun 'in the middle' is awkward at best.[30]

[26] Later sources ascribe both alternatives to Archimedes; Neugebauer, *A History of Ancient Mathematical Astronomy*, pp. 647, 691–92.

[27] In the *Planetary Hypotheses*, Ptolemy also emphasizes the uncertain order of every planet but the Moon. He treats the subsolar positions of Venus and Mercury as assumptions and gives optical reasons for our inability to perceive solar transits of Venus and Mercury: some partial solar eclipses are not noticed even though portions of the Moon far larger than the apparent planets obscure it. Goldstein, *The Arabic Version*, pp. 6–7.

[28] Toomer, *Ptolemy's Almagest*, pp. 419–20.

[29] Heiberg, *Syntaxis mathematica*, vol. II, p. 207 line 18.

[30] Jābir ibn Aflaḥ (Geber) would also point this out, arguing that the Moon should then be grouped with the superior planets; Lorch, 'The Astronomy of Jābir ibn Aflaḥ', esp. p. 98.

The final phrase in the quotation adds a notable qualification: it rules out a position for Mercury or Venus that would give either planet a parallax, which the Moon *does* display. This condition seems to exclude the lowest planet's contiguity to the Moon's sphere — certainly at lower conjunction (unobservable), but arguably even at observable elongations, as Ptolemy apparently looked for a parallax when Mercury was visible.[31]

The astrological *Tetrabiblos* — Ptolemy's second book about the heavens — was available in Latin; unlike the *Almagest*, it explicitly placed Mercury 'next above the sphere of the Moon' (*Tetrabiblos* 1.4). Consistent with this picture, Ptolemy explicitly identified the seven periods of human life with the seven planetary 'spheres' in the following order: Moon-Mercury-Venus-Sun-Mars-Jupiter-Saturn (*Tetrabiblos* 3.10). In short, this list is not random, but represents an ascending order.[32]

Later, in his *Planetary Hypotheses*, Ptolemy laid out his nesting hypothesis, discussed planetary order, and computed planetary distances. In doing so, he abandoned the restriction on locating the 'inferior' planets in the zone of observable parallax. He used the mathematical models of the *Almagest* to compute minimum planetary distances from the Earth, constructed a cosmology of nested spheres, and advanced suggestions for building a 3-dimensional model of the universe. This construction made Mercury's perigee equal at minimum to the Moon's apogee, even though Mercury had no detectable parallax and the Moon did. He focused his doubts on the place of the Sun in the sequence.[33]

1.3. After Ptolemy[34]

Planetary order was debated in late antique Latin astronomy (notably the heliocentric arrangement of Mercury and Venus in Martianus Capella) and even more so in the Carolingian era, when scholars had to cope with the sketchy

[31] Assuming the Ptolemaic order, Jābir ibn Aflaḥ went to the trouble of computing the presumed parallaxes of Venus and Mercury at greatest elongation. At 6′ and 4′, respectively, they were more than the 2′51″ Ptolemy had computed for the Sun. This was why Jābir rejected the Ptolemaic order and placed Venus and Mercury above the Sun; see Lorch, 'The Astronomy of Jābir ibn Aflaḥ', p. 97.

[32] Robbins, *Ptolemy. Tetrabiblos*, pp. 39, 443–47. Although Neugebauer (*A History of Ancient Mathematical Astronomy*, p. 690) warns about the hazards of identifying a list of planets in a text with an ordering in space, Ptolemy's lists — sometimes with, sometimes without the luminaries — are consistent (e.g., *Tetrabiblos* 1.4, 2.8, 3.11, 3.13, 4.9).

[33] Indeed, after conceding that his values might merely be minimal ones and making the correctness of his distances hypothetical, he inferred that parallaxes should be observable from Mercury to Mars; Goldstein, *The Arabic Version*, pp. 6–7, 9; Neugebauer, *A History of Ancient Mathematical Astronomy*, p. 919; summary in Van Helden, *Measuring the Universe*, pp. 21–27.

[34] See Swerdlow, *Ptolemy's Theory of the Distances*, chap. 3; Van Helden, *Measuring the Universe*, ch. 4; Lerner, *Le Monde des sphères*, vol. I, chap. 4–6.

reports of multiple arrangements they had inherited.[35] By the ninth-tenth centuries in Islamic civilization, however, the assumptions and methods behind the nesting hypothesis and planetary sizes in Ptolemy's *Planetary Hypotheses* had anonymously become almost a commonplace. They undergirded almost all later computations of planetary distances, even though their source was known to only a few scholars in Hellenistic and Islamic civilization. Authors such as Thābit ibn Qurra (d. 901), al-Farghānī (d. 870), and al-Battānī (d. 929) discussed the order and sizes of the planetary spheres, computing distances from the *Almagest*'s models, without suggesting that they knew of the *Planetary Hypotheses* or ascribing the procedures to Ptolemy.[36]

When this Arabic material was translated into Latin in the twelfth century, scholars likewise adopted (and sometimes modified) the values in it. They did so without suspecting that these had originated in Ptolemy's project of computing the dimensions of the cosmic system in the *Planetary Hypotheses*. In this way, computations of planetary spheres and cosmic dimensions found a permanent home in Latin astronomical literature, including Campanus of Novara's *Theorica planetarum*.[37]

2. Regiomontanus presents the *Commentaria* on *Almagest* 9.1

The preceding sketch may serve as the formal textual background for the fifteenth-century altercation mentioned above. The fragment discussed in this article focuses, despite its length, only on *Almagest* 9.1. There is no space to analyze the responses of George of Trebizond and Regiomontanus to all of book 9: the *Defensio* alone devotes 35 folios (fols 151–86) to George's treatment of that book. The brevity of *Almagest* 9.1 makes conspicuous how much non-*Almagest* material George brought to bear in commenting on this passage, from Aristotelian cosmological material through the covert legacy of the *Planetary Hypotheses* to thirteenth-century Latin astronomical literature.

[35] For example, Eastwood, *Ordering the Heavens*, ch. 2.3–4, 3.3, passim; for background, see Obrist, *La cosmologie médiévale*, vol. I, pp. 72–76, 231–38, 253–55. Buridan mentions as probable an idea that resolves a tension with Averroes, namely that Mercury, Venus, and the Sun 'are fixed in the same sphere, although they have different epicycles and eccentrics within it'. Grant, *Planets, Stars and Orbs*, p. 313.

[36] Thābit describes the nesting hypothesis and assigns figures to the sizes of the orbs in his *De his que indigent expositione antequam legatur Almagesti*. Pseudo-Thābit's *De quantitatibus stellarum* is derived from al-Farghānī and, like him, presents the computation of planetary distances not as Ptolemy's, but as an extension of Ptolemy's procedure for the luminaries. See Carmody, *The Astronomical Works of Thābit b. Qurra*, pp. 128–30, 133, 136–37; pp. 145–48, esp. 146–47. Al-Birūnī was aware of the *Planetary Hypotheses*; see Nallino's discussion in *Al-Battānī sive Albatenii opus astronomicum*, vol. I, pp. 287–89; Swerdlow, *Ptolemy's Theory of the Distances*, pp. 137–62.

[37] Van Helden, *Measuring the Universe*, pp. 29, 31; Benjamin and Toomer, *Campanus of Novara*.

Regiomontanus's response illustrates his general approach to George of Trebizond's commentary throughout the *Defensio*. First, he typically quotes from the *Commentaria* (at length, in this instance). He then proceeds to attack specific points in the cited text. For simplicity's sake, I discuss the views of both men only from the material in the *Defensio*. Indeed, Regiomontanus quotes very reliably from George's commentary. He not only cared about such things, but also evidently had a good manuscript of that work.[38]

2.1. Preliminaries: history and the Sun-heart analogy

Regiomontanus begins by quoting the opening remarks from George's own commentary on *Almagest* 9.1 (in italics below). To this, he appends his own brief remarks (in roman).[39] Regiomontanus's response to this passage typifies the tone of the larger controversy as it articulates some of his general objections to George's stance in the commentary:

> *Plato,* he says, *places the Moon first, then the Sun; Aristotle also says this, but they had followed what the Greek mathematicians thought in their day. For myself, however, I will not hesitate to say that, on this issue, we have been sent to Ptolemy by Aristotle himself. For the man who in* <u>*De caelo et mundo*</u> *enjoins [one] to turn to those devoted to such things (since, according to him [Aristotle], the arts and the sciences are increased by additions [to them])*[40] *undoubtedly intends that one start with the later teachers of these disciplines, who added much to them.*[41]

[38] My spot-checks of the *Commentaria* in Vienna, ÖNB cod. 3106 have matched the *Defensio* quotations exactly; on that ms, see Monfasani, *Collectanea Trapezuntiana*, p. 670.

[39] George gave the first chapter of his commentary on book 9 the heading 'On the order of the globes/spheres of the Sun, Moon and other wandering stars' (*De ordine globorum solis, lune ceterarumque stellarum erraticarum*); Vienna, ÖNB, cod. 3106, 162r.

[40] This allusion resonates less with *De caelo* than with both Aristotle, *Metaphysics* 12 (1073b11–17) and Simplicius's *Commentary on Aristotle's De caelo*, which paraphrases the *Metaphysics* passage; see Mueller, *Simplicius*, pp. 44–45; and Bowen, *Simplicius on the Planets*, pp. 168–69. (I thank Noel Swerdlow and Nick Jacobson, respectively, for pointing out these parallels). See also Simplicius's commentary on *De caelo* II.10: Bowen, *Simplicius on the Planets*, p. 99, lines 10–13. In addition, Averroes's commentary on the *De caelo* explicitly appeals to Ptolemy to supplement Aristotle: see the *quaesitum quartum* in *De caelo*, book 2. Marcantonio Zimara's metacommentary on the same passage ascribes to Aristotle the view that Venus and Mercury are above the Sun, contrasting the approach of the natural philosopher to that of the astronomer/astrologer and commenting on the problems that Venus and Mercury present for the speed-distance rule; Zimara, *Aristotelis opera cum Averrois commentariis* vol. V: *De caelo...*, 136va-137va.

[41] 'Lunam', inquit, 'primum Plato deinde solem collocat. Id Aristoteles etiam dicit; sed secuti fuerant quod temporibus suis grecorum mathematici opinabantur. Ipse autem non dubitabo dicere ad Ptolemeum nos quantum ad hec pertinet ab ipso Aristotele mitti. Nam qui in libris De celo et mundo ad eos ire iubet quibus hec cure erant cum artes atque scientie additionibus colligantur secundum ipsum, is profecto posteriores etiam ad disciplinarum pro-

In this curious passage, George identifies Aristotle's view with that of Plato, characterized only by the contiguity of Moon and Sun. In *Timaeus* 38C-D, Plato had proposed the following ascending order: Moon, Sun, Venus, Mercury, omitting the order of the remaining planets. Ever since Macrobius, who associated both permutations of Venus and Mercury above the Sun with Plato or Platonists, the supra-solar position of the so-called 'inferior' planets was named that of the 'Platonists and the Egyptians'.[42] More notable is George's association of such a view with Aristotle, since the latter's genuine books (*De caelo*, *Metaphysics*, or even *Meteorology*) do not discuss planetary order explicitly. As George's reference hints, his likely source is the pseudo-Aristotelian *De mundo*, which in Latin translation sometimes circulated with the *De caelo*. Despite skepticism about its authorship in the late Middle Ages, the *De mundo* in the fifteenth century was considered genuine by such leading scholars as Bessarion, Ficino, and the two Picos.[43] To this list, George of Trebizond's name must now be added. The *De mundo* (first century) is familiar with Stoic sources and endorses the so-called 'Platonic/Egyptian' order of Moon, Sun, Venus, Mercury, etc.[44]

George's conflation of Aristotle's planetary order with Plato's is noteworthy because he believes it to be erroneous. In later years, he would express vehement opposition to Plato and strong partisanship for Aristotle. In 1451, however, George commends Aristotle for endorsing not a correct view, but a correct principle about the growth of knowledge in the arts and sciences. George implicitly turns the Philosopher's deference to specialists into an expectation of scientific obsolescence: Aristotle anticipated that his views would be revised, as indeed they were, notably by Ptolemy's 'additions'. Although George does not discuss subsequent progress, he evidently saw himself as a participant in it, witness his claims later in the commentary.

After quickly turning a compliment against George, Regiomontanus attacks the latter's inconsistency and his representation of early astronomy and its subsequent development:

> He nicely reviews as much about Aristotle, who determines the foremost men to be consulted in any faculty if something obscure should come up, as about the continuous additions of the arts and sciences, which discrete items increase every day.

fessores qui multum eis addiderunt proficisci vult'. George of Trebizond, quoted in Regiomontanus, *Defensio*, fol. 151v.

[42] Stahl, *Roman Science*, pp. 158–59.

[43] Pseudo-Aristotle, *On the Cosmos*, ed. and transl. in Johan C. Thom (ed.), *Cosmic Order*, pp. 23–26; Burri, 'The Geography of *De mundo*', p. 96; Kraye, 'Disputes over the Authorship', esp. pp. 182–83.

[44] See [Pseudo-Aristotle], *De mundo*, 392a20–31 (more ambiguously at 399a1–12), transl. Thom *et al.* in Thom, *Cosmic Order*, pp. 22–25, 46–47; see also Maguire, 'The Sources of Pseudo-Aristotle's *De mundo*', esp. pp. 121–22.

> But this snappish man is utterly unaware of his fickleness. Throughout, he attacks Theon harshly as outside the mathematicians' ranks, often accuses other Greeks of ignorance, and does not spare the Arabs, especially the Spaniard Geber, whom he falsely accuses of inexperience. But he now thinks that one should start with the later professors of these disciplines, who added much to the latter on Aristotle's advice; and he now assails all the more recent ones as basically lazy and completely mediocre; and finally, he brags that, with himself as the laughable exception, no one has [ever] existed from whom Ptolemy has shone forth and who understood his teaching.[45]

Here Regiomontanus ridicules George for his contradictory beliefs about progress in astronomy. George recommends starting from the masters of astronomy even though his commentary has berated leading specialists in the subject (Theon, Geber) and dismissed everyone but himself. Regiomontanus goes on to list key contributors to astronomy whom George omits:

> He certainly disregards or conceals what has been discovered about the extent to which the Greeks' talents were increased by these Chaldean discoveries, and how much the Arabs became illustrious in the science of the stars: their intervention in all the mathematical sciences has abundantly enriched the Latin world. Is it not the case that, in our era, nearly everything that Roman [= Latin] mathematicians have usually taught on this subject flows from the Arabs, albeit mediated by poor and crude translation? Finally, I cheerfully pass in silence over the Latin men who, very illustrious in mathematics and abstaining from almost everything else, brought to light so many things from the most profound treasury of nature, lest we, in urging praise on our own, seem immodestly to give ourselves credit for them [= their discoveries].[46]

[45] 'Hec bene quidem recenset tam de Aristotele qui primores in quacumque facultate viros consulendos esse arbitratur si quid paulo obscurius occurrat, quam de continuis artium scientiarumque additamentis quibus singule queque dietim coalescunt. Verum suam mordax homo haudquaquam persentiscit. Nam qui Theonem non parum mathematicis addentem extreme insectatur, aliosque grecos ignorantie sepenumero accusat, qui que arabibus et presertim Gebro hispalensi non parcit, sed imperitiam falso obiectat nunc ad posteros disciplinarum professores qui multum eis addiderunt monitu aristotelico proficiscendum censet, nunc contra iuniores universos quasi inertes ac prorsus negligendos lacessit et postremo neminem ex quo Ptolemeus claruit, se uno dempto, extitisse qui doctrinam eius perceperit ridiculo gloriatur'; Regiomontanus, *Defensio*, fol. 151r.

[46] 'ignorans utique aut rescire dissimulans quantopere grecorum ingeniis adaucte sint ille inventiones chaldaice; quantum que arabes in siderali presertim disciplina claruerint, quorum interventu omnibus etiam mathematicis studiis latinitas abunde locupletata est. Nempe quicquid ad hec ferme nostra secula romani didicere mathematici, ab arabibus pene totum quamvis scabro et exili admodum interpretamento profluxit. Qui demum latini viri per quam clari in mathematicis plerisque omnibus evaserint/evituerint quantasque res penitissimo e thesauro nature rompserint, silentio preterire libet ne gentiles nostros laudaturi/laudando nobisipsis non nihil paulo arrogantius/immodestius tribuere videamur'. *Defensio*, fol. 151r-v.

In this partial echo of his 1464 inaugural lecture on al-Farghānī at Padua, Regiomontanus approaches astronomy with a consciously historical point of view and an implicit notion of progress.[47] The members of four peoples — we would say 'cultures' — have contributed to the enterprise: Chaldeans, Greeks, Arabs, and Latins.[48] His praise of these predecessors is fulsome but mostly anonymous (Theon and Geber appear, but not al-Battānī, whom he greatly admired). Also, he modestly refrains from naming the Latin individuals, among whom he surely included Peuerbach, himself, and perhaps (despite earlier criticisms) Bianchini (mentioned in the margin).[49] In short, the tradition is improving thanks to the 'additions' to it. By implication, Ptolemy is not the final word.

In short, George here has inconsistently diverged from his self-aggrandizing program of belittling most of traditional astronomy:

> This hopeless manikin indiscriminately tries to disparage, ruin, destroy, and demolish completely so many men of this sort, mocking them as blind and deluded wasters of time and labor, so that he might persuade some stupid and idiotic neophytes in astronomy that he is the sole commentator of Ptolemy, of whom he has not even taken the first bite, although he implies that he has seen into the order of the spheres more keenly than Ptolemy himself.[50]

These opening paragraphs of book 9 illustrate Regiomontanus's polemical strategy throughout the *Defensio*. He characterizes George as denigrating post-Ptolemaic astronomy and touting not only his own mastery of the *Almagest*, but also his improvement upon it by allegedly demonstrating the planetary order. Regiomontanus emphasizes this last point, so that George later will seem to fall from an even greater height.

When George finally turns to the content of *Almagest* 9 itself, he embroiders on Ptolemy's minimalist discussion of the order of the planetary spheres

[47] 'Oratio Johannis de Monteregio...', in Schmeidler, *Joannis Regiomontani opera collectanea*, pp. 43–53. These remarks strongly suggest, as does his Paduan Oration, that Regiomontanus did not believe in cycles, but in progress, pace Byrne, 'A Humanist History of Mathematics'; and Špelda, 'From Closed Cycles to Infinite Progress', esp. p. 210.

[48] Aristotle had referred to Chaldeans in *Metaphysics* 12.8 and *De caelo* 2.12, also discussed by both Simplicius and Averroes in their commentaries on *De caelo*. See respectively Bowen, *Simplicius on the Planets*, pp. 121, 225, and Zimara, *Aristotelis opera cum Averrois commentariis*, vol. V, 138vb.

[49] In the top margin of 151v, Regiomontanus has written the intriguing note: 'Quando de figura sideris scribes, Blanchini mentionem si videtur facias'.

[50] 'Tot tantosque viros perinde quasi cecos et oleum ut aiunt impensamque ludificatos homuncio futilis passim floccifacere pessundare, obterere ac prorsus abolere conatur ut stolidis quibusdam ac stupidis astronomie persuadeat tirunculis sese unicum esse Ptolemei illustratorem, cuius ne prima quidem ipse libamenta gustavit; quamvis ordinem spherarum acutius se animadvertisse subostentet quam Ptolemeum ipsum'. *Defensio*, fol. 151v.

with a long paraphrase that goes far beyond the text. Regiomontanus introduces the quotation with a rebuke:

> To protect the truth faithfully and to defend men under attack while we censure the expositor's conceited little [quest for] glory,[51] which he now hopes to attain with his very loose talk, let his words be broadcast for all to see: *Therefore it is necessary,* he says, *that the part/organ of the universe that is proportioned to the heart and is, as it were, the seat of its soul also hold the middle position of the same universe. Therefore, since the lunar parallax can demonstrate that, of all the stars that, owing to the magnitude of their distance, have either no parallax or a minimal one, the Moon seems to be close to the center and middle of the Earth, two other planets must therefore be sought that are more plausibly located below the Sun, so that three are below it, three above it; for* [the Sun] *itself is in the middle of all things.*[52]

Straying from the *Almagest*'s content, George's arguments start with controversial natural-philosophical assumptions, which he nevertheless calls necessary. The macrocosm-microcosm analogy underlies his assumption of a 'proportionality' between the position of the heart — the seat of the (presumably rational) soul, implicitly located in the middle of the (presumably human) body — and the position of the universe's heart and soul, implicitly the Sun. The 'proportionality' in this analogy allegedly guarantees both the position of the Sun in the 'middle of all' and the necessity that George attributes to his conclusions. Since the Sun is in the middle and three planets are above it, three must therefore also be below it. Being unique in having a measurable parallax, the Moon is closest to the Earth. Two other planets — Mercury and Venus — therefore must join the Moon below the Sun. Only his conclusion clarifies that, by 'middle', George means 'ordinally intermediate' (4th of 7).

Since *medium* can signify 'middle', 'intermediate', 'mean', 'central', etc., George's language is ambiguous, and his argument is loose. Yet he evidently

[51] I add 'quest' to translate *gloriuncula* both to capture Regiomontanus's pejorative intention and to meet the relative clause's need for an antecedent with a positive connotation from George's point of view. Pace Edward Rosen, Copernicus clearly did not coin *gloriuncula* in translating Theophylactus Simocatta's Letter 76 from the Greek: Czartoryski, *Nicolaus Copernicus. Minor Works*, pp. 69–70, with trans. and comm. by Edward Rosen. Jean Gerson was already using the word a century earlier; see Yule, *The Statistical Study*, p. 243.

[52] 'Quo autem fidelius veritatem tueamur et lacessitos quoscumque viros defendamus simul que gloriolam expositoris taxemus quam laxiori sermonis filo impresentiarum se adipisci/adepturum sperat, verba eius in medium proferantur: Quare necesse est, inquit, ut membrum quoque mundi quod cordi proportionatur et quasi anime ipsius sedes est medium ipsius mundi locum obtineat. Cum igitur lunaris diversitatis aspectu demonstretur lunam omnium stellarum que propter distantie magnitudinem diversitatis aspectum vel non habent vel minimum habent, proximam terre centro ac medio esse pateat querendi sunt alii duo planete qui verisimilius sub sole collocantur, ut tres sub ipso, tres super ipsum; ipse vero in omnium medio sit'. George of Trebizond, quoted in Regiomontanus, *Defensio*, fol. 151v.

thinks — erroneously, as Regiomontanus will show — that he is rehearsing uncontroversial claims. George now takes up the details:

> *He [= Ptolemy] therefore selected Venus and Mercury, first, since everyone straightforwardly locates Saturn, Jupiter, and Mars above the Sun, which cannot be denied since it is perceived in their conjunctions; and second, since he thought it more natural to place below the Sun [both] Mercury and Venus, which are always carried around near the Sun; finally, since he says that the ancients, namely the Chaldeans or the Egyptians, also thought thus, although some say that Thales of Miletus also posited this order. Thus far Ptolemy.*[53]

George of Trebizond's argument for treating Mercury and Venus quickly glosses over the order of the superior planets. His cryptic reference to their uncontroversial order being seen or perceived 'in their conjunctions' presumably means the *periods* between the superior planets' conjunctions with the Sun, i.e., an order determined by their speed in the ecliptic (a reasonable interpretation of Ptolemy's truncated discussion).[54] As his last three words show, George believes he is paraphrasing Ptolemy. *Almagest* 9.1, however, mentions neither an analogy between the Sun and the heart, nor Thales, nor the Chaldeans, nor the Egyptians. Indeed, George's conflation of Chaldeans and Egyptians is odd, as most sources contrast them, claiming that the latter make the Moon and Sun the two lowest planets.[55]

Without naming anyone, Ptolemy had inclined toward an order that he associated with 'older astronomers' and qualified as 'more in accordance with nature'. This arrangement put the Sun 'in the middle' to separate the five planets by the range of their elongation from the Sun: oppositions for Saturn, Jupiter, and Mars; but no opposition for Venus and Mercury. Ptolemy had nothing to say about the heart of the universe.

[53] 'Venerem igitur et Mercurium cepit tum quia omnes simpliciter Saturnum Iovem Martemque super solem locant, nec negari hoc potest cum in coniunctionibus ipsorum perspectum sit; tum quia naturalius esse putavit Mercurium et Venerem qui prope solem semper feruntur sub ipso ponere; tum denique quia priscos ita putasse ait Chaldeos forsan aut Egyptios, quamvis nonnulli Thaleta quoque Milesium hunc ordinem posuisse dicunt. Hec Ptolemeus'. George of Trebizond, quoted in Regiomontanus, *Defensio*, fol. 151v.

[54] Campanus of Novara (on whom George seems to draw; see below) discusses in detail correlations between conjunctions with the Sun; Benjamin and Toomer, *Campanus of Novara*, pp. 302–07.

[55] The historically fanciful nomenclature of 'Chaldean' and 'Egyptian' planetary order often surfaces in the Carolingian tradition, which drew heavily on Pliny, Macrobius, Martianus Capella; see Eastwood, *Ordering the Heavens*, e.g, p. 50. Following Macrobius's *Commentary on the Dream of Scipio*, Campanus associates Plato and the Egyptians; Benjamin and Toomer, *Campanus of Novara*, pp. 333, 436. Neugebauer cautions against assuming that the order of planets in a text corresponds to the author's understanding of their spatial arrangement: Neugebauer, *A History of Ancient Mathematical Astronomy*, p. 690.

Arguing against the Pythagoreans in *De caelo* 293b5–8, Aristotle had drawn an analogy between the center of the universe and the center of the animal, emphasizing the difference between their 'natural' and geometrical centers. He did not explicitly name the heart as the natural center of the animal, but that point was thoroughly Aristotelian (see *De partibus animalium* 3.4, 665 b21–24). George surely knew this, having translated both works into Latin.[56]

2.2. 'Demonstrating' planetary order and distances in the Sun-Moon interval

In the *Commentaria*, the heart-Sun analogy precedes George's lengthy treatment and computation of the sizes of the planetary spheres, which marks yet another significant departure from the spare text of the *Almagest*: '*We can/shall*, he says, *produce a demonstration from the distances taken proportionally if we first attain some natural necessity*'.[57] This demonstrative language promises precisely what *Almagest* 9.1 denies: without known parallaxes for the bodies beyond the Moon, planetary order *cannot* be determined with certainty. George's boast is one of the 'improvements' that Regiomontanus mocked as seeing 'into the order of the spheres more keenly than Ptolemy himself'.

The 'natural necessity', by which George presumably means 'incontrovertible facts' on which to erect his demonstration, evidently includes the absolute distance of the lunar apogee:

> *In Book 5, therefore, it has already been demonstrated that the greatest distance of the Moon is 64;10 Earth radii* [*e.r.*], *whereas the distance of the Sun is approximately 1210 e.r.; the distance of the Moon he* [= *Ptolemy*] *demonstrated from the parallax angle. For the Sun, this angle, no matter where it is measured, is not perceived to vary much, and therefore he necessarily used a single distance for the entire eccentric of the Sun. For the Moon, however, the parallax angle is perceived to vary much, in relation to the distances of the Moon itself. He therefore demonstrated the various distances of the Moon by means of the variation of this angle. Therefore, he posited a unique solar distance of 1210* [*e.r.*]. *If from this you subtract 64;10* [*e.r.*] *for the Moon's apogee, approximately 1146 e.r. remain between the Sun and the Moon.*[58]

[56] Monfasani, *Collectanea Trapezuntiana*, pp. 703–07. Other possible sources of this analogy are Macrobius and Chalcidius; Grant, *Planets, Stars and Orbs*, p. 227 n. 28.

[57] 'Nos a distantiis, inquit, proportionaliter captis demonstrationem afferemus si prius naturalem quandam necessitatem attingamus'. George of Trebizond, quoted in Regiomontanus, *Defensio*, fol. 151v.

[58] 'In quinto igitur libro iam demonstratum est maximam lune distantiam esse 64 10' talium, qualis est unius semidiameter terre. Solis autem distantiam indifferenter captam 1210 earundem partium; quam distantiam lune demonstravit per angulum diversitatis aspectus. Is angulus in sole ubivis capto non multum variatur sensibiliter; et ideo unicam necessario per totum eccentricum cepit distantiam solis. In luna vero angulus diversitatis aspectuum multum variatur sensibiliter secundum ipsius lune distantias; ideo demonstrate sunt ab ipso diverse distantie lune per diversitatem huiusmodi anguli; unica ergo solaris distantia ab ipso posita est 1210: a qua, si auferas 64 10' maximam lune, relinquentur 1146 proxime inter solem et lunam

Starting from Ptolemy's procedure for calculating the distances of the luminaries in *Almagest* 5, George lays out the standard post-*Almagest* framework of assumptions for computing planetary distances (ultimately rooted in the *Planetary Hypotheses*).[59] For the distance between the lunar apogee and the Sun, he uses the *Almagest*'s values. Ptolemy had obtained 64;10 e.r. for the lunar apogee by using two ancient Babylonian lunar eclipse observations (*Almagest* 5.14). From this figure and his ignorance of annular eclipses (i.e., he made total solar eclipses at lunar apogee his limiting case), he computed the distance of the Sun to be 1210 e.r.[60] As George notes, Ptolemy could detect no solar parallax and therefore treated this distance as fixed.[61] Indeed, the *Almagest*'s planetary models focus on velocities and positions against the zodiac, not distances from the Earth, as its lunar theory's tolerance for wild variation in distance makes abundantly clear.

Significant for George's argument is that this solar distance derived from measurements and calculations is independent of Ptolemy's models for the Sun and Moon (the underlying observations are Babylonian). The difference between this (fixed) solar distance and the lunar apogee yields a gap of 1146 e.r.:

> *This is why, since everyone concedes from the phenomena that Saturn, Jupiter, and Mars are above the Sun, if we did not place Venus and Mercury between the Sun and the Moon, that space would be empty. Indeed, because of the magnitude of this space, we cannot believe that* [*its size*] *could come from the solar and lunar distances being approximate rather than precise. If therefore a vacuum is altogether impossible, it is necessary that Venus and Mercury be located between the Sun and the Moon.*[62]

partes tales, qualis est unius semidiameter terre'. George of Trebizond, quoted in Regiomontanus, *Defensio*, fols 151v–152r. For brevity's sake, I translate *maxima distantia* and *minima distantia* as 'apogee' and 'perigee', and *diversitas aspectuum* as 'parallax'.

[59] Van Helden, *Measuring the Universe*, p. 31.

[60] In the geometry of Ptolemy's procedure, small angular changes (including errors) in the apparent solar diameter translate into large variations in distance. This constraint helps to explain the great compression of the *Almagest*'s Earth-Sun distance, which is too small by a factor of 20; *Almagest*, 5.14–15; Toomer, *Ptolemy's Almagest*, pp. 252–57; Swerdlow, 'Hipparchus on the Distance of the Sun', esp. p. 294; Neugebauer, *A History of Ancient Mathematical Astronomy*, pp. 104–06, 109–10.

[61] At *Almagest* 5.11, Ptolemy doubted that the Sun had a parallax (Toomer, *Ptolemy's Almagest*, p. 244). He effectively treated its apparent diameter as a constant (Neugebauer, *A History of Ancient Mathematical Astronomy*, pp. 104, 112).

[62] 'Quare cum omnes ex iis que apparent coacti concedant Saturnum, Iovem, et Martem sole superiores esse, nisi Venerem et Mercurium inter solem et lunam collocaverimus, erit illud spacium vacuum. Non enim possumus propter magnitudinem tanti spacii suspicari quod hoc accidat quia distantie solaris atque lunaris proxime non exquisite omnino capte sunt. Si ergo impossible omnino est vacuum dari, necesse est inter solem et lunam Mercurii et Veneris situm esse'. George of Trebizond, quoted in Regiomontanus, *Defensio*, fols 151v–152r.

Without explaining how the planets' conjunctions with the Sun undergird it, George takes the consensus about the order of Saturn, Jupiter, and Mars above the Sun to be uncontroversial. The gap of 1146 e.r. between Sun and Moon, however, can be neither an error nor empty: the two remaining planets must therefore fill it. Recall that in *Almagest* 9.1, Ptolemy had neither discussed in detail the order of the superior planets, nor mentioned the need to fill a gap with the 'inferior' ones.

> *We will now consider the very same matter by means of distances, not apparent parallax. For, since the parallaxes of Venus and Mercury are effectively imperceptible and make an angle much less than perceptible, we cannot easily discover the distances by these means.*[63]

Here George distinguishes ambiguously between two different methods of computing the Moon-Sun distance, but again digresses from Ptolemy. He presents computation 'by means of distances' as an alternative to the parallax method (arguably also a 'distance' method). Used to calculate the lunar distance, the parallax technique should also work theoretically for the planet immediately above the Moon, since its perigee is assumed to be contiguous to the Moon's apogee. In fact, none of the other planets, including the Sun, exhibits naked-eye parallax. *Pace* George, Ptolemy in *Almagest* 9.1 thought that imperceptible parallaxes made measurements of distance not difficult, but impossible. This was just why he treated the order of the Sun, Mercury, and Venus as uncertain.[64]

The burden of the next section of the *Commentaria* is therefore to show that 'by means of distances', as George calls this method, one can obtain a solar distance independently of the eclipse measurement. The procedure is simply to add the sizes of all the intervening spheres, working upward from the Moon's apogee. George here follows in outline the time-honored approach from the *Planetary Hypotheses* to al-Battānī and beyond. Given the lunar apogee in earth radii and assuming its equality to the perigee of the next closest planet, one computes in sequence the absolute distances for the radial range of motion of each of these two planets using the perigee-to-apogee ratio in the *Almagest*'s relevant planetary model. If these two different approaches--from parallax (and eclipses), and 'from distances' computed from the planetary models — yield very nearly the same result for the Moon-Sun distance, the converging

[63] 'Nunc per distantias, non per apparentem diversitatem idipsum consideremus. Nam cum aspectus diversitas Veneris et Mercurii pene insensibilis sit, ac angulum multo minus sensibilem faciat, non facile possumus ab ipso distantias invenire'. George of Trebizond, quoted in Regiomontanus, *Defensio*, fol. 152r.

[64] Ptolemy changed his mind in the *Planetary Hypotheses*; see note 35.

results would seem to indicate that the distances and the assumed order are correct.[65] But with which of the two planets does one start?

> *Transposing proportionally from the distances that he [= Ptolemy] demonstrated, however, we will demonstrate that Mercury can be located, in immediately ascending [order], after the Moon, and Venus after Mercury, and then the Sun. It is therefore necessary, lest there be a vacuum, that the apogee of the Moon be the perigee of either Venus or Mercury; not Venus, however, therefore Mercury. For Mercury is recorded as having passed below Venus, and is faster than Venus. And it is necessary that, in circular motion, the inferior be faster.*[66]

George's first argument for locating Mercury below Venus is empirical. It adduces an undocumented observation of Mercury 'passing below' — i.e., presumably in front of — Venus. If true, this alleged observation removes all ambiguity about the relative orders of Mercury and Venus. This is arguably the second item of 'natural necessity' to which George has appealed above. George's unnamed source here is perhaps Simplicius's commentary on the *De caelo* II.10, which reports such an observation (also uncredited, vague, and in similar language).[67] Secondly, George claims that the order Moon-Mercury-Venus is also justified by the principle that concludes the paragraph above: 'in circular motion, the inferior must be faster'. Recall that George presents his argument

[65] This is no demonstration, of course, but an affirmation of the consequent. The coincidence of apparently consistent results reached by two different procedures also conferred great reliability on Ptolemy's solar distance (erroneously, as it happened). Neugebauer, *A History of Ancient Mathematical Astronomy*, p. 112.

[66] 'sed proportionaliter traducentes a distantiis quas ipse demonstravit, demonstrabimus Mercurium post lunam statim ascendendo et Venerem post Mercurium collocari; deinde solem. Necesse igitur est, ne vacuum detur, ut maxima distantia lune minima sit aut Veneris aut Mercurii. Sed non Veneris, Mercurii ergo. Nam et subiisse Mercurius Venerem scribitur; et velocior est quam Venus: necessariumque est ut velociores in circulari motu inferiores sint'. George of Trebizond, quoted by Regiomontanus, *Defensio*, fol. 152r.

[67] Bowen, *Simplicius on the Planets*, p. 105, lines 19–21: 'observations in which the star of Mercury is reported running beneath the [star] of Venus make it clear in fact that Mercury is found *below* Venus'. One Latin edition reads: 'Quod autem Mercurius sub Venere deprehenditur significant observationes in quibus Mercurii stella subivisse Veneris stellam narratur'. Simplicius, *Commentaria in quatuor libros de Caelo Aristotelis* (Venice: Hieronymus Scotus, 1555), fol. 72rb, which is close to George's '*Nam et subiisse Mercurius Venerem scribitur*'. Simplicius's teachers had been students of Proclus, whose *Hypotyposis astronomicarum positionum*, ed. Manitius, *Procli Diadochi hypotyposis*, ch. 7.22–23, pp. 222–25, in a passage later used by Copernicus, also reports such an observation in vague terms (Venus was observed below Mars, just as Mercury was observed below Venus); see Giorgio Valla's Latin translation of Proclus cited in Lerner et al., *Nicolas Copernic. De revolutionibus*, vol. III, pp. 127–28. Although the phrasing makes Simplicius the most likely source, George of Trebizond evidently once had access to the Greek *Hypotyposis*; Monfasani, *Collectanea Trapezuntiana*, pp. 685–86.

for this planetary order as a demonstration founded on a necessary principle and an observation. It is therefore perhaps no coincidence that Simplicius had also used demonstrative language for the order of Venus and Mercury: he not only adduced the observation, but immediately thereafter claimed (wrongly) that the *Almagest* proved this order from the two planets' apogees and perigees.[68] Even more telling, George phrases his principle of motion in relation to the Earth, as Simplicius does (the closer to the Earth is faster), whereas Aristotle's argument was dynamic, starting from the motion of the stellar sphere (the outermost sphere is fastest and dominates the other motions, so that the planets nearest to it take the longest time to 'go through their own circles').[69] In short, Simplicius's *Commentary on De caelo* looks suspiciously like one of George's sources.

[68] Referring to the observation of Mercury below Venus, Simpicius writes (in Bowen's translation): 'This fact is *proven* as well from the account of the distance of their apogees and perigees, since the greatest distance of Venus is *proven* somehow to be the same as the distance of the Sun (so that Venus is close to the Sun), and the greatest [distance] of Mercury is [proven] somehow [to be] near the least [distance] of Venus, and the greatest [distance] of the Moon [to be] near the least [distance] of Mercury. Certainly, these facts are *proven* in Ptolemy's *Syntaxis* if the account of the eccentricity of the planets in transformed into an account of their [eccentricity] from the center of the Earth'. Bowen, *Simplicius on the Planets*, p. 105 [my italics; the verb translated by *proven* is δείκνυμι]. Ptolemy does no such thing in the *Almagest*. For this and other problems in this passage, see Bowen, *Simplicius on the Planets*, pp. 211–13.

[69] Simplicius writes: 'the motions are in proportion to their distances because [*planets*] *that are nearer the Earth, like the Moon, move faster*, whereas those that are farther move more slowly in the proportion of their distances' (Bowen, *Simplicius on the Planets*, p. 99). For Aristotle, however (*De caelo* 2.10; Bowen translation): 'Let us theorize on the basis of [works] on astronomy about the ordering of the [heavenly bodies] — the way in which each moves in that some are prior and others posterior — and how they are related to one another in their distances, since it is discussed [in these works] sufficiently. It turns out that the motions of each are in proportion to their distances in that some [motions] are faster and some slower. That is to say, since it is supposed that the outermost revolution of the heavens is simple and the fastest, and that the [motions] of the others are slower and more numerous — for each moves in a direction opposite to the heavens along its own circle — it is actually reasonable that *the* [*body*] *nearest the simple and primary revolution goes through its own circle in the longest time*, that the one that is farthest away in the least time, and that of the others the nearer always [goes through its own circle] in more time and the farther in less time. The reason is that the one that is nearest [the outermost revolution] is dominated [by it] most of all whereas the one farthest [is dominated] least of all on account of its distance, and the intermediate [bodies are] actually [dominated] in the ratio of their distances, just as the astronomers in fact prove' (Bowen, *Simplicius on the Planets*, p. 97). On this passage, see Pellegrin, 'The Argument for the Sphericity', esp. 163–64. See also Goldstein, 'Copernicus and the Origin', which sees Vitruvius and Martianus Capella behind Copernicus's path to heliocentrism and is pertinent to the controversy about ordering discussed here.

Having settled the problem of relative planetary order to his satisfaction, George outlines the framework underlying his computation of the absolute distances (in earth radii) of Mercury and Venus from the Moon, beginning with Mercury:

> *The maximum distance of the Moon in e. r. is 64;10 from the center of the Earth. The body of any planet cannot stick out of its sphere; therefore, to the maximum distance of the Moon, which goes from the center of the ecliptic to the center of its [= the Moon's] body, it is necessary to add the radii of the Moon and Mercury using the same units, lest the body of some planet stick out of its orb. But the Moon's radius is 0;17,32 (e.r.), as Ptolemy demonstrated. He did not set down [the radius] of Mercury, no more than those of the four other planets, as it was not his purpose to investigate the diameters, except [those] of the Sun and Moon on account of eclipses. All [of these diameters] can easily be measured with instruments, however. We therefore list them as they are recorded, for in fact we have not used instruments, nor made any observations, nor do we think it worth worrying much about such a small difference.*[70]

As George correctly notes, the *Almagest* does not discuss the radii of the five classical planets. It was in the *Planetary Hypotheses* that Ptolemy had given estimates for them.[71] Thanks to the few astronomers in the Hellenistic and Islamic worlds with access to that work, Ptolemy's figures became canonical in both Arabic and Latin treatises on cosmic dimensions. George is cavalier about the alleged ease of measuring planetary diameters — well-nigh impossible with the naked eye. After stating that he has neither used instruments, nor tried to measure the planetary diameters, George lists the traditional numbers. He justifies this move by asserting a negligible difference between the latter and the measurements he 'easily' could have carried out.[72]

[70] 'Est autem maxima distantia lune secundum quod semidiameter terre est unius partium 64 10' a centro terre; verum quoniam planetarum corpora nulla ex parte possunt spheras suas excedere, ad maximam distantiam lune que est a centro orbis signorum ad centrum corporis sui semidiametros suam et Mercurii secundum easdem partes addere necesse est, ne corpus stelle alicuius orbem excedat suum. Est autem lune semidiameter 0 17' 32" earundem a Ptolemeo demonstrata. Mercurii vero sicut et aliarum quatuor stellarum non ponitur ab ipso; non enim erat opus ad negocium suum nisi solis et lune, propter eclipses, diametros investigare. Capi tamen omnes possunt per instrumenta facile; eas igitur sicut capte instrumentis scribuntur sic ponemus. Nam nos quidem nec instrumentis nec ullis observationibus usi sumus, nec tamen multum curandum censemus de tam parva differentia'. George of Trebizond, quoted by Regiomontanus, *Defensio*, fol. 152r.

[71] Goldstein, *The Arabic Version*, pp. 8–9; Van Helden, *Measuring the Universe*, p. 27. He adopted the diameter of Venus, and perhaps others, from Hipparchus; Neugebauer, *A History of Ancient Mathematical Astronomy*, pp. 21–22.

[72] In the *Planetary Hypotheses*, the planetary diameters are credited to Hipparchus, suggesting that Ptolemy did not measure them either; Goldstein, *The Arabic Version*, p. 8. See also Bowen, *Simplicius on the Planets*, pp. 288–89.

2.3. Computing planetary distances

Next, George carries out a string of computations of planetary distances. Assuming that the planetary spheres are contiguous, he works his way upward from the lunar apogee using the *Almagest*'s perigee-to-apogee ratios for each planetary model. Since *Almagest* 9.1 precedes Ptolemy's detailed discussions of these models, George's commentary is here curiously premature. George's first calculation will illustrate his procedure:

> *The maximum [distance] of the Moon therefore is 64;10 (e.r.); its radius 0;17, 32 (e.r.). Converted to the same units, the radius of Mercury is approximately 0;2,8 (e.r.). It is recorded as being approximately 1/28 e.r.*[73] *Adding them together [equals] 64;29,40 [e.r.], the perigee of Mercury's center, where the radius of the lunar deferent is 60 parts.*[74]

This procedure builds on assumptions (e.g., no empty spaces) and practices standard in Arabic and Latin astronomy.[75] The most notable exception is the inclusion of planetary radii in the calculations. Behind this move lies a physical concern: the planets, like the luminaries, are bodies, not points. Theoretically, therefore, their dimensions matter when computing the thicknesses of the planetary spheres that constitute the cosmos. Treating the planets as points, as the *Almagest* and most subsequent astronomers do, is an inappropriate simplification that implicitly concedes physical impossibilities such as planets protruding into adjoining spherical shells and therefore colliding. The radius of each physical spherical planetary shell must therefore be larger by the planet's radius than that assumed in the *Almagest*'s geometrical models.

To the *Almagest*'s apogee of the Moon's center, George thus adds one lunar radius in order to obtain what might be called the 'physical' apogee of its sphere, to distinguish it from its ordinary (geometrical) apogee, reckoned from the planet's center. The point of the Moon's body farthest from the Earth's center thus marks the boundary of its sphere.

[73] This figure, like the absolute dimensions of the other planets given here and in the earlier Latin literature, goes back to the Arabic tradition (primarily via Thābit ibn Qurra and al-Farghānī) and indirectly to Ptolemy's *Planetary Hypotheses*. Swerdlow, *Ptolemy's Theory of the Distances*, pp. 137–56; Van Helden, *Measuring the Universe*, pp. 27, 30–37. As we shall see below, one of the likeliest sources of George's figures is Campanus's *Theorica planetarum*: Benjamin and Toomer, *Campanus of Novara*, p. 55.

[74] 'Lune igitur maxima 64 10'; eius semidiameter 0 17' 32": Mercurii semidiameter in easdem partes traducta 0 2' 8" proxime. Vigesima enim et octava proxime pars diametri terre conscribitum esse. Simul, 64 29' 40", minima distantia centri Mercurii secundum quod semidiameter deferentis lunam est 60 partium'. George of Trebizond, quoted by Regiomontanus, *Defensio*, fol. 152r-v.

[75] See the summary in Benjamin and Toomer, *Campanus of Novara*, p. 530.

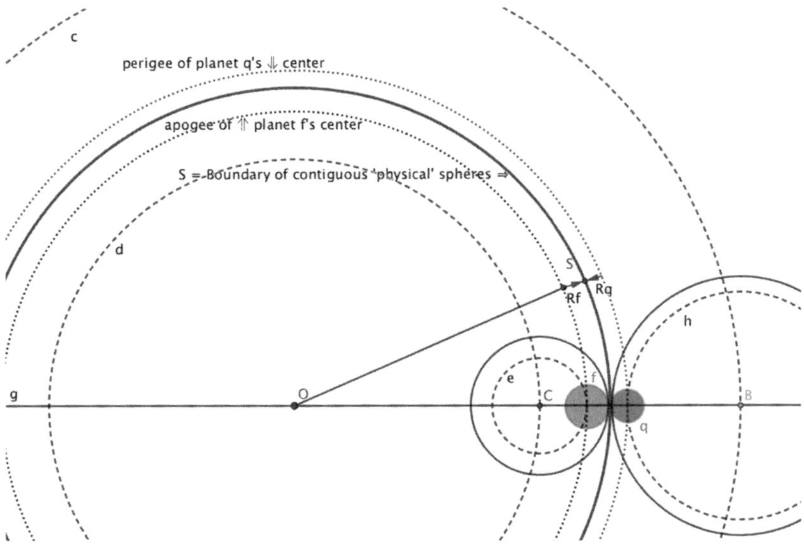

Figure 1. The concept behind George of Trebizond's computation of cosmic dimensions. The observer is at **O**, the grey circles represent the bodies of 2 arbitrary planets f and q. The *Almagest*-like models are represented as dashed-line circles: simplified concentric deferents of radii OC and OB carry the epicycles with centers C and B, respectively, bearing planets with radii R_f and R_q. The 2 dotted-line circles concentric to O show boundaries of the spheres determined respectively by the 'ordinary' apogee of **f** and the 'ordinary' perigee of **q**, when the planets are treated as dimensionless points.

Circle **S** represents the contiguous physical boundaries of the two planets' spheres, with dimensions corrected by adding the planetary radii to obtain, respectively, planet **f**'s 'physical' apogee, and planet **q**'s 'physical' perigee.

The same reasoning applies to the dimensions of the next planet's sphere (Mercury, as George argues). Again, assuming that the spheres are contiguous (i.e., no empty spaces or interpenetrations of bodies), the 'physical' lunar apogee will be equal not to the perigee of Mercury's center (as in the *Almagest* model), but to the part of Mercury's body nearest to the Earth — one Mercury radius beyond the center of the planet. Adding the radius of Mercury to the Moon's physical apogee now gives the absolute distance of the perigee of the center of Mercury's body. Only after this correction can the *Almagest*'s values of Mercury's relative perigee and apogee properly enter the computations. The Rule of Three that gives the absolute distance of a new planet's apogee ($p_{rel}:p_{abs}::a_{rel}:a_{abs}$, where a_{abs} is the unknown) always uses *Almagest* values for the planet's center. Since the result of the computation gives the new 'ordinary' apogee, it must always be corrected to obtain the new physical apogee.

Curiously, George of Trebizond shares this unusual procedure with the *Theorica planetarum* of Campanus of Novara (d. 1296).[76] Their computations of

[76] Benjamin and Toomer, *Campanus of Novara*, p. 56.

planetary distances have more in common than the usual parameters they draw from the *Almagest* and the Hellenistic and Arabic traditions on cosmic dimensions; they also share near-identical numbers for the distances and sizes of the planetary spheres up to, and including, the Sun. The hypothesis that George drew on Campanus's *Theorica planetarum* for his exposition will grow stronger as we proceed.

Since George believes (on unspecified empirical grounds) that Mercury is below Venus and necessarily closest to the Moon, he next computes the dimensions of Mercury's planetary sphere, starting from its perigee. George first gives the answer, then the computation:

> *According as the radius of Mercury's deferent is 60 parts, its perigee is 33;04. For 57 parts of this sort is the epicycle center's perigee that is opposite to the apogee, which [57 parts] are decreased by approximately 1;26 in the signs of Gemini and Aquarius. Indeed, the epicycle radius is 22;30 parts of this sort. The perigee of Mercury's center therefore is the remainder of 33;04, from which, by translation* [= of proportions], *we find the apogee of the same planet in the following manner.*[77]

This calculation of Mercury's perigee confirms the oddity of George's decision to treat planetary distances here, since the *Almagest* so far has discussed only the models for the Sun and Moon. Whereas the Moon had already been discussed in books 4–6, the five classical planets are all discussed after 9.1, even though the parameters of their models are needed to compute planetary distances.

Although the details of the *Almagest*'s Mercury model do not enter into George's computation, its main feature matters. Ptolemy constructed this model to generate the one apogee and the *two* perigees that he had found.[78] For the other planets, the apogee and perigee of the epicycle center are 180° apart and define the line of apsides. The perigee of Mercury's epicycle center, however, is not on the line from the apogee through the Earth. (In Figure 2, the straight vertical line that cuts the model in half, passing through the centers of the deferent and the Earth, 'C mundi', at the intersection of the 2 lines forming the X in the figure). The epicycle center has 2 perigees approximately 120° from the apogee on either side of the line of apsides (not shown, but approximately at 4 o'clock and 8 o'clock on the inside of the white ring). Ptolemy generates this configuration with a 'crank mechanism' (the smallest circle, with center

[77] 'secundum autem quod semidiameter ipsius Mercurii est 60, minima eius est 33 04'. Est enim 57 partium huiusmodi minima centri epicycli distantia que maxime opponitur; qua minores fiunt in Geminis et Aquario gradibus 1 26' proxime. Semidiameter vero epicycli est partium huiusmodi 22 30'. Quare relinquitur minima distantia centri Mercurii 33 4', ex qua per traductionem maximam distantiam eiusdem stelle invenimus hoc modo'. George of Trebizond, quoted by Regiomontanus, *Defensio*, fol. 152v.

[78] The lunar model also generates two perigees, but only the apogee mattered as the starting point for George's computations of the sizes of the other planetary spheres.

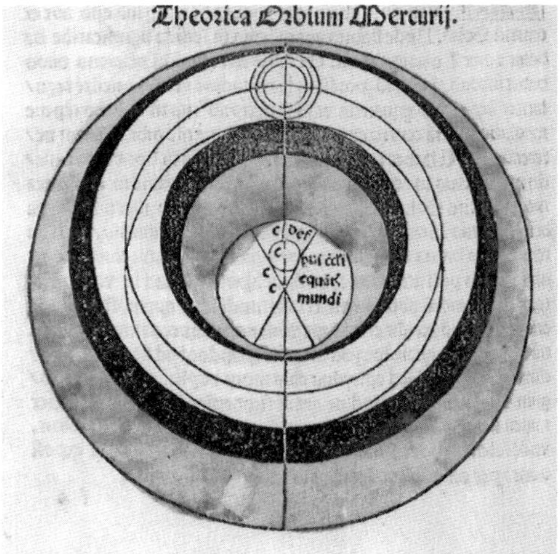

Figure 2: The Mercury model from Georg Peuerbach's *Theoricae novae planetarum* in Ratdolt's first *Sphaera mundi* compendium (Venice, 1482) and based on Regiomontanus's first edition. Since Mercury is not depicted, the lines inside the epicyclic orb presumably represent the latter's 3 components of longitude and latitude. By courtesy of the Department of Special Collections, Memorial Library, University of Wisconsin-Madison.

'C parvi circuli') that carries the center of Mercury's deferent ('C def[rentis]'), which in turn carries the epicycle center. By moving the epicycle center on the deferent's rim now towards, now away from, the Earth, the device varies the apparent size of the epicycle for the observer and generates 2 perigees.[79]

George's numbers derive from this model. He begins by giving the perigee's value as 33;04 units (60ths of the deferent radius). Instead of proceeding, he generates this number with a retroactive computation. Subtracting the eccentricity of the Earth (3) from the deferent radius (60) equals 57 units. Subtracting the epicycle radius (22;30) from this figure yields 34;30, which would be Mercury's perigee according to a standard *Almagest* model with apogee and perigee on the same diameter. But Mercury has two perigees, which are not on the line of apsides. By subtracting 1;26 from 34;30, George gets 33;04, the correct 'relative' perigee of Mercury's center (in 60ths of the deferent radius). The arithmetic works [57- (22;30 + 1;26) = 33;04], but George does not say where he got his correction of 1;26, which looks suspiciously like the complement needed for the right answer. Oddly, he has neither carried out the requisite full computation, nor even computed Mercury's perigee straightforwardly from the values for the epicycle center's perigee and the epicycle radius in *Almagest* 9.9.[80]

[79] Hartner, 'The Mercury Horoscope', esp. 110–18.

[80] In that chapter, Ptolemy had calculated the parameters of the Mercury model and obtained 55;34 for the perigee of the epicycle center. Although Ptolemy did not do so himself, subtracting the epicycle radius from this figure yields Mercury's perigee directly (55;34–22;30 = 33;04). See Neugebauer, *A History of Ancient Mathematical Astronomy*, p. 164. This

Although obscured by George's terminology, the schema is a simple Rule of Three that he will apply repeatedly in series, from the Moon's apogee in e.r. to the sphere of the fixed stars. The problem with computations beyond the Moon, however, is precisely that the planetary order is unknown and that George claims to be demonstrating it. He thus justifies the placement of Mercury next to the Moon with a vague, undocumented claim for a sighting of Mercury below Venus. Based on this alleged observation, George treats Mercury as necessarily next to the Moon, with Venus immediately thereafter. To confirm this order, he will now show that Mercury and Venus, in ascending order, fit almost perfectly between the Moon and the Sun, thus vindicating the traditional order.

> *The first* [*number is*] *33;4,30 approximately for the same* [= *least*] *distance, where the terrestrial radius is one* [*sic*!].[81] [The second is 64;29,40 e.r., evidently silently rounded up to 64;30]. *The third* [*is*] *91;30 for Mercury's apogee,*[82] *where the radius of its deferent is 60 parts. Having multiplied the second by the third and divided the result by the first, one obtains 178;28* (*where the Earth's radius is one*) *from the Earth's center to Mercury's center in its distance 'simply', that is, when the epicycle is at the maximum of the eccentric and the planet at the maximum of the epicycle.*[83]

George shares this general procedure with many predecessors. The 'first number' (1) is the planet's perigee in relative terms (60ths of its deferent radius), but it is written here with an additional 30″ for which I cannot account. The second (2) is the planet's perigee in absolute terms (e.r.); the third (3) is the planet's apogee in relative terms (in 60ths of the deferent radius). The fourth number (4), not named as such, is the sought quantity, the planet's apogee in absolute terms (e.r.). Thus, for a given planet:

is also the operation by which Campanus of Novara obtains the perigee. With the exception of 1;26, the other numbers match those of Campanus; Benjamin and Toomer, *Campanus of Novara*, pp. 238–39.

[81] In 60ths, not Earth radii. The error may be the result of a confusion of the very similar numbers for Mercury's perigee in 60ths (33;4) and the Moon's perigee in *e.r.* (33;33). Regiomontanus will pounce on this slip (*Defensio*, fol. 158v).

[82] One can visualize this apogee as the result of stretching end-to-end (in a straight line) all the elements of Ptolemy's 'crank mechanism'. Although not computed in the *Almagest*, this figure is the sum of the epicycle radius (22;30) and the apogee of the epicycle center (69), which is in turn the sum of the deferent radius (60), the diameter of the 'crank' circle (6), and the equant-Earth distance (3); Toomer, *Ptolemy's Almagest*, pp. 459–60. The sum also appears in Campanus (Benjamin and Toomer, *Campanus of Novara*, pp. 238–39).

[83] 'Primus 33 4' 30" proxime eiusdem distantie secundum quod terre semidiameter est unius; tertius 91 30' maxime distantie Mercurii secundum quod semidiameter sui deferentis est 60 partium; diviso per primum numero producto ex multiplicatione secundi in tertium fiunt 178 28' partes secundum quod semidiameter terre est unius a centro terre ad centrum Mercurii in distantia eius simpliciter; hoc est quando epicyclus est in maxima eccentrici et stella in maxima epicycli'. George of Trebizond, quoted by Regiomontanus, *Defensio*, fol. 152v.

$$\frac{(1)\ \text{perigee (60ths)}}{(2)\ \text{perigee (e.r.)}} = \frac{(3)\ \text{apogee (60ths)}}{(4)\ \text{apogee (e.r.)}}$$

This proportion underlies George's effective formula:

(4) apogee (e.r.) = (3) apogee (60ths) × (2) perigee (e.r.) / (1) perigee (60ths),

which he repeatedly expresses as 'having multiplied the second by the third and divided the result by the first, one obtains' the planet's apogee in e.r. The adverb 'simply' indicates that it must still be corrected for the size of the planet itself.

Walking through the Mercury computation will not only illustrate George's application of this (standard) procedure and his foibles, but also shed some light on his sources and aids. The apogee of Mercury's center in e.r. should be the product of the second (not specified in George's description of the operation) and third numbers, divided by the first.

Using the numbers George gives in describing the operation and supplying the omitted second number, 64;29,40 (his computation of the perigee of Mercury's center), one obtains:

(64;29,40 × 91;30) / 33;04,30 = 178;25,12

This is not the result he gives. Using the rounded value of 33;04 sixtieths that he had given earlier for the denominator, one obtains

(64;29,40 × 91;30) / 33;04 = 178;27,53, still off the mark.

His result comes from yet another (silent) rounding, this time of Mercury's perigee. Thus: (64;30 × 91.30) /33;04 = 178;28. As it happens, both roundings and this answer also appear in Campanus's *Theorica planetarum*, the beginning of a pattern.[84]

Whether relative or absolute, the perigees and apogees in this proportion are the distances from the Earth to the *center* of the planet. Thus George reminds the reader: '*Lest there be a vacuum, this apogee of Mercury becomes Venus's perigee, having added the radius of each, lest the planet stick out of its sphere*'.[85]

To the apogee of the planet's center used in the *Almagest*, one must add the planet's radius. Likewise for Venus, which is also a finite body.

Again, to every apogee in e.r. derived from the *Almagest*'s proportions, one must add the radii of both the planet and the next higher planet. Only this

[84] Benjamin and Toomer, *Campanus of Novara*, pp. 238–34.

[85] 'Hec Mercurii maxima, ne vacuum detur, minima fit Veneris, semidiametro utriusque addita ne stella spheram suam excedat'. George of Trebizond, quoted by Regiomontanus, *Defensio*, fol. 152v. Copernicus made a similar move in the *Commentariolus*; Swerdlow, 'The Derivation and First Draft', pp. 466–67.

corrected figure can be set equal to the *Almagest*'s perigee for the next higher planet and used in the Rule of Three to compute the corresponding new apogee.

As we have said, however, Mercury's radius is 0;2,8; that of Venus, 0;26,40: [added all] together, 178;56,48 [e.r.]. This second number is placed in conversion. We convert to these parts [= e.r.]; the first is the number of the same distance, where the radius of Venus's deferent is 60, and he [Ptolemy] has demonstrated it to be 15;35 in book 10; the third, the apogee of Venus, in 60ths of Venus's deferent, is 104;25. Having multiplied the second by the third, and divided by the first, one obtains 1199;2,18, the distance in e.r. of Venus's center simply.[86]

George's figure for Mercury's diameter is standard, but that for Venus has no counterpart in the earliest literature on planetary distances. It matches, however, Campanus of Novara's number, which apparently came from misinterpreting the Latin translation of al-Farghānī's figure for Venus's diameter.[87] The sum that George gives for Venus's 'physical' perigee is, like his other numbers, identical to that in Campanus, since both men take the rare step of including planetary radii in the sizes of the planetary spheres. This approach, combined with the duplication of the tell-tale error for Venus's diameter, strongly suggests that George of Trebizond relied on Campanus's *Theorica planetarum* when computing his cosmic dimensions.

Using the same procedure as above:

(4) apogee (e.r.) = (2) perigee (e.r.) × (3) apogee (60ths)/ (1) perigee (60ths)

For the case of Venus:

apogee of Venus = 178;56,48 × 104;25 / 15;35 = 1199;2,18,

George's figure, which is also identical to Campanus's.

Adding to these the radii of Venus and the Sun in the same units, one obtains the perigee of the Sun's center. As we have said, however, the radius of Venus measured by instruments is 0;26,40, as handed down; that of the Sun is 5;30 of the same parts [e.r.], as Ptolemy demonstrated in book 5, for a total of 1204;58,58 [e.r.]. This is the Sun's perigee, from the Earth's center to its own center, [when] situated in the perigee of its

[86] 'Est autem semidiameter Mercurii, ut diximus, 0 2' 8"; Veneris, 0 26' 40"; simul, 178 56'48". Hic numerus secundus ponitur in traductione; ad has enim partes traducimus. Primus est eiusdem distantie numerus secundum quod semidiameter deferentis Venerem est 60 et est 15 35' in decimo libro demonstratus ab ipso; tertius, maxima Veneris secundum easdem partes deferentis Venerem distantia est 104 25'; et ab eodem decimo libro colligitur. Multiplicato secundo in tertium et producto partito per primum, haberetur 1199 2' 18", maxima distantia centri Veneris simpliciter secundum partes de quibus semidiameter terre est unius'. George of Trebizond, quoted by Regiomontanus, *Defensio*, fol. 152v.

[87] Benjamin and Toomer, *Campanus of Novara*, pp. 326–27, 433. The sexagesimal equivalent of al-Farghānī's value is 0;18.

eccentric, where the Earth's radius is one part, through which [distance] its maximum is easily converted into similar units.[88]

To calculate the 'physical' perigee of the Sun (the edge of its body closest to the Earth), George of Trebizond, as usual, adds the radii of both Venus and the Sun to the apogee of Venus's center. Again, the computations match those of Campanus.[89]

The first set down is the Sun's perigee demonstrated in book 3, where the radius of its eccentric has 60 parts; and it [= the perigee] is 57;30; the second, the same minimum distance in e.r., is 1204;58,58; the third, the maximum in 60ths of the eccentric's radius is 62;30. After multiplication and division, they yield 1309;45,50 — this is the Sun's apogee in e.r.[90] *And thus, from perigee to apogee, there are approximately 105 parts of this sort [e.r.],*[91] *from which it is clear that the Sun's distance of 1210 similar parts [e.r.] (which he computed from the angle that does not vary sensibly and is subtended by the body's diameter) is approximately minimum.*[92]

From these computations, George concludes that Ptolemy's figure for the solar distance, reached using eclipses, was near perigee. He might be paraphrasing Campanus, who says as much and whose parameters and computations match George's exactly.[93]

[88] 'His si addideris semidiametros Veneris et solis secundum easdem partes, habebitur minima distantia centri solis. Est autem semidiameter Veneris, ut diximus, 0 26' 40" instrumentis capta, ut traditur; solis, 5 30' earundem partium, sicut Ptolemeus in libro quinto demonstravit; simul, 1204 58' 58". Hec est distantia solis minima a centro terre ad centrum ipsius in minima sui eccentrici distantia collocati secundum quod semidiameter terre est partis unius per quam maxima eius ad partes similes facile traducitur'. George of Trebizond, quoted by Regiomontanus, *Defensio*, fols 152v–153r.

[89] Benjamin and Toomer, *Campanus of Novara*, pp. 326–27, 332–33.

[90] (1) perigee (60ths) : (2) perigee (e.r.) :: (3) apogee (60ths) : (4) apogee (e.r.)

57;30 : 1204;58,58 :: 62;30: apogee of the Sun. Therefore the Sun's apogee = 1204;58,58 × 62;30 / 57;30 = 1309;45,50, which matches George's figure.

[91] 1309;45,50–1204;58,58 = 104;46,52, rounded up to 105, as in Campanus; Benjamin and Toomer, *Campanus of Novara*, pp. 332–33.

[92] 'Primus ponitur minima distantia solis demonstrata in libro tertio secundum quod semidiameter sui eccentrici habet 60 partes; et est 57 30'. Secundus, eadem minima secundum quod semidiameter terre est unius, 1204 58' 58"; tertius, maxima secundum partes semidiametri eccentrici et est 62 30'. Faciunt post multiplicationem et partitionem 1309 45' 50"; hec est maxima solis distantia secundum partes de quibus semidiameter terre est unius. Et sic a minima ad maximam sunt partes similes 105 proxime. Unde patet solis distantiam 1210 similium partium quam collegit ex angulo qui non variatur sensibiliter subtenso a diametro corporis eius minimam esse proxime'. George of Trebizond, quoted by Regiomontanus, *Defensio*, fol. 153r.

[93] cf. Campanus: 'the distance between the center of the Sun's body and the center of the earth stated by Ptolemy [1210 e.r.] ... is, strictly speaking, not the distance of the point of greatest distance or of the point of least distance, but of a point on the [the Sun's] eccentric not far removed from the least distance' (Benjamin and Toomer, *Campanus of Novara*, p. 333). He emphasizes his 'own very exact calculations' vs. Ptolemy's 'rough' ones, but notes

> *He [Ptolemy] did not want to inquire by proportions into these various easy matters, since he was focused on providing the foundations for all the items from which he who understands him [= Ptolemy] could easily infer other things. To this [point], it is also clear that the 5 sixtieths of the solar eccentric's radius (the apogee's excess over the perigee) equal approximately 105 parts in e.r. And thus 1/120th of the eccentric's diameter is found to contain the entire Earth diameter approximately 21 times.*[94]

George is pleased with these results, which he sees as supplying the demonstration of planetary order missing in the *Almagest*:

> *We have therefore demonstrated that the order of the planets is the one that Ptolemy posited. For this reason, those who place Mercury and Venus above the Sun do nothing but make a vacuum of the whole space between the lunar apogee and the solar perigee, which is approximately 1140;48,58 e.r.*[95] *The distances of Venus and Mercury fill this astonishing space to a T, which the demonstrator himself can see from the thickness of the globes. For the Moon's apogee is approximately 64;30 [e.r.],*[96] *which is the perigee of Mercury, the apogee of which is approximately 178;56 in the same units [e.r.]. After subtraction, the remainder is the thickness of Mercury's globe, 114;26 in the same units [e.r.]. In turn, since Venus's perigee is 178;56, if you subtract these parts from approximately 1204;59[e.r.; the solar perigee], the remainder is the thickness of Venus's globe, approximately 1026;03 in the same units [e.r.]. Together, the two orbs of Mercury and Venus therefore make a thickness of approximately 1140;29 of these same parts [e.r.], which space we also approximately computed between the Moon's apogee and the Sun's perigee. Thus, by demonstrations, everything squares everywhere.*[97]

that the difference between them is negligible; Benjamin and Toomer, *Campanus of Novara*, pp. 332–33.

[94] 'Noluit autem hec facilia et varia proportionaliter inquirere quoniam radices solummodo ad omnia tradere studuit; quibus qui eum intelligit, facile cetera consequentur. Huic etiam patet quod quinque partes de 60 semidiametri eccentrici solis quibus maxima distantia minimam excedit, faciunt 105 partes proxime(?) de partibus de quibus semidiameter terre est unius. Et sic pars una de 120 diametri eccentrici solis invenitur continere totam diametrum terre vicibus xxi proxime'. George of Trebizond, quoted by Regiomontanus, *Defensio*, fol. 153r.

[95] This number 1140;48,58 (64;10 subtracted from 1204;58,58) shows that George, despite many other numerical identities and conceptual similarities with Campanus, did not uncritically copy verbatim from him, but recomputed the latter's figures. For reasons that remain unclear (see Benjamin and Toomer, *Campanus of Novara*, pp. 436–37), Campanus somehow got 1138, a number consistently repeated in the entire manuscript tradition of his *Theorica planetarum*. George's other numbers are those of Campanus, sometimes rounded; Benjamin and Toomer, *Campanus of Novara*, pp. 326–27.

[96] This is significantly rounded up from 64;27,32 — the apogee of the lunar body's center, to which one should, on George's account, add the lunar radius in e.r.

[97] 'Demonstratus ergo nobis ordo erraticarum stellarum est qui a Ptolemeo ponitur. Quare qui Mercurium et Venerem supra solem collocant, nihil aliud agunt quam ut totum spacium a maxima lune ad minimam solis, quod est 1140 48' 58" proxime, secundum quod semidiameter terre est unius, vacuum sit; quod spacium mirum in modum distantie Veneris atque Mercurii ab ipso demonstrante ad unguem replent quod a grossitie globorum perspicuum sit. Nam lune maxima est 64 30' proxime; et eadem est minima Mercurii, cuius maxima est 178 56' proxime

Arguing from the great fit that he sees in his computations, George claims to have demonstrated the traditional order of Moon, Mercury, Venus, and Sun, in ascending order from the Earth's center.

2.4. The shadow of Campanus of Novara

Here we must briefly take stock of the parallelisms between Campanus and George so far. The identity of many of the fundamental numbers in the two works is to be expected, for they come directly (or are one arithmetic operation away) from those in the *Almagest*'s models, and the computations are straightforward. More unusual are the parallel decisions of both Campanus and George to compute the planetary spheres' size not from the planets' centers, but from their bodies' outermost edges.[98] Almost all of George's results and procedures are thus identical to those of Campanus. Even more significant is George's apparent use of Campanus's error for the diameter of Venus.[99] In the computations of the interval between the Moon and the Sun, the sole difference between George of Trebizond's values and those of Campanus is the calculation of the space between lunar apogee and solar perigee. Here, whereas the entire Campanus manuscript tradition seems to have propagated a copying error, George produced the correct figure, apparently perhaps because he computed or recomputed it.[100]

The emerging hypothesis of George's reliance on Campanus of Novara gains more plausibility in the following passage, in which George brings up an unusual argument for the traditional planetary order:

> *This is why I cannot stop marveling at this man Ptolemy and proclaim his sagacity and discretion; for him in such a matter, the authority of the ancients sufficed who named the seven days, by the revolution of which all time endures, for the seven planets. Thus, wherever you start, if you assign one hour to* [*each of*] *the individual planets and* [*if*] *you do this according to the planetary order, the analogous hour of the following day will belong to the planet that names the day. This planet is the one that follows the*

earundem; facta subtractione, remanet grossities globi mercurialis 114 26' earundem. Rursus quia minima Veneris est 178 56' si has partes subtraxeris a 1204 59' proxime minima solis, remanet grossities globi Veneris 1026 3' proxime earundem. Duo autem orbes Mercurii atque Veneris simul faciunt grossitiem earundem partium 1140 29' proxime, quod spatium etiam proxime colligebamus inter maximam lune et minimam solis. Ita undique omnia demonstrationibus coacta quadrant'. George of Trebizond, quoted by Regiomontanus, *Defensio*, fol. 153r.

[98] Al-ʿUrḍī (d. 1266) had done so as well; his discussion of planetary distances and sizes appears anonymously in Goldstein and Swerdlow, 'Planetary Distances and Sizes', esp. ca. p. 148. Subsequent evidence suggests that ʿUrḍī was the author: Saliba, 'The First Non-Ptolemaic Astronomy', pp. 571–76; see also Van Helden, *Measuring the Universe*, pp. 32–33.

[99] Benjamin and Toomer, *Campanus of Novara*, pp. 326–27, 433.

[100] George correctly obtained 1140;29 whereas the Campanus manuscript tradition consistently propagated 1138; Benjamin and Toomer, *Campanus of Novara*, pp. 436–37.

two intervening ones.[101] *Thus were the days denominated in the customs of nearly all peoples.*[102]

This passage seems to paraphrase a similar argument in Campanus's *Theorica planetarum*, in which the planetary order 'assumed by Ptolemy' (in George's words) is justified by the fact that it underlies the 'ancient' system of 'regent' hours that names the weekdays. Any other planetary order would disrupt that universal schema. George congratulates Ptolemy for following the ancients and for not needing more than their authority, a point that will provoke Regiomontanus to outrage.[103] This discussion adds another plausibility argument for George's dependence on Campanus: both discuss the weekly sequence after settling the order and distances of the spheres of the Sun and inferior planets.

After this digression, George returns briefly to *Almagest* 9.1:

> *It is also said to be most natural to separate with the intermediate Sun the three planets that can be any distance from the Sun from the two that cannot, so that three planets are above the Sun and three below it. Thus has he referred very secretly to the Sun's being like the heart of the universe and to the reason/explanation of the [planetary] speed. Those that are faster than the Sun sometimes follow, sometimes precede. Those that are faster in circular motion are lower. These things were sufficient for him [= Ptolemy], since he did not doubt that this matter could most certainly be demonstrated by means of things that would be demonstrated later. I am therefore surprised that*

[101] This astrological schema starts below the primum mobile, descending from Saturn to the Moon using the 'traditional' 7-planet sequence (here credited to Ptolemy). Following that sequence, one assigns a planet to each of the 24 hours of the day, with the first hour both 'ruling' and naming the day. This sequence of hours will generate the weekly sequence that we still use, thus — so the argument goes — attesting to the correctness of the planetary order underlying the sequence. For example, on Day 1 ('Satur[n]day'), the 24 hours will run through 3 complete planetary sequences (3 × 7 = 21 hours) and begin the 4th sequence (with Saturn, Jupiter, and Mars naming the last 3 hours of day 1). The first hour of day 2 will be the next in the hourly planetary sequence: the Sun will thus name both the next hour and the entire next day (Sunday); and so on. A simple rule can generate the weekly sequence of planetary days without all the enumeration: to find the name of the next day (and its first hour), skip over two planets in the descending 'Ptolemaic' sequence: Saturn (day 1 = Saturday) [skip Jupiter and Mars], Sun (day 2 = Sunday) [skip Venus and Mercury], Moon (day 3 = Monday) [skip Saturn and Jupiter], Mars (day 4), etc.

[102] 'Unde mirari soleo et prudentiam et modestiam hiuis viri Ptolemei dico; suffecit sibi ad tantam rem priscorum autoritas qui dies septem quorum revolutione cuncta tempora constant septem planetarum nominibus nominarunt; ita ut undecumque incipias si singulis planetis horam unam attribuas idque ordine dicto planetarum facias erit sequentis diei similis hora planete a quo dies denominata est. Is planeta est qui sequitur duobus interiectis. Sic in institutione omnium fere gentium dies denominati fuerunt'. George of Trebizond, quoted by Regiomontanus, *Defensio*, fol. 153r-v.

[103] Benjamin and Toomer, *Campanus of Novara*, pp. 332–35, lines 538–48. At *Defensio*, fols 158v–159r, Regiomontanus attacks this imputation of subservience to authority.

> *some mathematicians who comment [on the issue] have dared to grunt against a truth that is altogether clear.*[104]

Here, George starts from the plausibility argument in *Almagest* 9.1. The rationale for planetary order is the Sun's separation of the three planets visible in opposition from the two that never are. Into this criterion, George reads Ptolemy's most secret (*occultius*) endorsement of the analogy of the heart to the Sun and the principle of a linkage between a planet's speed and its distance from the center of the universe. Implicitly conceding that Ptolemy did not demonstrate the planetary order, George suggests that no proof appears at the beginning of *Almagest* 9.1 for reasons of presentation, not because such a proof is impossible. The lacuna is a concession to demonstrative sequence. In George's opinion, the premises required for such demonstration appear only later in the *Almagest*. As George's concluding sentence shows, this interpretation has critics, left unidentified.

These final remarks highlight several oddities about George's opening commentary on book 9. Why does he present here a long and detailed argument for planetary order and spacing? Why does he insist on calling it a demonstration, when commenting on a passage in which Ptolemy states explicitly that planetary order must remain uncertain without evidence of parallax? George apparently thought that he could improve on the *Almagest*'s presentation by moving into his commentary on book 9 a full discussion and putative demonstration of planetary order, both of which Ptolemy had allegedly omitted to discuss more pressing matters.

George says neither what, nor where, are the 'things that would be demonstrated later'. Most obviously, by *Almagest* 9.1, the use of relative apogees and perigees had been determined only for the two luminaries. The models for the five classical planets are all discussed thereafter. The opening of book 9, however, is Ptolemy's last word on planetary order in the *Almagest*. He postponed such a discussion to the *Planetary Hypotheses*, in which he computed the thicknesses of the planetary spheres and assigned values to the planets' diameters. The full text of the *Planetary Hypotheses* had been directly accessible to only a few scholars in Islamic civilization and to none in the Latin world.[105] Never-

[104] 'Naturalius etiam esse dicitur separare tres stellas que per omnem distantiam possunt removeri a sole a duabus que id facere non possunt, per medium solem ut tres super solem et tres infra solem planete sunt. Ita occultius quasi cor mundi solem esse et velocitatis rationem tetigit. Velociores enim sole sunt qui eum modo sequuntur, modo precedunt. Velociores vero in circulari motu inferiores sunt. Hec illi suffecerunt quia non dubitavit ab iis que postea demonstraturus erat posse rem istam certius demonstrari. Quare miror nonnullos quos inter mathematicos ennarant grunnire adversus veritatem sic undique apertam fuisse ausos'. George of Trebizond, quoted by Regiomontanus, *Defensio*, fol. 153v.

[105] Bessarion owned three partial Greek manuscripts of the work (Venice, Bibilioteca Nazionale Marciana, Z314, Z323, and Z324), which Heiberg used for the Teubner edition: Hei-

theless, the numerical values for planetary size and the strategies for computing planetary distances in the *Planetary Hypotheses* diffused widely. Its values for the planets' apparent diameters turn up in the treatises of al-Farghānī' and al-Battānī, among others, but without an attribution to Ptolemy. In fact, Ptolemy was so far from demonstrating a planetary order in *Almagest* 9.1 that he only alluded to a plausible one without spelling out its details. This reticence became the springboard for George's attempt to 'demonstrate' both the traditional order and sizes of the planetary spheres. Oddly, despite the importance George ascribed to the alleged observation of Mercury passing below Venus, he did not comment on the difference it would have made to Ptolemy's discussion.

Strikingly, at the page break at the top of fol. 153v of the *Defensio*, Regiomontanus inserted the following marginalium:

> the three superiors are linked to the Sun by their epicycle; the three inferiors, not by the epicycle, but by the motion in longitude; this can be one rhetorical reason for the position and order of the planets. Why did Nature not place the five retrogrades in one region? Why did she not notice the rationale of sex?[106]

This reading note occurs in the middle of the multi-page quotation from George, which makes its force ambiguous. Is Regiomontanus talking to himself? addressing George? both? The first point — the various linkages between the planets and the Sun — is familiar and straightforward. Campanus of Novara had made it in his *Theorica*, as had Johannes de Fundis and Peuerbach in their own *Theoricae novae planetarum*.[107] Indeed, Regiomontanus himself had made this objection in his copy of al-Biṭrūjī's *De motibus celorum*.[108] His first comment here takes the general structure of a near-traditional order for granted. He does not, however, take a position on the relative planetary order within each category. Saturn, Jupiter, and Mars are linked to the Sun by their epicycle (each planet revolves about the center of its epicycle in lock-step with the mean Sun). Next comes the Sun itself — in an intermediate position, as three bodies remain. Finally, Regiomontanus groups Venus and Mercury (in an

berg, *Claudii Ptolemaei opera*, vol. II, pp. viii, xi, clxvi. See also Mioni, *Codices graeci*, vol. II, pp. 27, 43, 46. All three manuscripts end before the cosmological second part of book 1, which survives only in Arabic and Hebrew and was mistakenly omitted from Heiberg's edition; Goldstein, *The Arabic Version*, p. 3.

106 'Tres superiores per epicyclum soli colligantur; tres vero inferiores, non per epicyclum sed per motum longitudinalem; hec esse potest una ratio rethorica situs planetarum et ordinis. Cur non quinque retrogrados in una parte natura locavit? Cur non animadvertit sexus rationem ~~qualitatem~~?' Regiomontanus, *Defensio*, fol. 153v, top margin.

107 Benjamin and Toomer, *Campanus of Novara*, pp. 306–07; Pedersen, 'The *Theorica planetarum* and its Progeny', esp. pp. 75–77; Peuerbach, *Theoricae novae planetarum*, [9v] in Schmeidler, *Joannis Regiomontani opera collectanea*, p. 772; Aiton, 'Peurbach's *Theoricae novae planetarum*', p. 23.

108 Nuremberg, Stadtbibliothek, Cent V 53; see Zinner, *Leben und Wirken*, pp. 61–62.

unspecified order) with the Moon as the 'three inferiors' that share a longitudinal linkage with the Sun. He does not spell out his reasoning, but a solar component famously links the deferents of Venus and Mercury (they move with the mean Sun), the lunar phases, and the 'crank mechanism' of the refined lunar model. This rationale for the traditional taxonomy arguably improves on Ptolemy's argument, which lumped the Moon with Venus and Mercury despite its unbounded elongation.

After identifying this first argument as 'rhetorical', Regiomontanus uses questions to propose two equally rhetorical organizational criteria.[109] The language seems counterfactual, as the negation and the perfect tense suggest: why did Nature *not* do x or notice y? As we shall see, however, Regiomontanus believes that the planetary order is genuinely uncertain. Nature's rationale is precisely what must be established. The first alternative groups all the planets with retrograde motion together. Although the second argument also appears counterfactual, Regiomontanus will later explore it as a possible order (i.e., Venus, contiguous to the Moon, followed by the hermaphroditic Mercury, and the 4 male planets, from the Sun to Saturn). Despite its ambiguities, the marginalium clearly shows that Regiomontanus finds the reasons underlying Nature's choice puzzling, and the standard rationale for the order of Sun, Mercury, and Venus unconvincing.

Returning to the text, George of Trebizond continues:

> *I am, however, altogether baffled by Theon [of Alexandria], whose exposition of Ptolemy's demonstrations perhaps follows [that of] others without understanding. I do not see — nor does he say — what else, indeed, could lead him to posit the order of the spheres according to Plato by ignoring demonstrations? Indeed, if someone wanted to demonstrate these things by parallax, he should simply use an instrument to get the parallax of Mercury at apogee or of Venus at perigee, for it is nearly the same; and everything relevant to this matter will be demonstrated since, from the first [measurement], one can get the lunar apogee and, from the other, the solar perigee.*[110]

In this passage, George categorizes Theon as a follower of Plato's planetary order rather than Ptolemy's. His reading suggests that the manuscript he used, presumably also borrowed from Bessarion,[111] belonged to the 'Byzantine recen-

[109] I take *una* here to be emphatic and enumerative.

[110] 'De Theone autem omnino stupesco qui demonstrationes Ptolemei exponit non intelligens forsan aliorum secutus. Quid enim aliud inducere ipsum potuit ut spherarum ordinem secundum Platonem spretis demonstrationibus poneret, nec ipse dicit, nec ego video. Verum si quis per diversitatem aspectus demonstrare ista voluerit, studeat habere diversitatem aspectus Mercurii instrumento in maxima distantia aut Veneris in minima simpliciter, idem enim pene est; et omnia que ad rem pertinent demonstrabuntur, cum hinc maxima lune illinc minima solis habeatur'. George of Trebizond, quoted by Regiomontanus, *Defensio*, fol. 153v.

[111] Monfasani, *George of Trebizond*, p. 74.

sion' of Theon's commentary. Ioanna Skoura's work shows that the mischief comes from a missing privative alpha in the manuscript tradition of book 9.1: βέβαιον vs. the correct ἀβέβαιον). The meaning of ἀβέβαιον ('unsteady', 'unconfirmed') is consistent with the skepticism that Theon expresses earlier about Plato's order (Venus and Mercury above the Sun). Forming βέβαιον by dropping the alpha, however, turns Theon's negation into an assertion: the word ('steady', 'firm', 'confirmed') seems to signal Theon's endorsement of Plato, contradicting his earlier endorsement of Ptolemy.[112]

Happy to find fault with Theon by letting the contradiction stand, George thus proposes a refutation of the latter's alleged Platonic order. He encourages the reader to make 'simple' (!) parallax measurements for Mercury's apogee or Venus's perigee — not so easy when staring into the Sun.[113] From these putative measurements, which are assumed to converge on one value, one theoretically can get the lunar apogee and the solar perigee (presumably by computation from the *Almagest*'s models for Mercury and Venus). In this argument, both the order of the two planets and their subsolar location are assumed.

This is surely an odd commentary on *Almagest* 9.1, in which Ptolemy had denied that planetary parallaxes could be detected (he had evidently tried). Given this lack of evidence for supralunar parallax, Ptolemy had even warned that the subsolar spheres of Venus and Mercury should not be located 'close enough to the Earth that there can result a parallax of any size'.[114] Undeterred by either sunlight, or Ptolemy's attempted measurements, or his opposition to any planet's contiguity with the Moon, George was recommending 'simple' parallax measurements at Mercury's apogee (therefore away from presumed contiguity with the Moon's sphere).

[112] The two readings are ἀβέβαιον (only in Vatican City, Bibl. Vat., Vat. gr. 1087) vs. βέβαιον (the 'Byzantine recension'). I thank Ioanna Skoura for this valuable information from her research on Theon's commentary (personal communication). Regiomontanus apparently found George's remarks on Theon's alleged position puzzling. On 159r, he wrote in the margin: 'See about Theon in what way he followed Plato on the order of the planets'. If he ever checked, his puzzlement may not have subsided: the version of Theon's commentary in his later possession (Nuremberg, Stadtbibl., Cent V, app. 8) also belongs to the Byzantine recension. Thanks to Swerdlow for noting that Giovanni Pico della Mirandola's *Disputationes* 10.4 (1496) drew from that same tradition: 'Verum non modo Ieber acutissimus mathematicus, sed et ipse Theon, interpres graecus Ptolemaei, dissentiens a magistro putat [Solem] supra Lunam statim collocandum'. See Garin, *Giovanni Pico della Mirandola*, vol. II, pp. 372–73, which translates as: 'not only Jābir, the acute mathematician, but also Theon himself, the Greek interpreter of Ptolemy, dissenting from the master, believe the sun is to be placed immediately above the moon, which Plato and Aristotle also affirm'. (Appendix to his translation in Swerdlow, 'Copernicus and Astrology').

[113] Regiomontanus makes this very point later in the *Defensio* (159v).

[114] Toomer, *Ptolemy's Almagest*, p. 420 (also quoted above); Neugebauer, *A History of Ancient Mathematical Astronomy*, pp. 148–49. He changed his mind in the *Planetary Hypotheses* (see note 35 above).

2.5. The superior planets

Having thus settled to his own satisfaction the order and sizes of the planetary spheres from the Moon to the Sun, George now turns to the sizes of the spheres above the Sun's. It is to them that he now applies Ptolemy's proviso about a lack of parallax:

> *If someone wants also to investigate the distances of the superiors in Earth radii, since they have no parallax, let him convert proportionally, as we have done for the inferiors, having added the [planetary] radii, as stated; and let him take, for the first and the second [numbers in the Rule of Three], the perigee of the planet whose distance he seeks or [the apogee] of the one below it–it is the same [distance]; for the first, [the perigee] in parts of its eccentric's radius; for the second, [the perigee] in e.r.; for the third, the same planet's apogee in [60th] parts of its eccentric; and let these parts [60ths] be converted into those parts [e.r.], as we show in fact. We give here, however, the numbers assumed by Ptolemy and the divisions of multiplications, estimated rather than exact, as anyone can easily get the exact ones. For we want to present the procedure and manner of understanding these things, not to show them [in detail].*[115]

Strangely, George now professes to forego exact calculation in favor of estimates, placing the burden of precision on the reader. He does not justify this unnecessary move, which turns out to be revealing. The 'numbers assumed by Ptolemy' are obviously the radii of the epicycles and the relative perigees and apogees of the epicycle centers in his models (in 60ths of the deferent radius). It is from these figures, all summarized in *Almagest* 11.10, that the absolute perigees and apogees of the planets' centers can be calculated.[116] The expression 'divisions of multiplications' designates the Rule of Three for a given planet: the product — in his now-familiar terminology — of the second by the third, divided by the first.

What George of Trebizond provides, however, are not estimates, but computations with bizarre results:

> *We found most exactly that the Sun's apogee from the Earth's center, where the Earth's radius is one, is approximately 1309;46. Add 5;30 for the Sun's radius in the same units, as demonstrated in book 5, and 1;10 for the radius of Mars's body. Together, this*

[115] 'Quod si quis superiorum etiam distantias ad partes de quibus semidiameter terre est unius rimari velit, quoniam nulla in illis diversitas est, traducat proportionaliter sicut nos in inferioribus fecimus, additis semper semidiametris, ut dictum est; et eius minimam cuius distantiam querit vel inferioris proximi — idem enim est — capiat pro primo et secundo; pro primo, partes semidiametri eccentrici sui; pro secundo, secundum quod semidiameter terre est unius; maxima vero eiusdem planete pro tertio secundum partes eccentrici sui, et traducantur partes iste ad partes illas sicut re ipsa ostendimus. Ponemus autem hic numeros sumptos a Ptolemeo et partitiones multiplicationum estimative magis quam exacte ut facilius uniusquisque possit ad exacta pervenire; modum enim et viam intelligendarum harum rerum dare non ostentare nos volumus'. George of Trebizond, quoted by Regiomontanus, *Defensio*, fol. 153v.

[116] Toomer, *Ptolemy's Almagest*, p. 546.

yields 1316;26 for the second number; for the third, let Mars's apogee demonstrated in the tenth [*book*] *be 105;30 in 60ths of the deferent's radius. Carrying out the approximate multiplication of the second and the third, and likewise the division by 14;30 of Mars's perigee simply, the result is 9923, and this is Mars's apogee 'simply' in e.r., from its own center to the Earth's center.*[117]

The language and final result are puzzling. Computing Mars's absolute apogee from George's numbers yields 9578 e.r., not 9923.[118] This is no scribal error. The relative values for Mars's apogee (105;30) and perigee (14;30) are straightforwardly Ptolemaic. Regiomontanus evidently copied George's numbers correctly for they match those in the Vienna manuscript of George's commentary (ÖNB, cod. 3106). The language of 'approximate multiplication... and division likewise' is strange, for the 'exact' operation (urged on the reader) is easy. Moreover, the result shows no rounding, as expected in a guess or estimate.

A leading clue about this puzzle lies in George's heavy use of Campanus's *Theorica planetarum*, which suggests the following hypothesis about his reasoning. With Campanus at hand, George may have been confused when, using relative numbers consistent with the *Almagest*'s for Mars's apogee and perigee, his results differed, as they do, from those in the *Theorica planetarum*. Uncharacteristically, Campanus made a half-degree mistake when recording Mars's eccentricity as 6;30 (instead of 6 in Ptolemy). When this error entered into his computations, Campanus's value for Mars's apogee (106) became too large by 30′ and the perigee (14) correspondingly too small by the same amount. These unexplained deviations from the Ptolemaic values, together with the divergence of George's own presumed computations from those in Campanus, no doubt puzzled him greatly. To this point, George's results had matched those of Campanus. With his erroneous values of Mars's apogee and perigee in the Rule of Three, however, Campanus got 9967;15,36 e.r. for the apogee of Mars's center.[119]

[117] 'Maxima solis a centro terre distantia prout semidiameter terre est unius inventa nobis exactius fuit 1309 46' proxime; adde 5 30' semidiametri solis similium partium, ut in libro quinto demonstratur, et 1 10' pro semidiametro corporis Martis; simul, 1316 26' pro numero secundo; pro tertio, ponatur maxima Martis simpliciter demonstrata in decimo partium 105 30' de partibus semidiametri sui deferentis 60; facta multiplicatione secundi et tertii estimativa et partitione similiter per 14 30' minime distantie Martis simpliciter proveniunt 9923 et hec est distantia Martis maxima simpliciter a centro terre ad centrum suum secundum quod semidiameter terre est unius'. George of Trebizond, quoted by Regiomontanus, *Defensio*, fols 153v–154r.

[118] Mars apogee (e.r.) = Mars apogee (60ths) × Mars perigee (e.r.) /Mars perigee (60ths) = 105;30 × 1316;26 / 14;30 = 9578;11 e.r.

[119] Campanus's computation gives 106 × 1316;25,50 / 14 = 9967;15,35 (rounded up to 36). This erroneous eccentricity was picked up in the fourteenth century by Jean de Lignères and the author of the *Equatorie of the planetis*, possibly Chaucer; Benjamin and Toomer, *Campanus of Novara*, pp. 337, 438; de Solla Price, *The Equatorie of the Planetis*, pp. 69, 126–27.

Apparently, George neither trusted his own computations, nor tried to find the source of the discrepancy with Campanus. Using nearly the same physical apogee of the Sun as Campanus (rounded up by 10″) and the Ptolemaic values for Mars's perigee and apogee, George should have got 9578, as noted above. The 389 e.r. difference between this number and Campanus's result dwarfed by at least an order of magnitude the sum of all the planetary radii that George, like Campanus, so painstakingly had added to the traditional dimensions of the planetary spheres. George's own figure of 9923 usefully decreased his difference by an order of magnitude, but I see no complimentary explanation of his result.[120]

The preceding hypothesis would explain the odd preamble to his calculation, with its unnecessary estimates and approximations that erase completely the hyper-precision of the added planetary radii. Indeed, George's qualifications and swerves make little sense apart from deference to a standard. It was Campanus of Novara's unmentioned but omnipresent *Theorica planetarum*, not the *Commentaria*'s arithmetic, that shaped George's results and the size of his cosmos. Whatever its precise origin, the mysterious figure of 9923 for Mars's apogee served to compute the dimensions of the remaining spheres:

> *For Jupiter, likewise. For the first, 45;45 or approximately 46 (it does not matter) in 60ths of the deferent radius; for the second, 9928;44, having added to the apogee [of Mars, 9923] 1;10 and 4;34 for the radii of the bodies of Mars and Jupiter. Let these [parts] be multiplied by the third, that is, 74;15,*[121] *for Jupiter's apogee simply in 60ths of the deferent radius, and having divided [this product] by the first, [the operation] yields 16125, Jupiter's apogee simply, reduced to the aforementioned units [e.r.].*[122]

[120] There is no matching solution for the eccentricity X in integers and minutes (George omits seconds):

9923 = [(60 + X + 39;30) × 1316;26]/(60 − X − 39;30)
= (99;30 × 1316;26) + (1316;26X)/ (20;30 − X)

Thus

9923 (20;30 − X) = (99;30 × 1316;26) + 203421;30 − 9923 X
= 130985;06 + 1316;26X

so that 72436;24 = 11239;26 X

and X = 6;26

This eccentricity yields values closest to George's: (60−6;26) − 39;30 = 14;04 (perigee) and (60 + 6;26) + 39;30 = 105;56 (apogee). The Rule of Three gives a result 10 e.r. short of George's; using the next closest pair, the answer is too high by 3 e.r.

[121] The relative proportions and the planets' sizes also appear in Campanus's *Theorica planetarum*; Benjamin and Toomer, *Campanus of Novara*, pp. 243, 333, 339–41.

[122] 'In Iove similiter pro primo 45 45', aut 46 proxime, nihil enim curamus secundum partes semidiametri terre unius additis ad maximam Martis 1 10' et 4 34' semidiametrorum corporis Martis et Iovis 9928 44' pro secundo hec in tertium, hoc est in 74 15', maxime simpliciter Iovis distantie secundum partes semidiametri deferentis 60 multiplicentur factaque par-

Again, George uses the standard Ptolemaic values for Jupiter's relative perigee and apogee, which should yield 16026 e.r., that is, 99 e.r. short of his result of 16125.[123] Here is an additional odd 'estimate' that also compensates for errors he was disinclined to track down.

> *To this [16125], add the radii of Jupiter and of Saturn, that is, 4;34 and 4;30, and you will get 16134 for the second; for the third, however, approximately 70 for Saturn's apogee in 60ths of its deferent radius. Having carried out the multiplication [of the second by the third] and division by the first, that is, by approximately 50 for Saturn's perigee,*[124] *one gets its apogee] of 22587 e.r.; if you add to this Saturn's radius of 4;30, there will be 22591;30 [e.r.] from the Earth's center to the concave surface of [the sphere] of the fixed [stars].*[125]

Uncharacteristically for the superior planets, the arithmetic for Saturn's apogee is correct (although it still builds on erroneous values for the two preceding planets). By keeping the minutes of the Ptolemaic values, George would have obtained 22523;10′ e.r. By rounding to the nearest unit, George decreased the numerator and increased the denominator slightly. After adding Saturn's radius, the Rule of Three gave him 22591;30 e.r. for the distance to the fixed stars, a figure that comfortingly approximates Campanus's 22607;58,16 e.r. for Saturn's apogee.[126] The original error of 389 e.r. for Mars should have been compounded; miraculously, it has shrunk to 16 upon reaching Saturn.

titione per primum exeunt 16125 maxima Iovis simpliciter ad partes ut dictum est reducta...' George of Trebizond, quoted by Regiomontanus, *Defensio*, fol. 154r.

[123] The Ptolemaic values also appear in Campanus; Benjamin and Toomer, *Campanus of Novara*, p. 339. Using George's numbers, the computation for Jupiter's apogee (e.r.) should yield (9928;44 × 74;15) /46 = 16026;16. Various roundings fail to generate George's number of 16125.

[124] Here George has got his 'approximate' values of 50 and 70 for the relative apogee and perigee of Saturn by rounding by 5′ (up and down, respectively) the Ptolemaic values, which are also those of Campanus; Benjamin and Toomer, *Campanus of Novara*, p. 341. Without rounding the minutes, George would have obtained 22523;10.

[125] '...cui numero [= 16125], adde Iovis et Saturni semidiametrum, hoc est 4 34' et 4 30', et habebis 16134 pro secundo; pro tertio, autem, 70 proxime maxime Saturni distantie secundum partes semidiametri sui deferentis 60; factaque multiplicatione ac partitione per primum, hoc est per 50 proxime, minime Saturni distantie habetur maxima eius prout semidiameter terre est unius partium 22587 quibus si addideris semidiametrum Saturni 4 30' erit a centro terre ad concavam superficiem fixarum 22591 30". George of Trebizond, quoted by Regiomontanus, *Defensio*, fol. 154r.

[126] George seems not to have noticed that this figure — the last one Campanus gives in e.r. — should be larger by 4;30 e.r., since Campanus omitted the final step of adding the radius of Saturn's body in e.r. to the apogee of Saturn. Immediately after this passage, however, Campanus did include this radius when calculating the distance of the sphere of fixed stars in miles. Benjamin and Toomer, *Campanus of Novara*, p. 343.

And thus it is clear that the distance of the fixed stars contains the Sun's mean distance approximately 18 times.[127] *From here, it will be easy to find the circumference and the area of these same orbs and the circumference of the epicycles and the sizes of individual planets reduced to similar parts, that is, insofar as the radius of the Earth is one, by the proportion of diameters reduced to cubes. For example, the Earth's diameter is to that of Saturn roughly as 2 to 9, the cubes of each are 8 and 729, and the latter contains the former 91;22' times approximately.*[128] *Likewise, the diameters of the Earth and Sun are as 2 to 11, their cubes 8 and 1331, and the latter contains the former 162;22' times approximately. Likewise, the proportion of the diameters of the Earth and Jupiter is as 2 to 9;8, the cubes of which numbers are 8 and 762 approximately, which is why the planet Jupiter contains the Earth 95;15 times;*[129] *likewise for the others. And this is by* Elements *12.29, for the proportion of a globe to a globe, that is, of a ball to a ball, is the proportion of the diameters cubed. This is partly because we can easily go from diameters to circumferences and, from both of these, to areas, and thence to globes. Indeed, from here, insofar as circles are magnitudes, it is not difficult to find the paths and the circumferences and the circles of epicycles and likewise of eccentric deferents.*[130]

Although he does not cite Euclid, Campanus makes comments similar to these after his own computations of planetary distances. Here, George also works out the Earth-to-Saturn ratio, as he had earlier the Earth-to-Jupiter ratio, while noting that the other values can be calculated in the same way.

For since we can get the epicycle diameters in terms of the parts of eccentrics, that is, diameters that I might call eccentrical, and the parts of eccentrical diameters [*expressed*] *as e.r., here reduced by us, in this investigation of distances, it is laborious to carry out,*

[127] 22591;30 / 1257 = 18;04 [George's mean Sun computed from the apogee and perigee figures given at 153v].

[128] An error, since 729/8 = 91;07,30. This is perhaps a scribal error caused by the eye skipping to the 22 minutes imbedded in similar phraseology two lines below.

[129] See Benjamin and Toomer, *Campanus of Novara*, p. 341, where Campanus's value is equivalent to 95;14,07.

[130] 'Et sic patet quod media distantia solis decies octies proxime continetur a distantia stellarum fixarum. Hinc ambitum ac superficiem ipsorum orbium et epicyclorum circuitum et planetarum singulorum magnitudines ad partes similes, hoc est prout semidiameter terre est unius redactas facile erit invenire partim a proportione diametrorum in cubos redacta. Verbi gratia, diameter terre ad Saturni diametrum est sicut 2 ad 9 proxime, cubi utriusque 8 et 729 et continet hic illum vicibus 91 22' proxime. Item terre ac solis diametri sunt sicut 2 ad xi cubi 8 et 1331 et continet hic illum 166 22' proxime. Item proportio diametrorum terre ac Iovis est sicut 2 ad 9' 8 horum numerorum cubi sunt 8 et 762 proxime. Quare stella Iovis continet terram 95 15' vicibus et similiter in ceteris. Idque per 29m duodecimi Elementorum. Proportio enim globi ad globum, hoc est pile ad pilam, est proportio diametrorum triplicata; partim quoniam facile a diametris ad circumferentias et ab utrisque ad superficies et tandem ad globos solemus pervenire. Ambitus vero et circuitus ac circulos epicyclorum et similiter deferentium eccentricorum prout circuli sunt magnitudines, non est difficile hinc invenire'. George of Trebizond, quoted by Regiomontanus, *Defensio*, fol. 154r-v.

> *but easy to understand, first how the epicycle diameters are reduced to e.r., and next how the circles are investigated by means of their diameters.*[131]

With this wordy statement, Regiomontanus concludes his very long quotation from George of Trebizond's commentary.

3. Regiomontanus's criticisms of the *Commentaria*

Regiomontanus's response is characteristically detailed and scathing. It is also surprising in content. The most general part of his attack on George's commentary refutes the boast of having demonstrated the order of the planets, both superior and inferior. George had argued that an observation and a 'good fit' justified the traditional order of the inferior planetary spheres. Regiomontanus now undermines this argument by showing that one obtains a better fit by inverting their traditional order (i.e., placing Venus next to the Moon and Mercury next to the Sun). While showcasing his dialectical prowess, Regiomontanus will also offer glimpses of fundamental physical assumptions that he shares with contemporaries, including George (e.g., the rejection of vacua between the spheres). More importantly, he will sharply criticize traditional assumptions, notably the project of computing cosmic distances and the principle of uniform circular motion.

For now, he proceeds by dismantling George's claims one by one:

> We have transcribed in full the expositor's more-prolix-than-useful narration so that it will be clear to which passages our arguments pertain. We introduce these arguments not to impugn Ptolemy's attempt at a planetary order, but to show that the reasons the expositor deems tough as diamonds are very weak. Indeed the things admitted [as premises] by analogy or with other loci, whether dialectical or rhetorical, we may quickly judge to be unworthy means of making astronomical laws solid.[132]

Regiomontanus approaches George as if he were in an academic disputation. Highlighting the opponent's bad argumentation by any means available

[131] 'Nam cum habeamus diametros epicyclorum secundum partes eccentricorum, id est, diametrorum — ut sic dicam — eccentricalium, et partes eccentricalium diametrorum ad partes prout semidiameter terre est unius hic nobis redactas, in ipsa investigatione distantiarum laboriosum quidem factu sed facile intellectu est quomodo primum diametros [read diametri?] epicyclorum ad partes prout semidiameter terre est unius reducuntur, deinde a diametris circuli investigantur'. George of Trebizond, quoted by Regiomontanus, *Defensio*, fol. 154v.

[132] 'Hanc expositoris prolixam magis quam utilem narrationem integre transscripsimus ut ~~adnotationes~~ rationes nostre quibus locis accommodentur manifestum fiat, quas quidem introducimus/afferimus non qua planetarum ordinem a Ptolemeo probatum impugnemus, sed qua rationes expositoris quas ipse adamantinas putat, infirmas plerumque esse ostendamus. Verum que per similitudines quasdam aut alios sive dialecticos sive rhetoricos locos sumuntur cursim attingemus tanquam non satis dignas quibus astronomica roborentur decreta'. Regiomontanus, *Defensio*, fol. 154v.

is sometimes more effective than finding the truth of the matter, a stance that sometimes complicates our assessment of Regiomontanus's conclusions. Although the latter certainly undermine George of Trebizond's claim to have demonstrated planetary order, we cannot always be sure that his polemical conclusions double as positive theses that Regiomontanus endorses. Nevertheless, he minimizes the ambiguity in principle, if not in practice, by professing to exclude dialectical and rhetorical arguments from astronomy, in effect everything except demonstrative ones.

3.1. The 'middle' in the heart-Sun analogy

For this reason, at the outset, Regiomontanus cannot resist confronting contradictions and ambiguities in George's analogy between the heart, the Sun, and the middle of the universe. He begins by setting up a *reductio ad absurdum*:

> First, he affirms, it is necessary that the organ of the world that is proportioned to the heart occupies the middle position in that universe. That fellow does not notice the discrepancy lurking in this point. Let it be granted that the Sun has a similarity to the heart and is for that reason proportioned to it, that is, that it occupies the middle position in the universe. One must now see if the Sun, being in the middle of the planets, might have a similar location in the celestial region. It is evident, however, that if we existed as the whole depth[133] of the Plinian/Ptolemaic heavens and adapted it, for example, to the height of a human body, the Sun by its location occupies the position of the foot rather than the heart. For the thickness of the human foot near the heels is close to the twentieth part of the entire human stature. And the Sun's distance from the elementary region where the superior world begins is almost 1/20th of the entire celestial height. He who compares the Sun to the heart and therefore affirms that the place it now holds in the heavens is rightly assigned to it, implies either that the human heart in such a place is located at the feet, or that the Sun is to be placed in Jupiter's sphere so that it might take the middle place in the height of the heavens, neither of which Nature surely allows. <margin: examine this place more precisely>[134]

[133] In this unusual construction, I take *profunditatem* to be an accusative of space predicated on an existential sense of *esse*.

[134] 'In primis itaque necessarium esse affirmat ut membrum mundi quod cordi proportionatur medium ipsius mundi locum obtineat. Non advertit homo ille latentem in hac re discrepantiam. Detur enim solem habere similitudinem cordis idcircoque proportionatum, id est medium in mundo locum obtinere. Iam videndum est si sol medius planetarum existens similem in regione celesti situm habeat. Constat autem si totam celi profunditatem plinensi <margin: ptolemaici> fuerimus et eam proceritati, verbi gratia humani corporis adaptaverimus, solem ipsum situ suo pedis locum potius quam cordis occupare. Crassitudo enim/siquidem pedis humani iuxta talos a vigesima parte totius proceritatis humane haud milium discrepat. Et remotio solis ab elementari regione unde mundus superior initium sumit vigesima ferme pars est totius celestis altitudinis. <margin: vide per calculum> Qui ergo solem cordi assimilat ideoque ei locum quem nunc in celo habet iuste tributum esse affirmat aut cor humanum in loco tali vel pedis ponendum, aut solem in sphera Iovis statuendum esse insinuat ut medium

To undermine the Sun-heart analogy so central to the microcosm-macrocosm parallelism, Regiomontanus examines the relative spatial proportions of its two terms. The Sun is about 1000 e.r. from the Earth, and the fixed stars roughly 20,000 e.r. Transferring this 20:1 ratio to the human body puts the heart near the feet.[135] Conversely, if the Sun held approximately the middle position in the cosmos at some 10,000 e.r., it would be inside Jupiter's (partial) sphere. But Nature allows neither a heart in the feet, nor a Sun in Jupiter's sphere. Regiomontanus was perhaps working from memory here: his marginalium suggests that he wanted to check the proportions.

Regiomontanus anticipates the rebuttal by examining another possible construal of this ambiguous analogy:

> And if someone should say that the Sun has an intermediate place in terms of, not continuous spatial equidistance, but rather a discrete [= ordinal] one, such that it has three planets below it and three above, why therefore does he exclude the entire multitude of fixed stars that make up the eighth sphere, as if they were not members of the celestial body? Yet these [stars] have the greatest power over this inferior world, if we believe Ptolemy himself <margin: as Ptolemy himself determined, they pour out [the greatest power] on terrestrial bodies>. This is why, if the Sun is to be given the middle position among the stars, it surely cannot be below Saturn, lest it have more [stars] above than below.[136]

For the Sun, the alternative ordinal meaning of 'middle' makes no more sense than the linear or geometrical one, since it arbitrarily omits the fixed stars from the count. On Ptolemy's authority, however, the stars have the greatest astrological effects on the lower world (Regiomontanus reinforces the point in the margin) and should therefore be counted. Placing the Sun in the middle of this ordinal series would have the counterfactual consequence of locating the Sun above Saturn. On neither mathematical understanding of 'middle', then, is George's (widely shared) analogy tenable. Since Regiomontanus's target is George, he does not mention that his argument also strikes Ptolemy who, without mentioning the heart, placed the Sun 'in the middle' of a taxonomy of celestial bodies that also excluded the fixed stars.

in altitudine celi locum possideat <margin: vide exactius locum hunc>, quorum certe neutrum natura permittit'. Regiomontanus, *Defensio*, fols 154v–155r.

[135] Regiomontanus's figure precisely matches the proportion for the top of the instep to the full height (3 units out of 60) in the *De statua* of Alberti, whom Regiomontanus knew; see Aiken, 'Leon Battista Alberti', pp. 68–96, esp. 94.

[136] 'At si quis dicat solem possidere medium locum non quidem spaciali et continua sed discreta quadam equidistantia quod tres sub se et totidem supra se habeat planetas, cur ergo excludet tantam stellarum multitudinem que in octava sphera consistunt quasi non sint membra celestis corporis? Que tamen maximam vim in mundum istum inferiorem habent si Ptolemeo ipsi credimus [*margin*: terrenis corporibus Ptolemeo ipso decernente infundunt]. Quare si medius inter stellas soli tribuendus est locus sub Saturno utique esse non poterit ne plures supra quam infra habeat'. Regiomontanus, *Defensio*, fol. 155r.

After this opening banter, Regiomontanus turns to the more substantive issue of the Sun-Moon interval:

> But setting aside this lame argumentation, we may attack another argument, in the preliminaries of which he relates that, for the Moon, the parallax angles vary much and appreciably, and therefore that Ptolemy had demonstrated the diverse distances of the Moon via the difference of angles of this sort [= parallax]. We would forgive this trivial difficulty as a slip-up if [George] had not shown himself more careful in this passage than in others, which one is surely allowed to infer from his more diffuse exposition [elsewhere].[137]

Characteristically, Regiomontanus also addresses secondary issues if he can turn them into evidence of George's misunderstanding or ignorance:[138]

> Indeed Ptolemy did not investigate the various distances of the Moon from the difference of angles that are known from the parallaxes. Rather, from the various distances of the Moon from the Earth, he inferred the various parallaxes and the differences of angles pertinent to them, and put them in the parallax table. He found a single parallax with the parallactic ruler, from which he derived the Moon's distance from the Earth in Earth radii. And since the Moon in that position also had a known distance in eccentric radii, having converted (as usual) the proportions into the new terms, he also knew, in Earth radii, the remaining distances of the Moon from the Earth, which earlier had been known in terms of the eccentric radius; from here, then, [the distances] revealed the remaining parallaxes, together with the angles associated with them.[139]

In short, George has misunderstood Ptolemy's procedure. As Regiomontanus states, summarizing *Almagest* 5.11ff, Ptolemy did not determine the distances of the Moon from parallaxes, but the other way around. Ptolemy reported using

[137] 'Sed dimissa hac fragili argumentatione in aliam aggrediamur in cuius apparatu commemorat angulos diversitatis aspectuum in luna multipliciter ac sensibiliter variari atque idcirco per diversitatem huiusmodi angulorum a Ptolemeo diversas lune distantias esse demonstratas. Hunc levem honum condonaremus errorem nisi in hoc passu circumspectiorem se ostentaret quam in aliis locis, quod quidem ex diffusiori narratione sua coniectari datur'. Regiomontanus, *Defensio*, fol. 155r.

[138] This approach also appears in his *Disputationes*; see Pedersen, 'The Decline and Fall', esp. p. 185.

[139] 'Ptolemeus quidem non per diversitatem angulorum qui ex diversitate aspectus innotescunt varias lune investigavit distantias, sed econtra, per varias lune a terra distantias, varias diversitates aspectuum et angulorum ad eas attinentium elicuit et in tabula diversitatis aspectus collocavit. Unicam enim de aspectibus diversitatem instrumento regularum didicit, unde remotionem lune a terra conclusit respectu semidiametri terrestris; cumque in eo situ luna notam quoque respectu semidiametri ecentrici haberet distantiam traductis, ut assolet, proportionibus ad novos terminos reliquas lune a terra distantias que prius ad semidiametrum eccentrici note fuerunt, ad semidiametrum quoque terrestrem cognovit; hinc demum reliquas aspectuum diversitates una cum angulis se respicientibus latere non poterant'. Regiomontanus, *Defensio*, fol. 155r-v.

the parallactic ruler for *one* measurement of the Moon on the meridian. From the difference between this measurement and the value predicted by his lunar theory (reckoned from the Earth's center), he calculated the lunar parallax. He then used this value to express the relative sizes of the lunar parameters in absolute distances (e.r.).[140] Since Regiomontanus was keenly aware that Ptolemy's lunar theory implied nearly a 2:1 variation in distance (hence 4:1 in area) from quadrature to syzygy (*Epitome* 5), he was particularly sensitive to the discrepancy between phenomenon and computation for the Moon in particular.[141]

> Next, during this long argumentation, [George] insinuates that the parallaxes of Mercury and Venus are all but imperceptible. If he means imperceptible, that is, [in the sense of] incomprehensible/ungraspable, his statement is tolerable, for only with difficulty can the parallax be understood for these planets, as their true motions cannot be found exactly, since one cannot find the parallax without first knowing the true place of the planet [margin: Mercury alone].[142] Each of the preceding planets is carried through so many circuits that one might properly suspect that their motions are beyond human understanding, especially that of Mercury, which is rarely visible.[143]

In seeking an acceptable meaning of 'imperceptible', Regiomontanus's first concession to George does not focus on the empirical failure to find parallaxes for the inferior planets (imperceptible in the first sense, that of *Almagest* 9.1). Rather, the problem lies in the 'incomprehensibility' of the motions of Venus and especially Mercury. Although an empirical consideration surfaces briefly in the difficulty of seeing Mercury, the passage highlights primarily the inaccuracies and complexities of the theories of the inferior planets, and of Mercury in particular. Inevitably, Regiomontanus turns next to the unacceptable meaning:

> If by 'imperceptible', he means that it does not reach 2′ or 3′, as is usually said to be the case for the parallax of Mars or Jupiter, we say: this man's skill is already obvi-

140 Toomer, *Ptolemy's Almagest*, pp. 243–44; Pedersen, *A Survey of the Almagest*, pp. 203–07; Neugebauer, *A History of Ancient Mathematical Astronomy*, pp. 100–01.

141 First noticed in the fourteenth-century by Ibn al-Shāṭir and Levi ben Gerson, and picked up by Henry of Langenstein and Regiomontanus (see below).

142 This marginal addition looks like a later editorial change (the ink is lighter and the hand more cursive than those of the text). The marginalium suggests that the original sentence, which discusses both planets, should pertain to 'Mercury alone'. The multiplicity of circuits describes the complex Mercury model better than that of Venus. Regiomontanus returns to this point at the end of his critique (159v).

143 'Deinde in processu huius longe argumentationis insinuat Mercurii Venerisque diversitatem aspectus pene insensibilem esse. Si quidem insensibilem dicit, hoc est incomprehensibilem, tolerandum est hoc verbum. Difficiliter enim valde in iis stellis comprehendi potest aspectus diversitas quarum ne veri quidem motus ad unguem perquiri possunt; quandoquidem diversitas aspectus non nisi precognito vero stelle loco inveniri potest. <margin: Mercurius solus> Uterque autem predictorum planetarum tantis fertur ambagibus ut suspicari quis haud iniuria possit motus eorum humanitus esse incomprehensibiles et presertim Mercurii perraro ad visum apparentis'. Regiomontanus, *Defensio*, fol. 155v.

ous. [Whether] you suppose that either Mercury or Venus is [immediately] above the Moon, its perigee is nearly the lunar apogee. Therefore, when either of these [planets] is found in such a perigee, it will necessarily show a parallax like the Moon's at apogee, which, when it is investigated seriously, can in no way be called imperceptible.[144]

In this apt criticism of George (who here strays far from the *Almagest*),[145] Regiomontanus highlights an abiding tension in the nesting hypothesis, with its rejection of vacua or useless gaps. Whichever inferior planet is next to the Moon must, at perigee, have the same parallax as the Moon at apogee. He gives no figure, but the 1;7° derived from Ptolemy's meridian measurement of the Moon in *Almagest* 5.13 gives the order of magnitude. Equally striking is the specification of an order of magnitude for the parallax of Mars or Jupiter. For the planet closest to the Moon, therefore, an 'imperceptible' parallax cannot mean 'too small to be seen' for the Moon at apogee has a visible parallax.

Regiomontanus is, however, far from seeking to undermine the subsolar location of Venus and Mercury, whatever their order may be:

We do not deny that it is reasonable to offer the space inserted between the luminaries to some other planets, particularly to Venus and Mercury, lest the universe have either an empty space or an immense celestial structure devoid of all function. Still unexplored, however, is which of the preceding planets should be placed immediately above the Moon.[146]

As this important passage shows, Regiomontanus shares key physical principles with George and others: the universe has neither a vacuum, nor a useless plenum. Venus and Mercury, in a still unspecified order, are thus presumed to fit in the large gap between the Sun and the Moon.[147] With this physically moti-

[144] 'Si vero insensibilem significat ut que duas aut tres sexagesimas unius gradus non attingat quemadmodum de diversitate aspectus Martis aut Iovis <margin: dici solitum est> quantulacumque sit, dicimus: iam patet hominis peritia. Nam sive Mercurium lune sive Venerem supponas, constat minimam eius a terra distantiam esse maximam lune proxime. Quando igitur alter eorum in minima tali distantia reperitur, necessario diversitatem aspectus sortietur quantam luna habere potest in maxima sui remotione que cum magnopere operepretium investigetur insensibilis dici non poterit'. Regiomontanus, *Defensio*, fol. 155v.

[145] Recall that, in the *Planetary Hypotheses*, Ptolemy dropped the *Almagest*'s requirement that the inferior planets be arranged to conform with the absence of parallax.

[146] 'Quod igitur spatium luminaribus interiectum rationabiliter quibusdam aliis vendicetur planetis presertim autem Veneri et Mercurio ne aut vacuus in mundo locus aut ingens celestis moles omnis officii exors inveniatur, non diffitemur. Uter autem dictorum planetarum statim supra lunam poni debeat nondum exploratum est'. Regiomontanus, *Defensio*, fol. 155v.

[147] In *Epitome* 9.1 ('Sphere celestes quo ordine habende sint ostendere' — 'To show what order the celestial spheres should have'), Regiomontanus argues that the space between the Moon and the Sun should be full because 'nature will not allow this space to be empty; therefore necessarily some celestial body will occupy it' (Swerdlow's translation; original in Schmeidler, *Joannis Regiomontani opera collectanea*, pp. 192–93). Venus and Mercury will fill this large space *commoditate naturali* — 'by natural fitness/suitability' [cf. 'natural proportion'

vated concession, Regiomontanus proceeds to undercut George's allegation of an empirical basis for the traditional order (Moon, Mercury, Venus):

> What this fellow implies — that Mercury is recorded as having gone below Venus [*Defensio* 152r] — has the air of a dream, since neither can he name the author of this claim, nor can vision adequately detect a lower position owing to the smallness of the planet Mercury, which also has its own light.[148]

Regiomontanus gives two reasons for dismissing the alleged sighting of Mercury below Venus.[149] First, it is undocumented. Second, and more fundamentally, Regiomontanus doubts that this alleged phenomenon could have been perceived: the planet is not only too small (difficult to see in the best of times), but also self-luminous (a point left unsubstantiated). The argument attacks the implicit assumption of a partial occultation of Venus by Mercury. If both planets are self-luminous, however, there can be no occultation, and therefore no way of telling which planet is lower (what we call transits would all be impossible to see).[150]

Regiomontanus now considers the speed-distance rule in George's final argument for 'demonstrating' a planetary order (key marginal additions are italicized):

> <Greater> speed *of revolution*, however, cannot warrant a lower position [for Mercury], *as the expositor claims to be necessary, unless we want to place the Sun below Venus, along with al-Biṭrūjī, the sole author of this delirious opinion*. Indeed, the Sun has a longitudinal motion equal to that of Venus; but the circuit of the anomaly,

in Swerdlow, 'Copernicus and Astrology']. Emphasizing the uncertain order of Venus and Mercury, Regiomontanus focuses on Mercury, noting that the most favorable positions for seeing it (at 'mean distances') are not optimal for parallax observations: 'But which of these two [Venus and Mercury] is located above the other cannot be discovered with certainty. For in most climates Mercury appears very rarely and, if it appears, it does so when it is near mean distances of the epicycle; and although it then has a parallax, the parallax is much smaller than what it would have, were it in the perigee of the epicycle. Therefore such a parallax [of Mercury] cannot be found exactly since neither in instruments necessary to this matter nor in computing the motions of Mercury can we obtain complete precision. And the same will be proper to hold concerning Venus'. Schmeidler, *Joannis Regiomontani opera collectanea*, translated by Swerdlow, 'Copernicus and Astrology'.

148 'Nam quod ille scribi autumat, Veneri Mercurium successisse/succurrisse, somnii speciem habet cum neque autorem eius rei nominet, neque subiectio huiuscemodi visu satis comprehendi possit propter parvitatem stelle mercurialis, lumen etiam proprium habentis'. Regiomontanus, *Defensio*, fol. 155v.

149 Regiomontanus uses 'scribi' to paraphrase George, implying a record of some sort (reinforced by the past infinitive in the synonyms *successisse/succurrisse*).

150 Averroes claimed to have seen a double transit when Venus and Mercury were both in conjunction with the Sun, whereas Levi ben Gerson used an argument similar to that of Regiomontanus against the possibility of observing transits. Goldstein, 'Some Medieval Reports', esp. pp. 53–54.

> which is measured carefully in the epicycle, is faster. For, in one year, it [= Mercury] traverses the zodiac as well as the epicycle more than 3 times whereas Venus covers only 5/8ths of its epicycle.[151]

Early on, George had given allegiance to the quasi-Aristotelian principle that '*in circular motion, it is necessary that the inferior be faster*' (152r). Regiomontanus here attacks not the speed-distance rule itself, but the contradiction between it and George's endorsement of the traditional planetary order. A consistent application of the principle yields not the order that George defends, but that of al-Biṭrūjī (Venus above, and Mercury below, the Sun).

Regiomontanus's heavy revisions — substitutions in the left margin and addenda in the lower margin — show that he went over this passage several times. Since they reflect shifts in his thinking, their flow therefore deserves examination. Indeed, the marginalium *vide diligentius hoc* ('examine this very carefully'), now crossed out, suggests additional research. When he first composed the main text, Regiomontanus presented the Mercury-Sun-Venus sequence as an unpalatable consequence of George's commitment to the speed-distance rule, without mentioning al-Biṭrūjī. In the revisions underlined in the quotation above, however, Regiomontanus identified this order with the 'delirious opinion' of al-Biṭrūjī.

In the margin at the foot of the page, he also added: 'See Alpetragius who places Venus ~~and Mercury~~ above the Sun. Likewise Abraham Ibn Ezra. Likewise Martianus. Likewise Pliny'.[152] In this note to himself, he originally (and erroneously) thought al-Biṭrūjī placed both Venus and Mercury above the Sun. Regiomontanus eventually corrected the lapse, whether momentary or not. He most probably crossed out 'and Mercury'[153] after listing the names

[151] 'Velocitas autem *circuitionis* inferiorem situm *veluti expositor tamquam necessarium autumat inferiorem situm prestare non potest nisi Veneri solem subiicere velimus cum Alpetragio huius opinionis unico autore delirantis*, quippe motum longitudinalem sol habet equalem cum Venere, circuitum autem diversitatis qui in epicyclo perpenditur longe celeriorem. In anno enim suo tam zodiacum quam epicyclum emetitur ter et amplius semel dum Venus quinque solum octavas partes epicycli sui perambulat'. Regiomontanus, *Defensio*, fols 155v–156r.

[152] 'Vide Alpetragium qui Venerem ~~etiam et Mercurium~~ supra solem sistit. Item Abraham avenezre. Item Martianum. Item ~~…um~~ Plinium'. Regiomontanus, *Defensio*, fol. 155v, lower margin. More than 3 dozen manuscripts of Pliny's astronomical excerpts survive, so that the Carolingian tradition associates him with the order of Moon-Mercury-Venus-Sun, against the Platonic/Egyptian order (Moon-Sun-Mercury-Venus); Eastwood, *Ordering the Heavens*, pp. 36–43, 250. On the controverted question of Ibn Ezra's views on Venus and Mercury, see Rodríguez-Arribas, 'Did Ibn Ezra Maintain a Circumsolar Arrangement', esp. pp. 202–12.

[153] In his manuscript of al-Biṭrūjī (Nuremberg, Stadtbibl., Cent V 53), Regiomontanus wrote at the very beginning of the work: 'Geber Mercurium et Venerem supra solem posuit' (74r); and later about al-Biṭrūjī: 'Venus supra solem' (101v) but also 'Quod Venus et Mercurius non obscurant solem non esse sufficiens indicium sue supra solem collocationis'; see also Zinner, *Leben und Wirken*, pp. 61–62.

of Ibn Ezra, Martianus Capella, and Pliny (the ink and script are consistent with one sitting). All three authors described non-traditional planetary orders, and the first two subscribed to a circumsolar arrangement for Venus and Mercury.[154] When interpreting Regiomontanus's note, much hinges on his word 'item'. Did he intend merely to check the 3 authors, just as he had done with al-Biṭrūjī? 'Item' then pertains to his examination ('vide'). Or was he grouping the 3 authors with al-Biṭrūjī because they placed at least one of the 'inferior' planets above the Sun, even if temporarily? If so, the 'item' pertains to their shared non-traditional arrangement. One argument for this last reading is a note on the facing page. In the upper margin of 156r (for lack of space on 155v?), Regiomontanus wrote 'Cicero in *On the Nature of the Gods* [= book 2.20] placed Mercury below Mars; see that [passage]'.[155] In short, Regiomontanus was reminding himself to re-read arguments for alternative planetary arrangements in the older literature.

The most likely sequence of his revisions thus seems to be the following. He first reminded himself to 'check this out very carefully'. Next came the note in the bottom margin (probably continued with the Cicero reference on the facing page). After checking his sources, Regiomontanus then crossed out the first reminder and added the notes in the left margin (italicized in the quotation above). These remarks are arguably the last because, unlike the bottom marginalium, they correctly identify al-Biṭrūjī's planetary order with the one that Regiomontanus's logic had forced upon George as a consequence of the speed-distance rule. Strikingly, the passage in which al-Biṭrūjī discusses planetary order is critical of Ptolemy's reasoning in *Almagest* 9.1, notably the matter of Venus and Mercury passing before the Sun.[156]

It is exclusively with al-Biṭrūjī that Regiomontanus's addition identifies the placement of Venus above the Sun. Al-Biṭrūjī rejected not only eccentrics and epicycles, but also explanations premised on the planets' 'contrary' motions in the heavens. On his account, there is only one celestial mover: the daily rotation of the outermost stellar sphere, which is the fastest and the cause of all other celestial motions. In relation to this frame and power, the speed of each of the seven planets, from Saturn to the Moon, gradually 'drops back' (*accurtat*) from the daily rate the farther it is from the stellar sphere. The Earth, being most distant at the center, is stationary. Greater retardation (decreasing speed) — a function of greater distance from the mover — therefore serves as an ordering principle. With this speed-distance rule, al-Biṭrūjī thus argues that Venus lags less than Sun, which in turn lags less than Mercury.

[154] For a recent overview, see Eastwood, *Ordering the Heavens*, pp. 103–09 (Pliny), 238–44 (Martianus Capella).

[155] 'Cicero in *De natura deorum* sub Marte Mercurium statuit. Vide illic'. Regiomontanus, *Defensio*, fol. 156r.

[156] Carmody, *Al-Biṭrūjī*, pp. 127–29.

Regiomontanus's critique of George translates al-Biṭrūjī's anti-epicyclic account into Ptolemaic terms. Whereas their sidereal periods may serve to rank the superior planets and the Sun, this criterion does not distinguish the inferior planets from the Sun, which all share the Sun's mean motion. If the 'circuit of the anomaly, measured carefully in the epicycle' (the synodic period) becomes the criterion for the so-called 'inferior' planets, their motions spread out nicely in descending order: Venus (584 days), Sun (365 days) and Mercury below it (just under 116 days),[157] as al-Biṭrūjī had proposed.

Regiomontanus, however, dismisses this position as absurd without giving an argument. One source of absurdity may derive from his allegiance to the nesting principle: assuming its contiguity with the lunar sphere, the Mercury model alone will not fill the space between Sun and Moon.[158] Here, as elsewhere in the *Defensio*, Regiomontanus is harsh on al-Biṭrūjī.[159] In the *Epitome*, by contrast, he had not only mentioned al-Biṭrūjī's alternative order without invective, but also explained the rationale for it.[160] In sum, George has no demonstration of the traditional planetary order, leaving the relative order of Venus and Mercury unsettled.[161]

3.2. Inverting Mercury and Venus

Regiomontanus's critique now takes a surprising turn:

> Since neither of these reasons sufficiently indicates the position of Venus and Mercury, one must find out by trial and error if, by placing Venus immediately above the

[157] Neugebauer, *A History of Ancient Mathematical Astronomy*, pp. 157, 167.

[158] al-ʿUrḍī also had an alternative order, of which Regiomontanus was evidently unaware; see Goldstein and Swerdlow, 'Planetary Distances and Sizes', e.g. p. 148; Van Helden, *Measuring the Universe*, pp. 32–34.

[159] Likewise, in the marginalia of his own manuscript of al-Biṭrūjī; Zinner, *Leben und Wirken*, pp. 61–62. Later in the *Defensio* (159v), Regiomontanus explains al-Biṭrūjī's reasoning as classifying the speeds in relation to the epicyclic motion. This can be seen most clearly if one converts the Sun's eccentric into its equivalent epicyclic model. All three deferents will then share the motion of the mean Sun, and the three epicycles will fall into a sequence of increasing speeds from Mercury (the fastest) through the Sun to Venus (the slowest). Here is an additional reason for Regiomontanus to explore the equivalence of eccentric and epicyclic models.

[160] 'However al-Biṭrūjī, who believed that the inequalities of motions and apparent velocities of the planets occur through a kind of falling behind (*quadam incurtatione*), placed under Mars Venus, under which the Sun, then Mercury, for Venus falls behind (*incurtat*) from the first motion less than the Sun, as he said, in fact on account of the epicycle, but Mercury more than the Sun'. See Swerdlow's translation of Regiomontanus, *Epitome* 9.1, as an appendix to his review of Robert Westman's *The Copernican Question*, in 'Copernicus and Astrology'.

[161] At the end of his diatribe, Regiomontanus will blame George for claiming he has demonstrated the planetary order even though he has said nothing about the order of the superior planets; *Defensio*, fol. 158v.

> Moon and Mercury below the Sun, the thicknesses of the two spheres in this order would fill the gap between the luminaries; if this is found to be so, one will boast in vain of being able to demonstrate that, in ascending [order], Mercury should hold the second place and Venus the third.[162]

The argument here sets George's promise of demonstrating the order of Venus and Mercury against both the contingency that Regiomontanus emphasizes and the need for testing both options. After trying only the traditional order, George found a decent fit and ended his search. Regiomontanus now examines the second possibility (Venus below Mercury):

> To throw the expositor's javelins back at him, we will use the pattern of his [own] computation with 'superadditions' of radii that he adopted from somewhere, although this cannot serve as a foundation if someone should try to ascend to the heavens' highest peak by a logical sequence of the planets' distances individually referred back to their eccentrics' radii and then compared, once the proportions have been converted to the Earth's radius, as is his wont. Indeed, as will be clear below, in this journey many orbs are skipped that, by their thickness, add not a little to the height of the heavenly space.[163]

Again, the flavor is that of an academic disputation, in which the strategy is to draw untoward consequences from the opponent's premises and explore contradictions in his conclusions. For polemical purposes, Regiomontanus's argument for inverting the order of Venus and Mercury will rest — despite his own skepticism — on George's procedure and numbers. The pejorative phrase 'the superadditions he got from somewhere' seems disingenuous, as Regiomontanus almost surely knew that Campanus had added planetary radii to the traditional computations of planetary spheres.[164] Indeed, and significantly, Regiomontanus criticizes George not for being uselessly detailed, but for omitting

[162] 'Cum itaque neutra harum rationum Mercurii et Veneris situm satis indicet, experiundum est, si Venerem statim immediate supra lunam et Mercurium sub sole posuerimus, crassitudo duarum spherarum hoc ordine sitarum intercapedinem luminarium expleat. Quod si ita compertum fuerit, iam deinceps frustra quis se demonstraturum iactitabit quod mercuriale sidus secundam inter planetas ascendendo sedem obtineat et Venus tertiam'. Regiomontanus, *Defensio*, fol. 156r.

[163] 'Ut ergo sua in expositorem tela reiaculemur, servabimus formam <margin: de pila(?) fac mentionem post calculum> calculi sui cum superadditionibus semidiametrorum quas ipse undecumque transsumpsit quamvis ne id quidem sufficiat si quis per consequentes stellarum distantias ad semidiametros eccentricorum suorum singulatim relatas et deinde traductis, ut assolet, proportionibus ad semidiametrum terrestrem comparatas, summa celorum fastigia scandere conetur. Plurimi enim in hoc transitu orbes pretereuntur qui crassitudine sua non parum augent celestis spacii altitudinem quemadmodum inferius aperietur'. Regiomontanus, *Defensio*, fol. 156r.

[164] Regiomontanus's copy of Campanus's *Theorica planetarum* is Nuremberg, Stadtbibliothek, Cent V 58; Zinner, *Leben und Wirken*, p. 73; Benjamin and Toomer, *Campanus of Novara*, pp. 114–16.

known spheres that would add appreciably to his computations of distances, a point to which he will return. Assuming Venus is below Mercury, Regiomontanus recomputes the distances:

> Thus, the apogee of the Moon from the center of the world, which will be the perigee of Venus, will be, after adding the radii of the Moon and Venus, 64;54,12 e.r.[165] The perigee of Venus in 60ths of the eccentric radius is 15;35, and the apogee in the same units is 104;25. Converting the proportion yields an apogee of 434;53,13 e.r., to which are added the Venus radius of 0;26,40 and the Mercury radius of 0;2,8, so that Mercury's perigee reaches 435;22 e.r.[166] This perigee of Mercury based on an eccentric radius of 60 parts is 33;4 and its apogee in the same units is 91;30. This number 33;4 is to 91;30 as 435;22 is to 1204;43,12.[167] Such would be the apogee of Mercury, to which one should add the two radii of Mercury, 0;2,8, and the Sun, 5;30. And the Sun's perigee from Earth will come out as 1210;15, which quantity[168] Ptolemy assigns to the Sun indifferently, wherever it may be located.[169]

The last sentence is particularly important, albeit slightly jarring. Regiomontanus effectively privileges the measurement of 1210 e.r., presented here as fixed regardless of longitude, over the variable distance implied from the eccentric model. In short, Regiomontanus takes the eccentric model for the Sun to be a solution to the problem of variable speed. The variable distance that follows from it is an awkward consequence that contradicts the fixed measured distance.

Regiomontanus completes his computation with the apogee of Mercury, which, following George's procedure, he corrects by adding the radii for the

[165] 64;10 + 0;17,32 + 0;26,40 = 64;54,12.

[166] (64;54,12 × 104;25) / 15;35 = 434;53,13 // 434;53,13 + 0;26,40 + 0;02,08 = 435;22,01 (Regiomontanus drops the single second in the text, but silently retains it in his next computation).

[167] 33;04 : 91;30 :: 435;22 : x, so that x = (435;22 × 91;30) / 33;04 = 1204;43,09. Regiomontanus got his slightly larger answer (1204;43,12) by silently retaining the single second in Mercury's perigee instead of using the rounded-down number given in the text.

[168] (1204;43,12 + 0;02,08 + 5;30) = 1210;15,20. Regiomontanus carried out this addition in the margin.

[169] 'Maxima itaque lune a centro mundano distantia que et minima Veneris erit adiunctis lune Venerisque semidiametris habet 64 54' 12", ut semidiameter terre est pars una. Minima autem Veneris remotio ad semidiametrum eccentrici sui 60 partium relata est 15 35' et maxima eodem respectu 104 25', factaque traductione proportionis ambos terminos notos habentis proveniet maxima Veneris remotio ad semidiametrum terre collata 434 53' 13" cui adiungantur semidiametri Veneris quidem 0 26' 40", Mercurii autem 0 2' 8", ut minima distantia Mercurii colligitur 435 22' prout terrestris semidiameter est una; hec autem minima Mercurii distantia secundum partes semidiametri eccentrici sui 60, est 33 4' et maxima eius distantia hoc respectu 91 30'; est autem numerus ille 33 4' ad 91 30' sicut 435 22' ad 1204 43' 12". Quare tanta haberetur maxima Mercurii remotio cui superaddantur due semidiametri Mercurii quidem 0 2' 8", solis autem 5 30'. Et emerget minima distantia solis a terra 1210 15', quantam videlicet Ptolemeus soli ubicumque posito indifferenter tribuit'. Regiomontanus, *Defensio*, fol. 156r-v.

bodies of Mercury and the Sun. Unlike George's computed value, which falls short of Ptolemy's solar distance of 1210 e.r., Regiomontanus's computation matches it:

> One concludes therefore that the two spheres of Venus and Mercury, in this [ascending] order, exactly fill the gap between the two luminaries, which is most wonderful, since the aforementioned gap is obtained not by a continuous conversion of proportions, but from the angle under which the luminaries are determined, together with the lunar parallax and the shadow's radius, as Ptolemy reports it. The distances of Venus and Mercury, however, can be investigated separately in relation to the diameters of their eccentrics as if they had almost nothing in common with the luminaries. Since therefore these can square so exactly that they seem established by compact, who will dare to bellow in reply that these two planets must not be interposed in this order between the luminaries, if we do not blush to change the decrees of our elders, who ranked them [= the planets]?[170]

With this conclusion, Regiomontanus shows that George carelessly overstated his case and certainly did not demonstrate the planetary order. It is by dismissing George's alleged empirical evidence for the occultation of Venus by Mercury that two possible options remained, not one. George's 'good fit' argument is no longer a supplemental quantitative confirmation of the one secure arrangement determined empirically. Instead, it becomes George's *sole* argument for the traditional order. Having granted for purposes of debate George's procedure for computing planetary order, Regiomontanus effectively treats the latter's computations as only the first of two options. Turning George's enthusiasm for the persuasive power of a good fit against him, Regiomontanus completes the trial by checking the second option. Inverting the traditional order of Mercury and Venus yields an even better fit. George's own criteria thus undercut his argument and justified in principle Regiomontanus's departure from the ancients, whether or not he put stock in it.

To strengthen his case, Regiomontanus adds a taxonomic rhetorical argument based on the rank order of the sexes:

> Also, extending this further, if there is room for rhetorical arguments, it is very appropriate and most natural that the two feminine planets together be inferior to

[170] 'Constat igitur duas spheras Veneris et Mercurii hoc ordine sitas intercapedinem duorum luminarium ad unguem explere, quod multo maxime admirandum est, cum intercapedo memorata non per continuam proportionum traductionem, sed per angulum sub quo luminaria cernuntur, una cum diversitate aspectus lunari, ac semidiametro umbre quemadmodum Ptolemeus tradit, elicita sit. Distantie autem Veneris atque Mercurii ad semidiametros eccentricorum suorum relate seorsum quasi nihil ferme cum luminaribus habeant communitatis, investigate sint. Cum ergo hec ad amussim ita quadrent ut ex composito instituta videantur, quis remugire audebit quin hoc ordine memorati duo planete luminaribus interponi debeant si maiorum decreta eos aliter ordinantium immutare non vereamur/erubescimus'. Regiomontanus, *Defensio*, fol. 156v.

> all the masculine ones and that the changeable and promiscuous Mercury should sponge from each sex. For in our earlier inquiry, we rejected the explanation of the more rapid circuit [derived] from the order of the planets.[171]

The criterion of sex is thus consistent with a planetary taxonomy that inverts the traditional order of Venus and Mercury. Although it is useless to order the 4 highest planets, it crudely groups them and also 'explains' why the ambiguous Mercury, contiguous to both male and female, now replaces the Sun as the boundary marker that Ptolemy had identified. Since Regiomontanus had disparaged rhetorical arguments in astronomy when dealing with the heart-Sun analogy, his labeling of this sex argument as rhetorical identifies the level of seriousness with which he treated it. His marginalium on 153v (above) now makes sense.

Regiomontanus was needling his opponent while displaying his skills in argumentation, but his disputation was not pro forma. Stimulated by the controversy, he was also thinking critically about plausible, implausible, and impossible criteria of planetary order. In the tradition of *secundum imaginationem* arguments that were the stock-in-trade of late-medieval natural philosophy, Regiomontanus's exploration of alternative cosmic arrangements also tested propositions for their contingency or necessity, usually forcing a shift from the latter category into the former. What is not necessarily true or false, remains possible until shown to be otherwise.[172]

3.3. Concentric spheres with non-uniform motion and the Viennese tradition

Remarkably, the uncertainty of both the planetary order and the sizes of the planetary spheres that George had calculated now introduces a radical reassessment of traditional astronomical tools:

> It may be possible to save Ptolemy's demonstrated constant distance of the Sun from Earth [= 1210 e.r.] indifferently, in whatever part of its orb the Sun may be. This will be the case a fortiori if we can also understand that the Sun is carried on a concentric without an epicycle, such that, for example, although it can move non-uniformly about the center of the world, yet it is deemed to be carried uniformly about some other point. In this way, every apparent anomaly in its motion can nevertheless

[171] 'Hoc etiam amplius/rursum attento, si rhetoricis locus datur argumentis, quod convenientius est atque naturalius duos quidem planetas femineos masculinis cunctis esse inferiores: Mercurium autem versipellem atque promiscuum utrique sexui interparasitari. Nam celerioris rationem circuitus ab ordine planetarum explorando antehac reiecimus'. Regiomontanus, *Defensio*, fol. 156v.

[172] Albert of Saxony's commentary on *De caelo* 2.10 lists 4 possible options, but Venus below Mercury is not one of them; Grant, *Planets, Stars and Orbs*, pp. 310–11. There is second-hand evidence that Archimedes had proposed a schema in which Venus was below Mercury, and both below the Sun; Neugebauer, *A History of Ancient Mathematical Astronomy*, pp. 647–50.

> be conveniently saved. Indeed, what they assume [as a justification] for introducing eccentrics and epicycles is not necessary, namely, that any celestial body is carried uniformly about the center of its proper orb, since Ptolemy himself could not safeguard this assumed principle.[173]

In this shocking passage, Regiomontanus proposes to eliminate a fundamental feature of Ptolemaic astronomy: the solar eccentric. The replacement he proposes is clearly not the *Almagest*'s geometrical equivalent to it (the deferent-epicycle model), since he specifies not only an Earth-centered 'concentric without an epicycle', but also a concentric that revolves non-uniformly about the Earth. The center of its uniform motion must therefore be 'some other point', in effect a solar equant point. Equally remarkable, this single-concentric proposal is a departure from Regiomontanus's own earlier homocentric theory of the Sun sketched in the 'Letter to Vitéz' (1460), which combined two uniformly revolving concentrics.[174] The stated motive for his new proposal is the preservation of 'Ptolemy's demonstrated constant distance of the Sun from the Earth indifferently'.[175] Recall that the point of the solar eccentricity in Ptolemy's model was to account for the Sun's measurable velocity, not its distance. Conveniently, he could treat as negligible the parallax that follows from that eccentricity.[176]

Regiomontanus's rationale is sufficient for favoring *some kind* of concentric model, but does not explain why he is abandoning his own two-concentric solar model from 1460: all concentric models will preserve equidistance. Better prediction cannot justify the shift either, as the new model offers no improvement on the equivalent classic one. Since the old model in the 'Letter to Vitéz' was not only complex, but also left key mechanical problems unsolved, the motivation for abandoning it arguably derives from a concern for simplicity, merely implied here but stated explicitly below.

173 'Quod autem distantiam solis a terra Ptolemeo demonstratam indifferenter in quacumque orbis sui parte sol fuerit servari liceat; hinc maxime declarabitur si solem ipsum in concentrico ferri absque epicyclo etiam intelligamus; ita videlicet ut, quamvis inequaliter circa mundi centrum moveatur, equaliter tamen circa aliud quoddam punctum circumduci estimetur. Sic enim nihilominus omnis diversitas que in motu eius apparet commode salvabitur. Non enim quod ad introducendum eccentricos atque epicyclos supponuntur necesse est celeste quodlibet corpus equaliter circa centrum orbis proprii circumferri cum ne Ptolemeus quidem illud sumptum principium custodire possit'. Regiomontanus, *Defensio*, fol. 156v.

174 Swerdlow, 'Regiomontanus's Concentric-Sphere Models', esp. pp. 6–11.

175 Regiomontanus's inclination to treat Ptolemy's 1210 e.r. as a single, constant solar distance may have taken some comfort from al-Battānī's recomputation of the minimum lunar diameter as 29′ 30″ (which allowed annular eclipses without changing the apparent solar diameter; see Nallino, *Al-Battānī sive Albatenii opus astronomicum*, vol. I, pp. 58, 236; Swerdlow, 'Al-Battānī's Determination of the Solar Distance', esp. p. 100.

176 Neugebauer, *A History of Ancient Mathematical Astronomy*, pp. 104, 112.

What evidently matters most are the empirical basis and the physical characteristics of the model, particularly saving Ptolemy's fixed solar distance by the simplest means possible — the equidistance of a concentric orb that carries the Sun. Implicit in this project is the notion that the variable distances of the Sun in the *Almagest* model are inconvenient features of that model. They do not describe the world, since no parallax is observed: the solar apogee and perigee are theoretical constructs without empirical warrant. They are consequences of an eccentric model contrived to save variation in velocity while retaining uniform motion about the center of the solar sphere. In effect, this critique extends to the Sun an argument analogous to Levi ben Gerson's critique of the unseen 2-fold variation in the apparent lunar diameter predicted by Ptolemy's second lunar model. The price for saving lunar velocity was a 4-fold variation in area, unseen though easy to detect in principle. Incidentally, Regiomontanus's shift in critical focus corresponds to a longstanding pattern of conceptual transfers from lunar to other theories that Neugebauer has noted.[177]

From Regiomontanus's point of view, assigning a fixed distance to both the Moon and the Sun made more sense on observational grounds than did accepting the variations in distance implied by Ptolemaic theory.[178] For the Sun, a single concentric could do the job — if only one was willing to take the momentous step of giving up uniform motion.

Alone, however, the goal of preserving a fixed solar distance does not explain why Regiomontanus was prepared, even for polemical purposes, to abandon uniform circular motion. This was a principle of astronomy so fundamental that it was almost universally shared by the astronomers and natural philosophers of the Greek, Arabic, Hebrew, and Latin worlds. It had drawn the consensus of very different intellectual camps ever since Eudoxus's homocentric spheres and Aristotle's *De caelo* (1.2; 269a3–30) in the fourth century BC. Indeed, it continued to be endorsed even though Hipparchus's pioneering use of epicycles and eccentrics, their flowering in the *Almagest*, and Ptolemy's introduction of the equant point were arguably partial violations of the principle's integrity. The revival of homocentric schemes (from al-Biṭrūjī onward) and the 'Marāgha School's' critiques of the equant constituted deliberate attempts to reaffirm that integrity. The principle was also endorsed in the critical medieval Aristotelian natural philosophy in at least five languages, and the Latin revival

[177] Neugebauer, *A History of Ancient Mathematical Astronomy*, p. 86.

[178] To assign concentric models to both Sun and Moon, Regiomontanus had to ignore, disbelieve, or explain away claims about observed annular eclipses in works he knew, notably Battānī's *Opus astronomicum* (Nallino, *Al-Battānī sive Albatenii opus astronomicum*, vol. I, pp. 58, 236–37) and a text that Regiomontanus himself summarized, notably in *Epitome* 5.21 (Schmeidler, *Joannis Regiomontani opera collectanea*, p. 143); see also Shank, 'The Notes on al-Biṭrūjī', esp. pp. 17–19.

of the *Almagest*. Regiomontanus himself favored it elsewhere.[179] As late as 1543, Copernicus took this principle as self-evident, expressing it as follows: 'it is impossible for a heavenly body that is simple to move irregularly in a single sphere' (*De revolutionibus*, 1.4). Soon thereafter, Erasmus Reinhold named its positive form 'the astronomical axiom' and inscribed it on the title page of his copy of Copernicus.[180] In short, Regiomontanus's proposal was going against the grains of both a venerable tradition and an overwhelming consensus that only Kepler would bring to an end.

Regiomontanus's proposal was not original. Although he did not say so, he had at least two late-fourteenth-century predecessors whose views eventually made it to the University of Vienna. Henry of Langenstein (d. 1397), who had studied and taught at Paris until 1382, then settled in Vienna after 1383, had first proposed a non-uniformly rotating concentric sphere for the Sun in his Parisian *De reprobatione ecentricorum et epicyclorum* (1364). An otherwise unknown Magister Julmann emulated him in his own *De reprobationibus epiciclorum et eccentricorum*, which cites several observations from 1377 and knows of Langenstein's work (the careers of the two men evidently overlapped in Paris). Both treatises traveled to Central Europe, including Vienna.[181] Regiomontanus copied Langenstein's *De reprobatione* into his Viennese notebook (Vienna, ÖNB, cod. 5203). Beyond criticizing the anonymous *Theorica planetarum communis* used in the universities, Langenstein took on Ptolemaic modeling itself: he sketched a non-uniformly moving concentric alternative to the solar eccentric and also a concentric model for the Moon, responding to Levi ben Gerson's criticism of what the crank mechanism did to apparent disk size. Langenstein used the natural-philosophical language of the intension and remission of forms to discuss this 'difform' motion.[182] This conceptual framework evidently smoothed the path to his acceptance and advocacy of non-uniform circular motion. Langenstein made his peace with this proposal by using contemporary natural philosophy: he drew a clever distinction between the motion of a sphere, which could be difform, and its mover (i.e., the intelli-

[179] e.g., Regiomontanus, *Defensio*, fols 210v, 211v; see Shank, 'Regiomontanus on Ptolemy', pp. 193–94, 199.

[180] North, *Cosmos*, pp. 73–77, 92–94, 203–07; Grant, *Planets, Stars, and Orbs*, pp. 488–93; Copernicus: 'Quoniam fieri nequit, ut coeleste corpus uno orbe inaequaliter moveatur', in Lerner et al., *Copernic. De revolutionibus*, vol. II, p. 21; vol. III, p. 85; Gingerich, 'From Copernicus to Kepler', esp. p. 515.

[181] Zinner, *Entstehung und Ausbreitung*, pp. 82–84. Julmann's work survives in two manuscripts (Munich, BSB, Clm 26667, 109v–116r; and Vienna, ÖNB, cod. 5292, 180r–197v), Langenstein's *De reprobatione ecentricorum et epicyclorum* survives in at least 8 manuscripts, now found in Prague, Vienna, and Melk, among other locations.

[182] Kren, 'Homocentric Astronomy in the Latin West', esp. pp. 271–74, 278; Zinner, *Leben und Wirken*, pp. 64, 307.

gence moving it), which could not. The apparent paradox was solved by arguing that the latter could act uniformly relative to a non-central point, such as the equant point.[183]

Consistent with his solar proposal and its inspiration from Langenstein, Regiomontanus denies that eccentrics and epicycles are necessary for astronomical theory.[184] The rationale for positing these devices in the first place was adherence to the principle that, as Regiomontanus phrases it, 'any celestial body is carried uniformly about the center of its proper orb'. But, he argues, Ptolemy himself could not follow this principle. Why then should his successors be bound by it, and especially by the epicycles and eccentrics that were contrived to uphold it? Since the principle does not hold, why not simply postulate (physical) concentric spheres that move non-uniformly? This is roughly what one observes for the Sun and Moon.

When noting Ptolemy's failure to safeguard (*custodire*) the principle of uniform motion, Regiomontanus is alluding to compromises associated with the equant point, which the *Almagest* had introduced as the center about which the epicycle center moved uniformly. Invented in simpler times, the principle of uniform motion had found an ideal embodiment in Eudoxus's nonquantitative astronomy, in which all combinations of spheres moved both uniformly and concentrically about the Earth's center. In this one point coincided the center of the universe, the centers of the spheres, and the centers of their uniform motion.

The work of Eudoxus (and Aristotle's use of it) largely did not use numerical data and did not aspire to quantitative prediction. Beginning with the inequality of the seasons, 'anomalies' or 'inequalities' in the motions of the celestial bodies proved incompatible with the concentric model. After making contact with Babylonian quantitative astronomy and its program of prediction, Hipparchus and others turned to astronomical models using eccentric deferents and epicycles. These devices multiplied centers of uniform motion away from

[183] 'Est etiam ista difformitas tanto maior quanto punctus reducibilis fuerit a centro fuerit distantior. Patet ergo quod contingit spheram difformiter uniformiter moveri, scilicet super centro proprio vel super puncto preter eius centrum, quem punctum astronomi centrum equantis vocant. Hoc tamen est differenter (read: dupliciter), quoniam uniformitas super centro proprio est semper tam in velocitate motus quam in [R: + in] velocitate circuitionis. Sed uniformitas super punctum preter centrum solum est unius, scilicet circuitionis; et attenditur penes descriptionem angulorum equalium in equalibus temporibus. Et igitur credo omnem motorem superiorem suum orbem uniformiter movere aliquo dictorum modorum ita quod conatus cuiuslibet uniformiter movendi vel est respectu centri proprii orbis quem movet, vel respectu cuiusdem puncti a centro distanter; alias enim videretur sequi mutatio voluntatis movendi in intelligentia, quod non diceret philosophus'. Henry of Langenstein, *De reprobatione ecentricorum et epicyclorum*, Vienna, ÖNB, cod. 5203, 102r.

[184] Regiomontanus had made this point earlier in the *Defensio* (Shank, 'Regiomontanus on Ptolemy', pp. 198–99) and would return to it in books 12 and 13.

the center of the universe — the price to be paid for greater compatibility with observational records and better predictions. Ptolemy built on the work of his predecessors. His models gave predictions in even better tune with observations if the epicycle center moved uniformly about a point other than the deferent center. In the case of the improved lunar theory, that point was the eccentric Earth. In the case of the planetary theories, Ptolemy invented a 'circle of uniform motion' (sketched in *Almagest* 9.2), the center of which had no physical significance. The Latin translators dubbed these features *circulus aequans* and *punctum aequans*, respectively, to signify that the line from the equant point to the epicycle center sweeps out equal angles in equal times.[185]

3.4. Ptolemy's violations of uniform motion

Note that Regiomontanus's criticism above began by rebuking George of Trebizond's use of a solar perigee and apogee instead of Ptolemy's single measured fixed distance (*Almagest* 5.11).[186] It continues now by addressing directly contradictions in the *Almagest* itself, namely that Ptolemy's theories of the Moon and the five planets all violate the principle of uniform motion:

> For [1] he who shows the lunar epicycle revolving uniformly about the center of the world must admit that the very same epicycle is necessarily carried non-uniformly about the eccentric's center. The rigor of geometry confirms this point, since it is impossible for the same body to be carried uniformly about several centers in the same plane figure. [2] Next, this is the case for the epicycles of all the retrograde planets, in that those that individually are carried uniformly about the centers of their equants, move non-uniformly about the centers of their deferent orbs. But [3] the same thing necessarily obtains also for the planets themselves; for whenever one of them moves [on the epicycle] away from the epicycle's continually changing mean apogee (as they call it), [the planet] necessarily describes unequal angles on the center of its epicycle and is therefore carried around unequally about its most proper orb.[187]

[185] Neugebauer, *A History of Ancient Mathematical Astronomy*, pp. 86, 154–55; Benjamin and Toomer, *Campanus of Novara*, pp. 212, 214, passim; and note 3 on p. 404; Pedersen, *A Survey of the Almagest*, pp. 273–83.

[186] Toomer, *Ptolemy's Almagest*, p. 244.

[187] 'Quippe qui lunarem epicyclum equaliter in centro mundano rotari ostendens, ipsum eundem inequaliter in centro eccentrici circumduci confiteatur necesse est. Id enim geometrica roboratur firmitudine cum sit impossibile idem corpus equaliter ferri circa centra plura in eodem plano signato. Hoc denique omnium retrogradorum epicyclis obtingit ut qui equaliter in equantium suorum centris singuli ferantur, inequaliter circa centra orbium se deferentium moveantur. Sed planetis quoque ipsis idem obtingere necesse est, cum enim equaliter quivis eorum ab auge, ut vocant media, epicycli sui que continue mutatur, discedere soleat, necesse est ut in centro ipsius epicycli inequales describat angulos atque idcirco in orbe suo propriissimo inequaliter circumferatur'. Regiomontanus, *Defensio*, fols 156v–157r.

These examples are awkward if one professes adherence to the principle that a sphere or circle must rotate uniformly about its 'proper center'. In argument [1], Regiomontanus considers the second lunar model, in which Ptolemy makes the center of the Moon's epicycle move uniformly about the Earth. It therefore necessarily moves non-uniformly about every other point,[188] most problematically about the center of the deferent sphere that carries the epicycle center. In this lunar model, the eccentric Earth is functionally an equant point (for the first time in the *Almagest*): the lunar model dissociates the geometrical center of the deferent sphere from the center of its uniform motion (the Earth in this case).[189] What Ptolemy has covertly allowed here for the Moon — the non-uniform motion of a (deferent) sphere about its center — is just what Regiomontanus, following Langenstein, openly proposes for the Sun.[190] Eccentrics and epicycles were invented to preserve the principle of uniform motion, yet the *Almagest* violates it often. Why not simplify matters by rejecting the principle outright?

A similar problem [2] also plagues the epicycle centers of the five retrograde planets. This is the archetypical illustration of the equant problem. On this account, the epicycle center is carried equidistantly but non-uniformly around the center of the deferent sphere, whereas it moves uniformly (sweeping out equal angles in equal times) about the equant point, while varying its distance from the latter. This arrangement is incompatible with the notion of the deferent sphere rotating uniformly about its diameter, as the principle of uniform motion prescribes.

Regiomontanus's third example [3] is another corollary of the equant problem in that the planet's motion about the epicycle center is uniform with respect to a shifting equant-related reference point: the mean apogee — the 'mean aux, as they call it', says Regiomontanus with little enthusiam. The planet should move uniformly around the epicycle center, but Regiomontanus argues that it does not. His brief critique builds on Peuerbach's claim that the motion of the planet is 'irregular about the center of the epicycle' but 'irregular according to this rule: that the distance of the body of the planet from the point of the mean apogee of the epicycle, whatever it may be, follows a regular pattern'.[191]

[188] Peuerbach's *Theoricae novae planetarum* [3r] mentions these items in sequence but does not emphasize their consequences for the principle of uniform motion; Schmeidler, *Joannis Regiomontani opera collectanea*, p. 759; Aiton, 'Peurbach's *Theoricae novae planetarum*', esp. p. 12.

[189] Neugebauer, *A History of Ancient Mathematical Astronomy*, p. 86; Toomer, *Ptolemy's Almagest*, pp. 221–22; Pedersen, *A Survey of the Almagest*, pp. 186–87. Regiomontanus returns to this point at *Defensio* 12 (224v).

[190] Copernicus too would later complain about this problem in Ptolemy's lunar theory in *De Revolutionibus* 4.2–3.

[191] Peuerbach, *Theoricae novae planetarum*, [7r] in Schmeidler, *Regiomontani opera collectanea*, p. 767; Aiton, Peurbach's *Theoricae novae planetarum*', pp. 18–19.

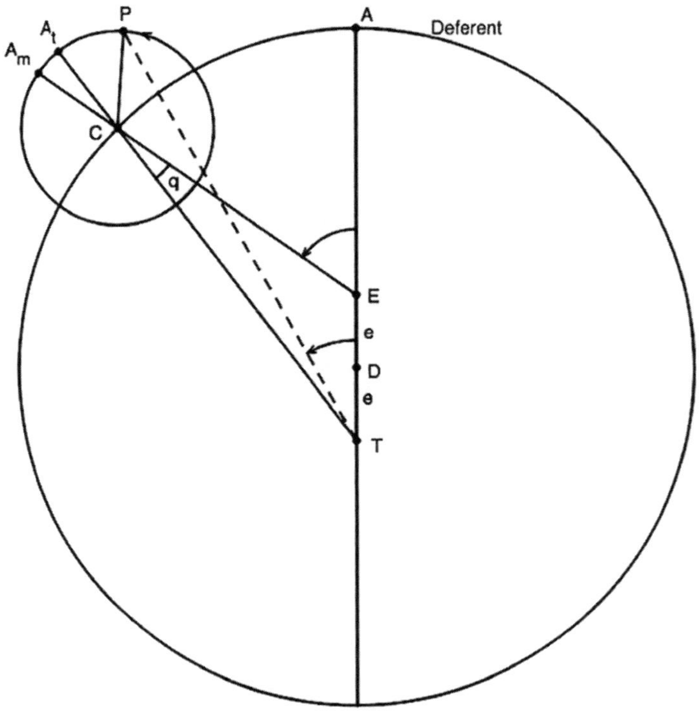

Figure 3. The true apogee A_t and mean apogee A_m of the epicycle with center C. The Earth/observer is T, the deferent center D, the equant point E, each separated by the eccentricity e. The planet is P. The true apogee A_v is the prolongation of TC to the epicycle, the mean apogee A_m the prolongation of EC to the epicycle. After Pedersen. Compliments of Nick Jacobson.

The mean apogee, from which the planet's mean epicyclic anomaly is reckoned, is the prolongation of line EC to A_m on the far side of the epicycle.[192] As the epicycle center is carried around the deferent, A_m oscillates on either side of the true apogee A_t, coinciding with it on the line of apsides. This is why Regiomontanus qualifies the mean apogee as 'continually changing', as does Peuerbach, who states explicitly that the epicycle rotates faster through the upper half of the deferent, but slower through the lower half.[193] But a planet moving in this way cannot sweep out equal angles in equal times about the epicycle center, the 'most proper center' of the planet's motion — another violation of the principle of uniform motion.

Like Regiomontanus's proposal for a non-uniformly moving solar concentric, the first and third criticisms parallel those of Henry of Langenstein's *De repro-*

[192] Pedersen, *A Survey of the Almagest*, pp. 283–85.

[193] Peuerbach, *Theoricae novae planetarum*, [7r] in Schmeidler, *Joannis Regiomontani opera collectanea*, p. 767; Aiton, 'Peurbach's *Theoricae novae planetarum*', p. 19. See also Malpangotto, 'The Original Motivation', section 6c.

batione ecentricorum et epiciclorum, a text that includes 'a detailed refutation of Ptolemy's theory of planetary distances and sizes'.[194] Regiomontanus's first criticism parallels one of Langenstein's criticisms of the lunar theory,[195] whereas the third briefly summarizes the *reductio ad absurdum* with which Langenstein argues that the oscillation of the mean apogee makes the motion of the planet on the epicycle non-uniform.[196] In other words, Copernicus's three criticisms of Ptolemy's lunar theory in *De revolutionibus* 4.2 appear in both Langenstein and Regiomontanus.[197]

Returning to the main topic from this radical digression into fundamental principles, Regiomontanus cuts short his critique, now restricting his argument to the goal of preserving a fixed solar distance:

> We discussed these things extensively elsewhere; we recall them now, not so that we might destroy the approach of epicycles and eccentrics, but so that we may establish the solar distance posited indifferently by Ptolemy [= 1210 e.r.] and then show, with the help of our superior computation, the greater suitability <of placing> Mercury always below the Sun, if one is allowed to dissent from the ancients, than of placing it above the Moon; and, even more, that we may display the lameness of the argument that the expositor holds up as wonderful and most certain, which from another angle can be demonstrated to be sloppy and fragile.[198]

The seemingly straightforward statement raises a host of intriguing questions, even as Regiomontanus's qualifications leave his own views ambiguous. First, Regiomontanus claims that his criticisms are not new: he has treated them at length 'elsewhere', i.e., not in the *Defensio* itself, which he usually cross-ref-

[194] Kren, 'A Medieval Objection to Ptolemy', esp. pp. 379–83; Mancha, 'Ibn al-Haytham's Homocentric Epicycles', esp. pp. 73–75.

[195] I transcribe the conclusion of Langenstein's argument from the copy in Regiomontanus's hand: 'Ex quibus evidenter sequitur centrum epicycli lune difformiter moveri in ecliptica secundum ponentes ecentricos, quoniam ponunt centrum ecentrici lune uniformiter moveri super centro mundi et ipsum declinare a via solis, ut patet in omnibus theoricis eorum. At tamen ponunt lineam exeuntem a centro mundi per centrum epicycli esse lineam medii motus lune; ergo secundum eos movetur uniformiter in ecliptica; sequitur ergo implicatio contradictionis ex isto modo ponendi. Et similiter in aliis planetis quorum eccentrici declinant a via solis', Vienna, ÖNB, cod. 5203, 102r.

[196] Kren, 'A Medieval Objection to Ptolemy', pp. 379–81; Mancha, 'Ibn al-Haytham's Homocentric Epicycles', pp. 74–77.

[197] This is an unappreciated part of the Latin background of Copernicus; Lerner et al., *Copernic. De revolutionibus*, vol. III, pp. 493–96.

[198] 'Hec alibi diffusius a nobis declarata, commemoramus impresentiarum non quo ianuam eccentricorum atque epicyclorum infringamus, sed quo distantiam solarem a Ptolemeo indifferenter sumptam adstruamus ac deinceps calculo nostro superiori attestante ostendamus convenientius Mercurium sub sole statim si a priscis dissentire licet quam supra lunam ordinari; imo potius ut manifestemus infirmitatem rationis quam expositor pro mirabili atque certissima ducit; quam aliunde etiam fluxam atque fragilem esse demonstrari potest'. Regiomontanus, *Defensio*, fol. 157r.

erences. He says more here, however, than he does in his other best known discussion of this topic, namely, the letters to Johannes Vitéz (1460) and to Giovanni Bianchini (*c.* 1463–64), which circulated neither much, nor at all, respectively. In the 'Letter to Vitéz', which sketched homocentric theories for the Sun and the Moon, he expressed his interest in eliminating eccentrics and epicycles from planetary theory as well.[199] But these letters are not much more detailed than is this passage of the *Defensio*. In short, the fulsome critique Regiomontanus cross-references apparently refers to another work, perhaps the mysterious *Problemata Almagesti*, most of which has yet to be found.[200]

Citing a fuller treatment of the problems with eccentrics and epicycles elsewhere, Regiomontanus cuts short his discussion here, a distraction from the main point — the unique solar distance that Ptolemy determined from eclipse observations. This reiterated goal suggests that we should take seriously his proposal about the concentric of the Sun, since this is the simplest way of accounting for that fixed solar distance.[201] Next, Regiomontanus argues that the inversion of the traditional order of Venus and Mercury is more likely because, using George's numbers, it fits better in the space between the Sun and Moon. Regiomontanus's final comment in the quotation above relativizes that remark, however: his overarching goal is to undermine George of Trebizond's purportedly demonstrative arguments. The perfect fit of the inverted order thus displaces George's computations and, using his own criteria, completely undermines his boast of having 'demonstrated' the traditional planetary order.

3.4.1. Skepticism about computing cosmic dimensions

The inverted order clearly tells us little about Regiomontanus's own views. Indeed, in the comment that follows, he doubts George's numbers, which he has just used polemically, to compute the 'better-fitting' planetary order. The first reason for distrusting George's numbers is that he has undercounted large orbs that affect the sizes of the planetary spheres:

> Indeed, if it is necessary to assign to each celestial motion its [own] orb (which is right and to be assiduously worked out by the philosopher, unless we wish to believe that an accident subsists without a subject), one must first add to the Moon an orb concentric to the world, which is carried westward uniformly nearly 0;3,11° per day around the axis of the zodiac so that, by this motion [= the daily longitudinal ret-

[199] Swerdlow, 'Regiomontanus's Concentric-Sphere Models', pp. 6–9.

[200] Zinner, *Leben und Wirken*, pp. 118–20, 324–25; and the updated summary of fragmentary loci in Malpangotto, *Regiomontano e il rinnovamento*, pp. 205–07.

[201] Regiomontanus returns to this point in his concluding remarks, where he criticizes George for computing distances using the solar perigee and apogee instead of Ptolemy's single distance, thus gratuitously adding 100 e.r. to the sphere of the Sun; *Defensio*, fols 159v–160r.

rogression of the lunar nodes],[202] the oblique circles of the luminaries may cut each other continuously in points ever farther westward.[203]

Regiomontanus here restates approvingly the traditional Aristotelian view that each motion must have one mover. He also justifies it on technical philosophical grounds: 'accidents' (i.e., quantities, qualities, relations) exist not independently, but only insofar as they inhere in substances (the only entities with independent existence). The motions of the planets are 'accidents' in this technical sense, with consequences for astronomy. If a planet appears to have several motions, several movers must cause them — in this case, one orb for each motion.[204] The task of assigning these orbs is what 'is to be assiduously worked out by the philosopher', suggesting that the work is both unfinished and difficult.

Regiomontanus thus proceeds to enumerate — beyond the basics of the epicycle, deferent, and crank mechanism — some of the orbs needed to power the other lunar motions. His point is that, since George's planetary distances include tiny planetary radii, he should be counting the (non-zero thickness) orbs responsible for all motions, not merely those responsible for longitude and apogee. First, an additional sphere is therefore necessary to move the lunar nodes (i.e., the intersections of the Moon's inclined path with the ecliptic) in their slow retrogression around the ecliptic. The marginalium 'About thickness' highlights the main theme of the next argument in the text:

Not even the younger students doubt the two orbs that, with an uneven and interchanged [= complementary] thickness, surround the lunar eccentric so that the entire sphere may remain concentric to the world, filling every vacuum and excluding the penetration of bodies. However, some philosophers vociferously affirm, as if it were a great novelty, that the lunar globe also is imbedded in some small orb, by the rotation of which the spotted face of the Moon always faces our eyes.[205]

[202] Neugebauer, *A History of Ancient Mathematical Astronomy*, p. 83.

[203] 'Si enim, quod iustum est et a philosopho summopere excogitandum, suum cuique celesti motui orbem tribui oportet (ne sine subiecto accidens quodpiam subsistere opinemur), lune in primis adiungendus est orbis mundo concentricus qui circa axem zodiaci spatio diurno trium primarum ac undecim fere secundarum sexagesimarum equaliter ad occasum feratur: ut hoc motu obliqui luminarium circuli in aliis continue atque alii se invicem punctis occasum versus secent'. Regiomontanus, *Defensio*, fol. 157r.

[204] Grant, *Planets, Stars and Orbs*, pp. 488–95.

[205] '<margin: De spissitudine> De orbibus autem duobus qui impari ac permutata crassitudine ecentricum lunarem ambiunt ut tota sphera mundo reddatur concentrica indeque vacuum omne impleatur et corporum penetratio excludatur, ne discipuli quidem recentiores ambigere solent. Quod autem lunaris etiam globus orbiculo quodam implicitus sit cuius contra epicyclum volutione maculosa illa lune effigies oculis nostris semper obvertatur, etsi inusitatum sit additamentum, a nonnullis tamen argutie philosophantibus affirmatum'. Regiomontanus, *Defensio*, fol. 157r-v.

These first two orbs are the partial ones that frame the eccentric path of the Moon's epicycle about the center of the Earth, and are familiar from the illustrations in Regiomontanus's edition of Peuerbach's *Theoricae novae planetarum* (*c.* 1472).[206]

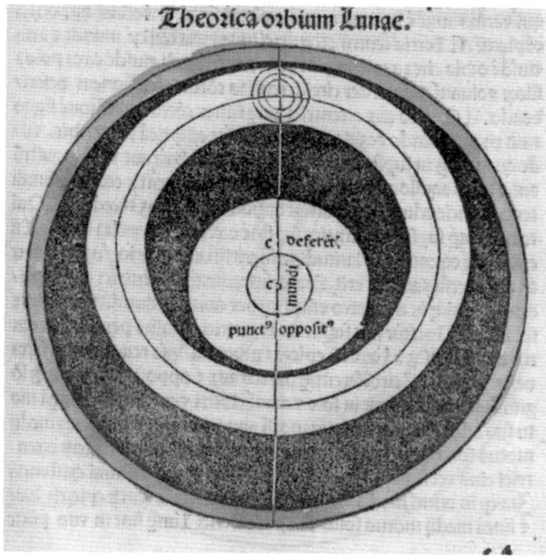

Figure 4. The lunar model, from Peuerbach's *Theoricae novae planetarum* in Ratdolt's first *Sphaera mundi* compendium (Venice, 1482), based on Regiomontanus's first edition. Note that the partial eccentric orbs have a non-zero thickness at their thinnest point. (By courtesy of the Department of Special Collections, Memorial Library, University of Wisconsin-Madison).

Here Regiomontanus explicitly mentions the vacuum-filling role of the orbs, about which Peuerbach was silent. Finally, he brings up as controversial a point debated by fourteenth-century natural philosophers. If the Moon were imbedded in its epicycle, we should see now the face of the Moon, now its backside as the epicycle makes a full rotation. But the Moon only shows us one face. We must therefore postulate an additional sphere around the Moon that counteracts the rotation of the epicycle by rotating at the same rate and in the opposite direction.[207]

> How many orbs, then, do you think must be added to Venus as well as Mercury, if you correctly consider the 3-fold latitudinal motion of each? At minimum both [planets] surely must be surrounded by epicycles, from the variation of which are derived the

[206] Peuerbach, *Theoricae novae planetarum* [3r–4r], in Schmeidler, *Joannis Regiomontani opera collectanea*, pp. 759–61.

[207] John Buridan got the principle of the argument right, but not the details. He also considered dispensing with the additional sphere (as Regiomontanus did not, probably for deliberate polemical reasons) by simply giving the body of the Moon a rotation equal, but opposite, to that of the epicycle; Grant, *Planets, Stars and Orbs*, pp. 299–302.

> twin latitudes of inclinations (*inclinationum*) and slants (*obliquationum*)[208] in the globe of the epicycle and then in the planet itself. To make a celestial structure of this sort more perfect, especially if all celestial motions can agreeably be reduced to circular uniformity (*circularem equalitatem*), other [spheres] also will have to be superimposed on the aforementioned ones surrounding the epicycle. As to the rest, for the third motion in latitude, which they call 'deviation' (*deviationem*), an immense orb will have to be adapted to both planets, an orb that can surround the other joined orbs of the sphere and represent this latitude of deviation with its variable motion.[209]

The details of the intricate latitude theory for the inferior planets are not the issue here (they are discussed in book 13).[210] Instead, Regiomontanus's damaging question focuses on George's failure to include in his computations the dimensions of the three orbs that, in combination, produce the inferior planets' motions in latitude.

The first two latitudinal motions that Regiomontanus mentions are the inclination (*inclinatio*) and the slant (*obliquatio*) of the epicycle with respect to the plane of the deferent.

In the *Almagest*, Ptolemy analyzes the inferior planets' changes in latitude into three variable components, two for the epicycle (*inclination* ι_1 and *slant* ι_2) and one for the deferent (*deviation* ι_3). Measured in the direction the observer's line of sight, the epicycle's *inclination* ι_1 has maximum and minimum values when the epicycle center is 90° from the line of apsides (at 90° and 270° in Figure 5) and is zero when on the line of apsides (with the epicycle center at

208 I take *latitudines* and *geminas* to be case errors (for *latitudinis* and *geminae*, respectively). Although *obliquatio* is sometimes transliterated, 'slant' typically translates the corresponding term in the *Almagest* and the Latin synonym of *reflectio* in Peuerbach's *Theoricae novae planetarum*. Toomer, *Ptolemy' s Almagest*, p. 599 (passim); Aiton, 'Peurbach's *Theoricae novae planetarum*', p. 34. Twice in *De revolutionibus* 6, Copernicus uses *inflexio* and *inclinatio* as equivalents to *obliquatio*; Lerner et al., *Copernic, De revolutionibus*, vol. II, p. 459, lines 5–6; vol. III, p. 435, and the useful chart at vol. III, p. 425. *Obliquatio* can also signify the obliquity of the ecliptic, in the *Defensio* and elsewhere.

209 'Quot denique orbes tam Veneri quam Mercurio adiiciendos esse arbitraberis si triplicem utriusque latitudines [*read* latitudinis?] motum rite mediteris? Nempe epicyclis quidem binos ad minimum circumplecti oportet, quorum varia mutatione geminas [*read* geminae?] inclinationum atque obliquationum latitudines in epicycli globum et deinceps in stellam ipsam <ss: planetam ipsum> deriventur. Quo autem completior fiat huiusmodi celestis structura, presertim si universos celi motus ad circularem reducere equalitatem libeat, alii quoque prefatis epicyclum ambientibus superimponendi erunt. Ceterum pro tertio latitudinis motui quem vocant deviationem adaptandus erit utrique planete orbis ingens qui reliquos sphere orbes cunctos ambiat motuque suo mutabundo hanc deviationis latitudinem effingat'. Regiomontanus, *Defensio*, fol. 157v.

210 Swerdlow, 'Ptolemy's Theories of the Latitude', pp. 41–71, esp. 50–56; see also Pedersen, *A Survey of the Almagest*, ch. 12; Neugebauer, *A History of Ancient Mathematical Astronomy*, pp. 206–26, 1006–16.

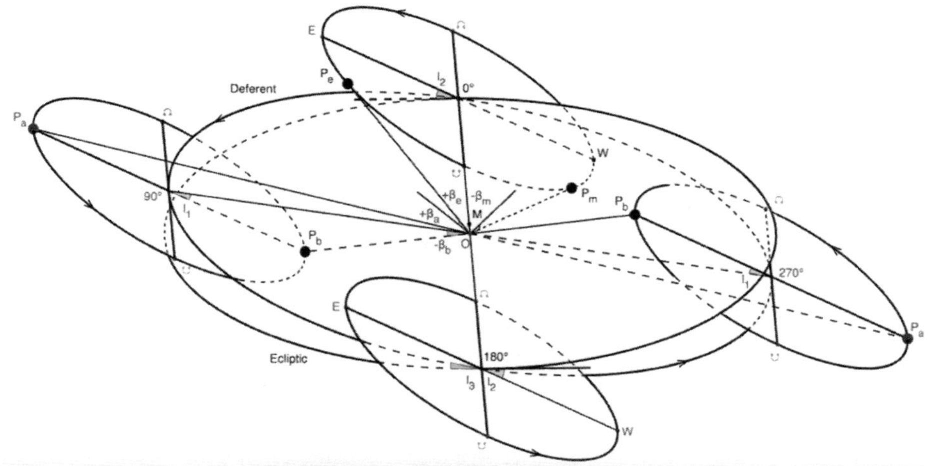

Figure 5. Simplified inner planet latitude theory. The observer is at O, the center of the ecliptic and universe, but eccentric to the model, in which M is the center of the deferent. OM therefore defines the line of apsides. The deferent has a variable *deviation* ι_3 from the ecliptic. The epicycle is inclined in two ways to the plane of the deferent. The *inclination* ι_1 is measured in the direction of sight (i.e., radially from O toward the epicycle center and the mean Sun), shows maxima and minima (β_a, β_b) at 90° and 270° from OM for positions P_a and P_b, and is null at 0° and 180° (when the line of sight coincides with the deferent). The *slant* ι_2 is measured orthogonally to the line of sight, reaches maxima and minima along EW at 0° and 180° and is null at 90° and 270° (when the epicycle diameter orthogonal to the line of sight is in the deferent plane). β_a and β_b are the latitudes for planetary positions P_e and P_m (maximum elongation) (after Swerdlow, not to scale) Courtesy of Nick Jacobson.

0° and 180°). Inversely, the epicycle's *slant* ι_2 displays maximum and minimum values perpendicularly to the line of sight, when the epicycle center is on the line of apsides (at 0° and 180°). The epicycle's slant ι_2 is zero when on the normal through the Earth to the line of apsides.[211] To each of these motions, Regiomontanus assigns a sphere (implicitly of non-zero thickness). Peuerbach's discussion in the *Theoricae novae* is more tentative: 'on account of these inclinations and slants, *some assume* small orbs holding the epicycles within themselves according to the motion of which the same [phenomena] happen'.[212]

When Regiomontanus mentions the unmet goal of a 'more perfect structure' that fits the criteria of uniform motion, he also states without elaboration

[211] For the geometrical challenge that Ptolemy had to solve, see Riddell, 'The Latitudes of Venus'.

[212] 'Propter dictas epicyclorum inclinationes atque reflectiones orbes parvi epicyclos intra se locantes a quibusdam ponuntur ad quorum motum eaedem contingunt'. Peuerbach, *Theoricae novae planetarum* [17r], in Schmeidler, *Joannis Regiomontani opera collectanea*, p. 787; Aiton, 'Peurbach's *Theoricae novae planetarum*', pp. 34–36, esp. p. 36, adding italics and modifying the end of Aiton's translation.

that the epicycle will need yet more spheres to attain this higher perfection: the two spherical shells for inclination and slant enclose the epicyclic sphere for longitude theory. This picture beautifully suits his polemic: if the shells for the latitude theory are external to the planet, George's computations of the cosmos should add them to the planet's radius.

Not least is the *deviation*, the transliterated Latin term for the small variable pitch of the deferent in relation to the ecliptic (angle ι_3 in Figures 5 and 6).[213] Regiomontanus emphasizes the enormous size of the sphere needed to account for that motion: it evidently encompasses both the deferent for longitude theory and the aforementioned congeries of epicyclic spheres. Again, in his computations, George has omitted the huge spheres, implicitly of non-zero thickness, that account for the deviations of Mercury and Venus, although he included the tiny dimensions of the inferior planets themselves. Here too, the *Defensio* echoes Peuerbach, who treated the deviations of Mercury and Venus by assigning to each a large sphere 'concentric to the world and enclosing all those mentioned before'.[214]

Pushing his case to the limit by moving beyond the Sun-Moon interval, Regiomontanus mentions also the orbs needed to explain the motions of the Sun and the superior planets:

> Finally, for their latitudinal properties, the three superiors also require orbs of this sort, albeit fewer. We do not intend — lest we go on at length — to detail them precisely, since we have not even reviewed all the inferior [planets'] orbs, having completely skipped the Sun, which, since the center of its eccentric stands at a significant distance from the center of the world, as discerned by the observations of the illustrious astronomers Ptolemy and al-Battānī, and of other trustworthy individuals, begs for other orbs by the motion of which such an accident may be produced.[215]

Also omitted by George, these additional spheres throw off even more his final computations of cosmic dimensions. Although the superior planets have no deviation, they nevertheless will each require around the epicycle two additional latitude spheres for the slant and inclination. Commenting on the Sun, Regiomontanus notes that Ptolemy's and al-Battānī's observations point to a

[213] In the *Handy Tables* and *Planetary Hypotheses*, Ptolemy made the deviation constant; Swerdlow, 'Ptolemy's Theories of the Latitude', p. 57 and passim.

[214] The intersection of the planes of the ecliptic and of the deferent thus produces a fixed nodal line that runs through the eccentric Earth and therefore cuts the deferent asymmetrically; see Pedersen, *A Survey of the Almagest*, pp. 358–59.

[215] 'Tales demum orbes sed pauciores pro latitudinum qualitate tres quoque superiores exposcunt; quos, ne longiores simus, exacte prosequi non est consilium, cum ne inferiorum quidem universos orbes recensuerimus; sole penitus/prorsus preterito, qui quoniam centrum eccentrici sui multifariam a centro mundano removet quemadmodum observationibus illustrium astronomorum Ptolemei et Albategnii atque aliorum fide non carentium dinoscitur, orbes alios exposcit quorum motu tale accidens emergat'. Regiomontanus, *Defensio*, fol. 157v.

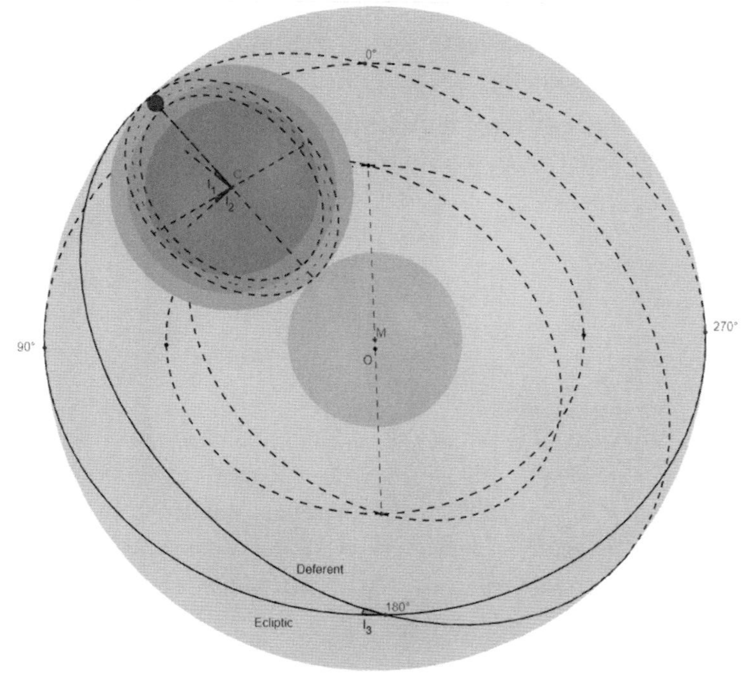

Figure 6. The large sphere, centered at M, that governs the variable deviation (ι_3) of the deferent in the *Almagest*'s latitude theory of an inferior planet. The earth is at O; the combined spheres that produce the inclination (ι_3) and slant (ι_2) are centered on C and carried about M. (Not to scale; compliments of Nick Jacobson).

solar eccentric, curiously, since George included the Sun's perigee and apogee in his computations. Most significant is his treatment of the Sun as an unsolved problem that demands a multi-spherical solution, a point consistent with his 'Letter to Vitéz', but in clear tension with the non-uniform homocentric sphere he had suggested above. The overarching message is clear, however:

> We have noted all these things for the following reason: since, whatever their thicknesses, orbs of this sort very recently superimposed upon the spheres increase [the thickness] of the celestial region, they are effectively laboring in vain who get excited about climbing the celestial heights on stairs cut out here and there. Indeed, although individual planets may have [known] eccentric and epicyclic radii and known eccentricities with their own individual radii, and [although] the Moon's apogee from the center of the universe in e.r. has been discovered, nevertheless the conversion of proportions necessary for this work must be kept in check by so many orbs of unknown thickness inserted here and there.[216]

[216] 'Hec itaque pluscula adnotavimus ut, quoniam orbes huiusmodi nuperrime spheris superimpositi crassitudinem quantamcumque habent, celesti regionis augent, operam ludere videantur quincumque celorum culimam scandere gestiunt scalis passim intersectis. Fieri enim oportet ut quamvis eccentricorum atque epicyclorum semidiametros ipsasque eccentricitates

In this arresting paragraph, Regiomontanus dismisses as a waste of time the computation of planetary distances solely from the perigee-to-apogee ratios of planetary models. Although the parameters of the models are well established, even a hypothetical complete enumeration of all orbs, some very recent,[217] cannot give this project a secure foundation, since the orbs' thicknesses are unknown, a point that returns a folio later.[218]

Significantly, Regiomontanus juxtaposes his proposal for a non-uniformly rotating solar concentric to his critique of George's discussion of the planetary spheres' order and sizes. Planetary distances and order do in fact bring up notable tensions in the physical implications of Ptolemaic astronomy. Not coincidentally, such tensions surface soon after the translation of several Levi ben Gerson works into Latin in the early 1340s.[219] It is likely through him that Latin astronomers became aware of the mismatch between observation and Ptolemy's mature lunar theory. Critiques similar to Levi's appear in the astronomical contexts of mid- and late-fourteenth-century Paris (Langenstein, Julmann), and mid-fifteenth-century Vienna (Regiomontanus). Ptolemy's lunar theory worked well enough to track and predict the velocity and position of the Moon as a point source. But the 'crank mechanism' used to yield these results also produced a nearly 2-fold variation in the Moon's distance from the Earth. Irrelevant for a point source, this effect on a body with a half-degree apparent diameter should produce a visible 4-fold variation in area.

In addition, the Moon at apogee does exhibit parallax (*Almagest* 5.11–19). Yet, on the nesting hypothesis, the lunar apogee equals the perigee of the next contiguous inferior planet, traditionally Mercury. Whether for the sake of argument or for more serious reasons, however, Regiomontanus had made a hypothetical case for Venus with 'good fit' criteria that he knew to be non-demonstrative.

Not least, there was the problem of the solar distance, to which Regiomontanus kept returning. To account for the variation in solar speed, Ptolemy's theory relies on an eccentric that carries the Sun with a minimum apparent diameter of 0;31,20°. The change in distance computed from this model yielded an angular variation in diameter of almost 3′, likewise unobserved. Unlike Ptolemy, al-Battānī emphasized the nearly 3-minute difference, all the

cum semidiametris propriis habeant notas singuli planete inventaque sit maxima lune a centro mundi remotio ad terre semidiametrum collata, traductio tamen proportionum ad hoc opus necessaria prohibeatur tot passim orbibus ignota spissitudine interiectis'. Regiomontanus, *Defensio*, fols 157v–158r.

[217] I take *nuperrime* in the previous note to modify *superimpositi* and to signify more plausibly a contemporary proposal (e.g., Peuerbach's *Theoricae novae*) than Regiomontanus's mention of spheres in his own previous sentence.

[218] *Defensio*, fol. 159v.

[219] Mancha, 'The Latin Translation'.

more so as he used two of his own eclipse observations (including an annular one) to modify Ptolemy's parameters.[220]

It should now be clearer why Regiomontanus, a mathematical astronomer who also cared about physical data, could, in the 'Letter to Vitéz', bother not only to elaborate a homocentric hypothesis for the Sun and Moon, but also to anticipate extending it to the other planets. It made sense to give the Sun and Moon combinations of concentric spheres that used fixed distances to account for *both* the inequalities in their motions *and* their observed sizes and parallaxes (none in the case of the Sun, measurable in the case of the Moon).

For modern critics, homocentric theories are rooted in Aristotelian dogma and have as their Achilles' heel the constant distances of planets and luminaries from the Earth at the center of the universe. But these are not the concerns that we see here. As the arguments of Regiomontanus (and also Henry of Langenstein) show, constancy in distance is the very feature that makes a homocentric model attractive — on observational rather than dogmatic natural philosophical grounds. This is particularly so when thinking critically about theories that predict *many* unseen variations in apparent size: small ones for the Sun, large ones (4:1) for the Moon, and even greater ones for Venus (45:1) and Mars (52:1), as Regiomontanus had complained in a letter to Bianchini.[221] For astronomers newly attuned to predicted variations in the physical sizes of the luminaries and planets, the appeal of concentric models lay partly in observation. As further evidence that traditional natural philosophy was not the deciding factor, the fundamental principle of uniform circular motion was precisely what Langenstein was willing to give up, and Regiomontanus at least willing to consider.

Space limitations do not permit a full analysis of the last two folios of this discussion, which returns to earlier material. They can be summarized as follows. Regiomontanus attacks George for conceding that he has made no astronomical observations, and for failing to understand how difficult it is to measure planetary diameters. Regiomontanus derides his opponent for claiming to have demonstrated the planetary order without either addressing the superior planets or having shown whether Venus or Mercury is closer to the Moon (158r-v). He assails George for praising Ptolemy's deference to the ancients, arguing that Ptolemy did no such thing, but reasoned about the order he preferred (158v–159r).

Regiomontanus criticizes once again George's Averroistic heart-Sun trope (he mentions the Commentator by name): Nature would never have 'sunk' the Sun

[220] Nallino, *Al-Battānī sive Albatenii opus astronomicum*, vol. I, pp. 58, 236–37; Regiomontanus summarized this material in *Epitome* 5 (esp. prop. 21) and in the *Defensio* called it a monster (*monstrum*; 218v–219r).

[221] See the translation by Swerdlow, 'Regiomontanus on the Critical Problems', esp. pp. 173–74.

to such a lowly ordinal location (with only 3 small 'stars' below it and implicitly the whole star catalogue above it). He goes on to challenge the consistency of George's speed-distance principle with the sub-solar position for Venus he favors, now using al-Biṭrūjī's criterion of synodic period polemically to put that planet above the Sun (159r).

Finally, he turns to parallax problems. He criticizes George of Trebizond for absurdly claiming that one can simply use an instrument to measure the parallaxes of Mercury's apogee or Venus's perigee: the planets are invisible in conjunction with the Sun (159v).[222] He accuses George of failing to understand that the planets' true motions must be computed first, before finding their parallaxes: 'Therefore, whoever investigates parallax while neglecting the true position of the planet (*stella*) is trying in vain to build an unsupported roof, suspended in air'.[223] Regiomontanus once again emphasizes the greater reliability of Ptolemy's unique measured solar distance (which he takes to be constant) while deprecating the attempt by George (and presumably others) to obtain it by computing the dimensions of the spheres (160r). Accordingly, he then attacks George for 'insulting' Ptolemy by treating 1309;46 as the perigee of Mars ('he strays from the truth by 100 e.r.'; *Defensio*, fol. 160r). Again, he effectively rejects the eccentricity of the Sun (a feature of the model, not of measurement) and implying that Mars's perigee should take the (fixed) 'maximum solar distance as 1210 e.r.' (160r), a measurement far more reliable than the assumptions George needed for his computations.

4. Conclusion

The *Defensio*'s attack on George of Trebizond's *Commentaria* 9.1 confirms that the order of the planetary spheres was hotly contested in the second half of the fifteenth century. In his commentary on *Almagest* 9.1, George chose to discuss neither the uncertainties that Ptolemy had briefly noted in his treatment of celestial order, nor the various post-Ptolemaic alternatives. Instead, he used book 9 as the springboard to detailed computations of the dimensions of the planetary spheres and the cosmos. He presented his own efforts as empirically grounded (the alleged record of Mercury passing below Venus) and as offering a demonstration that Ptolemy might have carried out, but omitted. His computations built on, and reinforced, the traditional order that Ptolemy preferred despite the uncertainties in it.

As the preceding analysis suggests, George's *Commentaria* did not produce cosmic dimensions *de novo*. Instead, I infer that he silently took comfort from one work in the tradition of such computations, most likely Campanus of

[222] Regiomontanus makes a similar point in *Epitome* 9.1; see note 148.

[223] 'Qui ergo aspectus diversitatem vestigat vero loco stelle neglecto tectum in aere pendulum nullo sustentaculo instituere frustra conatur'. Regiomontanus, *Defensio*, fol. 159v.

Novara's *Theorica planetarum*. The two works share several tell-tale idiosyncrasies in procedure and emphasis. The most obvious similarities are (1) their insistence on adding planetary radii to the radii of the spheres inferred from the *Almagest*'s models, and (2) their justification of the traditional planetary order by its ability to generate the proper sequence of weekdays. Less uniquely, both marvel at the convergence of independent approaches (eclipses, proportions, days of the week), which they take in the aggregate to confirm their planetary order and distances.

The most compelling argument for George's direct dependence on Campanus, however, rests on errors that are otherwise very hard to explain. George behaves strangely in the *Commentaria* precisely when errors surface in the Campanus manuscript tradition. For the superior planets, George suddenly introduces, without warrant, talk of approximations when his own results diverge from both the numbers he should have got from straightforward computation and the erroneous ones given in Campanus manuscripts. A probable explanation of this swerve is the worry of straying too far from the (erroneous) answers in Campanus, whom he seemingly trusted more than his own arithmetic.

For his part, Regiomontanus was following the *Almagest* itself when he insisted that the order of the planetary spheres could not be settled without evidence of parallax. Historically, this uncertainty clouded primarily the positions of Venus and Mercury, exceptions to the rough speed-distance rule that failed to undermine its appeal for the remaining planets (arguably including the Sun). As the *Defensio* shows, Regiomontanus's concise statement of this uncertainty at the end of *Epitome* 9.1[224] is merely the tip of Regiomontanus's argumentative iceberg.

In the *Defensio*, Regiomontanus's long case for the uncertainty in the order and location of Venus and Mercury was primarily destructive. It consisted, first, in showing that George's alleged demonstrations failed. The *Commentaria*'s case for the necessity of the traditional order rested on a supposed observation of Mercury below Venus, conjoined with a ban on vacua and useless space. George thus treated the 'good fit' of the computed distances of the inferior planets as establishing a necessary truth. Like George, Regiomontanus rejected empty space between the Sun and Moon: Venus and Mercury definitely belonged there. Unlike George, however, Regiomontanus rejected as undocumented and unbelievable the reported observation of Mercury below Venus. He thereby downgraded George's 'good fit' argument for this order to merely the computation of one possible arrangement of the four lowest planets.

[224] This passage is contrasted with the *Commentariolus*'s discussion by Lerner et al., *Copernic, De revolutionibus*, vol. I, pp. 242–44.

The highpoint of Regiomontanus's argument was his exploration of the other alternative. After hypothetically inverting the received order of Mercury and Venus, he computed the sub-solar planetary distances based on George's numbers and assumptions. By showing that his newly postulated order displaced George's good fit with a better one, Regiomontanus undermined both George's alleged demonstration and his second, weaker line of defense. In short, George's 'proof' of the traditional order of Venus and Mercury was neither necessary nor sufficient.

Regiomontanus's argumentative approach echoes that of the university disputation, in which masters dissected their opponents' arguments from every angle, ruthlessly searching for contradictory assumptions, fallacious reasoning, and any other exploitable weakness. For this polemical reason alone, we cannot be sure that Regiomontanus gave more credence to his own hypothesis than to George's. His primary goal was not to defend a planetary order that inverted Venus and Mercury, but rather to highlight his opponent's incompetence by showing that the alternative order met George's own criteria better. Question marks replaced certainty.

Since George almost certainly never saw them, Regiomontanus's sharp analyses and *secundum imaginationem* arguments left their main target untouched. Conveyed orally, however, their gist may have met another, perhaps more important goal, that of undermining George's bid for patronage at the court of Hungary. Their most profound effect was surely on Regiomontanus himself. They not only sharpened in detail the unsettled state of key questions in astronomy and cosmology (a favorite theme of his), but also stimulated his thinking about alternatives and arguments for them. Thus, in the opening folios of *Defensio* 9, Regiomontanus's discussions of planetary distances and order quickly turned to substantive criticisms of contemporary astronomy and to hopes for a simpler astronomy. In Regiomontanus's mind, the uncertainties surrounding the received planetary order and distances were intimately tied to the complexity of the planetary models themselves, witness his inclusion of latitude theory in cosmic dimensions.

Whatever he believed about the other celestial bodies, Regiomontanus was prepared to streamline the physical theory of the Sun, apparently on empirical grounds. Echoing Ptolemy's remarks on the absence of solar parallax, Regiomontanus repeatedly endorsed a fixed solar distance as computed from measurements in the *Almagest*. On this basis, he therefore proposed a single concentric sphere for the Sun. Such a picture could make sense only by severing the age-old necessary connection between uniform motion and spherical motion. In driving a wedge between these two properties of celestial spheres, Regiomontanus was echoing suggestions found in his own copy of *De reprobatione eccentricorum et epicyclorum* that Henry of Langenstein brought from Paris to Vienna in the late fourteenth century. Regiomontanus's proposal and its link to Langenstein point to a substantive Viennese thread in late-medieval astronomy, one notably critical of fundamental tools in the field.

Regiomontanus concludes his critique of George's computations of planetary distances with remarks that undermine the standard program of calculating cosmic dimensions. At the core of his argument, the focus on longitude theory vastly oversimplifies the sizes of the spheres by omitting many relevant orbs — most notably those responsible for the planets' motions in latitude. These excluded orbs are not abstract entities, but physical bodies with unknown thicknesses. When computing cosmic distances, adding only the orbs with 'known' thicknesses while neglecting the others can yield no reliable information about either the sizes or the order of the planetary spheres. Although Regiomontanus had little to say here about his own positions, his criticisms challenged many a long-held assumption. When Copernicus revisited this territory a generation later, he would see in the apparent disarray the seeds of a different order.

Bibliography

Aiken, Jane A., 'Leon Battista Alberti's System of Human Proportions', *Journal of the Warburg and Courtauld Institutes* 43 (1980), pp. 68–96.

Aiton, Eric J., 'Peurbach's *Theoricae novae planetarum*. A Translation with Commentary', *Osiris* n.s. 3 (1987), pp. 5–43.

d'Alessandro, Paolo, and Piero D. Napolitani, *Archimede latino. Iacopo da San Cassiano et il corpus archimedeo alla metà del quattrocento con edizione della* Circuli dimensio *et della* Quadratura parabolae, Paris: Les Belles Lettres, 2012.

Benjamin Jr., Francis S., and Gerald J. Toomer, *Campanus of Novara and Medieval Planetary Theory. Theorica Planetarum*, Madison: University of Wisconsin Press, 1971.

Bowen, Alan C., *Simplicius on the Planets and their Motions. In Defense of a Heresy*, Leiden: Brill, 2013.

Burri, Renate, 'The Geography of *De mundo*', in Johan C. Thom (ed.), *Cosmic Order and Divine Power. Pseudo-Aristotle, On the Cosmos*, Tübingen: Mohr Siebeck, 2014, pp. 89–106.

Byrne, James, 'A Humanist History of Mathematics? Regiomontanus's Padua Oration in Context', *Journal for the History of Ideas* 67 (2006), pp. 41–61.

Carmody, Francis J., *Al-Biṭrūjī. De motibus celorum. Critical Edition of the Latin Translation by Michael Scot*, Berkeley: University of California Press, 1952.

Carmody, Francis J., *The Astronomical Works of Thābit b. Qurra*, Berkeley: University of California Press, 1960.

Czartoryski, Paweł, *Nicolaus Copernicus. Minor Works*, Warsaw: Polish Scientific Publishers, 1985.

Duhem, Pierre, *To Save the Phenomena. An Essay on the Idea of Physical Theory from Plato to Galileo*, trans. E. Doland and C. Maschler, Chicago: University of Chicago Press, 1969.

Eastwood, Bruce, *Ordering the Heavens. Roman Astrology and Cosmology in the Carolingian Renaissance*, Leiden: Brill, 2007.

Garin, Eugenio, *Giovanni Pico della Mirandola. Disputationes adversus astrologiam divinatricem*, 2 vols, Firenze: Vallecchi, 1952.

Gingerich, Owen, 'From Copernicus to Kepler. Heliocentrism as Model and Reality', *Proceedings of the American Philosophical Society* 117 (1973), pp. 513–22.

Goldstein, Bernard R., *The Arabic Version of Ptolemy's Planetary Hypotheses*, Philadelphia: American Philosophical Society, 1967 [*Transactions of the American Philosophical Society* n.s. 57, 4].

Goldstein, Bernard R., 'Some Medieval Reports of Venus and Mercury Transits', *Centaurus* 14 (1969), pp. 49–59.

Goldstein, Bernard R., and Noel M. Swerdlow, 'Planetary Distances and Sizes in an Anonymous Arabic Treatise Preserved in Bodleian MS Marsh 621', *Centaurus* 15 (1971), pp. 135–70.

Goldstein, Bernard R., 'Copernicus and the Origin of his Heliocentric System', *Journal for the History of Astronomy* 33 (2002), pp. 219–35.

Grant, Edward, *Planets, Stars and Orbs. The Medieval Cosmos, 1200–1687*, Cambridge: Cambridge University Press, 1994.

Hankins, James, *Plato in the Renaissance*, 2 vols, Leiden: Brill, 1990.

Hartner, Willy, 'The Mercury Horoscope of Marcantonio Michiel of Venice', *Vistas in Astronomy* 1 (1955), pp. 84–138 [reprinted in Id., *Oriens et Occidens. Ausgewählte Schriften zur Wissenschafts- und Kulturgeschichte. Festschrift zum 60. Geburtstag*, Hildesheim: Olms, 1968, pp. 440–95].

Heiberg, Johan L., *Syntaxis mathematica*, 2 vols, Leipzig: Teubner, 1898-1903 [*Claudii Ptolemaei opera quae exstant omnia*, vol. I].

Heiberg, Johan L., *Opera astronomica minora*, Leipzig: Teubner, 1907 [*Claudii Ptolemaei opera quae exstant omnia*, vol. II].

Van Helden, Albert, *Measuring the Universe. Cosmic Dimensions from Aristarchus to Halley*, Chicago: University of Chicago Press, 1985.

Juste, David, 'MS Escorial, Real Biblioteca del Monasterio de San Lorenzo, d.II.5' (updated 01.06.2017), at *Ptolemaeus Arabus et Latinus. Manuscripts*, URL = http://ptolemaeus.badw.de/ms/219.

Kraye, Jill, 'The Philosophy of the Italian Renaissance', in George H. R. Parkinson (ed.), *The Renaissance and 17th Century Rationalism*, London: Routledge, 1993, pp. 16–69.

Kraye, Jill, 'Disputes over the Authorship of *De mundo* between Humanism and Altertumswissenschaft', in Johan C. Thom (ed.), *Cosmic Order and Divine Power. Pseudo-Aristotle, On the Cosmos*, Tübingen: Mohr Siebeck, 2014, pp. 181–97.

Kremer, Richard L., 'Text to Trophy. Shifting Representations of Regiomontanus's Library', in James Raven (ed.), *Lost Libraries. The Destruction of Great Book Collections since Antiquity*, New York: Palgrave Macmillan, 2004, pp. 75–90.

Kren, Claudia, 'Homocentric Astronomy in the Latin West. The *De reprobatione ecentricorum et epicyclorum* of Henry of Hesse', *Isis* 59 (1968), pp. 269–81.

Kren, Claudia, 'A Medieval Objection to "Ptolemy"', *British Journal for the History of Science* 4 (1969), pp. 378–93.

Labowsky, Lotte, *Bessarion's Library and the Biblioteca Marciana. Six Early Inventories*, Roma: Edizioni di Storia e Letteratura, 1979.

Lerner, Michel-Pierre, *Le Monde des sphères*, 2 vols, 2nd ed., Paris: Les Belles Lettres, 2008.

Lerner, Michel-Pierre, Alain-Philippe Segonds and Jean-Pierre Verdet, *Nicolas Copernic. De revolutionibus orbium coelestium / Des révolutions des orbes célestes. Édition critique, traduction et notes*, 3 vols, Paris: Les Belles Lettres, 2015.

Lorch, Richard, 'The Astronomy of Jābir ibn Aflaḥ', *Centaurus* 19 (1975), pp. 85–107.

Maguire, Joseph P., 'The Sources of Pseudo-Aristotle's *De mundo*', *Yale Classical Studies* 6 (1939), pp. 111–67.

Malpangotto, Michela, *Regiomontano e il rinnovamento del sapere matematico e astronomico nel Quattrocento*, Bari: Cacucci, 2008.

Malpangotto, Michela, 'The Original Motivation for Copernicus's Research. Albert of Brudzewo's *Commentariolum super Theoricas novas Georgii Purbachii*', *Archive for History of Exact Sciences* 70 (2016), pp. 361–411.

Mancha, José Luis, 'Ibn al-Haytham's Homocentric Epicycles in Latin Astronomical Texts of the XIVth and XVth Centuries', *Centaurus* 33 (1990), pp. 70–89.

Mancha, José Luis, 'The Latin Translation of Levi ben Gerson's Astronomy', in Gad Freudenthal (ed.), *Studies on Gersonides. A Fourteenth-Century Jewish Philosopher-Scientist*, Leiden: Brill, 1992, pp. 21–46 [reprinted in Id., *Studies in Medieval Astronomy and Optics*, Aldershot: Ashgate/Variorum, 2006, article III].

Manitius, Karl, *Procli Diadochi hypotyposis astronomicarum positionum*, Leipzig: Teubner, 1909 [reprinted Stuttgart, 1974].

Mioni, Elpidio, *Codices graeci manuscripti bibliothecae Divi Marci Venetiarum. Thesaurus antiquus*, vol. II, Roma: Istituto Poligrafico e Zecca dello Stato, 1985.

Mohler, Ludwig, *Kardinal Bessarion als Theologe, Humanist und Staatsmann. Funde und Forschungen*, 3 vols, Paderborn: F. Schöningh, 1923–1942 [reprinted Aalen: Scientia Verlag, 1967].

Monfasani, John, *George of Trebizond. A Biography and Study of his Rhetoric and Logic*, Leiden: Brill, 1976.

Monfasani, John, *Collectanea Trapezuntiana. Texts, Documents, and Bibliographies of George of Trebizond*, Binghamton, N. Y.: Medieval and Renaissance Texts and Studies, 1984.

Mueller, Ian, *Simplicius. On Aristotle's* On the Heavens *2.10–14*, Ithaca, N. Y.: Cornell University Press, 2005.

Nallino, Carlo, *al-Battānī sive Albatenii opus astronomicum. Ad fidem codicis Escurialensis arabice editum. Latine versum, adnotationibus instructum*, 3 vols, Milano: Ulrico Hoepli, 1899–1907 [reprinted Hildesheim: Olms, 1977].

Neugebauer, Otto, *A History of Ancient Mathematical Astronomy*, 3 vols, Berlin: Springer, 1975.

North, John D., *Cosmos*, Chicago: University of Chicago Press, 2008.

Obrist, Barbara, *La cosmologie médiévale. Textes et images*, vol. I: *Les fondements antiques*, Firenze: SISMEL, Edizioni del Galluzzo, 2004.

Pedersen, Olaf, 'The Decline and Fall of the *Theorica planetarum*. Renaissance Astronomy and the Art of Printing', in Erna Hilfstein, Paweł Czartoryski and Frank Grande (eds), *Science and History. Studies in Honor of Edward Rosen*, Wrocław: Ossolineum, 1971, pp. 156–85.

Pedersen, Olaf, *A Survey of the Almagest*, Odense: Odense University Press, 1974 [revised edition by Alexander Jones, New York: Springer, 2011].

Pedersen, Olaf, 'The *Theorica planetarum* and its Progeny', in Graziella Federici Vescovini and Francesco Barocelli (eds), *Filosofia, Scienza, e Astrologia nel Trecento Europeo. Biagio Pelacani Parmense*, Padua: Il Poligrafo, 1992, pp. 53–78.

Pellegrin, Pierre, 'The Argument for the Sphericity of the Universe in Aristotle's *De caelo*. Astronomy and Physics', in Allan Bowen and Christian Wildberg (eds), *New Perspectives on Aristotle's De caelo*, Leiden: Brill, 2009, pp. 163–85.

Riddell, R. C., 'The Latitudes of Venus and Mercury in the Almagest', *Archive for History of Exact Sciences* 19 (1978), pp. 95–111.

Robbins, Frank E., *Ptolemy. Tetrabiblos*, Cambridge (Mass.): Harvard University Press, 1940.

Rodríguez-Arribas, Josefina, 'Did Ibn Ezra Maintain a Circumsolar Arrangement of Venus and Mercury?', *Nuncius* 25 (2010), pp. 191–212.

Saliba, George, 'The First Non-Ptolemaic Astronomy at the Maraghah School', *Isis* 70 (1979), pp. 571–76 [reprinted in Id., *A History of Arabic Astronomy. Planetary Theories during the Golden Age of Islam*, New York: New York University Press, 1994, pp. 113–18].

Schmeidler, Felix, *Joannis Regiomontani opera collectanea. Faksimiledrucke von neun Schriften Regiomontans und einer von ihm gedruckten Schrift seines Lehrers Purbach*, Osnabrück: Otto Zeller Verlag, 1972.

Shank, Michael H., 'The *Notes on al-Biṭrūjī* Attributed to Regiomontanus. Second Thoughts', *Journal for the History of Astronomy* 23 (1992), pp. 15–30.

Shank, Michael H., 'Regiomontanus on Ptolemy, Physical Orbs, and Astronomical Fictionalism. Goldsteinian Themes in the *Defense of Theon against George of Trebizond*', *Perspectives on Science* 10 (2002), pp. 179–207.

Shank, Michael H., 'Regiomontanus as a Physical Astronomer. Samplings from the *Defence of Theon against George of Trebizond*', *Journal for the History of Astronomy* 36 (2007), pp. 325–49.

Shank, Michael H., 'Made to Order', *Isis* 105 (2014), pp. 167–76.

Shank, Michael H., 'The *Almagest*, Politics, and Apocalypticism in the Conflict between George of Trebizond and Cardinal Bessarion', *Almagest: International Journal for the History of Scientific Ideas* 8, 2 (2017), pp. 49–83.

de Solla Price, Derek J., *The Equatorie of the Planetis*, Cambridge: Cambridge University Press, 1955.

Špelda, Daniel, 'From Closed Cycles to Infinite Progress. Early Modern Historiography of Astronomy', *History of Science* 53 (2015), pp. 209–33.

Stahl, William H., *Roman Science. Origins, Development and Influence to the Later Middle Ages*, Madison: University of Wisconsin Press, 1962.

Swerdlow, Noel M., *Ptolemy's Theory of the Distances and Sizes of the Planets. A Study of the Scientific Foundations of Medieval Cosmology*, Ph.D. Dissertation, Yale University, 1968.

Swerdlow, Noel M., 'Hipparchus on the Distance of the Sun', *Centaurus* 14 (1969), pp. 287–305.

Swerdlow, Noel M., 'The Derivation and First Draft of Copernicus's Planetary Theory. A Translation of the *Commentariolus* with Commentary', *Proceedings of the American Philosophical Society* 117 (1973), pp. 423–512.

Swerdlow, Noel M., 'Al-Battānī's Determination of the Solar Distance', *Centaurus* 17 (1973), pp. 97–105.

Swerdlow, Noel M., 'Regiomontanus on the Critical Problems of Astronomy', in Trevor H. Levere and William R. Shea (eds), *Nature, Experiment, and the Sciences. Essays on Galileo and the History of Science in Honour of Stillman Drake*, Dordrecht: Kluwer, 1990, pp. 165–95.

Swerdlow, Noel M., 'Regiomontanus's Concentric-Sphere Models for the Sun and Moon', *Journal for the History of Astronomy* 30 (1999), pp. 1–23.

Swerdlow, Noel M., 'Ptolemy's Theories of the Latitude of the Planets in the *Almagest*, *Handy Tables*, and *Planetary Hypotheses*', in Jed Z. Buchwald and Allan Franklin (eds), *Wrong for the Right Reasons*, Dordrecht: Springer, 2005, pp. 41–71.

Swerdlow, Noel M., 'Copernicus and Astrology, with an Appendix of Translations of Primary Sources', *Perspectives on Science* 20 (2012), pp. 353–78.

Thom, Johan C. (ed.), *Cosmic Order and Divine Power. Pseudo-Aristotle, On the Cosmos*, Tübingen: Mohr Siebeck, 2014.

Toomer, Gerald J., *Ptolemy's Almagest*, Berlin: Springer, 1984.

Westman, Robert S., *The Copernican Question. Prognostication, Skepticism, and Celestial Order*, Berkeley: University of California Press, 2011.

Yule, George Udny, *The Statistical Study of Latin Vocabulary*, Cambridge: Cambridge University Press, 1944.

Zepeda, Henry, 'Euclidization in the *Almagestum parvum*', *Early Science and Medicine* 20 (2015), pp. 48–76.

Zimara, Marcantonio, *Aristotelis opera cum Averrois commentariis*, 5 vols, Venezia: Giunti, 1562–1574.

Zinner, Ernst, *Entstehung und Ausbreitung der Coppernicanischen Lehre. Zum 200 jährigen Jubiläum der Friedrich-Alexander-Universität zu Erlangen*, Erlangen: Mencke, 1943.

Zinner, Ernst, *Leben und Wirken des Joh. Müller von Königsberg, genannt Regiomontanus*, 2nd ed. Osnabrück: O. Zeller Verlag, 1968.

Optimus Malorum: Giovanni Pico della Mirandola's Complex and Highly Interested Use of Ptolemy in the *Disputationes adversus astrologiam divinatricem* (1496), A Preliminary Survey

H. Darrel RUTKIN[1]

In this essay I will survey a broad range of Giovanni Pico della Mirandola's explicit mentions of Ptolemy in his long, dense and influential attack on astrology, the *Disputationes adversus astrologiam divinatricem* of 1496.[2] Searching the text, I have found that Pico explicitly mentioned Ptolemy's name 376 times in various contexts. We know fully well, however, that Pico was not always explicit or straightforward in how he used Ptolemy, as I have shown elsewhere.[3] Furthermore, Pico's use of Ptolemy is located at the intersection of two larger issues: [1] The full range of Pico's complex and interested use of authorities overall (mainly astrological, philosophical and theological) in the *Disputations*,[4] and [2] the increasing knowledge of Ptolemy's Greek text in the Renaissance.

Although Pico was long dead by the time that the brilliant humanist scholar, Joachim Camerarius, published the *Editio Princeps* of the Greek text of Ptolemy's *Tetrabiblos* in 1535, we know that Pico was one of the first scholars to systematically use the Greek manuscripts that Lorenzo de' Medici had collected to philologically critique the earlier Arabo-Latin translations and their associated commentaries.[5] In this essay, I will explore Pico's explicit mentions of Ptolemy in the *Disputations* in relation to the authentic *Tetrabiblos* and *Almagest*, and the pseudonymous *Centiloquium*, which Pico thought was authentic. These are

[1] I would like to acknowledge that this article was completed as part of a project that has received funding from the European Union's Horizon 2020 Research and Innovation Programme (GA n. 725883 EarlyModernCosmology), as well as support from the University of Sydney while I was an Honorary Associate in History of Science at its School of History and Philosophy of Science.

[2] I use the Latin text that Eugenio Garin edited for the National Edition of Pico's works: Garin, *Giovanni Pico*. The translation is mine, and will ultimately appear in the I Tatti Renaissance Library. My thanks to the organizers of this marvelous and memorable conference, and especially to David Juste and Dag Nikolaus Hasse for their very helpful responses to a range of queries, and for their valuable comments on the submitted first draft of this essay.

[3] Rutkin, 'The Use and Abuse'.

[4] See (e.g.) Caroti, 'Le fonti medievali'.

[5] See Gentile, 'Pico e la biblioteca'.

Ptolemy's Science of the Stars in the Middle Ages, ed. by David Juste, Benno van Dalen, Dag Nikolaus Hasse and Charles Burnett, PALS 1 (Turnhout, 2020), pp. 387–406
© BREPOLS PUBLISHERS 10.1484/M.PALS-EB.5.120185

the only three Ptolemaic texts that Pico mentions. These three texts were all well known, moreover, and were all, in fact, normal texts of university education, as we can see in the detailed 1405 statutes for the University of Bologna.[6] After briefly surveying and categorizing Pico's explicit mentions, I will discuss a few examples in greater depth.

*

Pico was up against tremendous odds in his passionate endeavor to undermine and ultimately eradicate astrology, which was still very much 'normal science' at the time he wrote in the early 1490s.[7] Mounting such an attack may seem obvious to us from an early twenty-first-century perspective, but from a late fifteenth-century Renaissance or Early Modern perspective, it would have been a hugely daunting prospect. A resonant analogy would be of someone trying to criticize and destroy Newtonian mechanics in either Cambridge in the middle of the nineteenth century, with respect to both its overall epistemic authority as well as its institutional establishment at the finest universities and learned academies. Pico had his work cut out for him![8]

Here is another pointed analogy: Despite his profound respect for Ptolemy's work in astronomy, geography, harmonics and optics, Pico criticizing Ptolemy for his astrological writings would be similar to a hypothetical nineteenth-century critic of Newton's respecting him for his work in mathematics, mechanics and optics, but objecting to his work in alchemy. The major relevant difference, however, is that Newton did not publicize his alchemical passions.[9] Pico was painfully aware of how solidly established and deeply rooted astrology was both conceptually and institutionally, at both the universities — including those he had attended at Bologna, Padua and Ferrara — and throughout a broad spectrum of society, politics and culture.[10] Astrology was not marginal in any respect. Rather, Pico's attack itself would have been considered profoundly marginal in its time.[11]

[6] See (e.g.) Federici Vescovini, 'I programmi degli insegnamenti'.

[7] See in particular, Boudet, *Entre science et nigromance*, and in a much shorter compass, my 'Astrology'.

[8] I refer here to Cambridge, UK at Cambridge University and in Cambridge, MA at both Harvard and MIT.

[9] For the most up-to-date information on Newton's alchemy, see William R. Newman, *Newton the Alchemist*, and his Indiana University website, 'The Chymistry of Isaac Newton': http://webapp1.dlib.indiana.edu/newton/index.jsp.

[10] For a lively recent study offering many examples contemporary with Pico, see Azzolini, *The Duke and the Stars*.

[11] Nevertheless, Pico is part of a long ancient, Arabic and medieval Latin tradition of critics of astrology, including the well-documented cases of Sextus Empiricus, Moses Maimonides, Nicole Oresme and Henry of Langenstein. See most recently Nothaft, '*Vanitas vanitatum*'.

In this context, Pico wrote the *Disputations against Divinatory Astrology* at the very end of his short but passionate life.[12] In it, he tried to undermine, destroy and indeed wholly eradicate astrology from the cultural landscape by any means necessary — using many different skills and strategies — in a long and difficult work that has not yet been fully understood in modern scholarship.[13] Towards this end, in my 2002 Indiana University PhD thesis, I focused primarily on Pico's attack on astrology's natural philosophical foundations in *Disputations* Book III.[14] Here I will focus on articulating the contours of Pico's various uses of Ptolemy — astrology's principle ancient authority — towards the very same aim of undermining and destroying astrology.[15]

The *Disputations* is thus an extremely ambitious work, one part of Pico's larger unfinished project attacking the seven major enemies of the Church.[16] The only part he certainly wrote (and that still exists) is the *Disputations*, which Pico did not live to complete, leaving behind a fragmentary manuscript that no longer exists, despite Robert Westman's recent statement to the contrary.[17] In the *Disputations*, Pico attacked astrology from many different perspectives, including its natural philosophical foundations and its foundations for practice, *inter alia*, the doctrines of signs, houses and dignities, all of which were considered essential for astrological prediction.[18] Here Pico was famously followed a century later by Johannes Kepler in *his* attempts to reform (not reject) astrology.[19] Signs, houses and dignities were all employed in the four canonical types of astrological practice: general astrology or revolutions, nativities, interrogations and elections.[20]

Among many other things, Pico was keen to point out that perfectly legitimate mathematical devices otherwise useful for astronomical calculation were

[12] See (i.a.) Garin, *Giovanni Pico*, pp. 3–17. For a splendid evocation of Pico's life, times and works, see Grafton, 'Giovanni Pico'.

[13] For some valuable recent studies, see the essays collected in Bertozzi, *Nello specchio del cielo*.

[14] Rutkin, *Astrology, Natural Philosophy*. I also treat this more fully in volume II of my soon-to-be-forthcoming monograph: Rutkin, *Sapientia Astrologica*, vol. II.

[15] This essay draws on and further develops arguments I made in a memorable conference on Ptolemy at Caltech in 2007 organized by Alexander Jones: Rutkin, 'The Use and Abuse'.

[16] Garin, *Giovanni Pico*, p. 3.

[17] Garin, *Giovanni Pico*, in his introduction, informs us of the state of the manuscript, and Franco Bacchelli provides further information in the Aragno reprint, 'Appunti per la storia'. On the contrary, see Westman, *The Copernican Question*, n. 55 (p. 528): 'Although the published value may contain a typographical error, Garin, who made a critical comparison with the original manuscript, makes no comment here'. There are many other misprisions in his treatment of Pico, who provides the fulcrum for his larger argument.

[18] See (e.g.) Book VI.

[19] See (e.g.) Simon, *Kepler*.

[20] For a valuable treatment of the range of astrological practices, see Bezza, *Arcana Mundi*.

often turned in various ways into astrological predictors, which he characterized as arbitrary signifiers with no foundations in nature. These included the 360 degrees of the zodiac, its twelve 30-degree subdivisions, and their further 60-minute subdivisions.[21] Pico contrasts these arbitrary man-made mathematical devices (however useful) with actual celestial entities that have real celestial influences, which he certainly believed in, but severely delimited in scope.[22] For Pico, these celestial-efficient causal factors act only by means of motion, light and heat, within a well articulated Aristotelian understanding of nature, as he discusses in depth in *Disputations*, Book III. This includes how generation works, as we can see, for example, in *De generatione et corruptione* II.10, and especially as developed by Aristotle's later commentators, including Albertus Magnus.[23] In Pico's radical reinterpretation, however, he eliminated the unique nature of each planet's light, and thus its unique influence, as found, for example, in the first two chapters of al-Kindi's deeply influential *De radiis stellarum*, and Albertus Magnus's commentary on Aristotle's *De caelo*, II.3.1 ff.[24] Pico thus attempted to wrench off the by-his-time deeply entrenched astrological superstructure from its still-solid Aristotelian foundations.[25]

*

One of Pico's main tactics to weaken astrology's epistemic authority was precisely to undermine faith in its major authorities, beginning with Ptolemy, whom he calls (*inter alia*) the best of the bad ('optimus malorum') and the most learned of the astrologers ('doctissimus astrologorum').[26] One way Pico does this is [1] to explicitly, directly and sometimes abusively attack a range of Ptolemy's positive astrological doctrines in the *Tetrabiblos* and *Centiloquium*. If Pico can fundamentally shake a pro-astrological reader's faith in Ptolemy, that would be a huge step forward for his project, especially in the Renaissance. By contrast, [2] where Ptolemy ignores — or *himself* explicitly criticizes or outright rejects — an astrological doctrine, Pico then appropriates his great authority, and thus transforms him, paradoxically, into an anti-astrological

[21] See (e.g.) *Disputations* VI. 4 and 11.

[22] See my PhD thesis *Astrology, Natural Philosophy*, chapter 6, and volume II of my monograph *Sapientia Astrologica*.

[23] Hossfeld, *Albertus Magnus*; and see my 'Astrology and Magic'.

[24] For the Latin text of al-Kindi, see d'Alverny and Hudry, 'Al-Kindi, *De Radiis*'; for a partial English translation, see Adamson and Porman, *The Philosophical Works*, pp. 217–34. For Albert's *De caelo*, see Hossfeld, *Albertus Magnus*.

[25] I made this argument in chapter 6 of my PhD thesis *Astrology, Natural Philosophy*. I support the claims in this paragraph in much greater depth in volumes I and II of my monograph *Sapientia Astrologica*.

[26] Book I (70, 8 and 6).

ally.[27] Another frequent tactic is [3] to shine a harsh and often ridiculing light on the innumerable outright disagreements or conflicting teachings ('pugnantia') between the main astrological authorities.[28] As by far the most important ancient authority, Pico often used Ptolemy in this context by comparing and contrasting his views with other astrological authorities, primarily ancient and medieval.

In these heated and often sarcastically abusive critical pursuits, Pico regularly deployed his highly developed philological skills to highlight and diagnose — among other things — influential misinterpretations of Ptolemy's doctrines, derived from inaccurate and thus misleading translations.[29] Sometimes Pico refers explicitly to the Greek manuscripts he knew at first hand,[30] as well as to Latin translations of Ptolemaic texts and commentaries. Sometimes he even offers his own corrective translations directly from the Greek. Before turning to specific examples, however, I should briefly recall Pico's significant methodological statement in Book II, Chapter 6: if reforming astrology and not suppressing it were his intention, he would have written his book very differently.[31] The primary purpose of Pico's criticisms, therefore, was to undermine astrology in every possible way, but especially by casting doubt on its foundational doctrines and authorities. His multifold and highly interested uses of Ptolemy played a major role in that process. To their deeper exploration we shall now turn. This essay should be considered a preliminary sounding in deep and richly complex culture-historical waters.

*

I would now like to analyze some of the ways Pico used Ptolemy by focusing on the controversial theme of astrology's relationship to religion — Pico's central concern in the *Disputations* — which he treats in some depth, but not systematically. Here I will build up an admittedly incomplete picture, drawn from several disparate chapters, to offer a taste of Pico's approach to this centrally important subject, while focusing on his various uses of Ptolemy. Exploding religion's perceived subordination to astrology was Pico's greatest concern.

[27] For Ptolemy as a critic of earlier astrology, see Grafton, 'Giovanni Pico'.

[28] One (of many) examples is in *Disputations*, VI.3, in which Pico attacks the astrologers for disagreeing among themselves ('inter se pugnent') about the doctrine of the terrestrial houses.

[29] On Pico as a high-level philologist, in addition to Grafton, 'Giovanni Pico', see Gentile, 'Pico filologo'.

[30] Some of these are described in Gentile, 'Pico e la biblioteca'.

[31] 'Quod si docere hic potius astrologiam quam confutare instituissem, funderem manum ad errata iuniorum profitentium hanc artem; sed non hoc meum consilium. Adnotare tamen fortasse aliqua fuerit operae precium, quo magis fiat manifestum non posse eos vera praedicere etiam si verissima essent dogmata astrologorum' (142, 19–24).

I begin with Book II, Chapter 5, which is entitled: 'How harmful and noxious astrology is to the Christian religion'.[32] Pico mentions Ptolemy himself once in this chapter. He begins by strongly stating his overall view of astrology's relationship to religion, and it is not a pretty picture:

> Truly, for me reviewing and exploring [...] the enemies of the Church, I do not see where more supplies and more arms are supplied to all of them equally against the truth than from this profession [namely, astrology]. For from this, the fall is easy and headlong into impiety, bad religion, heresies, vain superstition, lost morals and irrevocable evil. For whence will impiety arm itself against the spears of religion better than that divine miracles, by which every religion is primarily confirmed, be referred to the heavens?[33]

This is very much in line with Pico's statement in the overall Proem that astrology is the mother of all superstitions.[34]

The next passage has the one explicit mention of Ptolemy:

> But I have read none of the main writers of astrology who do not subject religion and all laws — and likewise the rest of human affairs — to the configurations of the stars. In the second book of the *Apotelesmaton*, Ptolemy understands the fact — that in this nation (*gens*), this god (*numen*), and in that nation, a different one is worshipped — arose from nowhere else than from the different natures of the stars and constellations that rule these peoples and nations.[35]

The passage in question comes from *Tetrabiblos* II.3, and here Pico neutrally describes Ptolemy's position. He also mentions Ptolemy's anonymous Greek commentator soon after.

*

[32] 'Quam noxia sit astrologia quamque pestifera christianae religioni' (126, 11).

[33] 'Sane lustranti mihi undique omnia et ecclesiae hostes exploranti, non video unde omnibus pariter plus copiarum, plus armorum, adversus veritatem suppeditetur, quam ex ista professione. Hinc enim ad impietatem, hinc ad malam religionem, hinc ad haereses, hinc ad vanam superstitionem, hinc ad perditos mores irrevocabilemque malitiam praeceps et facillimus lapsus. Unde enim se potius adversus tela religionis armabit impietas, quam ut divina miracula, quibus omnis potissimum religio confirmatur, ad caelum pertendat esse referenda?' (126, 13–22).

[34] 'Est autem haec propria labes omnium superstitionum, quarum non alia professio quam praecepta tradere insaniendi; sed in primis hunc sibi titulum vendicavit astrologia, sicut et inter ipsas superstitiones, quarum mater alumnaque merito existimatur, obtinet principatum' (38, 28–40, 1).

[35] 'Ego vero ex scriptoribus astrologiae praecipuis neminem legi qui religionem et leges omnes, ut reliquas res humanas, constellationibus siderum non subiciat. Ptolemaeus, in secundo libro *Apotelesmaton*, quod apud hanc gentem illud numen, apud aliam aliud coleretur, non aliunde natum intelligit, quam ex varia siderum imaginumque natura populis illis et gentibus imperitante' (128, 13–19).

The next passages to be examined are from Book IV, Chapter 10, which is entitled: 'That bad laws, just as good laws also, are not subjected to the heavens'.[36] Pico mentions Ptolemy five times in this chapter. The first passage develops the material just discussed:

> But in general, that both good and bad religions do not depend on the heavens, experience itself sufficiently demonstrates. For some refer the origins and variety of religions to the stars, which rule cities and provinces, as they believe, by their own law. Others refer the origins and variety of religions to what they call the Great Conjunctions of the superior planets, especially of Saturn and Jupiter. The Arabs and Latins follow this opinion, pursuing the tracks of the Arabs. It seems that Ptolemy had approached this, who, as we will declare afterwards, never mentioned these Great Conjunctions. But in the second book of the *Apotelesmaton* he says that those Asiatics situated to the East and South worship Venus and Saturn, since they are under a triplicity of an arid quality, that is, Virgo, Taurus and Capricorn, which he thinks Saturn and Venus rule. Again, whoever lives between the South and West has Venus and Mars for gods, since they are located under a moist triplicity, which sets Mercury and Venus over them along with Mars.[37]

Here Pico describes more fully what he had just claimed for Ptolemy, namely, that places and their celestial rulers determine which gods are worshipped where. Once again, Pico neutrally and accurately describes Ptolemy's position. This time, however, he does so to refute it.[38]

Now Pico directly attacks Ptolemy's position with an argument from experience:

> Now, that opinion of Ptolemy — in which different stars rule different places and peoples in the same manner, so that he would also think that different religious rites exist in different places — is strongly refuted by experience itself: the same stars still rule those provinces lying between the East and South that ruled them formerly, and Venus and Saturn are no longer worshipped there, as they were formerly. Why do these gods rule them? As Ptolemy himself writes, it is because the earthly triplicity, corresponding to Taurus, Virgo and Capricorn, rules those regions, and Venus and

[36] 'Malas leges, sicuti nec bonas, caelo non subici' (486).

[37] 'In universum vero tam bonas quam malas religiones a caelo non dependere, ipsa satis experientia demonstratur. Alii enim religionum ortus et varietatem ad sidera referunt, urbibus et provinciis suo quodam, ut ipsi credunt, iure dominantia; alii ad magnas quas vocant coniunctiones planetarum superiorum, Saturni praesertim atque Iovis; et hanc quidem sententiam sequuntur Arabes et Latini, vestigiis Arabum insistentes; illi accessisse videtur Ptolemaeus, qui, ut postea declarabimus, de magnis illis coniunctionibus nullam umquam habuit mentionem. Sed libro secundo *Apotelesmaton* eos ait ex Asiaticis, qui ad orientem vergunt et meridiem, colere Venerem et Saturnum, quoniam trigono subsint aridae qualitatis, hoc est Virgini, Tauro et Capricorno, quibus ipse putat Saturnum Veneremque dominari; rursus qui inter meridiem habitant et occasum, Venerem atque Martem habere pro numinibus, quoniam humidae triplicitati subiciantur, cui cum Marte, Mercurium et Venerem praeficit' (486, 7–488, 1).

[38] I will discuss the central doctrine of Great Conjunctions below.

> Saturn are in that triplicity. For so he thought. But this partition of the world with its regions distributed under different triplicities is perpetual, not temporary. Wherefore, what existed at some time will always be the same. Why, therefore, O Ptolemy, does the same religion not remain also in the same places today? Finally, let us conclude this entire chapter by thus inquiring of Ptolemy himself: if different stars were worshipped formerly, among different peoples, from a different rulership of the stars, by the force and power of what star is it effected that no stars today are worshipped in almost any region of the entire world?[39]

This chapter ends with Pico calling Ptolemy out and ridiculing him in a mildly sarcastic manner, another way that Pico used Ptolemy in the *Disputations*, especially when Ptolemy promoted a doctrine that Pico rejects.

*

Next I will discuss Book V, Chapter 14, which approaches the subject of religion differently. It is entitled: 'That that which is commonly said to be the true geniture of Jesus is not, and there is no indication in it that he would either die a violent death or be a great prophet'.[40] In this chapter, Pico only mentions Ptolemy once, and he uses him differently here than before:

> What these little diviners (*divinaculi*) trifle about concerning the geniture of Jesus himself is wont to disturb some people. For he was born, they say, while the first face of Virgo — thus they call decans — was rising, about which Albumasar writes that there is in it a lovely maiden holding two spikes of grain in her hand and nourishing a boy, whom a certain people call Jesus. Thus they think that the miracle of the embodied Word is confirmed by the science of astrology, which finds among the celestial images the Virgin and Jesus. For my part, I am not especially angry with them, since they are accustomed to confirm their religious teachings with such testimonies. Therefore, they think that we will willingly accept these things to cor-

[39] 'Iam illam Ptolemaei opinionem, quae perinde atque variis locis et gentibus varia sidera dominantur, ita varios etiam ritus religionum in locis existimat, longe magis ab ipsa experientia confutatur. Cum et illis provinciis quae inter orientem iacent et meridiem praesint eadem sidera quae olim illis praesidebant, nec tamen ut olim ibi Venus Saturnusque coluntur. Cur enim haec illis numina dominabantur? Utique, ut ipse scribit Ptolemaeus, quoniam terrena triplicitas, quae constat ex Tauro, Virgine et Capricorno, regionibus illis praeest; illi vero triplicitati Venus et Saturnus. Ita enim ipse existimavit. At partitio haec mundi et regionum, sub aliis atque aliis triplicitatibus distributarum, perpetua est, non temporaria; quare eadem erit semper quae aliquando fuit. Cur non igitur, o Ptolemaee, eadem etiam in eisdem locis hodieque religio durat? Denique totam istam disputationem ita concludamus, Ptolemaeum ipsum interrogantes, si a varia siderum praesidentia apud alias gentes olim alia sidera colebantur, cuiusnam sideris vi ac potestate efficitur ut nulla sidera hodie aliqua fere totius mundi regione colantur' (488, 21–490, 8).

[40] 'Eam quae vulgo fertur, veram esse Iesu genituram, nec ex ea indicari illum aut violenta morte moriturum, aut magnum esse prophetam' (604, 1–3). For this and other related material, see Pompeo Faracovi, *Gli oroscopi di Cristo.*

roborate our religious teachings. But Christian truth has no need for these fables and dreams, among which even the weighty sayings of the philosophers are all but fables. Who has seen these images while their senses were quiet, which an extremely obscure account indicated? Where did either Ptolemy or some ancient ever mention them? These are the tidiest bits of nonsense, these figments of the Arabs, although they refer them back to the Indians, against which we will dispute more broadly in the following book. But Lord Jesus was not born with that virgin ascending, as they say. Otherwise, he would have come into the light almost two hours before midnight, but we have received from Church tradition that he was born *at* midnight. Wherefore, neither the third degree of Libra, as a certain person says, but rather the tenth occupied the rising point when the most desired of all people arose.[41]

Thus, even if the doctrine of decans were sound, according to Pico's analysis, Jesus in fact had ten degrees of Libra rising and not Virgo at all, let alone its first ten degrees. This would mean that a different decan altogether would be on the ascendant, which would thereby wholly undercut their argument. In this way, Pico uses a properly astrological argument to refute the astrologers, something he said he would do in Book III.1.[42] As he also said, Pico treats the decans later in greater detail in Book VI, Chapter 16, and he uses Ptolemy there too in the same way, namely, by employing Ptolemy's silence concerning decans to make his own anti-astrological point. It is noteworthy that the counter-argument presupposes some good knowledge of astrology.

*

The last chapter I will discuss concerning Pico's attempts to decouple religion from astrology is the rather long and involved Book V, Chapter 5, in which

[41] 'Movere autem solent nonnullos quae de genitura ipsius Iesu nugantur isti divinaculi; natus enim est, inquiunt, prima facie virginis (sic decanos vocant) ascendente, de qua scribit Albumasar, esse in ea virginem formosam duas manu spicas gerentem puerumque nutrientem, quae gens quaedam vocat Iesum. Sic confirmari putant miraculum Verbi corporati per scientiam astrologiae, quae inter caeli imagines Virginem Iesumque repperit. Hic eis equidem non magnopere irascor quoniam solent ipsi sua dogmata talibus testimoniis confirmare, quia putant libenter nos etiam haec recepturos, quibus dogmata nostra corroboremus. Sed non eget his fabulis somniisque veritas christiana, apud quam etiam seria philosophorum paene fabulae sunt. Quis has vidit imagines cui, sensu tacente, ratio illas occultior indicavit? Ubi de illis, vel Ptolemaeus, vel antiquus aliquis umquam fecit mentionem? Meracissimae nugae sunt Arabumque figmenta, quamquam illa referant ad Indos, adversus quas libro sequenti latius disputabimus. At neque ista virgine, ut dicunt, ascendente natus est dominus Iesus, alioquin per duas ferme horas ante mediam noctem in lucem apparuisset, quem ex ecclesiae traditione media nocte natum accepimus; quare nec tertia, ut quidam dicit, Librae pars, sed decima potius, cum oriebatur natorum desideratissimus, horoscopum occupabat' (604, 4–606, 11).

[42] 'Atque ipsam hanc nostram opinionem non aliis magis, quam quibus utuntur astrologi contra nos argumentis, asseveratam probabilemque reddemus' (178, 19–21).

Pico directly confronts the central issue of Great Conjunctions.[43] Pico mentions Ptolemy 16 times in this chapter, whose title is: 'That the planets joined in Great Conjunctions have no more power than when divided, and that these Great Conjunctions have been found to be something new, born from a bad understanding ("malus intellectus") of Ptolemy':[44]

> But even if we grant that planets which have been joined do more than when they are separated toward the great transformations (*mutationes*) of this world — as he had just argued — nevertheless, we will not grant that this is to be referred back to a conjunction of Jupiter and Saturn, or Mars. We will prove, however — both by reason and the authority of the greatest astrologers — that other planets obtain their power first in these matters. For none of the ancients ever made their judgments on universal transformations of the world via what are called Great Conjunctions. [Firmicus] Maternus says nothing about these, although he was certainly the most curious investigator of astrology. Paulus [Alexandrinus] says nothing; Hephaestion [of Thebes] says nothing; Theophilus says nothing; Astaxarchus says nothing. Ptolemy himself says nothing, whose testimony here will suffice, so that we do not use witnesses against them who are too little known.[45]

Here Pico uses Ptolemy along with some other ancients — and as their spokesperson — to argue against one of contemporary astrology's major doctrines that subordinated religion to astrology. We will now explore how he does so.

First Pico neutrally describes Ptolemy's relevant doctrine:

> In the second book of the *Apotelesmaton* (II.4), teaching in what way general transformations of the world are foreseen, [Ptolemy] refers them all and only to eclipses of the sun and moon. Nothing can be said more rationally [sc. in favour of an astrological doctrine], for universal and great effects ought to be referred to these causes, which are the greatest, universal and efficacious. Moreover, it is admitted by everyone that among the planets two only are of universal efficient causality (*efficientia universalis*), namely, the sun, and the moon, whose light is none other than the sun's light borne to earth as by a mirror. Wherefore, if any celestial power (*virtus*) ought

[43] For more on the doctrine of Great Conjunctions, see (e.g.) North, 'Astrology and the Fortunes', and now Hasse, *Success and Suppression*, pp. 272–89, with a discussion of Pico's critique thereof at pp. 277–78.

[44] 'Planetas magnis coniunctionibus iunctos non plus posse quam divisos, magnasque istas coniuntiones novum esse inventum de malo Ptolemaei intellectu natum' (544, 18–20).

[45] 'Quod si iunctos planetas plus facere quam separatos ad magnas istius mundi mutationes illis dederimus, non tamen dabimus hoc ad Iovis Saturnique aut Martis coniunctionem referendum, sed obtinere vim primam in istis rebus alia sidera, et ratione et summorum astrologorum auctoritate probabimus. Neque enim umquam aliquis veterum per has, quas isti vocant magnas coniunctiones, de universalibus mundi mutationibus iudicarunt, nihil de his Maternus, quamquam curiosissimus utique astrologiae investigator, nihil Paulus, nihil Ephestion, nihil Theophilus, nihil Astaxarchus, nihil ipse Ptolemaeus, cuius hic nobis testimonium erit satis, ne parum eis notis testibus adversus eos utamur' (546, 22–548, 2).

to be thought the origin and cause of universal and great effects, none should be thought other than these.[46]

These views can be much more fully developed from Pico's extensive and penetrating natural philosophical analyses in *Disputations*, Book III.[47]

He continues:

But I know what they say about Ptolemy, that this was omitted by him in the *Apotelesmaton* in order not to expose a mystery. For thus Haly, his commentator, writes. But in the *Centiloquium*, many examples of this thing have been given when, in the 50th *verbum*, he directs that we should not forget the conjunctions of the planets, in which there is great efficacy; and likewise in the 58th *verbum*. Then, in the 65th, he reminds us of these same matters, transmitting the great teachings of these things in distinguishing the greatest, median and smallest conjunctions.[48]

Here Pico sets up a contrast between Ptolemy's approach in the *Tetrabiblos* and *Centiloquium*. He also mentions how Haly, Ptolemy's commentator,[49] frames the differences as Ptolemy's deliberate choice in the *Tetrabiblos* in order to protect this major doctrine from careless exposure to the 'hoi polloi'.

Pico now goes on the offensive, setting the tone, as so often, with biting sarcasm. This time, however, it is not directed against Ptolemy himself, but at a very influential — and extremely pernicious — misinterpretation (as he sees it):

I want nothing more than for them to respond to me that, from here on out, either their teachings, whichever seem greater and more admirable, become open to all, or

[46] 'Is igitur, secundo libro *Apotelesmaton*, docens qua via generales et mundi mutationes praevideantur, eas omnes refert solummodo in Solis Lunaeque defectus; nec potest dici aliquid rationabilius, nam debent effectus universales et magni in eas referri causas quae maximae, universales et efficaces sint. Est autem confessum apud omnes inter planetas duos esse tantummodo efficientiae universalis, Solem scilicet et Lunam, cuius lumen non aliud quam Solis lumen per eam quasi per speculum, ut sic dixerim, ad terram delatum. Quare, si qua debet caelestis virtus origo et causa existimari effectum universalium atque magnorum, nulla debet potius quam siderum istorum talis existimari' (548, 2–13).

[47] See the analysis in my PhD thesis *Astrology, Natural Philosophy*, chapter 6, and in volume II of my monograph *Sapientia Astrologica*.

[48] 'Sed scio quid dicent de Ptolemaeo, omissum hoc ab eo in libro *Apotelesmaton* ne mysterium proderet. Ita enim scribit Haly eius interpres. Sed in *Centiloquio* (sic enim vocant) multa eius rei dedisse documenta, cum verbo quinquagesimo eius libri iubeat ne planetarum coniunctiones obliviscamur, in quibus magna sit efficacia; et verbo item tum quinquagesimo octavo, tum quinto et sexagesimo, earundem rerum nos admonet, magna tradens de his praecepta, coniunctionem maximam, mediam minimamque distinguens' (548, 21–550, 2).

[49] The real name of the *Centiloquium*'s commentator is Abū Jaʿfar Aḥmad ibn Yūsuf, and 'Haly' is just a wrong inference by Plato of Tivoli, which contaminated the entire Latin tradition. See Lemay, 'Origin and Success', pp. 103–04. This 'Haly' has nothing to do with and should not be confused with Haly Abenrudian (ʿAlī ibn Riḍwān), the commentator of the *Tetrabiblos*. See now also Hasse, *Success and Suppression*, pp. 370–74, for a clarifying discussion of all three Halys.

> [to admit] that they have emerged from either vain opinions or a false understanding of ancient authors. For what they say — that Ptolemy kept this quiet in the *Apotelesmaton* to not bring forth a mystery — I am so far from denying this that I believe he also did not bring it forth in the *Centiloquium*, in which, certainly, he said nothing more about these conjunctions than what he also did not say about the death of Priam or the Trojan war![50]

Pico has now set the sarcastic tone for what follows. As we will see, Ptolemy's doctrine is not the problem here. In fact, Pico fully agrees with it. Rather, the problem arises with an extremely influential later misinterpretation of what Ptolemy wrote.

Pico then discusses each text from the *Centiloquium* in turn, beginning with *verbum* 50:

> Ptolemy's 50th *verbum* is thus among them, that is, in the common edition (*in vulgata editione*): 'You ought not to forget that there are 120 conjunctions among the planets. For in these there is a greater knowledge of things that come to be in this world receiving increase and decrease'. In Greek it is thus: 'We should not overlook 119 conjunctions. For among them has been placed a conjunction of those things which come to be in the world of generation and corruption'.[51]

Pico here gives the normal translation of this text in Plato of Tivoli's twelfth-century version from the Arabic,[52] and then offers his own slightly but significantly different translation, directly from the Greek.

Pico now has a basis for his own revisionist analysis:

> First of all, the barbarous interpretation attributes more to these conjunctions than Ptolemy does[.] [...] Let us ascribe the transformations of lower things, above all, to the greatest conjunctions. Ptolemy certainly did not say that it is of the superior planets, but rather, if what he wrote is both read and understood accurately, it will

[50] 'Ego vero ab eis responderi nihil potius vellem, ut vel hinc palam omnibus fiat, quaecumque eorum dogmata maiora admirabilioraque videntur, ea vel ex vanis opinionibus, vel ex falsa veterum auctorum intelligentia pullulasse. Nam quod dicunt tacuisse hoc Ptolemaeum in libro *Apotelesmaton*, ne mysterium proderet, tantum abest ut negem, ut nec in *Centiloquio* credam proditum ab eo, in quo certe tam nihil magis locutus est de istis coniunctionibus, quam nec de Priami morte aut bello Troianorum' (550, 2–11).

[51] 'Est quinquagesimus Ptolemaei verbum ita apud eos, hoc est in vulgata editione: "non obliviscaris esse centum viginti coniunctiones, quae sunt in stellis erraticis; in illis enim est maior scientia rerum quae fiunt in hoc mundo suscipientia incrementum et decrementum". Graece est ita: 'Ne praetermittamus centum et decem novem coniunctiones. In his enim posita est coniunctio eorum quae fiunt in mundo generationis et corruptionis' (550, 13–20).

[52] The text here is my transcription of Erhard Ratdolt's 1484 Venice edition, *Liber Ptholemei*. The pages in this edition are not numbered: 'non oblivisceris esse 120 coniunctiones, quae sunt in stellis erraticis; in illis enim est maior scientia eorum quae fiunt in hoc mundo suscipienti incrementum et decrementum'. These are the differences from Garin's text: *obliviscaris*; number written out (*centum viginti*); *rerum* for *eorum*; *suscipientia*. The *status quaestionis* on the various versions of the *Centiloquium* is Boudet, 'Nature et contre-nature'.

> indicate the contrary to us. For he did not say that there are 120 of these conjunctions, as is commonly said, but only 119 because he does not number the conjunction of the sun and moon with them, so that it would have peculiar privileges, and a singular prerogative, in great and universal transformations. He restores all the rest to a disorderly mass of number and order, so that, among them, one would not rule another in a particular situation. But the barbarous expositors, not paying attention, as if it had been omitted by the fault of the scribes, added the conjunction of the sun and moon. They thought that 120 is to be read, not 119, which all the Greek codices have, so that, not content with a deviation of the sense, they also corrupted our faith in the letter.[53]

Here Pico makes a philological-critical argument to revise our understanding of this crucial passage. He argues that Ptolemy did indeed have one particular conjunction in mind, but it was of the sun and moon, not Jupiter and Saturn. In fact, the translations from the Arabic all mention 120 conjunctions, and those from the Greek — Pico's, Pontano's and George of Trebizond's — all have 119.[54]

Pico further supports this radical reinterpretation with *verba* 58 and 65, by arguing that the Greek term 'σύνοδος', when unqualified — as we find it in pseudo-Ptolemy's Greek manuscripts — refers only to conjunctions of the sun with the moon (552, 19–554, 20), that is, to the new moon.[55] Finally, Pico

[53] 'Primum plus tribuit istis coniunctionibus barbara interpretatio quam tribuat Ptolemaeus [...]. Sed in quas potissimum maximas quasque rerum inferiorum mutationes referamus superiorum esse siderum, Ptolemaeus certe non dixit, sed potius, si eius dicta recte et legantur et intelligantur, contrariam nobis sententiam indicabunt. Non enim centum et viginti, ut vulgo legitur, has esse dixit coniunctiones, sed solum centum decem et novem, quia his scilicet Solis et Lunae coniunctionem non numerat, ut quae privilegia habeat peculiaria praerogativamque singularem in magnis universalibusque mutationibus; reliquas omnes acervatim in numerum ordinemque redigit, ut inter quas alia aliae singulari nulla conditione praestaret. Quod non advertentes barbari expositores, quasi omissum foret vitio librariorum, Solis et Lunae coniunctionem addiderunt, legendumque centum et viginti, non autem centum decem et novem, quod graeci omnes codices habent, putaverunt, ut non contenti sensus depravatione, litterae quoque fidem adulterarent' (550, 20–552, 18).

[54] My thanks to David Juste for this information on the *Centiloquium*.

[55] On the face of it, this is neither a sound nor a persuasive argument, since 'σύνοδος' = 'conjunction' can refer to a conjunction of the sun and moon as well as of the other planets, as we find in *verbum* 50 ('συνόδους τῶν πλανήτων'), the definition in LSJ II.2. Nevertheless, this is, in fact, Ptolemy's normal usage when 'σύνοδος' is unqualified in both the authentic *Tetrabiblos* and *Almagest*, as well as in the pseudo-Ptolemaic *Centiloquium*. In the 13 instances of 'σύνοδος' indicated by Hübner in his index nominum to the *Tetrabiblos* (see Hübner, *Ἀποτελεσματικά*), all of the usages are unqualified and refer to the new moon. This is likewise the case in the 50 instances in the *Almagest* identified by the TLG in which one can find three related usages: [1] to the new moon (often along with the full moon); [2] in the phrase 'mean conjunction', in these cases always referring to the sun and moon; and [3] the conjunction of the sun and moon in relation to determining the time of eclipses. My thanks to

reinterprets greatest, median and smallest conjunctions to mean, respectively, eclipses, the new moon before the sun enters the tropical signs (namely, those of the four seasons), and those for every other month (554, 20–558, 11), which is truly far-fetched, and has no support in the *Tetrabiblos*. For Pico, this is Ptolemy's true opinion. Pico attributed the normal view of these conjunctions as of Jupiter and Saturn, on the other hand, to a barbarian, that is, Arabic misinterpretation of Ptolemy's text. Here Pico shows off his philological skills to shine a critical light on a profoundly influential misinterpretation (as he sees it) of Ptolemy's doctrine in the *Centiloquium* that had hitherto provided an authoritative foundation for Great Conjunctions.

*

In a full treatment of this interesting and important topic, I would characterize each of Pico's 376 explicit mentions of Ptolemy. Some will fall clearly into well defined categories, others will not. These are some of the more significant categories I have thus far detected: [1] As an example of discord between authorities to undermine faith in an astrological doctrine, with Ptolemy's as one of the conflicting teachings. [2] Pico using Ptolemy's silence or explicit criticism as powerful ammunition to help him attack and/or delimit a range of astrological doctrines. Ptolemy was himself, of course, a major critic of earlier astrology in the *Tetrabiblos*. [3] Pico also overtly attacks and sometimes ridicules Ptolemy himself for holding a number of astrological doctrines, as well as for other positions, including his purported incompetence as a philosopher.[56] In short, with Ptolemy in particular, Pico wanted to play it both ways: to hold up Ptolemy as an authority when he supports Pico's position, and to undermine and criticize him when he does not.

Alexander Jones for his timely assistance in this matter. In addition to the several unqualified uses of 'σύνοδος' in the *Centiloquium*, the verbal form of 'σύνοδος' ('συνοδεύειν') is qualified explicitly in *verbum* 63 as conjoining Saturn and Jupiter ('ὅτε συνοδεύει ὁ Κρόνος καὶ ὁ Ζεύς'). In *verbum* 65, pseudo-Ptolemy refers to smallest ('minima'), middle ('media') and greatest ('maxima') conjunctions, but he does not further qualify them. Perhaps these two *verba* in close proximity inspired the misinterpretation that Pico is attempting to rectify. Finally, the only other usage I could find of a qualified use of 'σύνοδος' itself in the Greek of any of these three Ptolemaic texts is in *verbum* 50 in the Greek text of the *Centiloquium* in Boer, *Καρπός*: 'Μὴ παραδράμῃς τὰς ριθ συνόδους τῶν πλανήτων'. For whatever reason, though — perhaps he was using a different Greek manuscript — Pontano does not reflect this qualified usage in his translation: 'Ne praetermittas centum et decem novem coniunctiones'. I use Boer's 1952 Teubner edition of the Greek text of the *Centiloquium*, and Pontano's Latin translation in the 1531 Basel edition. I also use Heiberg's edition of the Greek and Toomer's English translation of the *Almagest*, and Hübner's Greek text of the *Tetrabiblos* and Robbins's English translation in the Loeb Classical Library, as well as a printout from the *Thesaurus Linguae Graecae* on the various instances of 'σύνοδος' in the *Almagest*.

[56] Pico treats this theme at Book I (70, 9 ff.).

Pico often used philological-critical arguments towards a range of anti-astrological ends. They were usually deployed with explicit quotations and their often penetrating (if deeply interested) analyses. Sometimes the translations were new and revisionary by Pico himself. He also regularly mentioned Ptolemy along with one or more of his commentaries and/or commentators, including Haly, the anonymous Greek and others, often to highlight their so-called barbarous misinterpretations. In this light, given the broad and impressive display of his much vaunted philological skills — and even though he several times pointed out striking doctrinal differences between the *Tetrabiblos* and the *Centiloquium* (as just above) — it is surprising that he never drew the conclusion that one might be spurious.

Furthermore, Pico attacked Ptolemaic doctrine in numerous ways, including by arguing that Ptolemy's own position was misunderstood, which is then not a direct attack on Ptolemy himself, but on a particular (sometimes influential) interpretation, including by his major commentators. This is a place where, if Pico were interested in reform — not rejection — he could have cleared the way back to a more pristine Ptolemaic astrology with later distorting accretions (including such misinterpretations) removed.[57] We can see this intention with contemporary medically-oriented humanist scholars, including Nicolò Leoniceno, and Giovanni Mainardi, Leoniceno's student and one of the editors of the *Disputations*, along with Pico's nephew Gianfrancesco.[58]

*

Especially in the Renaissance, neutralizing or diminishing Ptolemy's stature as an astrological authority would have taken Pico a very long way indeed towards realizing his quixotic goal of suppressing astrology. Success in rebranding a perfectly legitimate and by-his-time well-established scientific astrology as *divinatory* astrology — that is, as the mother of all superstition and thus the preeminent enemy of the Church — would have completed his overly ambitious goal, but in this he was profoundly unsuccessful, especially in the short term. His ultimate goal, I believe, was to entirely remove astrology from the prophetic airwaves, as it were, especially in an age of widely disseminated annual astrological prognostications that were increasingly available and affordable.[59] In this

[57] We see such a reforming orientation towards astrology in Girolamo Cardano. See Grafton, *Cardano's Cosmos*.

[58] For Leoniceno, see (e.g.) Mugnai Carrara, *La biblioteca di Nicolò Leoniceno*. For Mainardi and his role in the complex many-handed process of editing Pico's *Disputationes*, see Zambelli, 'Giovanni Mainardi', and Farmer, *Syncretism*. Although the evidence Farmer presents is intriguing, his conclusions should be treated with caution.

[59] For valuable recent scholarship on this important topic, see Green, *Printing and Prophecy*, and Tur, *Hora introitus solis*.

way, the voice of a true divinely-inspired prophet, Girolamo Savonarola, could be heard without so much noisy and persuasive competition.[60]

To characterize astrology as divinatory — as Pico does in the *Disputations* — seems perfectly unobjectionable to us, but would have been taken quite differently by most of Pico's readers. The appropriate context for understanding what I mean is an influential and authoritative thirteenth-century text, the *Summa Theologiae* by Thomas Aquinas. In Questions 92 to 95 of the *Secunda secundae*, Thomas sharply distinguished both of what we call astronomy and astrology from divination, in discussing legitimate and illegitimate modes of knowing and/or predicting the future.[61] Although astrology is conjectural and not certain — like astronomy is — they are both legitimate modes of knowing and predicting because they both rely on causal knowledge. This is decidedly not the case with what Thomas explicitly calls *divinatory* practices — including augury and geomancy — which have no causal foundations, and thus rely solely on demons.

In the *Disputations*, then, Pico implicitly responded to and rejected Thomas's influential analysis, collapsing his careful distinctions, and casting astrology wholly into the snakepit of divinatory practices, which he would never dignify with the term arts. In their famous anti-divinatory papal bulls of 1586 and 1631, Sixtus V and Urban VIII both followed Pico in this rebranding effort.[62] The equally influential Rule IX of the *Index of Prohibited Books* (1564, 1596 and later), however, followed and expanded Thomas's views, thus setting up a conflict — valuable for us — between these two sets of legally binding texts, whose debates we can now see fully articulated in recently edited documents from the archives of the Roman Congregations of the Holy Office and the Index.[63] Thomas Aquinas, then, was another major, fundamentally pro-astrological authority for Pico to co-opt, but also a complex one — as we can now more easily see — as was Thomas's distinguished teacher, Albertus Magnus; but these are topics for another occasion.[64]

Despite Pico's furious efforts at rebranding, then, astrology was still considered legitimate knowledge, and continued to be taught at the finest early mod-

[60] For Savonarola in context, see (e.g.) Weinstein, *Savonarola and Florence*, and Dall'Aglio, *Savonarola*, and for his relationship to both Giovanni and Gianfrancesco Pico, see Garfagnini, 'Savonarola tra Giovanni e Gianfrancesco Pico', pp. 237–79.

[61] I discuss this material in part 2 of volume I of my monograph *Sapientia Astrologica*, and more fully and in a broader context in my 'Is Astrology a Type of Divination?'.

[62] See (i.a.) Ernst, 'Dalla bolla *Coeli et terrae*', pp. 255–79.

[63] See Baldini and Spruit, *Catholic Church and Modern Science*, vol. I, tomes 1–4. Although the texts they publish are extremely valuable, their interpretations should be treated with caution.

[64] For both Thomas's and Albert's views on astrology — including in relation to theology — see volume I of my monograph *Sapientia Astrologica*.

ern European universities, until well into the seventeenth century, and sometimes beyond.[65] Nevertheless, Pico's *Disputations Against Divinatory Astrology* — with its complex and highly interested uses of Ptolemy — certainly played a significant cumulative role in astrology's eventual removal from the time-honored and well-established premodern maps of legitimate knowledge and practice. Further study should make that role more fully understood.

Bibliography

Adamson, Peter, and Peter E. Pormann, *The Philosophical Works of Al-Kindī*, Oxford: Oxford University Press, 2012.

Azzolini, Monica, *The Duke and the Stars. Astrology and Politics in Renaissance Milan*, Cambridge (Mass.): Harvard University Press, 2013.

Bacchelli, Franco, 'Appunti per la storia del testo delle *Disputationes adversus astrologiam divinatricem*', *Dianoia* 13 (2008), pp. 141–59.

Baldini, Ugo, and Leen Spruit (eds), *Catholic Church and Modern Science. Documents from the Archives of the Roman Congregations of the Holy Office and the Index*, vol. I, tomes 1–4: *Sixteenth Century Documents*, Roma: Libreria Editore Vaticana, 2009.

Bertozzi, Marco (ed.), *Nello specchio del cielo. Giovanni Pico della Mirandola e le* Disputationes *contro l'astrologia divinatoria*, Firenze: Olschki, 2008.

Bezza, Giuseppe, *Arcana Mundi. Antologia del pensiero astrologico antico*, 2 vols, Milano: Rizzoli, 1995.

Boer, Emilie, *Καρπός. Pseudo-Ptolemaei Fructus sive Centiloquium*, Leipzig: Teubner, 1952 [*Claudii Ptolemaei Opera quae exstant omnia*, vol. III.2].

Boudet, Jean-Patrice, *Entre science et nigromance. Astrologie, divination et magie dans l'Occident médiéval, XIIe-XVe siècle*, Paris: Publications de la Sorbonne, 2006.

Boudet, Jean-Patrice, 'Nature et contre-nature dans l'astrologie médiévale. Le cas du *Centiloquium* du pseudo-Ptolémée', in Maaike van der Lugt (ed.), *La nature comme source de la morale au Moyen Âge*, Firenze: SISMEL edizioni del Galluzzo, 2014, pp. 383-410.

Caroti, Stefano, 'Le fonti medievali delle *Disputationes adversus astrologiam divinatricem*', in Marco Bertozzi (ed.), *Nello specchio del cielo. Giovanni Pico della Mirandola e le Disputationes contro l'astrologia divinatoria*, Firenze: Olschki, 2008, pp. 67–94.

Dall'Aglio, Stefano, *Savonarola and Savonarolism*, tr. John Gagne, Toronto: Centre for Reformation and Renaissance Studies, 2010.

d'Alverny, Marie-Thérèse, and Françoise Hudry, 'Al-*Kindi, De Radiis*', *Archives d'Histoire Doctrinale et Litteraire du Moyen Age* 41 (1975), pp. 139–259.

[65] For the basic chronology, see my article 'Astrology' and now my 'How to Accurately Account'.

Ernst, Germana, 'Dalla bolla *Coeli et terrae* all'*Inscrutabilis*. L'astrologia tra religione, natura e politica nell'età della Contrariforma', in Germana Ernst (ed.), *Religione, ragione e natura. Ricerche su Tommaso Campanella e il tardo Rinascimento*, Milano: Franco Angeli, 1991, pp. 255–79.

Farmer, Stephen A., *Syncretism in the West: Pico's 900 theses (1486). The Evolution of Traditional, Religious, and Philosophical Systems*, Tempe: Medieval and Renaissance Texts and Studies, 1998.

Federici Vescovini, Graziella, 'I programmi degli insegnamenti del Collegio di medicina, filosofia e astrologia dello statuto dell'università di Bologna del 1405', in Jacqueline Hamesse (ed.), *Roma, magistra mundi. Itineraria culturae medievalis, Mélanges offerts au Père L. E. Boyle*, Louvain-la-Neuve: Fédération des Instituts d'Etudes Médiévales, 1998, vol. I, pp. 193–223.

Garfagnini, Gian Carlo, 'Savonarola tra Giovanni e Gianfrancesco Pico', in Gian Carlo Garfagnini (ed.), *Giovanni Pico della Mirandola. Convegno internazionale di studi nel 500° anniversario della morte (1494–1994), Mirandola, 4–8 ottobre 1994*, Firenze: Olschki, 1997, pp. 237–79.

Garin, Eugenio, *Giovanni Pico della Mirandola, Disputationes adversus astrologiam divinatricem*, 2 vols, Firenze: Vallecchi, 1946–1952.

Gentile, Sebastiano, 'Pico e la biblioteca Medicea privata', in Paolo Viti (ed.), *Pico, Poliziano e l'umanesimo di fine Quattrocento*, Firenze: Olschki, 1994, pp. 85–101.

Gentile, Sebastiano, 'Pico filologo', in Gian Carlo Garfagnini (ed.), *Giovanni Pico della Mirandola. Convegno internazionale di studi nel 500° anniversario della morte (1494–1994), Mirandola, 4–8 ottobre 1994*, Firenze: Olschki, 1997, pp. 465–90.

Grafton, Anthony, 'Giovanni Pico della Mirandola. Trials and Triumphs of an Omnivore', in Anthony Grafton (ed.), *Commerce with the Classics. Ancient Books and Renaissance Readers*, Ann Arbor: University of Michigan Press, 1997, pp. 93–134.

Grafton, Anthony, *Cardano's Cosmos. The Worlds and Works of a Renaissance Astrologer*, Cambridge (Mass.): Harvard University Press, 2001.

Green, Jonathan, *Printing and Prophecy. Prognostication and Media Change 1450–1550*, Ann Arbor: The University of Michigan Press, 2011.

Hasse, Dag Nikolaus, *Success and Suppression. Arabic Sciences and Philosophy in the Renaissance*, Cambridge (Mass.): Harvard University Press, 2016.

Heiberg, Johan L., *Syntaxis mathematica*, 2 vols, Leipzig: Teubner, 1898-1903 [*Claudii Ptolemaei opera quae exstant omnia*, vol. I].

Hossfeld, Paul (ed.), *Albertus Magnus, Opera Omnia*, vol. V.1: *De caelo et mundo*, Monasterii Westfalorum: Aschendorff, 1971; vol. V.2: *De generatione et corruptione*, Monasterii Westfalorum: Aschendorff, 1980.

Hübner, Wolfgang, *Ἀποτελεσματικά*, Stuttgart/Leipzig: Teubner, 1998 [*Claudii Ptolemaei opera quae exstant omnia*, vol. III.1].

Lemay, Richard, 'Origin and Success of the Kitāb Thamara of Abū Jaʿfar Aḥmad ibn Yūsuf ibn Ibrāhīm from the Tenth to the Seventeenth Century in the World of Islam and the Latin West', in Ahmad Y. al Hassan, Ghada Karmi and Nizar Namnum (eds), *Proceedings of the First International Symposium for the History of Arabic Science (Aleppo, April 5–12, 1976)*, Aleppo: Institute for the History of Arabic Science, 1978, vol. II, pp. 91–107.

Mugnai Carrara, Daniela, *La biblioteca di Nicolò Leoniceno. Tra Aristotele e Galeno: cultura e libri di un medico umanista*, Firenze: Olschki, 1991.

Newman, William R., 'The Chymistry of Isaac Newton' (http://webapp1.dlib.indiana.edu/newton/index.jsp).

Newman, William R., *Newton the Alchemist: Science, Enigma, and the Quest for Nature's Fire*, Chicago: University of Chicago Press, 2018.

North, John D., 'Astrology and the Fortunes of Churches', *Centaurus* 24 (1980), pp. 181–211.

Nothaft, Philipp, '*Vanitas vanitatum et super omnia vanitas*. The Astronomer Heinrich Selder and a Newly Discovered Fourteenth-Century Critique of Astrology', *Erudition and the Republic of Letters* 1 (2016), pp. 261–304.

Pompeo Faracovi, Ornella, *Gli oroscopi di Cristo*, Venezia: Marsilio, 1999.

Pontano, Giovanni, *Liber commentationum in Centum sententiis Ptolemaei*, Basel: Andreas Cratander, 1531.

Ptolemy, *Quadripartitum*, tr. Plato of Tivoli, Venice, Erhardus Ratdolt, 1484.

Robbins, Frank E., *Ptolemy. Tetrabiblos*, Cambridge (Mass.): Harvard University Press, 1940.

Rutkin, H. Darrel, 'The Use and Abuse of Ptolemy's *Tetrabiblos* in Renaissance and Early Modern Europe. Two Case Studies (Giovanni Pico della Mirandola and Filippo Fantoni)', in Alexander Jones (ed.), *Ptolemy in Perspective. Use and Criticism of his Work from Antiquity to the Nineteenth Century*, Dordrecht: Springer, 2010, pp. 135–49.

Rutkin, H. Darel, *Astrology, Natural Philosophy and the History of Science, c. 1250-1700: Studies toward an Interpretation of Giovanni Pico della Mirandola's* Disputationes adversus astrologiam divinatricem, PhD dissertation, Indiana University, 2002.

Rutkin, H. Darrel, 'Astrology', in Lorraine Daston and Katharine Park (eds), *The Cambridge History of Science*, vol. III, Cambridge: Cambridge University Press, 2006, pp. 541–61.

Rutkin, H. Darrel, 'Astrology and Magic', in Irven M. Resnick (ed.), *A Companion to Albert the Great. Theology, Philosophy and the Sciences*, Leiden: Brill, 2013, pp. 451–505.

Rutkin, H. Darrel, 'How to Accurately Account for Astrology's Marginalization in the History of Science and Culture: The Essential Importance of an Interpretive Framework', in a special issue of *Early Science and Medicine* edited by Hiro Hirai and Rienk Vermij, 23 (2018), pp. 217-43.

Rutkin, H. Darrel, 'Is Astrology a Type of Divination?: Thomas Aquinas, the Index of Prohibited Books and the Construction of a Legitimate Astrology in the Middle

Ages and the Renaissance', *International Journal of Divination & Prognostication* 1 (2019), pp. 36-74.

Rutkin, H. Darrel, *Sapientia Astrologica: Astrology, Magic and Natural Knowledge, ca. 1250–1800, Vol. I: Medieval Structures (1250–1500): Conceptual, Institutional, Socio-Political, Theologico-Religious and Cultural*, Dordrecht: Springer, 2019; vol. II: *Renaissance Structures (1480–1500): Continuities and Transformations*, in process.

Simon, Gérard, *Kepler astronome, astrologue*, Paris: Gallimard, 1979.

Toomer, Gerald J., *Ptolemy's Almagest*, London/New York: Duckworth / Springer, 1984.

Tur, Alexandre, *Hora introitus solis in Arietem. Les prédictions astrologiques annuelles latines dans l'Europe du XVe siècle (1405–1484)*, Thèse de Doctorat, Université d'Orléans, 2018.

Weinstein, Donald, *Savonarola and Florence. Prophecy and Patriotism in the Renaissance*, Princeton: Princeton University Press, 1970.

Westman, Robert S., *The Copernican Question. Prognostication, Skepticism and Celestial Order*, Berkeley: University of California Press, 2011.

Zambelli, Paola, 'Giovanni Mainardi e la polemica sull'astrologia', in *L'opera e il pensiero di Giovanni Pico della Mirandola nella storia dell'umanesimo. Convegno internazionale per il V Centenario della nascita di Giovanni Pico della Mirandola (Mirandola, 15–18 settembre 1963)*, vol. II, Firenze: Olschki, 1965, pp. 205–79 [reprinted in her *L'ambigua natura della magia: Filosofi, streghe, riti nel Rinascimento*, Venezia: Marsilio, 1996, pp. 76-118].

Longomontanus on Mars: The Last Ptolemaic Mathematical Astronomer Creates a Theory

Richard L. KREMER

1. Introduction

In his 1630 annual astrological prognostication, the Stettin city physician Lorenz Eichstad introduced his readers to the conflicting astronomical calculations that were buffeting contemporary calendar-makers. Following standard practice, Eichstad computed dates and times for the Sun's entry into the four cardinal points, cast horoscopes for those times, and by interpreting those charts predicted the weather and other significant events for the four quarters of the coming year. But how should the astronomer calculate planetary positions for the horoscopes? From tables in the *Astronomia danica* (1622), authored by Tycho Brahe's former assistant, Longomontanus, Eichstad determined the time of the winter solstice and the planetary longitudes for that time. But he also extracted these longitudes from the tables of another former assistant of Tycho's, Johannes Kepler, as they had appeared in the recently published ephemerides (1630) prepared by Jacob Bartsch. And for good measure, Eichstad took positions from the 1616 ephemerides of Giovanni Magini, computed from the Copernican Prutenic Tables.[1] As can be seen in Table 1, these computed positions could diverge by more than a degree.

	Eichstad	Longomon.	Kepler	Magini
Moon	15;31	15;31	15;36	[14;44][a]
Mercury	9;39	9;39	9;22	10;53
Venus	9;36	9;36	9;17	8;44
Mars	15;35	[15;32][b]	15;35	15;36[c]
Jupiter	8;15	8;15	8;17	8;09
Saturn	3;18	3;18	3;00	3;35

Table 1. Eichstad's predicted planetary longitudes for winter solstice, 11 December 1629, 12;43 p.m., meridian of Stettin (ignoring zodiacal signs).

[a] Eichstad did not include a Copernican lunar longitude; I list the value given in Magini's ephemerides.
[b] Eichstad's prognostication does not specify a Mars value for Longomontanus but notes that it 'nearly agrees' with Magini's value. Using the *Astronomia danica*, I compute a Mars longitude of 15;32 for this time. My computations for the other planets agree exactly with Eichstad's.
[c] Magini's (Copernican) ephemerides places Mars at 16;56.

[1] Eichstadt, *Prognosticon astrologicum*, sig. A2r-A3v; Longomontanus, *Astronomia danica*; Bartsch, *Ephemeridis Ioannis Kepleri*; Magini, *Ephemerides coelestium motuum*.

Ptolemy's Science of the Stars in the Middle Ages, ed. by David Juste, Benno van Dalen, Dag Nikolaus Hasse and Charles Burnett, PALS 1 (Turnhout, 2020), pp. 407–443
© BREPOLS PUBLISHERS 10.1484/M.PALS-EB.5.120186

Throughout his analysis of these discrepant data, Eichstad wrote only of the 'Keplerian calculation', the 'Copernican calculation' and the 'Longomontanian calculation'. Never did he mention heliocentrism, ellipses, spheres or physical forces. This calendar maker concerned himself primarily with predictive computation, not geometrical models or cosmological structure. He did wonder, however, why the calculations of Tycho's two assistants diverge, even though they rest on the same foundation of Tycho's observations. Vaguely alluding to their differing 'hypotheses' that yield variations of 'some minutes', greater in the inferior than in the superior planets, Eichstad chose to base his horoscope on Longomontanus's results (second and third columns in Table 1). Unlike Kepler, Eichstad wrote, Longomontanus had made observations at Hven with Tycho for ten years and had continued this work after Tycho's death. In his *Progymnasmata* (1602), Tycho had praised Longomontanus as 'ingenious and industrious'. For Eichstad, a mathematical astronomy based on hard empirical work apparently offered more reliable predictions than did one based on new 'hypotheses'.[2]

Over the next two decades, Eichstad continued publishing not only annual calendars but also ephemerides to aid other calendar makers. At first, he took the solar and lunar computations from Longomontanus's *Astronomia danica* (henceforth *AD*) and the planetary positions from Kepler's *Rudolphine Tables*. But in later editions, Eichstad's ephemerides increasingly feature only Longomontanus's computations. Like other seventeenth-century Baltic astronomers, Eichstad rejected Kepler's 'special hypotheses' and claimed that Longomontanus's predictions better matched easily observed planetary phenomena such as eclipses or conjunctions of planets and stars. In 1644, Eichstad quoted a letter from Longomontanus: Kepler had 'relied upon physical and too uncommon speculations ... I could not approve his *Rudolphine Tables* ... In fact, I am certain that astronomy rests on principles that are much loftier than such physical ones'.[3]

Indeed, Longomontanus's *AD* would remain a very influential text over the middle third of the seventeenth century. Despite Kepler's new astronomy of forces and ellipses, first presented in the *Astronomia nova* (1609) (henceforth *AN*), the *AD* remained widely read across Europe. In Peking, Jesuit and Chinese astronomers even produced Chinese translations.[4] Called the 'Tychonian Almagest' by K. P. Moesgaard, the *AD* was the last text that sought to offer a comprehensive mathematical astronomy in the tradition of Ptolemy. The 550-page tome includes trigonometric preliminaries (not employing logarithms), kinematic geometrical models comprised of circles, epicycles and eccentrics

[2] Eichstadt, *Schreibcalender*, sig. A2r-A3v.

[3] Translated in Omodeo, 'The Scientific Culture', p. 140.

[4] Hashimoto, 'Longomontanus's *Astronomia Danica* in China'.

and tables to compute planetary longitudes and latitudes, parameters derived from selected observations made by Brahe, Copernicus, al-Battani, Ptolemy and Longomontanus himself, eclipse computational tools, and very little about the physics of the heavens or about astrology. As N. M. Swerdlow has shown from an analysis of his lunar theory, Longomontanus's geometrical models are the most complex and convoluted ever constructed within the 1500-year Ptolemaic tradition.[5]

We cannot, in this paper, provide a full analysis of the *AD* and its planetary theories. Rather, we shall limit our focus to Longomontanus's work on Mars, the planet that would lead Kepler to write his *Astronomia nova*. We shall trace the last Ptolemaic astronomer, practicing his craft following pathways laid down in Ptolemy's *Almagest* and in the *Astronomia nova*. And we shall see him occasionally leaving those paths, when he rejects equants (like Copernicus) and builds some of his astronomical parameters on Euclid's perfect numbers, the first four of which are 6, 28, 496 and 8128 (numbers equal to the sum of their factors, including 1). Despite these occasional departures, however, the *AD* belongs to the genre of the *Almagest*, even if its topics do not always correspond exactly to those of the earlier text.

The son of peasants, Christian Sørensen Longberg (1562–1647), better known as Longomontanus, the Latinized name of his birthplace, had studied at a cathedral school and the University of Copenhagen before entering, around 1589, Tycho Brahe's household on Hven.[6] Apparently possessing keen eyesight, he observed stellar positions for Tycho's new star catalog. He soon became Tycho's favorite assistant and was given responsibility for developing Tycho's lunar theory. They worked together until 1597 when Tycho was forced to dismantle his observatory and leave Hven. Longomontanus then took up traveling, continuing study at the universities in Leipzig and Rostock, earning in 1598 his Magister at the latter institution. In Breslau, he met the well-known humanist Jacob Monavius, who introduced him to the sister of Paul Wittich (1546–86), another former assistant of Tycho's whom the latter would accuse of plagiarism. Through Longomontanus's contacts with the family, Tycho would attempt to purchase Wittich's library.[7]

[5] Moesgaard, 'Tychonian Observations', p. 84; Swerdlow, 'The Lunar Theories of Tycho Brahe'; Swerdlow, 'Tycho, Longomontanus, and Kepler'.

[6] For a recent summary of the extensive biographical literature, see Christianson, *On Tycho's Island*, pp. 313–19. For previous studies of Longomontanus's astronomical work, see Moesgaard, 'How Copernicanism Took Root'; Moesgaard, 'Tychonian Observations'; Moesgaard, 'Cosmology in the Wake of Tycho'; Swerdlow, 'The Lunar Theories of Tycho Brahe', pp. 8–10.

[7] Brahe to Longomontanus, 24 March 1598, in Dreyer, *Tychonis Brahe Dani opera omnia*, vol. VIII, pp. 34–35. See Gingerich and Westman, 'The Wittich Connection'.

In 1600, after Tycho settled in Prague, Longomontanus rejoined his former boss and began working on the theory for Mars.[8] When Kepler arrived several weeks later, however, Tycho gave him Mars, asking Longomontanus to return to the lunar theory, which had remained unfinished after an earlier version published in 1598 proved to be flawed. By June of 1600, Longomontanus completed the revised lunar theory (it would be published in 1602 in the *Progymnasmata*) and left Prague. He remained in correspondence with Kepler; according to Voelkel, their discussions of the theories for the Moon and Mars would eventually shape the rhetorical structure of Kepler's *AN*.[9]

After a stint of teaching at his alma mater, the cathedral school, Longomontanus in 1605 secured a teaching position at the University of Copenhagen, where he would remain until his death. Holding the Chair of Mathematics, he also taught astronomy and in the 1640s planned and became first director of an updated version of Tycho's observatory, the Round Tower in Copenhagen, a building that still stands today in the city's center. In addition to the *AD*, Longomontanus published several annual calendars with astrological predictions, a treatise on squaring the circle,[10] academic disputations on various topics, and in 1639 an introduction to observational astronomy (the telescope is not mentioned).[11] The only assistant of Tycho's to ascend to a university position, Longomontanus enjoyed a considerable reputation during the first half of the seventeenth century as author of the 'Tychonian Almagest' (it 'made Longomontanus famous throughout Europe', writes J. R. Christianson[12]), even if Kepler, eventually, would become Tycho's best-known assistant. Their respective works on Mars, as will be argued below, nicely demonstrate the contrast between Ptolemaic mathematical astronomy and Keplerian physical astronomy.

2. Longomontanus's first Mars theory (LM1)

We begin our story with Longomontanus as Tycho's assistant on the island of Hven. By 1589, Tycho had finished his solar theory. The lunar motion, however, remained intractable; Tycho would continue working on that theory until his death in 1600. Both theories, with tables for computing positions, would not be published until 1602 when the printing of the posthumous *Progymnasmata* finally was completed. But that volume did not include planetary theories. By the early 1590s, Tycho had assembled a large set of planetary observations

[8] For details of Tycho's extended efforts to woo Longomontanus to return to his service, see Thoren, *The Lord of Uraniborg*, pp. 407–08, 419, and Donahue, *Johannes Kepler. Astronomia nova*, p. 134.

[9] Voelkel, *The Composition of Kepler's* Astronomia nova, pp. 153–64.

[10] van Maanen, 'The Refutation of Longomontanus' Quadrature'.

[11] Longomontanus, *Introductio in theatrum astronomicum*.

[12] Christianson, *On Tycho's Island*, p. 318.

and had convinced himself that many of Copernicus's parameters for the planetary theories were incorrect by small amounts. As Thoren has argued, Tycho also realized that Copernican predictions for all the planets differed systematically from the observations and wondered whether an additional inequality, not known to Ptolemy or Copernicus, might affect their motions. Or perhaps planetary theories should be referred to the true rather than mean Sun, as Ptolemy had done, and the other inequality could be solved by modifying the solar eccentricity. Given the similarity of Mars's orbit to the Sun's, Tycho suspected that the red planet might provide a key for reworking all the planetary theories.[13]

Tycho would not live to construct these revised planetary theories.[14] After moving to Prague at the end of 1599, he persuaded his former assistant, Longomontanus, to join him in that city and to take up the theory of Mars. At this point, Kepler, who reached Prague several weeks later, becomes our story teller, crafting what would become a classic narrative in the history of European astronomy about the move from a mathematical astronomy of circles to a physical astronomy of ellipses.

According to Kepler's account, presented in the *AN*, Ch. 7 ('The circumstances under which I happened upon the theory of Mars'), Longomontanus had extracted from Tycho's planetary observations, made between 1580 and 1600, a set of ten Mars acronychal observations (i.e., at mean opposition, when the true planet is in opposition to the mean Sun) and had 'invented an hypothesis' [excogitata hypothesis] that could match the observed data to within 2 arcminutes of longitude. In Ptolemaic astronomy, at mean opposition a superior planet's epicycle (i.e., the second anomaly) is essentially eliminated from the computation so that one uses observations at those times to test a theory's predictions for the first anomaly (the work of the equant point for Ptolemy). In Copernican astronomy, at mean opposition the earth's orbit is essentially eliminated, so that again the first anomaly is being explored, i.e., the theory is being tested for its predicted heliocentric longitudes.

[13] Thoren, *The Lord of Uraniborg*, p. 448. For a short overview of LM1, see Swerdlow, 'The Lunar Theories of Tycho Brahe', pp. 8–10.

[14] It is difficult to determine how much of LM1 was completed before Tycho's death. In his *Astronomiae instauratae mechanica* (1598), Tycho wrote that he had done 'all that I could' for theories of the five planets. '... we have assembled ... the apogees as well as the eccentricities, and further the angular motions and the ratios of their orbits and periods, so that they no longer contain all the numerous errors of previous investigations'. He found that the apogees are subject to 'another inequality' and that the annual motion is 'subject to a variation'. He also revised the values of maximum latitudes and the places of their nodes at the ecliptic. Thus, 'with regard to all five planets there remains only one thing to do, namely to construct new and correct tables expressing by numbers all that has been established by more than 25 years of careful celestial observations ... thereby demonstrating the inaccuracy of the usual tables'. (Quoted from Raeder et al., *Tycho Brahe's Description of His Instruments*, pp. 115–16.) See Thoren, *The Lord of Uraniborg*, pp. 448–50.

In the *AN*, Kepler did not fully describe the Mars theory as developed by Longomontanus and Tycho. He indicated that they had begun with the framework of Copernicus's Mars theory and the parameters of the Prutenic Tables and had:

i) increased Mars' mean motion by 1½ to 1¾ arcminutes;
ii) decreased the longitude of the apogee by 5;02° for the epoch beginning 1585;
iii) increased the total amount of precession since the epoch of Christ by 6½ arcminutes;[15]
iv) increased Mars' total eccentricity from 0.1960 to 0.2016;
v) shifted the ratio of the eccentricities, $(e_1+r')/(e_1-r')$ or the ratio of the eccentricity for direction to the eccentricity for distance, from 2/1 to 8/5 (more on this below, see Figure 1).[16]

They also had invented a 'hypothesis and table' for latitudes that Kepler did not describe. With these parameters and hypotheses, Kepler informs us, Longomontanus/Tycho had computed tables of i) Martian equations at one-degree intervals; ii) mean motions for 40 years 'exactly as was done for the solar and lunar motions in Book I of the *Progymnasmata*';[17] and iii) Martian latitudes.[18]

Reaching the peroration of his narrative, Kepler claimed that their Mars theory had failed. Longomontanus and Brahe had 'proclaimed' that it could match the observations at acronychal positions to within 2 arcminutes; but 'Christian got stuck' (Kepler did not specify the size of the errors!) at the acronychal latitudes and the parallax of the annual orb, i.e., the second anomaly. 'This result was a problem for him, as he was about to brood over the lunar motions'. Mars had outwitted Longomontanus. 'I therefore', concluded Kepler in one of the more under-stated asides in the history of astronomy, 'began to investigate the certitude of their [Longomontanus's] operation. What success came out of that labor it would be boring and pointless to recount. I shall describe only so much of that labor of four years as will pertain to our methodical enquiry'.[19]

[15] Kepler listed the revised value for precession of 28;02,30; I have computed the Prutenic value for 1 January 1585, the epoch Kepler assigned to Longomontanus's revised Mars tables.

[16] For details of Copernicus's models for the superior planets, see Swerdlow and Neugebauer, *Mathematical Astronomy*, Ch. 5. I use here, and below, the notation of Swerdlow and Neugebauer.

[17] See Dreyer, *Tychonis Brahe Dani opera omnia*, vol. II, p. 46 for mean motion tables extending from 1560–1619 for the Sun; vol. II, pp. 103–06 for the Moon, for the years 1560–1660.

[18] Donahue, *Johannes Kepler. Astronomia nova*, pp. 134–35. I will not consider latitude theory in this chapter.

[19] Donahue, *Johannes Kepler. Astronomia nova*, p. 135.

To explore LM1, Kepler prepared a horizontal table of 10 columns (stretching across two pages, cleverly placed on the innermost sheet of the printed quire to guarantee registration across the gutter), listing for the ten acronychal dates the observed longitudes (reduced to the plane of the ecliptic) and latitudes and the mean longitudes, apogees, precessions, and true longitudes of Mars as computed from the revised theory. Kepler did not give a column for the computed latitudes. And he did not include a column directly comparing observed and computed longitudes, i.e., exploring the claim that the new theory could match acronychal observations to ±2 arcminutes of longitude. If we add this comparison (Table 2), using data Kepler presented in Ch. 8 (Tycho's reductions of the raw observations) and in Ch. 10 (Kepler's reductions), we find that the deviations can vary by up to ±9 arcminutes, nearly five times the result claimed by Longomontanus and Tycho. Did Kepler omit these columns to avoid accusing his former employer of exaggeration? In any case, Kepler undoubtedly realized that his predecessor's revised Mars theory performed less well than they had claimed.

Kepler also did not present the Prutenic (PT) longitudes for the times in question. My comparison of Prutenic longitudes against those generated by LM1 (Table 2, col. 4) shows that Tycho's and Longomontanus's tinkering could shift the predictions by more than 4 degrees and by so doing could fit the observed positions much better than could the PT.

Obser.	Obs (*AN*, Ch. 8) – LM1	Obs (*AN*, Ch. 10) – LM1	LM1 – PT
1	−5	−5	31
2	−5	−5	−107
3	1	1	−142
4	6	4	−133
5	4	1	−105
6	−8	−5	19
7	9	9	264
8	−1	2	93
9	−4	−2	−65
10	−2	−1	−133

Table 2. Tests of LM1 (first anomaly), in arcminutes of longitude

The other columns in Kepler's table reveal additional details about LM1. The computed mean sidereal motions for Mars (col. 6) are advanced by an average of about 1½ arcminutes from the Prutenic mean longitudes for the dates in question (exactly as Kepler had described the new theory), assuming that Uraniburg is 40 minutes of time west of Königsberg, the meridian of the PT.[20]

[20] Neither the Prutenic Tables nor Tycho's *Progymnasmata* include 'tables of places', i.e., longitudes of various towns and cities of Europe. The *AD*, however, provides such a table,

The computed Martian apogee (col. 7), measured from the first star of Aries, is decreased from the Prutenic values by an average of 5;02°, again exactly as Kepler had stated. The column for precession (col. 8) Kepler labeled as 'nostra' and indeed, rather than using the Prutenic precession (variable rate) Longomontanus used Tycho's precession (fixed rate of 51 arcsecs/year); my computations match those listed by Kepler to ±4 arcsecs.[21]

Col. 9 lists the Martian true longitudes as generated by LM1. As noted above, since at mean opposition, the second anomaly is eliminated, the true longitudes give the correction (i.e., the equation of center) for the first anomaly for the various dates. In their analysis of Copernican planetary theory, Swerdlow/Neugebauer label this correction c_3. At mean opposition:

$$\lambda_{♂} = \bar{\lambda}^*_{♂} + \Pi - c_3 \tag{1}$$

where Π is precession and the starred value is the sidereal mean Martian longitude. As Swerdlow/Neugebauer show, this correction can be computed as (see Figure 1):

$$\tan c_3 = \frac{(e_1 + r')\sin \bar{\kappa}}{R + (e_1 - r')\cos \bar{\kappa}}, \tag{2}$$

where, in Copernicus's model for a superior planet, e_1 is the eccentricity, r' the radius of the small epicycle, R the radius of the eccentric circle, and $\bar{\kappa}$ the mean eccentric anomaly.[22] For Mars, Copernicus had set $e_1 = 1460$, $r' = 500$, values LM1 modifies to 1638 and 378, respectively. That is, LM1 replaces Ptolemy's bisected eccentricity for Mars, $(e_1+r')/(e_1-r') = 8/4$ and Copernicus's slight deviation therefrom (8/3.92), shifting the ratio of the eccentricities to 8/5, a value close to that Kepler will derive in AN, Ch. 16, for his vicarious hypothesis.

Kepler offered no hints as to how Tycho and Longomontanus had formulated new values for the eccentricities; his chief concern at this point in the *AN* was not to further tweak these values but to argue that one should use true, not mean, oppositions, to study models for the first anomaly. We must note, however, that Y. Maeyama has shown, in an anachronistic analysis, that the optimum eccentricities for the vicarious hypothesis (i.e., those that allow the model to most closely match the 'correct' elliptical orbit) result in an 8/5

showing Uraniburg 40 time minutes west of Königsberg. Presumably this was the value used by Tycho and his assistants. The Rudolphine Tables lists the separation as 38 time minutes.

[21] The amounts of precession to the dates listed in Kepler's table, at a fixed rate of 51 arcsecs/Julian year, suggest an epoch of precession for Tycho and Longomontanus of −395. They presumaby copied this epoch from the mean precession in *De revolutionibus*. See Swerdlow and Neugebauer, *Mathematical Astronomy*, p. 543.

[22] For details, see Swerdlow and Neugebauer, *Mathematical Astronomy*, pp. 297–99.

ratio! Somehow, Tycho and Longomontanus had found the 'best' solution for the Copernican model even before Kepler explored the problem with his vicarious hypothesis (see below for more on this issue).[23]

Using these modified parameters to compute c_3 for the times of the ten mean oppositions, I can match the Martian true longitudes given in col. 9 to ±8 arcsecs in seven cases (but only to −108, 79 and −76 arcsecs in the other three cases, presumably typographical or computational errors in the production of the *AN*).

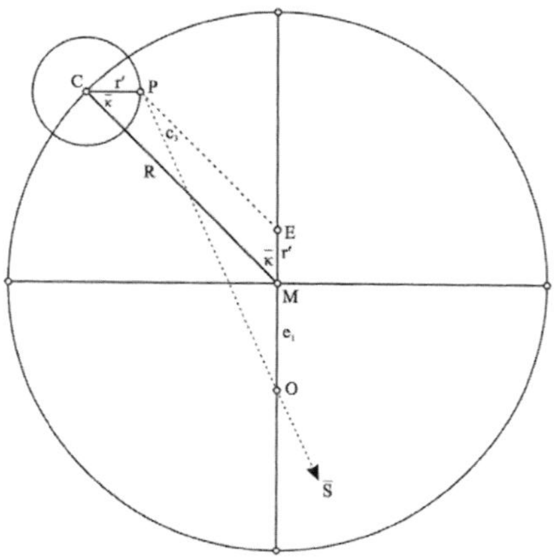

Figure 1. *De revolutionibus* model for superior planets at mean opposition, with the Earth at O, true planet at P and mean Sun in the direction of $\bar{S}$.

Kepler's description, although terse, does allow us to realize that Tycho and Longomontanus were not using the Martian model from Copernicus's *De revolutionibus*, with its eccentricity and a single small epicycle. Rather, wrote Kepler, their model expressed the 'maximum eccentricity' as the sum of 'the semidiameters of the two small circles', exactly the model Copernicus had used in his *Commentariolus* and had briefly mentioned in *De revolutionibus* V, 4 (see Figure 2).[24] We know that Tycho, by 1575, had acquired a manuscript copy of the *Commentariolus* and that Longomontanus himself had prepared an autograph copy of this text that, upon his departure from Prague in July 1600, he gave to his friend Johann Eriksen, another assistant of Tycho's.[25] In the *Progymnasmata*, Tycho briefly mentioned how he had acquired the *Commentari-*

[23] Maeyama, 'Kepler's *Hypothesis Vicaria*', p. 56.

[24] Donahue, *Johannes Kepler. Astronomia nova*, p. 134; see Swerdlow, 'The Derivation and First Draft', pp. 467–71. This model, of course, is nearly identical to that of the fourteenth-century Damascene astronomer, Ibn al-Shāṭir.

[25] Dreyer, *Tycho Brahe*, p. 83. Longomontanus's autograph copy of the *Commentariolus* is now in Vienna, ÖNB, lat. 10530. For the other two known copies, see Dobrzycki, 'The Aberdeen Copy'.

olus (from Thaddaeus Hagecius in Regensburg) and described its two-epicycle model for Saturn.

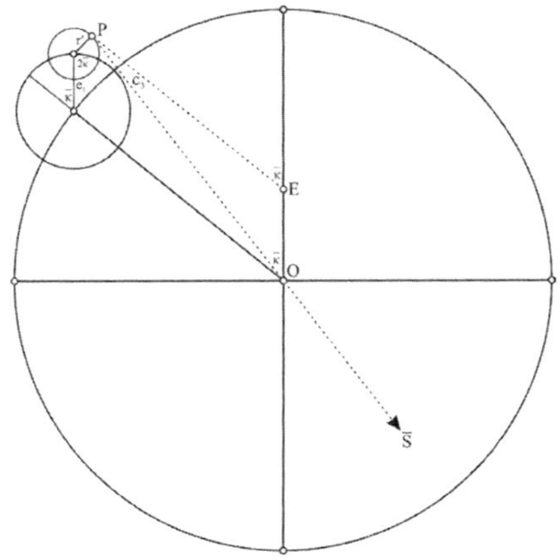

Figure 2. *Commentariolus* model for superior planets at mean opposition, with the Earth at *O*, the true planet at *P* and the mean Sun in the direction of $\bar{S}$.

Double-epicycle models also appear in planetary sketches drawn in 1578 by the itinerant mathematician, Paul Wittich, into several of his copies of the *De revolutionibus*. As Gingerich and Westman have shown, Wittich in 1580 visited Tycho in Hven, undoubtedly with his planetary sketches in tow. After Wittich's death in 1586, Tycho expended considerable effort trying to acquire Wittich's library and his Copernicus books. Gingerich and Westman suggested that Tycho may have been led to his geoheliocentric theory by Wittich's sketches; to protect his claim of originality, Tycho would thus have wanted to remove Wittich from the historical record. Be that as it may, Wittich's sketches might also be relevant to the birth of LM1. In 1598, Longomontanus managed to see Wittich's books in Breslau; by the summer of 1600, the books made their way to Tycho in Prague.[26] Good evidence thus suggests that, by 1600, Longomontanus (and of course Tycho) had seen both the *Commentariolus* (no diagrams) and Wittich's sketches.

Wittich supplied no parameters for his sketches. But his double-epicycle, heliocentric superior planet model is identical to Copernicus's superior planet model in the *Commentariolus*.[27] Longomontanus and Tycho had all this information in front of them as they created LM1. Surely Kepler knew about this potential provenance of LM1. But as far as I know, he discussed this matter neither in print nor his surviving correspondence. In any case, Equation 2 holds for the geometries of both Figure 1 and Figure 2.

[26] Gingerich and Westman, *The Wittich Connection*, pp. 21–22.

[27] cf. Swerdlow, 'The Derivation and First Draft', p. 481; Gingerich and Westman, *The Wittich Connection*, p. 119.

In the *AN*, Kepler then turned to his primary concern, the method of isolating the first anomaly by testing Mars' model against observations at mean opposition. To set the time of mean opposition, Tycho and Longomontanus needed to determine the place of the mean Sun and then to work with the observational data to get an 'observed' true Martian longitude for a time when true Mars was opposite the mean Sun. Since such times might occur during the day or cloudy evenings, such computation required considerable manipulation of the 'raw' observational data (correcting for precession, refraction, solar parallax, and shifting times by applying the Martian mean velocity in longitude). Comparing 'theory' against 'observation' was never straightforward in mathematical astronomy.

In the *AN*, Ch. 8, Kepler listed the mean solar longitudes, for the times of opposition, in a separate table alongside a column of the 'observed' true Martian longitudes taken from the earlier table (both to arcsecs). We might guess that Longomontanus/Tycho had used Tycho's solar theory, published in the *Progymnasmata* (1602), for this computation. Indeed my recomputation with that theory, for the listed times, shows agreement to ±1 arcsec of mean longitude. However, when Kepler then compared these mean solar longitudes against the 'observed' Martian longitudes, he found that only two of the 10 alleged mean oppositions were within 30 arcsecs of opposition; one varied by more than 13 arcminutes, another by more than 9 arcminutes. He concluded, rather laconically, that 'the exactness of their hypothesis [he means, of course, the inexactness] did not prevent my seeking another' [i.e., the launch of Kepler's war on Mars].[28] In Kepler's eyes, Tycho and Longomontanus had not cleanly separated the effects of the first from the second anomaly; they had not selected times when the observed true longitude of the planet was in exact opposition with the mean Sun. Thirteen arcminutes of mean Martian longitude translates into more than 10 hours from the time of exact opposition.

Trying one further intervention in LM1 before dropping that project and starting his own campaign on Mars, Kepler recomputed 'from the most recent table' of Tycho/Longomontanus the Martian mean motions for the times of mean opposition. Above, we found that LM1 had simply increased the Prutenic Mars mean motions by 1½ arcminutes. Kepler now appears to have reduced that increase by about ½ arcminutes. In a procedure I have not been able to reconstruct, he then recomputed values for c_3 and, using Equation 1 above, recomputed values for the true Martian longitude, finding that they agreed 'tolerably' with the true longitudes Tycho-Longomontanus had earlier computed. Kepler thus appears to have confirmed his predecessor's mathematics to about ±1 arcminute; he did not confirm their model and its parameters against

[28] Donahue, *Johannes Kepler. Astronomia nova*, p. 138.

the observed Martian positions.[29] And he did not try to save the model by tweaking their parameters or the geometrical arrangement of their Ptolemaic circles.

As noted above, Kepler reported that Tycho/Longomontanus thought their Mars theory had succeeded for the first anomaly; they admitted that it had failed for latitudes at mean oppositions and for longitudes for the second anomaly. We know nothing about their latitude theory. But if we assume they kept the Prutenic value for the radius of the Earth's (or Sun's) orbit in the double epicycle model (6577 in contrast with 6580 in *De revolutionibus*), we can evaluate the performance of LM1 for longitudes.[30] In Figure 3, I compare the predictions of LM1 and the Prutenic Tables against modern longitudes for the period spanning the January 1585 and March 1587 oppositions.[31] Thin vertical lines mark the dates of those oppositions. As noted above in Table 2, LM1 matches Tycho's observations at oppositions to about ±10 arcmins (it is known that Tycho's observations match modern positions to ±3 arcmins). So exactly at the mean oppositions, LM1 has reduced large Prutenic errors (reaching 2° of longitude) to a few minutes. However, at places other than mean oppositions, LM1 errors can reach 1½°. Tycho and Longomontanus must have noticed this as well, which may have prompted them to abandon work on LM1. Indeed, we know of its existence thanks only to Kepler's prolix writing style in the *AN*!

3. Longomontanus's second Mars theory (LM2)

After Tycho's death, Longomontanus eventually took up the chair for mathematics at the University of Copenhagen. It would take him more than twenty years after Tycho's death to complete the *AD*, printed in Amsterdam by William Caesius (with a second edition in 1640 printed in Amsterdam by Joan and Cornelius Blaeu, whose father, Willem, had worked with Tycho at Uraniburg in 1595–96).

[29] For the results of Kepler's recomputations, see Donahue, *Johannes Kepler. Astronomia nova*, p. 139, bottom table, rightmost column.

[30] For my computation with LM1, I use the Prutenic solar theory and Tycho's precession of 51 arcsecs/year with an epoch of –395, which matches the precessional values in the Tycho/Longomontanus data for the mean oppositions.

[31] Modern longitudes computed from JPL's on-line HORIZONS ephemeris, which uses the DE-431 models and parameters. I must emphasize, however, that graphs such as this one are anachronistic. Before the nineteenth-century no astronomer could have produced such comparisons over extended timespans; they could only check individual observations against the predictions of their models.

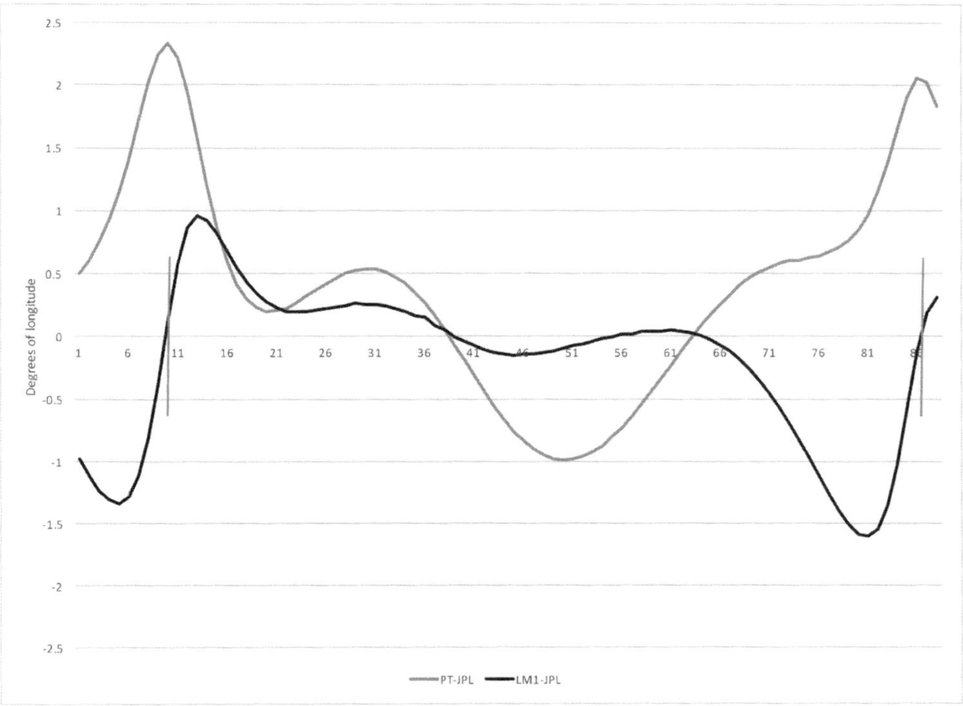

Figure 3. Longitude of Mars, Nov 1584 – Mar 1587 at 10-day intervals, comparing Prutenic and LM1 predictions against modern positions. Vertical lines mark dates of mean opposition.

The *AD*, a large volume in two parts plus an appendix on novae and comets, is a hybrid in genre between the mathematical astronomy of the *Almagest* or *De revolutionibus* and Tycho's more eclectic *Astronomiae instauratae progymnasmata* ('preliminary exercises of the restoration of astronomy'). Indeed, I can think of no other printed book before 1622 that could have served as a model for Longomontanus's project. The first part, named '*Prognorismaton astronomiae* ('preliminaries of astronomy') and clearly pointing to Tycho's earlier title, presents a 41-page introduction to plane and spherical trigonometry; a terse 7-page discussion of natural philosophy that reproduces, verbatim, Longomontanus's 1611 *Disputatio prima astronomica, de praecognitis*;[32] and a 100-page section on the geometry of the astronomical sphere. Longomontanus offered mostly Tychonic arguments against an annual motion for the Earth but accepted daily rotation. Since his notion of a luminiferous medium, a plenum with no boundaries between sub-lunar and celestial regions, denies any special role for the Earth, Moesgaard calls Longomontanus's cosmology 'Tycho-Copernican'. All bodies have their own gravity and move under divine power without material spheres or axes. Particular geometrical problems are solved, quan-

[32] See Moesgaard, 'How Copernicanism Took Root', pp. 126–34.

titatively, at the end of each chapter (e.g., find the distance, in miles, between Copenhagen and Jerusalem, given the longitude and latitude of each city), suggesting a pedagogical context for the book. A separately paginated 44-page appendix on the natural philosophy of comets and new stars rounds out the *AD*.

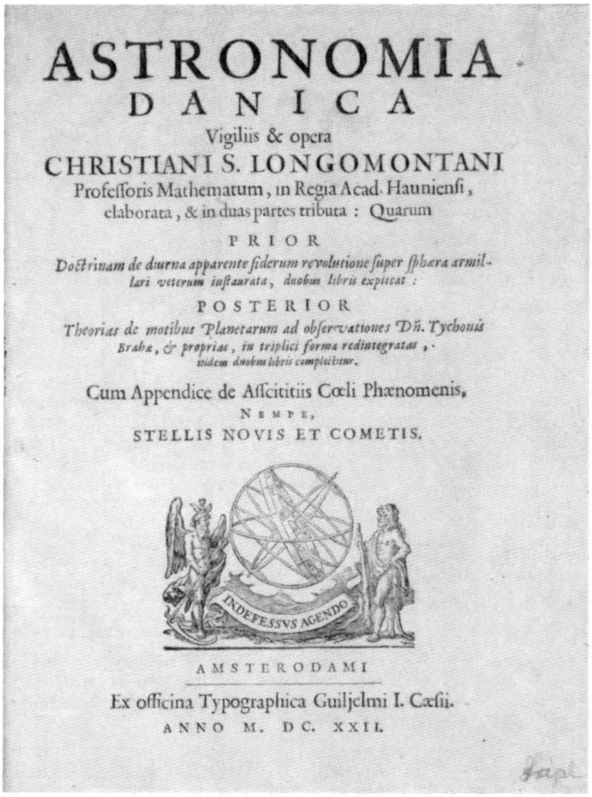
ASTRONOMIA
DANICA
Vigiliis & opera
CHRISTIANI S. LONGOMONTANI
Profeſſoris Mathematum, in Regia Acad. Haunienſi,
elaborata, & in duas partes tributa : Quarum
PRIOR
Doctrinam de diurna apparente ſiderum revolutione ſuper ſphæra armillari veterum inſtaurata, duobus libris explicat :
POSTERIOR
Theorias de motibus Planetarum ad obſervationes Dn. Tychonis Brahæ, & proprias, in triplici forma redintegratas, iisdem duobus libris complectitur.
Cum Appendice de Aſſcititiis Cœli Phænomenis,
NEMPE,
STELLIS NOVIS ET COMETIS.

AMSTERODAMI
Ex officina Typographica Guiljelmi I. Cæſii.
ANNO M. DC. XXII.

Figure 4. Longomontanus's *Astronomica danica*, 1622, title page

The second part (344 pages) has separate pagination and its own elaborate title page, which describes the content as follows:

> The second part of the Danish Astronomy, including theories of the planets restored in two books, of which the former, after a description and comparison of the three-fold hypothesis of the world, viz., the ancient Ptolemaic, the astonishing Copernican, the modern of Tycho Brahe, treats the apparent motions of the fixed stars, likewise of the Sun and Moon in the same way, re-established and adapted to all ages of the world, together with the entire theory of eclipses and besides this a special treatment of the Moon; the latter treats the motions of the other five planets, on the basis of the three-fold hypothesis, similarly restored to the appearances of the heavens in the same way.[33]

[33] Translation from Swerdlow, 'Tycho, Longomontanus, and Kepler', p. 172.

Swerdlow has fully explicated Longomontanus's historical investigation of earlier solar observations and theory and his construction of new theories of precession, obliquity of the ecliptic, and solar motion.[34] Swerdlow portrays Longomontanus as a fierce critic of the ancient observations, eager to correct errors he uncovers; an acute analyst of earlier theories and their hidden assumptions; a philosopher happy to build his solar parameters not only on observations but also on perfect numbers (Longomontanus used the first three of these numbers, 6, 28 and 496);[35] and as a careless, sometimes rather indifferent calculator whose math is filled with many small mistakes. This is the Longomontanus who created LM2, the Mars theory presented for the first time in the *AD*. Longomontanus did not tell readers when he had created LM2 or the other planetary theories published in the *AD*; Kepler did not mention LM2 in the *AN* so I shall assume that LM2 was made after Tycho's death and is thus the handiwork solely of Longomontanus.

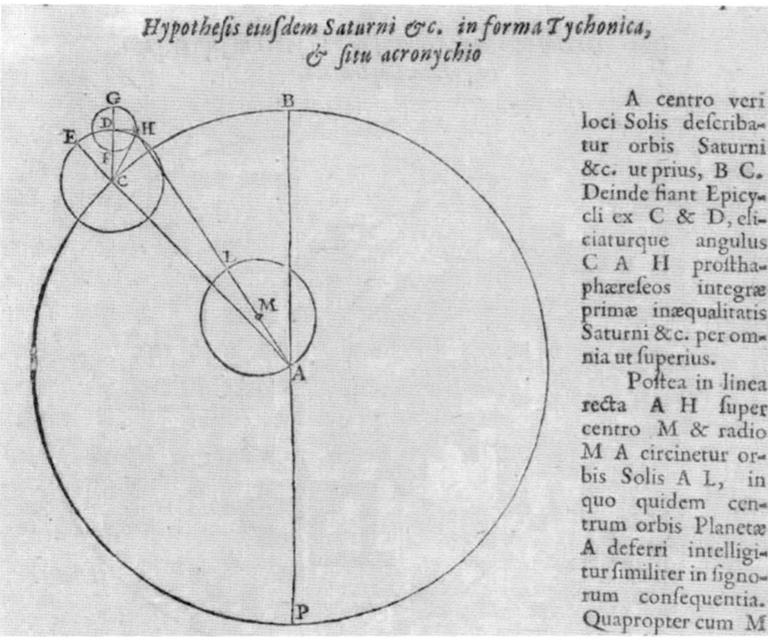

Figure 5. Longomontanus's hypothesis for Saturn in the 'Tychonic form', at an acronychal opposition of the true planet (*H*) and true Sun (*A*). The Earth is at *M*. *AD*, 2:204.

[34] Swerdlow, 'Tycho, Longomontanus, and Kepler', pp. 171–84; Moesgaard, 'Tychonian Observations'.

[35] As far as I know, Longomontanus was the only Ptolemaic astronomer who incorporated perfect numbers into his astronomical parameters. The rate of Longomontanus's anomaly of the equinox (used for the periodic portion of precession) is 6 arcmins/year; his solar eccentricity is 1/28; the radius of the small circle around which the pole of the actual ecliptic rotates is 1/496 of a quadrant. See Moesgaard, 'Tychonian Observations'.

Swerdlow did not discuss Longomontanus's planetary theories. Hence, I will briefly describe the model for Saturn and Jupiter before turning, in more detail, to Mars. For the superior planets, Longomontanus presented what he called the 'Tychonic form' of the hypothesis, with the true Sun (*A*) circling the fixed Earth (*M*) and the mean planets (*C*) circling the true Sun (see Figure 5).[36] A double epicycle corrects the first anomaly. The parameters of the circles for the first anomaly are based on a series of acronychal observations, this time however for oppositions with the true rather than the mean Sun, just as Kepler had done in the *AN*. Longomontanus's parameters for the eccentricities of Saturn and Jupiter vary only slightly from those of Copernicus (see Table 3); like Copernicus, he bisects the eccentricities. For the second anomaly, Longomontanus computed the Earth-Sun distance from two observations each for Jupiter and Saturn, assuming in each case a fixed radius or a circular orbit of the Sun around the Earth. For Mars, he would differently treat the second anomaly, in a theory that, to the best of my knowledge, had never appeared in mathematical astronomy, either before or after Longomontanus. We will analyze the components of LM2 in the order in which they are presented in the *AD*.

	e_1	r'	r	$(e_1+r')/(e_1-r')$
♄ De rev	8540	2850	10900	2/1
♄ PT	8540	2850	10911	2/1
♄ LS1	8721	2907	10426	2/1
♃ De rev	6870	2290	19160	2/1
♃ PT	6863	2287	19062	2/1
♃ LJ1	7155	2385	19349	2/1

Table 3. Parameters for the superior planets, in Swerdlow and Neugebauer's notation (e_1 = radius of the larger epicycle, r' = radius of the smaller epicycle, r = radius of the Earth's or Sun's orbit). Prutenic (PT) values from Savoie, 'La diffusion du copernicianisme', pp. 372–73.

First anomaly: To determine the heliocentric equation of center, what Longomontanus called the 'eccentric prosthaphaereses [correction]' or $\angle CAH$ in Figure 5, he proceeded similarly as he had done for Saturn and Jupiter. He assembled a list of acronychal observations since 1580, adjusted to times of opposition with the true Sun (unlike LM1 which used the mean Sun). But unlike the observations for Saturn and Jupiter, the Mars data were partly borrowed from Kepler's *AN*, Ch. 15 (Longomontanus added two final observations for 1608 and 1610 that do not appear in the *AN*). His times match Kepler's to ±3 minutes for 11 of the 12 cases (the other differs by 16 minutes), his reduced observed longitudes match Kepler's to within −1 arcminute for 11

[36] Following Kepler in the *AN*, Ch. 25, Longomontanus wrote of three 'forms' for astronomical hypotheses, the Ptolemaic, Copernican, and Tychonic.

of 12 cases (one differs by 0;10,05).[37] Of the five cases where Kepler listed the reduced observation to a precision of arcsecs, Longomontanus's values match to the nearest arcsec in four. This strongly suggests that Longomontanus took these data directly from Kepler rather than, on his own, completing the laborious reductions with spherical triangles and correcting for atmospheric refraction and precession of the reference stars.

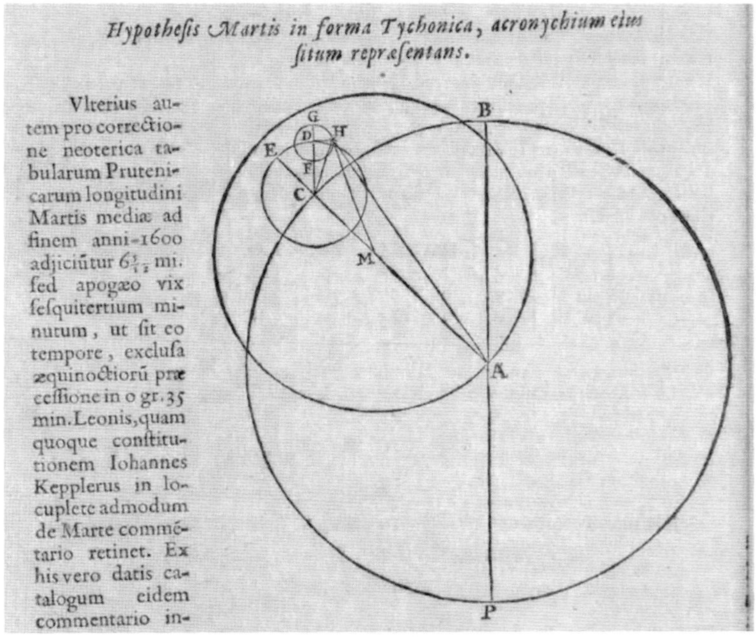

Figure 6. Longomontanus's hypothesis for Mars in the 'Tychonic form', at acronychal opposition of the mean planet and true Sun. The true planet is at H, mean planet at C, true Sun at A, Earth at M; $\angle ECD = \angle EAB$; $\angle CDH = 2\angle ECD$. *AD*, 2:220.

As is well known, Kepler in the *AN*, Ch. 16 used these acronychal data in iterative computations to determine the direction of the line of apsides and the radii of the two epicycles (3628 and 14988 units, respectively) for his so-called vicarious hypothesis.[38] Longomontanus says nothing about his procedures; he simply announces that his values for the epicycle radii (he calls them 'complicata quantitas'[39]) are 3710 and 14840 units, respectively. Longomontanus has

[37] Longomontanus did include in his table, however, his own 1608 and 1610 Mars acronychal observations, data he must have reduced himself.

[38] Gingerich, 'The Computer Versus Kepler'; Maeyama, 'Kepler's *Hypothesis Vicaria*'; Barker and Goldstein, 'Distance and Velocity'. For the systematic errors in these data, arising from Kepler's reliance on Tycho's solar theory, which in turn rested on an overly large value for the horizontal parallax of the Sun and an incorrect table of refraction, see Wilson, 'The Error in Kepler's Acronychal Data'.

[39] *AD*, 2:220.

slightly changed the ratio, $(r_1+r_2)/(r_1-r_2)$, from Kepler's 8/4.88 to exactly 8/5. Perhaps he simply wanted a 'cleaner' ratio in his Mars theory?

In the *AN*, Ch. 18, Kepler tested his vicarious hypothesis against the 12 observed Mars acronychal longitudes, finding all the differences to be less than ±2 arcminutes.[40] Longomontanus's differences, displayed in a similar table but listing different intermediate quantities than had Kepler, are less than ±3 arcminutes (see Figure 7, rightmost col.). Although he did not mention Kepler's table, Longomontanus surely must have had Kepler's table in mind as he drafted this presentation of LM2's treatment of the first anomaly. In any case, LM1 had matched the acronychal oppositions (to mean Sun) to about 9 arcminutes; LM2 improves the match by a factor of three.

	Ann.	Menſ.	Dies	H. M	Lõg. obſ. ♂	Long. ſimpl. ♂	Anomalia: ♂	Præc. æq.	Long. ſup. ♂	Differ.
					Sig.	Sex. 0 I II	Sex. 0 I II	0 I II	Sig. 0	
I	1580	Novẽ.	18	1 30	♊ 6 27 ½	0 27 54 35	4 27 11 10	27 57 50	♊ 6 27½	0
II	1582	Decẽ.	28	4 0	♋ 16 54 ½	1 11 28 40	5 10 44 20	27 59 10	♋ 16 55 ½	1 +
III	1585	Ianua.	30	19 15	♌ 21 35 ⅙	1 52 10 10	5 51 25 30	28 1 5	♌ 21 37 0	0⅙ +
IV	1587	Martii	6	7 20	♍ 25 42	2 32 47 35	0 32 2 40	28 3 0	♍ 25 42 ⅙	0⅙ +
V	1589	April.	14	6 20	♏ 4 23	3 16 16 20	1 15 31 10	28 5 0	♏ 4 26 ½	3½ +
VI	1591	Iunii	8	7 45	♐ 26 42	4 7 40 5	2 6 54 23	28 6 45	♐ 26 45 ⅓	3⅓ +
VII	1593	Augu.	25	17 30	♓ 12 15	5 11 49 20	3 11 3 20	28 8 40	♓ 12 14 ¾	0¼ −
VIII	1595	Octo.	31	0 40	♉ 17 30 ⅔	0 9 6 40	4 8 20 0	28 10 25	♉ 17 29 ⅙	1½ −
IX	1597	Decẽ.	13	16 0	♋ 2 27	0 55 2 30	4 54 15 25	28 12 20	♋ 2 26	1 −
X	1600	Iunii	18	14 0	♌ 8 37	1 36 24 45	5 35 37 10	28 14 5	♌ 8 37 ⅙	0⅙ +
XI	1602	Febr.	20	14 15	♍ 12 26	2 16 46 35	0 15 58 40	28 15 55	♍ 12 24	2 −
XII	1604	Marti.	28	16 20	♎ 18 36 ⅙	2 58 45 20	0 57 57 0	28 17 50	♎ 18 36 ½	⅓ +
XIII	1608	Iulii	24	2 0	♒ 11 10	4 45 53 15	2 45 14 30	28 21 45	♒ 11 13 ⅓	3⅓ +
XIV	1610	Octo.	8	16 50	♈ 25 30	5 48 35 0	3 47 55 20	28 23 50	♈ 25 26 ⅔	3⅓ −

Figure 7. Fourteen acronychal Mars observations, for testing the first anomaly in LM2. *AD*, 2:221.

As he had done for the acronychal observations used in LM1, Longomontanus in testing LM2 calculated the mean Mars longitudes (col. 7) from the Prutenic Tables, increasing each of the sidereal values by 6,25 arcminutes (he did not comment on this adjustment). Likewise, the values for precession (col. 9) are taken from the table used in LM1 but increased by 1 arcminute (again, uncommented). Values for the anomaly (col. 8) Longomontanus constructed by taking longitudes of the apogee (λ_a) directly from the *AN*, Ch. 18 for the dates in question and using other quantities from the table of Figure 7:

[40] Donahue, *Johannes Kepler. Astronomia nova*, pp. 206–07. Interestingly, Kepler, after employing his iterative computation to the acronychal observations, added 0;03,55 (p. 199) to all his mean longitudes that, he says, 'came from Tycho' (p. 183), perhaps a reference to LM1. In any case, LM2 will increase those mean longitudes by 0;05,04 (see below).

$$\bar{\varkappa} = \bar{\lambda}^* + \Pi - \lambda_a = \text{col}_7 + \text{col}_9 - \lambda_a. \quad (3)$$

Using Equation 3, I compute values of the anomaly that match those of Figure 7 to ±4 arcsecs. Finally, if I use Longomontanus's values for the anomaly (col. 8) and Longomontanus's stated values for the radii of the two epicycle circles and compute the Mars correction as per Equation 2 above, I match values extracted from Longomontanus's table by Equation 1 above to about ±30 arcsecs.[41] Hence, Longomontanus's test of LM2 for the first anomaly, i.e., his confirmation of the epicycle radii of 3710 and 14840, combines intermediate values taken from the Prutenic Tables and Kepler's *AN*. Longomontanus did not use the computational parameters and machinery of the finished LM2 to generate these tests for the first anomaly of that theory. He created LM2 by bootstrapping from previous theories.

Longomontanus nowhere commented on Kepler's long argument in the *AN* for why the true rather than mean Sun must be used for correcting the first anomaly; and he pointedly ignored the observed latitudes, showing that he was not persuaded by Kepler's central claim that a Martian theory must simultaneously yield longitudes and latitudes (Ptolemy and Copernicus had presented independent theories for longitudes and latitudes, as would Longomontanus). In other words, the very content of Longomontanus's table of acronychal Mars observations reveals his continued commitment to Ptolemaic (and Copernican) procedures for analyzing the first anomaly.

On the other hand, Longomontanus was quite willing to employ Kepler's vicarious hypothesis for treating the first anomaly of Mars. And he was willing to break with the earlier astronomers in treating the division of Mars eccentricity. Ptolemy had bisected the eccentricity for a ratio of 8/4; Copernicus slightly tweaked the ratio to 8/3.92; Kepler's vicarious hypothesis used 8/4.88, a value Magini slightly modified to 8/4.86. LM2 sets the ratio exactly to 8/5.[42] These are not perfect numbers. But I am not convinced that i) the constraints of geometry and Kepler's iterative method for treating the first anomaly, ii) the reduced acronychal observations, and iii) the mean motions cobbled together from the Prutenic Tables and the *AN* would have forced Longomontanus to land exactly on this ratio. Regardless of how he derived the parameters for the Martian

[41] In the first case, the difference is 115 arcsecs. As can be seen from Fig. 7, Longomontanus's table shows his Mars theory for the first date exactly matching the longitude extracted from Tycho's observations. This makes me suspect that for this date, Longomontanus fudged his 'computed' longitude to get a perfect match.

[42] cf. Swerdlow and Neugebauer, *Mathematical Astronomy*, pp. 356, 546; Voelkel and Gingerich, 'Giovanni Antonio Magini's "Keplerian" Tables'; Magini, *Supplementum ephemeridum*, pp. 174–76; Donahue, *Johannes Kepler. Astronomia nova*, pp. 200–01. I used Equation 2 above and least squares to extract from Magini's tabulated equation of center the values of 14948 and 3653 for the radii of the epicycles.

eccentricity, the rightmost column of Figure 7 shows Longomontanus's success; LM2 matches the acronychal observations to less than ±3 arcminutes, apparently Longomontanus's criterion of 'good enough' for astronomical theory. And his ratio of the eccentricities is 'cleaner' than Kepler's or Magini's.

Once he had determined the radii of Mars' two epicycles, Longomontanus could compute the table of corrections for the first anomaly (he called them the eccentric prostphaphaereses, I shall call them c_1 in LM2) using Equation 2 above. Given the maximal correction of 10;34,20, Longomontanus's values match mine to about ±30 arcsecs, with divergences reaching three times that amount in a pair of spikes centered around the argument of 112° (see Figure 8). I cannot explain how Longomontanus may have computed these corrections to generate this pattern of (small) differences.

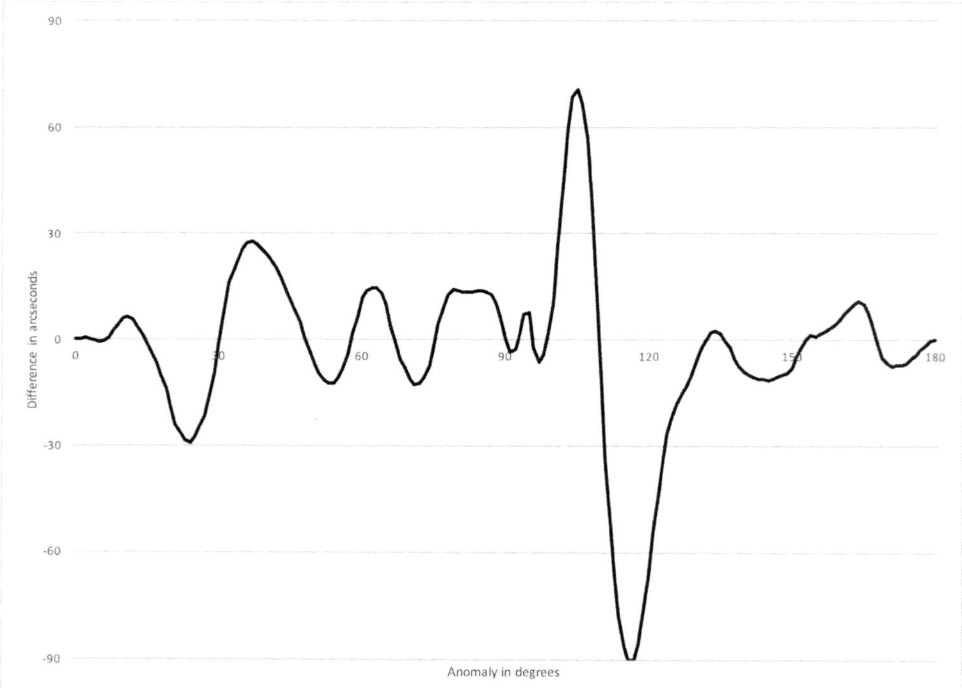

Figure 8. Differences between LM2's 'eccentric prosthaphaereses' (equation of center) or c_1 (i.e., column 1) and my recomputation with eccentricities of 0.1484 and 0.0371. See *AD*, 2:257–59.

Second anomaly: Longomontanus next treated the second anomaly arising from the annual motion of the Sun (Longomontanus usually refers to this simply as the 'annual orb' or the 'annual orb of Mars'[43]). LM2 breaks down the process of finding a geocentric longitude into three parts: i) compute Mars'

[43] See, for example, *AD*, 2:227.

heliocentric longitude and its distance from the Sun, ii) find the geocentric longitude of the Sun and the Earth-Sun distance, and iii) combine these results geometrically to obtain Mars' geocentric longitude. Exactly this technique, as noted by Bialas and explored in more detail by Voelkel and Gingerich, had been presented in 1614 by the Bolognese professor of astronomy and well-known table maker Giovanni Antonio Magini. Although no correspondence is extant between Magini and Longomontanus, the Italian astronomer had communicated with Tycho since 1590.[44] Several times they had discussed Mars, with Magini begging for Tycho's acronychal observations of that planet. But this correspondence does not elucidate the Mars theory Magini would use in his 1614 *Supplementum*, which implements Kepler's vicarious hypothesis. I find no mention of Magini in the *AD* and we cannot know, therefore, what Longomontanus might have known about the *Supplementum*. But it is clear that Magini and Longomontanus took similar computational approachs in their treatment of Mars.

Mars-Sun distance: As Voelkel and Gingerich have shown, Magini apparently took six computed distances from Kepler's *AN*, Ch. 56, based on the ellipse, and then filled in the intermediate distances by hand, smoothly but not using a rigorous function based on the geometry of the ellipse.[45] Longomontanus, on the other hand, simply computed the distances trigonometrically within the double-epicycle model of Figure 6, comprised only of circles. It can be shown that the distance of the planet from the Sun, 'Distantia a centro' in Longomontanus's tables or AH, can be computed as follows, where $r_1 = CD$ and $r_2 = DH$:

$$AH = \sqrt{1 + r_1^2 + r_2^2 + 2(r_1 - r_2)\cos\bar{\varkappa} - 2r_1r_2\cos 2\bar{\varkappa}}. \qquad (4)$$

With these equations and Longomontanus's values for the radii of the two epicycles, I can recompute his 'Distantia a centro' to ±60 parts out of a mean distance of 100,000 parts (see Figure 9). The maximal distance is $R + r_1 - r_2$, the minimal distance is $R - r_1 + r_2$. Magini had arranged his distances to emulate Kepler's ellipse; Longomontanus kept with the circles of Ptolemy, Copernicus and Tycho. But Longomontanus followed Magini's Kepler-based innovation of introducing distances into tabular computation of planetary longitudes, thereby taking into account the eccentric path of the Sun around the Earth (something ignored by both Ptolemy and Copernicus who worked with the mean Sun in designing planetary models). The 'Distantia a centro' are presented in c_2, the second column in Longomontanus's table of equations for Mars.

[44] Bialas et al., *Johannes Kepler. Gesammelte Werke*, vol. XI/1, p. 489; Voelkel and Gingerich, 'Giovanni Antonio Magini's "Keplerian" Tables'; Favaro, *Carteggio inedito di Ticone Brahe*.

[45] Voelkel and Gingerich, 'Giovanni Antonio Magini's "Keplerian" Tables', p. 246.

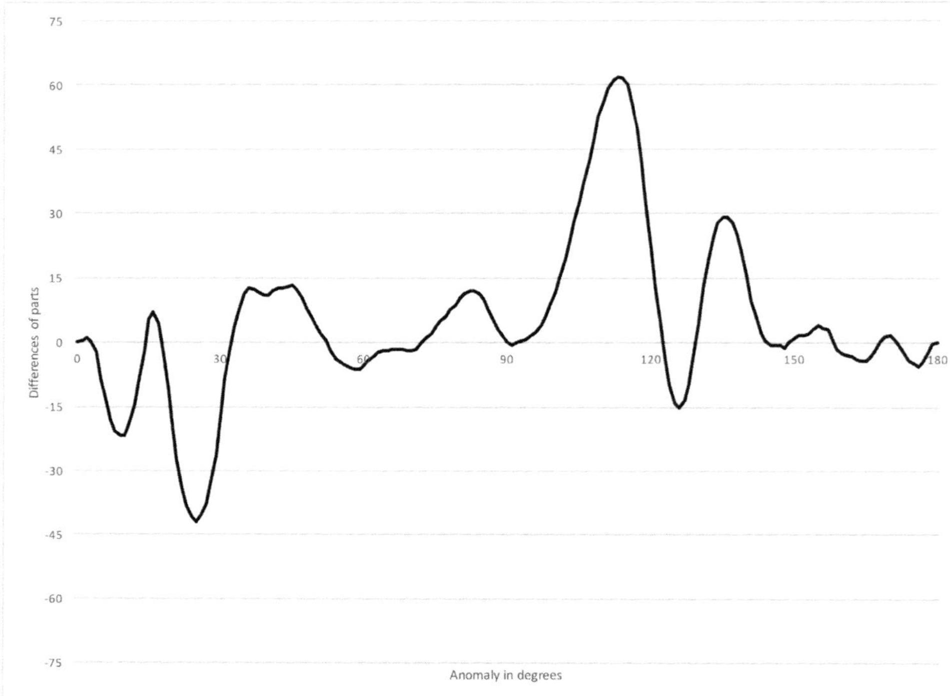

Figure 9. Differences between LM2's Mars-Sun distance (R = 100,000) or c_2 (i.e., column 2) and my recomputed values using Equation 4 and eccentricities of 0.1484 and 0.0371. See *AD* 2:257–59.

Earth-Sun distance: Copernicus, famously, had referred to the 'fixed symmetry' (certam symmetriam) of his heliocentric planetary models, in which the Earth-mean Sun distance provided a common measure for each model.[46] Longomontanus's challenge was to find the Earth-true Sun distance, i.e. to assume an eccentric rather than a circular path for the annual orbit. In the *AN*, Ch. 24, Kepler had solved this problem by selecting a series of Mars observations separated by intervals equal to the sidereal period of the planet. This meant that the location of Mars, relative to the true Sun, remained at the same eccentric position (i.e., relative to the Sun) and Kepler could then triangulate to find the places of the Earth in its eccentric orbit around the Sun.

Longomontanus cited Kepler, Ch. 24, in a marginal note but used a different procedure, selecting Mars observations when the planet was at its apogee or perigee.[47] His approach, the most complex feature of LM2, requires two steps: first, correct the distance between the Earth-mean Sun as a function of the

[46] Copernicus, *De Revolutionibus*, fol. iiiv.

[47] *AD*, 2:225. For two of these observations, Longomontanus borrowed reductions of the raw data to longitudes from Kepler (Donahue, *Johannes Kepler. Astronomia nova*, pp. 169, 212). The raw data for other two observations, on 1 January 1587 and 1 November 1589, can be

distance of true Sun from its apogee; second, correct the distance as a function of the distance of true Mars from its apogee. As far as I know, no astronomer, either before or after Longomontanus, used this method to tweak the geometry of the Ptolemaic planetary models. Tycho had added additional circles to Ptolemy's lunar model; Longomontanus, presumably inspired by Kepler's vicarious hypothesis, keeps the Copernicus – al-Shāṭir double epicycle model for Mars but corrects linear distances therein.

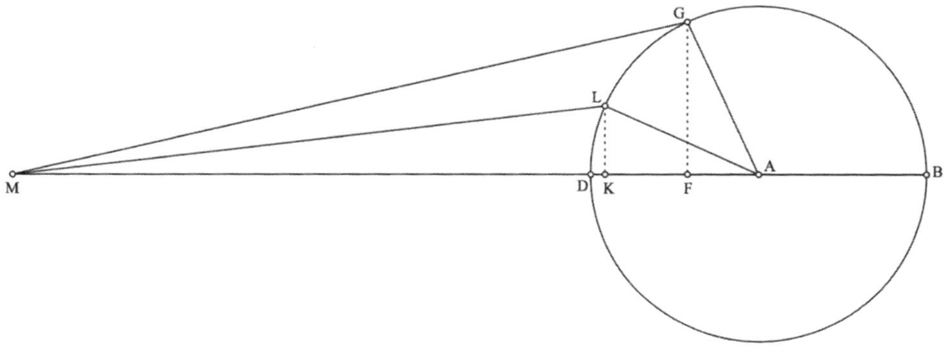

Figure 10. Longomontanus's solar epicycle, showing the location of the true Sun at two times (*G* and *L*) when Mars was at apogee. The Earth is at *M*, *MA* is the Earth-mean Sun distance, *MD* the minimum distance, *MB* the maximum distance.

Longomontanus created one table to use for both corrections, the 'Tabula analogiam distantiae solis ab apogaeo, in annuo orbe Martis, ad singulos binos gradus ostendens' (Table showing the proportion of the Sun's distance from the apogee in the annual orb of Mars at intervals of two degrees). We will first show how he derived this table and then will examine its use in correcting the Sun-Earth distance. He presented his derivation with an epicyclic solar model (Figure 10). The true Sun is at *G*, the mean Sun at *A*. As the true Sun rotates around the epicycle, the Earth-true Sun distance, *MG*, varies as a function of the true solar anomaly ($\angle BAG$). Since the radius of the solar epicycle is small compared to the dimensions of the Mars model, Longomontanus approximated the Earth-true Sun distance by varying the length of the distance through *MA* in harmonic motion between *MD* and *MB*. Hence, when the true Sun is at *G* (first observation), the Earth-Sun distance will be approximated by *MF*; when the true Sun is at *L* (second observation), the Earth-Sun distance will be *MK*.

found in Dreyer, *Tychonis Brahe Dani opera omnia*, vol. XI, pp. 177, 335. *AD*, 2:225, does not provide Longomontanus's reduction of these data, listing only the final times and longitudes.

Such approximation also means that the angular position of the Sun, when viewed from the earth, will be inexact by an amount equal to $\angle AMG$. In a separate section, Longomontanus showed that, given the distances in his Mars model, these angular discrepancies will always be less than 27 arcsecs when the planet is at apogee and always less than 81 arcsecs at perigee.[48] These small angles are ignored in LM2.

A further approximation arises in the derivation. The four observations were not set to the exact times of Mars being at apogee or perigee. Using the Mars model for the first anomaly, I compute the true anomalies for those times to be 0;43, 358;45, 185;46 and 181;19. Nonetheless, Longomontanus's derivations assume that the true anomaly, in each case, is zero or 180 degrees.

To define the parameters of the harmonic motion of the approximated Earth-true Sun distance, Longomontanus computed the actual Earth-true Sun distance from the Mars observations. Figure 11 shows LM2 for the first observation of the planet at apogee. $\angle ♈MH$, the geocentric longitude of Mars, is known from observation (119;18). The distance AH (1,112,970) and the heliocentric longitude of Mars ($\angle ♈AH$ = 149;32,59) are known from the preliminary Mars theory (first anomaly) already developed. The true geocentric longitude of the Sun ($\angle ♈MA$ = 356;37) is known from the solar theory (interestingly, Longomontanus used Tycho's solar theory from the *Progymnasmata* for this computation, not the solar theory of the *AD*).[49] These givens define a single location of the Earth (M) in the model and, by using the law in sines in triangle AMH, Longomontanus can determine the Earth-Sun distance (MA) in units of the Mars theory developed for the first anomaly (AC = 100,000 parts). For the case in Figure 11, he finds MA = 66,586 parts at the time of the first observation, a value I confirm by recomputation.[50]

For the second observation at apogee, he found MA = 65,691 parts, a difference of 895 parts resulting from the fact that the Sun was at a difference place in its epicycle. In Figure 10, this difference in Earth-Sun distance is represented by *KF*. Hence, Longomontanus could determine, by proportions, the maximal shift in earth-Sun distance when Mars is at apogee, i.e., the shift caused by the Sun's eccentric motion. At the first observation, the true solar anomaly (λ) was 260;57,55; at the second, it was 195;10,09 (Longomontanus's values). It can easily be shown that:

[48] *AD*, 2:243–44.

[49] The results differ here by about 1½ arcminutes.

[50] *AD*, 2:225. $\angle AMH = \lambda_{♂obs} + 360 - \lambda_{\odot theor} = 122;41$. $\angle HAM$ = 'anomalia orbis' – 180 = $\lambda_{\odot theor} - \lambda_{♂helio} - 180 = 27;05$. $\angle MHA = 180 - \angle AMH - \angle HAM = 30;14$. By the law of sines, $MA = AH \sin \angle MHA / AH = 66,586$.

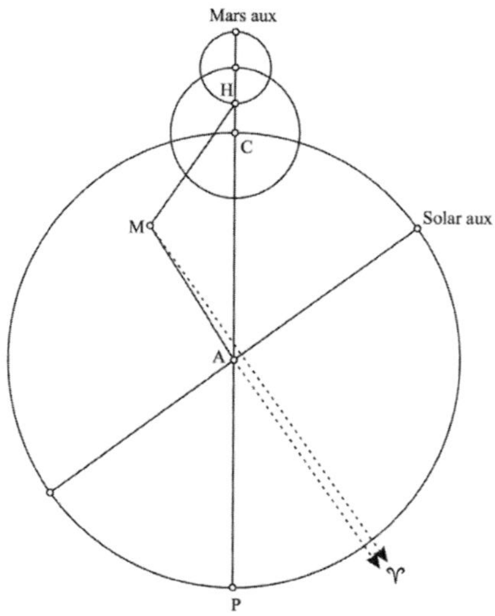

Figure 11. LM2 for 6 March 1600. Mars (*H*) at apogee, observed at longitude 29;18 ♋; Earth at *M*, true Sun at *A*.

$$KF = \cos(\alpha_{\text{obs2}}) - \cos(\alpha_{\text{obs1}}), \quad \text{and} \tag{5}$$

$$\frac{KF}{DB} = \frac{\text{Observed difference in Earth–Sun distances}}{\text{Maximal solar contribution at apogee}}. \tag{6}$$

For Longomontanus's values at apogee (he was using a cosine table with a radius of 10 million[51]):

$$\frac{8{,}080{,}836}{20{,}000{,}000} = \frac{895}{\text{maximal solar contribution}},$$

which yields a maximal solar contribution of 2215 parts. At perigee, using the same procedure, he found the maximal solar contribution to be 2415. Averaging the two values gives the maximal solar contribution of 2315 parts. To further refine this value, Longomontanus took two Mars observations from the *AN*, Ch. 27. Using Kepler's computation of the Earth-Sun distances, he found a maximal solar contribution of 2375. Roughly splitting the two figures, Longomontanus set maximal value for solar contribution to shifting Earth-Sun distances to 2350 (*DB* in Figure 10).[52] Crucially, the units for this value are the units of the Mars theory where the mean Mars-Sun distance is 100,000.

[51] Printed sine tables with 7 significant figures were readily available.

[52] Longomontanus then repeated Kepler's computation using 'our reduction' of the raw observational data, finding the same value of 2350. See *AD*, 2:226–27.

With this result, Longomontanus prepared the table that gives the solar contribution to the changing Earth-Sun distance, in units of the Martian theory, tabulated as excess over the minimum value, *MD*, in Figure 10. By solving the right triangle *GFM* in Figure 10 it can be shown that the excess over minimum, or *DF* in the case of anomaly ∠*BAG* is:

$$\text{excess over minimum} = c_3 = DB\cos^2(\tfrac{1}{2}\angle BAG). \tag{7}$$

With *DB* = 23,500 parts, Longomontanus computed his table of proportions (see Figure 13). Using the Equation 7, I can recompute Longomontanus's table to ±1 part (see Figure 12). In the *Progymnasmata*, Tycho had included a *Tabula distancia solis a terra*, similarly computed but giving the full distance of the Sun from the eccentric Earth, not the excess over the minimal distance.[53] Tycho's table was intended for determining the magnitude and duration of eclipses, not for correcting geocentric longitudes. As far as I know, Longomontanus's is the only such table ever designed for the latter purpose.

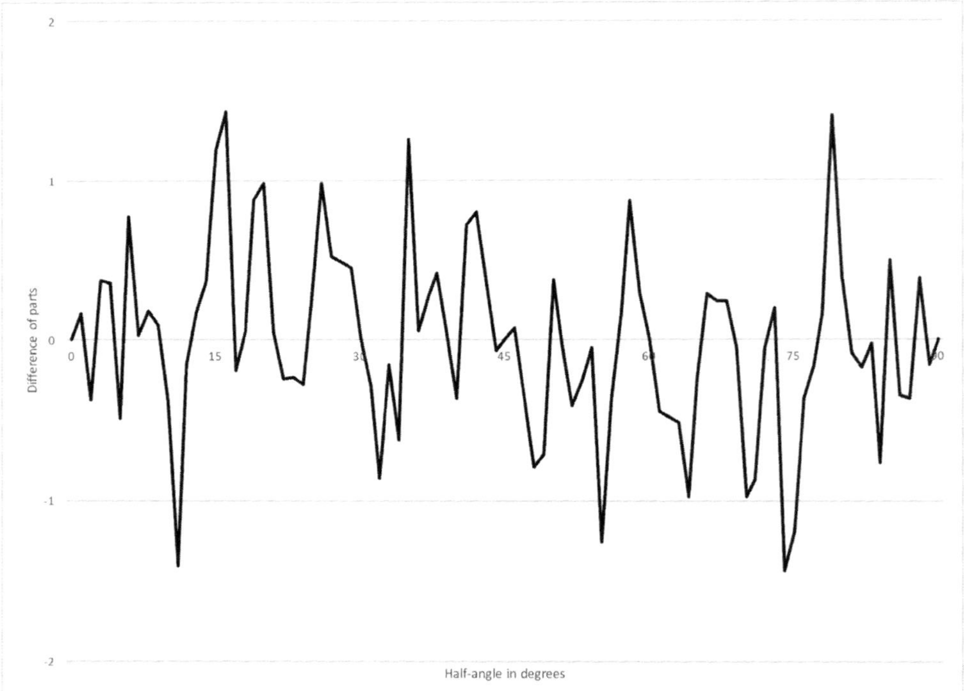

Figure 12. Differences between LM2's *Tabula analogiam ... ostendens* and my recomputation (*R* = 23,500 parts).

[53] Dreyer, *Tychonis Brahe Dani opera omnia*, vol. II, pp. 82–83.

Tabula Analogiam diſtantiæ Solis ab apogæo, in annuo orbe Martis, ad ſingulos binos gradus oſtendens.

	0 Sexagen.	Differē.	1 Sexagen.	Differē.	2 Sexagen.	Differē.	
0	23500		17625		5875		60
2	23493	7	17266	359	5523	352	58
4	23471	22	16900	366	5179	344	56
6	23436	35	16529	371	4843	336	54
8	23386	50	16152	377	4515	328	52
10	23321	65	15770	382	4197	318	50
12	23244	77	15381	389	3888	309	48
14	23151	93	14989	392	3588	300	46
16	23045	106	14593	396	3298	290	44
18	22925	120	14199	400	3018	280	42
20	22791	134	13790	403	3748	270	40
22	22643	148	13386	404	2490	258	38
24	22484	159	12979	407	2244	246	36
26	22311	173	12570	409	2009	235	34
28	22125	186	12160	410	1784	225	32
30	21927	198	11750	410	1573	211	30
32	21716	211	11340	410	1375	198	28
34	21491	225	10930	410	1189	186	26
36	21256	235	10521	409	1016	173	24
38	21010	246	10114	407	857	159	22
40	20752	258	9710	404	709	148	20
42	20482	270	9307	403	575	134	18
44	20202	280	8907	400	455	120	16
46	19912	290	8511	396	349	106	14
48	19612	300	8119	392	256	93	12
50	19303	309	7730	389	179	77	10
52	18985	318	7348	382	114	65	8
54	18657	328	6971	377	64	50	6
56	18321	336	6600	371	29	35	4
58	17977	344	6234	366	7	22	2
60	27625	352	5875	359	0	7	0
	5 Sexagen.		4 Sexagen.		3 Sexagen.		

Vſus Tabulæ præcedentis.

Data anomalia tam Solis quam Martis coæquata; primo Solis anomalia tabulam affixam ingrediendum eſt, cum ſexagenis ſcilicet ſupra infraque adſcriptis, & gradibus ad marginem utrinque integris proximeque minoribus; ſecundum quam in area

Figure 13. LM2's *Tabula analogiam ... ostendens*, used for the second anomaly. *AD*, 2:228.

In the *AD*, Longomontanus presented his derivation of the *Tabula analogiam ... ostendens* in the context of variations caused by the eccentric annual path of the Sun (or Earth), i.e., as a function of the true solar anomaly. However as noted above, the corrected Earth-Sun distance in LM2 is also a function of the Mars-Sun distance. As can be seen in Figure 11, as the Mars-Sun distance (*HA*) shifts, with the other angles remaining fixed, the Earth-Sun distance (*MA*) also shifts. To correct for this effect, Longomontanus used the same two apogee and two perigee Mars observations and computed the Earth-Sun distances at apogee and perigee, in units of the Mars theory, for when the Sun was at its mean distance from the Earth (i.e., *MA* in Figure 10). That is, he used Equation 6 with the mean solar contribution (2350/2 = 1175) and computed the Earth-Sun distance for that contribution.

Hence, from Observation 1 (Figure 11, at apogee of Mars) he had found an Earth-Sun distance (*MA*) of 66,586 units, when the true solar anomaly was 260;53,32 ($\angle BAG$ in Figure 10). Increasing the true solar anomaly to 270° (for the mean Earth-Sun distance) increases the Earth-Sun distance by {cos (260;53,32) – cos (270)} · 1175 or 186 units to 66,772 units. From Observation 2, also at apogee, he had found an Earth-Sun distance of 65,691 units, which, when adjusted to the mean Earth-Sun distance, becomes 66,825 units. Averaging these two values yields 66,799 units (Longomontanus set this distance to 66,788 units) when Mars is at apogee. A similar process for the perigee observations yields, by my computation, a distance of 64,192 units (Longomontanus found 64,202 units). Subtracting Longomontanus's apogee and perigee values yields 25,850 units, or the maximal shift in Earth-Sun distance resulting from the eccentric motion of Mars around the Sun. For the correction of the Earth-Sun distance resulting from the eccentric motion of the Sun around the Earth, Longomontanus had found a maximal shift of 23,500 units. Hence, the 'Mars correction' is 11/10 larger than the 'Sun correction'.

To use the *Tabula analogiam ... ostendens*, which gives the corrections as increases over the minimal Earth-Sun distance, Longomontanus had to determine that minimum distance. For this he used the two perigee observations, now adjusting them not to the mean but to the minimal Earth-Sun distance. From Observation 3 he found a distance of 63,045 units, from Observation 4, a distance of 62,990 units, which yields an average of 63,018 units (Longomontanus found 63,027 units).

Thus, to compute the Earth-Sun distance (*MA* in Figure 11) in LM2 requires two entries into the table:

$$MA = 630{,}275 + \varepsilon(\alpha_{\mathrm{Sun}}) + \tfrac{11}{10} \cdot \varepsilon(\alpha_{\mathrm{Mars}}), \tag{8}$$

where:

$$\varepsilon(\alpha) = 23{,}500 \cdot \cos^2(\tfrac{1}{2}\alpha). \tag{9}$$

These two corrections, derived from only four observations of Mars, allow LM2 to correct the Earth-Sun distance, a distance that had been fixed in Ptolemaic and Copernican astronomy. In his final theory, Kepler had corrected this distance with an ellipse; Magini, as shown by Voelkel and Gingerich, corrected the distance 'by hand', fitting six observations he extracted from the *AN*.[54] We will see, below, how well LM2 performs with its novel approach to solving the second anomaly.

Mean motions: Longomontanus treated the mean motions last in his presentation of LM2. I find inconsistencies and small computational errors in his

[54] Voelkel and Gingerich, 'Giovanni Antonio Magini's "Keplerian" Tables', pp. 245–46.

derivations, not unlike those Swerdlow found in Longomontanus's treatment of precession.[55] As in his study of precession, Longomontanus used observations from Ptolemy, corrected by Longomontanus's own solar theory and theory of precession, plus Tycho's modern observations (he provides no details of his reduction of Tycho's measurements). He tabulated two mean motions for each superior planet, motion in longitude and motion in anomaly (tropical longitude minus the apogee of the planet). The apogees move at different rates for each superior planet. By subtracting his mean longitudes from the mean anomalies, we find that Longomontanus gave Saturn's apogee a rate of 0;01,20,12°/yr, Jupiter's 0;00,57,52°/yr and Mars's 0;01,14,52°/yr. We cannot, therefore, call Longomontanus's apogees sidereal, as was the case in Ptolemaic and Alfonsine astronomy.[56] Nowhere does he discuss this feature of his mean motions.

To find the rate of change in Mars's heliocentric anomaly ($\angle BAC$ in Figure 6), Longomontanus computed the shift in its apogee between observations widely separated in time. Using three acronychal observations of the opposition of Mars and the mean Sun, made in AD 130, 135 and 139, Ptolemy in the *Almagest* X.7 had cleverly found the planet's apogee to be at 115;30°, employing his own theory of the Sun.[57] Longomontanus recomputed this value, using his revised solar theory and the true rather than mean Sun, finding Mars's apogee to be at 118;15 at the era of Antoninus Pius (which he defined as noon, 31 December 136). For a recent observation, Longomontanus simply stated that for 31 December 1600 'we find' (deprehendimus) Mars's apogee at 148;42, a shift of 30;27 in 1464 years, which yields an annual rate of change, noted above, of 0;01,14,52°/yr. Presumably, Longomontanus had found the modern apogee by deploying several acronychal Mars observations and the procedures of the *Almagest* X.7; but he did not inform his readers how he had set this important long-term parameter for LM2.[58]

For determining the mean motion in longitude, Longomontanus provided more details. Selecting the earliest Mars observation reported in the *Almagest*, an occultation of Mars and a star (β Sco) observed 'at dawn' on 18 January −271, Longomontanus drew a diagram to represent the arrangement of LM2 for that date. Taking as givens his solar theory and the radii and distances for Mars (the first anomaly) that he had earlier derived, Longomontanus found the mean tropical longitude of Mars at this date to be 182;56,30 (in contrast to Ptolemy's 184;12).[59] He then shifted this value back to the end of −272, finding

[55] Swerdlow, 'Tycho, Longomontanus, and Kepler', pp. 178–79.

[56] Pedersen, *A Survey of the Almagest*, p. 287. The linear portion of Longomontanus's theory of precession advances at a rate of 0;00,49,45°/yr.

[57] For an interesting analysis of how Ptolemy may have determined his mean motions, see Jones and Duke, 'Ptolemy's Planetary Mean Motions'.

[58] *AD*, 2:231–32.

[59] *AD*, 2:233; *Almagest* X.9.

174;11,08. For the end of 1600, he indicated (again offering no observational raw data) that the mean tropical longitude was 307;06,32. Hence, Mars's mean longitude had moved 132;55,30 plus 995 revolutions in 1872 years, which yields an annual rate of change of 191;17,10,06°/365 days or 0;31,26,39,28,13°/day. The medieval Alfonsine Tables had set this rate at 0;31,26,38,40,04°/day, Kepler's Rudolphine Tables would use 0;31,26,39,11,41°/day. Over 100 Julian years, Longomontanus's mean Mars would run ahead of the Alfonsine Mars by about 5 arcminutes, ahead of Kepler's Mars by about 3 arcminutes.

Using these rates, Longomontanus prepared a set of mean motion tables for the superior planets, listing the increments in mean longitude and mean anomaly for intervals of 20, 40 ... 100, 200 ... 1000, 1100 ... 2000, 2500 ... 6000, 6300 years. The tables have two radices, for the Creation of the World (−3963) and for the Birth of Christ (1 January 1, noon). Unlike Tycho, who had worried that his solar and lunar theories were reliable only for several hundred years, Longomontanus confidently presented his planetary models as a mathematical astronomy for the ages (Kepler's mean motions run from −4000 to +2100 years in the Rudolphine Tables).

Thus, to generate a geocentric longitude for Mars, LM2 computes three corrections:

1. For the correction of the first anomaly ($c_1(\bar{\varkappa})$ or $\angle CAH$ in Figure 6), enter the Table of Eccentric Prostphaphaereses with Mars' mean anomaly which is tabulated in the mean motion tables. This correction yields the true heliocentric longitude of Mars.
2. For the correction of the Mars-Sun distance (c_2), enter the Table of Distances from the Center, also with Mars' mean anomaly.
3. For the correction of the Earth-Sun distance (c_3), enter the Table Showing Proportion twice, once with the solar true anomaly, once with Mars' true anomaly, and compute the corrected distance using Equation 8.

With these results, the user must solve the Mars-Sun-Earth triangle (*AHM* in Figure 14) to find Mars' geocentric longitude. Corrections c_2 and c_3 yield the distances *AH* and *AM*, c_1 yields Mars' heliocentric longitude. The angle at *H* can be easily found from the figure with the constructed right triangles:

$$\tan H = \frac{\sin \varphi}{\frac{AH}{AM} - \cos \varphi}, \tag{10}$$

where

$$\varphi = \left| \pi - \lambda_{\text{Sun}} + \lambda_{\text{Mars heliocentric}} \right| . \tag{11}$$

It can then easily be seen that:

$$\lambda_{\text{Mars}} = H + \lambda_{\text{Mars heliocentric}}. \tag{12}$$

In his sole worked example, Longomontanus uses Equations 10–12 exactly as I have presented them to compute the geocentric longitude of Mars in LM2.[60] His computed longitude differs from mine by 2,37 arcminutes.

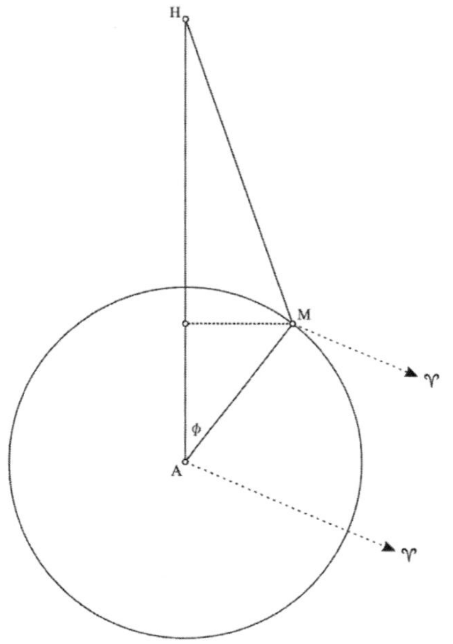

Figure 14. Converting Mars' heliocentric to geocentric longitude. The Sun is at *A*, Mars at *H*, the Earth at *M*.

4. Conclusion

Because of the ways they structured their algorithms, it is not easy to compare directly the Earth-Sun distances computed via Kepler's *Rudolphine Tables* and Longomontanus's LM2. We can, however, compare the final Martian longitudes generated by these tables (see Figure 15). As observers since Kepler have noted, Ptolemy's Mars model (and Copernicus's) generates predictions that give large errors when the planet is in opposition with the Sun, i.e., when the Mars-Earth distance is at the minimum. These errors at opposition could reach ±2°; every 32 years, as the geometry brings Mars even closer to the Earth, the errors could reach ±5°, creating the 'great Martian catastrophe' in Gingerich's colorful phrase.[61] By placing Mars in an elliptical orbit, Kepler reduced the errors at

[60] *AD*, 2:262–63.

[61] Gingerich, 'The Role of Erasmus Reinhold', p. 54; Gingerich, 'Early Copernican Ephemerides', p. 407; Gingerich, 'The Great Martian Catastrophe'.

opposition to about ±2 arcminutes (notice that Kepler's larger errors of about ±3 arcminutes occur not at opposition but when the Earth approaches Mars at octant positions in its orbit). Precisely at the time of opposition, the errors in LM2 are roughly equal to Kepler's errors. But before and after those times, Longomontanus's errors can rise to about ±30 arcminutes. Figure 15 thus suggests that, measured by the criterion of accuracy, Longomontanus's circles did not perform as well as did Kepler's ellipses. I have not explored to what extent LM2's error features, visible in Figure 15, might derive from errors in the direction of the apsidal line or the eccentricities, features unrelated, that is, to the geometry of the circle.

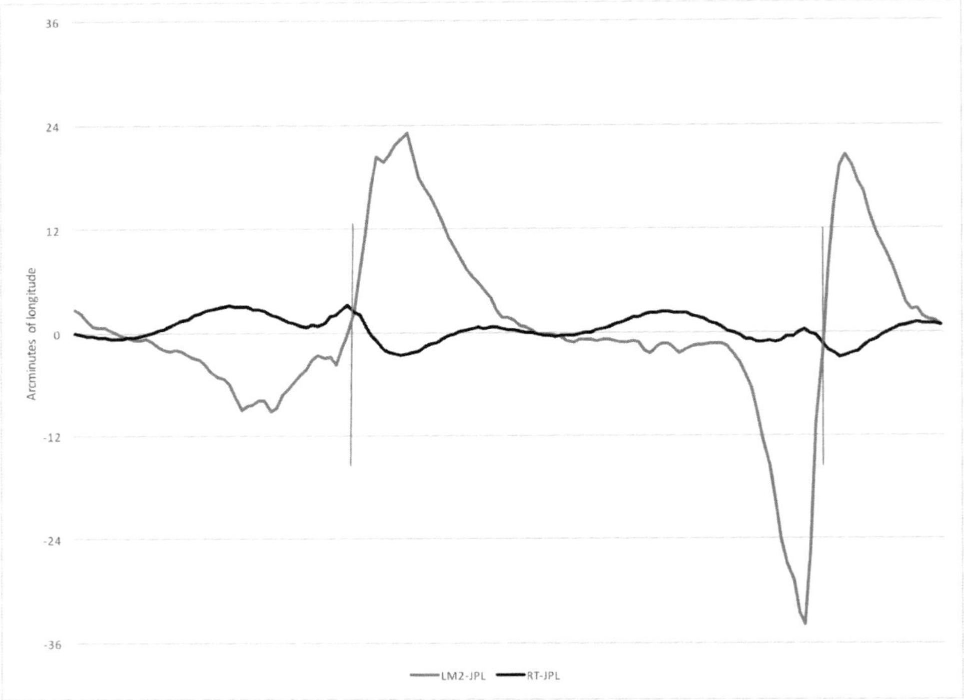

Figure 15. Longitude of Mars, Jun 1624 – May 1628 at 10-day intervals, comparing Rudolphine and LM2 predictions against modern positions. Vertical lines mark dates of true opposition.

As he introduced LM2 in the *AD*, Longomontanus noted that the observed shifts in Earth-Sun distances were not predicted by Ptolemy's equations, implying that precision alone (i.e., predicted positions matching observed positions) should govern astronomical theorizing. But he also commented, for the first and only time in this large book, on the approach taken by his former colleague, Kepler:

> Whence it happened that Johannes Kepler, wholly ascribing to the eccentric (as also seems [to be the case]) this discrepancy in the phenomena of the yearly orbit of Mars,

> had tried to think up other hypotheses instead of the usual eccentrics, indeed figures shy of the circle and oval or emulating ellipses, [figures] in which equal arcs of revolutions will not correspond precisely to equal times, namely with the Sun as director of all the motions, putting in motion the planets disposed around itself now more intensely, now more remissly. Moved by the weightiest causes, however, we almost alone uphold that Copernican axiom — that the perpetual motion of celestial bodies is uniform and circular or composed of circular motions — which in astronomy we value to the highest degree. Nor can we certainly think otherwise because of such anomalies of Mars and other [planets], nor must we [think otherwise] before necessity will have demanded it, until the circular or simple, or especially the complicated and compound [figures], by which nearly all curved figures of this sort and even straight lines can be described clearly, will have abandoned their role in relation to every kind of representation of the celestial phenomena. Why may we not assemble these materials? We do not hesitate to affirm that they [= the materials] are real, however, and have a kind of existence, which, armed with the force and power of centers, results in their commensurable conversions in equal times, just as these things are found more fully discussed by us elsewhere.[62]

In the margin at this point, Longomontanus referred to his 1611 public disputation, a text he reprinted in the first part of the *AD*.[63] He also cited *De revolutionibus* I.4 (the motions of heavenly bodies 'are circular or compounded of several circles') and Tycho's *Progymnasmata* I, p. 11 ('All celestial motion is essentially regular and uniform, unchanging and circulation, an axiom long accepted by all astronomers').[64] Obviously, Longomontanus had decided to

[62] 'Vnde evenerat quod Iohannes Kepplerus hanc in annui orbis Martis phaenomenis discrepantiam eccentrico (ut etiam videtur) in solidum adscribens, alias revolutionum hypotheses pro consuetis eccentricis excogitare conatus fuerit; quippe a circulo deficientes et figuras ovales vel ellipticas [reading for eclipticas] aemulantes; in quibus quoque aequales revolutionum arcus non prorsus [222/223] aequalibus temporibus respondebunt. Sole scilicet motuum omnium directore planetas circum se dispositos nunc intentius, nunc vero remissius ciente. Nos autem gravissimis causis moti, Copernicaeum illud, quod motus corporum coelestium sit aequalis et circularis perpetuus, vel e circularibus compositus, in astronomia maximi facimus, et quasi unice tuemur [in margin: Copernic. lib. I.c. 4. Tycho Brahe axioma astron. vocat lib. I. Progym. pag. 11.]; nec certe aliter propter tales Martias aut quorundam aliorum anomalias sentire possumus, neque debemus, antequam necessitas id imposuerit, ut circularia sive simplicia, sive, ut plurimum, complicata et composita, per quae omnes pene huiusmodi incurvatae figurationes, tum etiam rectae plane lineae describi possunt, officium suum circa omnimodam phaenomenωn coelestium repraesentationem deposuerint. Quin potius haec licet materialia non constituamus; realia tamen ac eiusmodi esse affirmare non dubitamus, quae centrorum vi ac virtute armata, conversiones suas temporibus aequalibus commensurabiles absolvunt; veluti haec alibi a nobis fusius disceptata reperiuntur'.

[63] Longomontanus, *Disputatio prima astronomica*; cf. *AD* 1:42–49. For a brief exposition of this treatise, see Moesgaard, 'How Copernicanism Took Root', pp. 126–34.

[64] 'Fateri nihilo minus oportet circulares esse motus, vel ex pluribus circulis compositos…' Copernicus, 1543, fol. 2v, translated in *Nicolai Copernici Opera omnia*, vol. II, p. 11. 'Motus autem omnes coelestes esse per se regulares et aequabiles, constantique lege circulariter ferri,

retain the Copernican axiom of circles and circular motion. Yet he also suggested that 'centers of force' might be 'real'; they simply do not belong in mathematical astronomy. Longomontanus remained a Ptolemaic astronomer.

This account of the development of LM1 and LM2 shows Longomontanus working as a bricoleur (or what Americans might call a 'Rube Goldberg'), combining existing models, parameters and observations from various sources, not worrying especially about consistency[65] or about natural philosophy (his diagrams flip between heliocentric and geocentric geometries). Yet he did hold to several commitments, to the Copernican axiom of circles and circular motion and to the ontological significance of perfect numbers. This approach allowed him to reduce the Ptolemaic/Copernican errors for Mars by a factor of about ten. Longomontanus could not match the precision of the Rudolphine Tables. But for many seventeenth-century astronomers, especially those in northern Europe like Eichstad in Stettin, Longomontanus's tables were 'good enough'; and they avoided the philosophical problems created by Kepler's centers of force and non-circular motions.

Finally, we must note that in addition to his bricolage, Longomontanus's solution to the second anomaly was quite original. Kepler in the *AN* had used observations of a sidereally fixed Mars to determine the eccentricity of the Earth's orbit. Longomontanus in the *AD* used observations at Mars's apogee and perigee to derive parameters for eccentric circles to represent that orbit. Tycho's two assistants each found new ways to extract data from their master's observations. But Longomontanus's Ptolemaic and Copernican commitments prevented him from accepting the ellipse and a mathematical astronomy involving the physics of forces. He was the last astronomer to create a new theory in the Ptolemaic tradition.

Acknowledgments

I thank participants at the 2015 London Ptolemaeus Arabus et Latinus conference for their stimulating questions and comments and Michael Shank, Glen van Brummelen, and Noel M. Swerdlow for their advice and comments on earlier drafts of this chapter. Henry Zepeda prepared the diagrams. Lars Gislén helped me reduce Longomontanus's astronomical models to spreadsheets. Gilsén's spreadsheets for the Rudolphine Tables, Prutenic Tables and Tycho's solar and lunar tables also proved invaluable for this study. Each of these

pro Axiomate ab Astronomis omnibus iamdudum receptum sit'. Dreyer, *Tychonis Brahe Dani opera omnia*, vol. II, p. 14.

[65] For example, in his worked examples, Longomontanus added precession after having corrected the mean longitudes. But in some of his derivations, he precessed the mean longitudes before adding the corrections. This ambiguity in procedure can affect final longitudes by up to ±2 arcminutes for Mars. See *AD*, 2:229, 261–63.

spreadsheets replicates the computational procedures, including linear interpolation, of the original tables.

Bibliography

Barker, Peter, and Bernard R. Goldstein, 'Distance and Velocity in Kepler's Astronomy', *Annals of Science* 51 (1994), pp. 59–73.

Bartsch, Jacob, *Ephemeridis Ioannis Kepleri pars secunda ... pars tertia [1621–1636]*, Sagan, 1630.

Bialas, Volker, et al., *Johannes Kepler. Gesammelte Werke*, vol. XI/1: *Ephemerides novae motuum coelestium*, München: C. H. Beck, 1983.

Brahe, Tycho, *Astronomiae instauratae mechanica*, Wadensburgi, 1598.

Christianson, John Robert, *On Tycho's Island. Tycho Brahe and his Assistants, 1570–1601*, Cambridge: Cambridge University Press, 2000.

Copernicus, Nicolaus, *De Revolutionibus orbium coelestium*, Nuremberg, 1543.

Dobrzycki, Jerzy, 'The Aberdeen Copy of Copernicus's *Commentariolus*', *Journal for the History of Astronomy* 4 (1973), pp. 124–27.

Donahue, William H., *Johannes Kepler. Astronomia nova*, Santa Fe: Green Lion Press, 2015.

Dreyer, John Louis Emil, *Tycho Brahe. A Picture of Scientific Life and Work in the Sixteenth Century*, Edinburgh: Adam and Charles Black, 1890.

Dreyer, John Louis Emil, *Tychonis Brahe Dani opera omnia*, 15 vols, Copenhagen: Libraria Gyldendaliana, 1913–1929.

Eichstadt, Lorenz, *Prognosticon astrologicum ... annum MDCXXX*, Stettin, 1630.

Eichstadt, Lorenz, *Schreibcalender auff das Jahr ... MDCXXX*, Stettin, 1630.

Favaro, Antonio, *Carteggio inedito di Ticone Brahe, Giovanni Keplero e di altri celebri astronomi e matematici dei secoli XVI. e XVII. con Giovanni Antonio Magini*, Bologna: Nicola Zanichelli, 1886.

Gingerich, Owen, 'The Computer Versus Kepler', *American Scientist* 52 (1964), pp. 218–26.

Gingerich, Owen, 'The Role of Erasmus Reinhold and the Prutenic Tables in the Dissemination of Copernican Theory', in Jerzy Dobrzycki (ed.), *Études sur l'audience de la théorie héliocentrique. Conférences du Symposium de l'UIHPS (Toruń 1973)*, Wroclaw-Warszawa: Ossolineum, 1973, pp. 43–62 and 123–25.

Gingerich, Owen, 'Early Copernican Ephemerides', in Erna Hilfstein, Pawel Czartoryski and Frank D. Grande (eds), *Science and History. Studies in Honor of Edward Rosen*, Wrocław: Ossolineum, 1978, pp. 403–17.

Gingerich, Owen, 'The Great Martian Catastrophe and How Kepler Fixed It', *Physics Today* 64/9 (2011), pp. 50–54.

Gingerich, Owen, and Robert S. Westman, *The Wittich Connection. Conflict and Priority in Late Sixteenth-Century Cosmology*, Philadelphia: The American Philosophical Society, 1988.

Hashimoto, Keizo, 'Longomontanus's *Astronomia Danica* in China', *Journal for the History of Astronomy* 18 (1987), pp. 95–110.

Jones, Alexander, and Dennis W. Duke, 'Ptolemy's Planetary Mean Motions Revisited', *Centaurus* 47 (2005), pp. 226–35.

Longomontanus, Christen Sørensen, *Disputatio prima astronomica, De praecognitis*, Copenhagen, 1611.

Longomontanus, Christen Sørensen, *Astronomia danica*, Amsterdam, 1622.

Longomontanus, Christen Sørensen, *Introductio in theatrum astronomicum*, Copenhagen, 1639.

Maeyama, Yasukatsu, 'Kepler's *Hypothesis Vicaria*', *Archive for History of Exact Sciences* 41 (1990/91), pp. 53–92.

Magini, Giovanni Antonio, *Supplementum ephemeridum ac tabularum secundorum mobilium*, Venice, 1614.

Magini, Giovanni Antonio, *Ephemerides coelestium motuum ... ab anno Domini 1611 usque ad annum 1630*, Venice, 1616.

Moesgaard, Kristian Peder, 'How Copernicanism Took Root in Denmark and Norway', in Jerzy Dobrzycki (ed), *The Reception of Copernicus' Heliocentric Theory*, Dordrecht: Reidel, 1972, pp. 117–51.

Moesgaard, Kristian Peder, 'Tychonian Observations, Perfect Numbers, and the Date of Creation: Longomontanus's Solar and Precessional Theories', *Journal for the History of Astronomy* 6 (1975), pp. 84–99.

Moesgaard, Kristian Peder, 'Cosmology in the Wake of Tycho Brahe's Astronomy', in Wolfgang Yourgrau and Allen D. Breck (eds), *Cosmology, History, and Theology*, New York: Springer, 1977, pp. 283–305.

Nicolai Copernici Opera omnia, vol. II, Warsaw: Officina Publica Libris Scientificis Edendis, 1975.

Omodeo, Pietro Daniel, 'The Scientific Culture of the Baltic Mathematician, Physician and Calendar-Maker Laurentius Eichstadt (1596–1660)', *Journal for the History of Astronomy* 48 (2017), pp. 135–59.

Pedersen, Olaf, *A Survey of the Almagest. With Annotation and New Commentary by Alexander Jones*, New York: Springer, 2010.

Raeder, Hans, Elis Strömgren and Bengt Strömgren, *Tycho Brahe's Description of His Instruments and Scientific Work as Given in* Astronomiae instauratae mechanica (*Wandesburgi 1598*), København: Munksgaard, 1946.

Savoie, Denis, *La diffusion du copernicianisme au XVIe siècle. Les Tables Pruténiques d'Erasme Reinhold*, PhD dissertation, École des hautes études en sciences sociales, 1996.

Swerdlow, Noel M., 'The Derivation and First Draft of Copernicus's Planetary Theory. A Translation of the Commentariolus with Commentary', *Proceedings of the American Philosophical Society* 117 (1973), pp. 423–512.

Swerdlow, Noel M., 'The Lunar Theories of Tycho Brahe and Christian Longomontanus in the *Progymnasmata* and *Astronomia danica*', *Annals of Science* 66 (2009), pp. 5–58.

Swerdlow, Noel M., 'Tycho, Longomontanus, and Kepler on Ptolemy's Solar Observations and Theory, Precession of the Equinoxes, and Obliquity of the Ecliptic', in Alexander Jones (ed.), *Ptolemy in Perspective. Use and Critisism of His Work from Antiquity to the Ninteenth Century*, Berlin: Springer, 2010, pp. 151–202.

Swerdlow, Noel M., and Otto Neugebauer, *Mathematical Astronomy in Copernicus's De revolutionibus*, 2 vols, New York/Berlin: Springer, 1984.

Thoren, Victor E., *The Lord of Uraniborg. A Biography of Tycho Brahe*, Cambridge: Cambridge University Press, 1990.

van Maanen, Jan A., 'The Refutation of Longomontanus' Quadrature by John Pell', *Annals of Science* 43 (1986), pp. 315–52.

Voelkel, James R., *The Composition of Kepler's* Astronomia nova, Princeton: Princeton University Press, 2001.

Voelkel, James R., and Owen Gingerich, 'Giovanni Antonio Magini's "Keplerian" Tables of 1614 and Their Implications for the Reception of Keplerian Astronomy in the Seventeenth Century', *Journal for the History of Astronomy* 32 (2001), pp. 237–62.

Wilson, Curtis, 'The Error in Kepler's Acronychal Data for Mars', *Centaurus* 13 (1969), pp. 263–68.

Indexes

Names and Work Titles

Modern Names

Manuscripts

List of Contributors

José Bellver
Research Fellow, University of Louvain-la-Neuve
josepbellver@gmail.com

Jean-Patrice Boudet
Professor, University of Orléans
boudetjp@gmail.com

Josep Casulleras
Associate Professor, University of Barcelona
jcasulleras@ub.edu

Bojidar Dimitrov
Research Fellow, University of Bologna
bojidar.dimitrov@unibo.it

Dirk Grupe
Independent researcher
dirk.grupe@gmx.de

Paul Hullmeine
Doctoral student, Ptolemaeus Arabus et Latinus / University of Munich
phullmeine@ptolemaeus.badw.de

Alexander Jones
Professor, New York University
alexander.jones@nyu.edu

Richard Kremer
Professor Emeritus, Dartmouth College
Richard.L.Kremer@dartmouth.edu

Tzvi Langermann
Professor, Bar-Ilan University
tlangermann@hotmail.com

H. Darrel Rutkin
Assistant Professor, University of Venice
drutkin2001@yahoo.com

Michael H. Shank
Professor Emeritus, University of Wisconsin-Madison
mhshank@wisc.edu

Nathan Sidoli
Professor, Waseda University
sidoli@waseda.jp

Carlos Steel
Professor Emeritus, University of Leuven
carlos.steel@hiw.kuleuven.be

Johannes Thomann
Researcher, University of Zurich
johannes.thomann@aoi.uzh.ch

Henry Zepeda
Assistant Professor, Wyoming Catholic College
henry.zepeda@wyomingcatholic.edu